FOURTH EDITION

An Invitation to Social Research

How It's Done

Emily Stier Adler

Rhode Island College

Roger Clark

Rhode Island College

WADSWORTH
CENGAGE Learning™

Australia • Brazil • Japan • Korea • Mexico • Singapore • Spain • United Kingdom • United States

WADSWORTH
CENGAGE Learning

An Invitation to Social Research: How It's Done, Fourth Edition

Emily Stier Adler, Roger Clark

Acquisitions Editor: Chris Caldeira

Assistant Editor: Erin Parkins

Editorial Assistant: Rachael Krapf

Publisher/Executive Editor: Linda Schreiber

Development Editor: Erin Parkins

Marketing Director: Kimberly Russell

Marketing Manager: Andrew Keay

Marketing Coordinator: Jillian Myers

Marketing Communications Manager: Laura Localio

Technology Production Manager: Lauren Keyes

Editorial Production Manager: Matt Ballantyne

Senior Art Director: Caryl Gorska

Manufacturing Director: Barbara Britton

Manufacturing Buyer: Rebecca Cross

Permissions Acq Manager, Image: Leitha Etheridge-Sims

Permissions Acq Manager, Text: Bob Kauser

Production Technology Analyst: Lori Johnson

Production Project Manager/CPM: PrePress PMG

Production Service: PrePress PMG

Cover Designer: Yvo Riezebos

Cover Printer: West Group

Compositor: PrePress PMG

For product information and technology assistance, contact us at
Cengage Learning Customer & Sales Support, 1-800-354-9706

For permission to use material from this text or product submit all requests online at **cengage.com/permissions**
Further permissions questions can be e-mailed to
permissionrequest@cengage.com

Library of Congress Control Number: 2009941183

ISBN-13: 978-0-495-81329-3

ISBN-10: 0-495-81329-X

Wadsworth
20 Davis Drive
Belmont, CA 94002-3098
USA

Cengage Learning is a leading provider of customized learning solutions with office locations around the globe, including Singapore, the United Kingdom, Australia, Mexico, Brazil, and Japan. Locate your local office at: **www.cengage.com/global.**

Cengage Learning products are represented in Canada by Nelson Education, Ltd.

To learn more about Wadsworth, visit **www.cengage.com/wadsworth.**

Purchase any of our products at your local college store or at our preferred online store **www.ichapters.com**

Printed in the United States of America
1 2 3 4 5 6 7 13 12 11 10 09

BRIEF CONTENTS

CONTENTS

6 Measurement   129

10 Qualitative Interviewing 251

11 Observational Techniques 291

*We dedicate this book to
Gracie and Jazzy*

PREFACE

We'd like to invite you to participate in one of the most exciting, exhilarating, and sometimes exasperating activities we know: social science research. We extend the invitation not only because we know, from personal experience, how rewarding and useful research can be, but also because we've seen what pleasure it can bring other students of the social world. Our invitation comes with some words of reassurance, especially for those of you who entertain a little self-doubt about your ability to do research. First, we think you'll be glad to discover, as you read *An Invitation to Social Research: How It's Done,* how much you already know about how social research is done. If you're like most people, native curiosity has been pushing you to do social research for much of your life. This book is meant simply to assist you in this natural activity by showing you some tried-and-true ways to enlightening and plausible insights about the social world.

SPECIAL FEATURES

Active Engagement in Research

Our second word of reassurance is that we've done everything we can to minimize your chances for exasperation and maximize your opportunities for excitement and exhilaration. Our philosophy is simple. We believe that honing one's skill in doing social research is analogous to honing one's skills in other enjoyable and rewarding human endeavors, like sport, art, or dance. The best way isn't simply to read about it. It's to do it and to watch experts do it. So, just as you'd hesitate to teach yourself tennis, ballet, or painting only by reading about them, we won't ask you to try learning the fine points of research methodology by reading alone. We'll encourage you to get out and practice the techniques we describe. We've designed exercises at the end of each chapter to help you work on the "ground strokes," "serve," "volleys," and "overheads" of social research. We don't think you'll need to do all the exercises at home. Your instructor might ask you to do some in class and might want you to ignore some altogether. In any case, we think that, by book's end, you should have enough control of the fundamentals to do the kind of on-the-job research that social science majors are increasingly asked to do, whether they find themselves in social service agencies, the justice system, business and industry, government, or graduate school.

The exercises reflect our conviction that we all learn best when we're actively engaged. Other features of the text also encourage such active engagement, including the "Stop and Think" questions that run through each chapter, which encourage you to actively respond to what you're reading.

Engaging Examples of Actual Research

Moreover, just as you might wish to gain inspiration and technical insight for ballet by studying the work of Anna Pavlova or Mikhail Baryshnikov, we'll encourage you to study the work of some accomplished researchers. Thus, we build most of our chapters around a research essay, what we call focal research, that is intended to make the research process transparent, rather than opaque. We have chosen these essays for their appeal and accessibility, and to tap what we hope are some of your varied interests: for instance, crime, gender, election polls, life in prison, attitudes towards environmentalism, immigrants' lives, and others.

Behind-the-Scene Glimpses of the Research Process

These focal research pieces are themselves a defining feature of our book. In addition to such exemplary "performances," however, we've included behind-the-scenes glimpses of the research process. We're able to provide these glimpses because many researchers have given generously of their time to answer our questions about what they've done, the special problems they've encountered, and the ways they've dealt with these problems. The glimpses should give you an idea of the kinds of choices and situations the researchers faced, where often the "real" is far from the "ideal." You'll see how they handled the choices and situations and hear them present their current thinking about the compromises they made. In short, we think you'll discover that good research is an achievable goal, and a very human enterprise.

Clear and Inviting Writing

We've also tried to minimize your chances for exasperation by writing as clearly as we can. A goal of all social science is to interpret social life, something you've all been doing for quite a while. We want to assist you in this endeavor, and we believe that an understanding of social science research methods can help. But unless we're clear in our presentation of those methods, your chances of gaining that understanding are not great. There are, of course, times when we'll introduce you to concepts that are commonly used in social science research that might be new to you. When we do, however, we will try to provide definitions to make the concepts as clear as possible. The definitions are highlighted in the margin of the text and in the glossary at the end of the text.

Focus on Ethics

Given the importance of doing research that is methodologically correct and practical as well as ethical, we've put a focus on ethical principals in each chapter. The "Thinking about Ethics" section of each chapter applies the ethical principles we cover in depth in Chapter 3 to research projects presented in the subsequent chapters.

Balance between Quantitative and Qualitative Approaches

We think you'll also appreciate the balance between quantitative and qualitative research methods presented here. Quantitative methods focus on things that are measured numerically. ("He glanced at her 42 times during the performance.") Qualitative methods focus on descriptions of the essence of things. ("She appeared annoyed at his constant glances.") We believe both methodological approaches are too useful to ignore. Emblematic of this belief is the inclusion of a chapter (Chapter 15), which devotes about as much space to the discussion of qualitative data analysis as it does to quantitative data analysis. The presence of such a chapter is another defining feature of the book.

Moreover, in addition to more conventional strategies, we will introduce you to some relatively new research strategies, such as using the Internet to refine ideas and collect data and visual methodologies. We cover the link between theory and research, compare research to other ways of knowing, and focus on basic and applied research.

Our aims, then, in writing this book have been (1) to give you firsthand experiences with the research process, (2) to provide you with engaging examples of social science research, (3) to offer behind-the-scenes glimpses of how professional researchers have done their work, (4) to keep our own presentation of the "nuts-and-bolts" of social science research as clear and inviting as possible, (5) to focus on doing research following ethical principles, (6) to give a balanced presentation of qualitative and quantitative research methods, and (7) to introduce recent technological innovations. Whether we succeed in these goals, and in the more important one of sharing our excitement about social research, remains to be seen. But rest assured, however, of our conviction that there is excitement to be had.

WHAT IS NEW IN THE FOURTH EDITION

The fourth edition represents a substantial revision of the third. Once again, we've rewritten major sections of every chapter to clarify the process of social research and to provide up-to-date material from the social research literature. In doing so, we've focused our presentation on the essentials of social research and covered some new material as well as classic sources. Throughout the text, you'll notice a new emphasis on visual sociology and methods that use this approach. In addition, we have expanded our data analysis chapter,

Chapter 15, to include a section on multivariate analysis and retained a thorough introduction to both quantitative and qualitative data analyses. We have added two new appendixes, one on comparing methods and one on using multiple methods, both by way of summarizing important issues in our discussion of sociological methods.

Our data analysis chapter (Chapter 15) also reflects, as does the rest of the current text, our belief that research, as practiced by social (and all other) scientists, is increasingly computer assisted and Internet based. So, for instance, in the data analysis chapter, we introduce students to data that they can analyze online. In other chapters we also present ways of finding research reports and data that can be accessed quickly online.

Themes from the first three editions have been retained here. This edition has 12 focal research pieces—7 of them new and 1 updated. While incorporating the new pieces, we have enhanced the balance between qualitative and quantitative research in the book. In Chapter 5, for instance, we present a new piece that considers calling cell phones and the implications for noncoverage bias. In Chapter 7, we present a new piece on how people make the transition to retirement, based on a series of interviews with workers before and after they retire from full-time employment. In Chapter 14, the new focal research describes a needs assessment done in Ecuador to evaluate the conditions that incarcerated women there face. We note that, as a result of their research, the researchers were able to influence policy and create social change. In all cases, our new contributors have volunteered important "behind-the-scenes" insights into the research process, insights that we gratefully share here.

ACKNOWLEDGMENTS

We cannot possibly thank all those who have contributed to the completion of this edition, but we can try to thank those whose help has been most indispensable and hope that others will forgive our neglect. We'd first like to thank all the students who have taken research methods courses with us at Rhode Island College for their general good-naturedness and patience as we've worked out ideas that are crystallized here, and then worked them out some more. We'd also like to thank our colleagues in the Sociology Department and the administration of the college, for many acts of encouragement and more tangible assistance, including released time.

We'd like to thank colleagues, near and far, who have permitted us to incorporate their writing as focal research or in text boxes, and, in many cases, then read how we've incorporated it, and told us how we might do better. They are Mikaila Mariel Lemonik Arthur, Nichole Banton, Susan Chase, Erica Chito Childs, Leah Christian, Desirée Ciambrone, Adam Clark, Michael Dimock, Danielle Dirks, Sandra Enos, Rachel Filinson, Joseph R. Ferrari, John Grady, Jill Harrison, Michele Hoffnung, Scott Keeter, Brandon Lang, Matthew T. Lee, Kristy Maher, Ramiro Martinez, Jennifer C. Mueller, Donald C. Naylor, Maureen Norton-Hawk, Leslie Houts Picaa, Chris Podeschi, Jennifer Racine, Matthew M. Reavy, and Jessica Holden Sherwood. These researchers include those who've given us the behind-the-scenes glimpses of the research process that we think distinguishes our book. We'd like to extend special thanks to Aaryn Ward at Louisiana State University who composed the excellent instructor's manual accompanying the book.

We're very grateful to Chris Caldeira and Erin Parkins, our editors at Wadsworth, Pre-Press PMG for their help with the fourth edition. We've benefited greatly from the comments of the social science colleagues who have reviewed our manuscripts at various stages of development. Mary Archibald; Cynthia Beall, Case Western Reserve University; Daniel Berg, Humboldt State University; Susan Brinkley, University of Tampa; Jeffrey A. Burr, State University of New York, Buffalo; Richard Butler, Benedict College; Daniel Cervi, University of New Hampshire; Charles Corley, Michigan State University; Norman Dolch, Louisiana State University, Shreveport; Craig Eckert, Eastern Illinois University; Pamela I. Jackson, Rhode Island College; Alice Kemp, University of New Orleans; Michael Kleinman, University of South Florida; James Marquart, Sam Houston State University; James Mathieu, Loyola Marymount University; Steven Meier, University of Idaho; Deborah Merrill, Clark University; Lawrence Rosen, Temple University; Josephine Ruggiero, Providence College; Kevin Thompson, University of North Dakota; Steven Vassar, Mankato State University; and Gennaro Vito, University of Louisville, offered important assistance for the first edition. For the second edition, Gai Berlage, Iowa College; Susan Chase, University of Tulsa; Dan Cooper, Hardin-Simmons University; William Faulkner, Western Illinois University; Lawrence Hazelrigg, Florida State University; Josephine Ruggiero, Providence College; Jeffrey Will, University of North Florida; and Don Williams, Hudsonville College, have been most helpful. For the third edition, we would like to thank Ronda Priest, University of Southern Indiana; Dennis Downey, University of Utah; Scott M. Myers, Montana State University; William Wagner, California State University, Bakersfield; and John Mitrano, Central Connecticut State University, for their valuable comments. For the fourth edition, we would like to thank Allan Hunchuck, Theil College; Aaryn Ward, Louisiana State University; Erin Leahey, University of Arizona; and Kevin Yoder, University of North Texas for their help.

Finally, we'd like to thank our spouses, George L. Adler and Beverly Lyon Clark, for providing many of the resources we've thanked others for: general good-naturedness and patience, editorial assistance, encouragement, and other tangible aids, including released time.

The Uses of Social Research

© Emily Stier Adler

INTRODUCTION

How do you see yourself 10 years from now? Will you be single? Will you be a parent? What about people you know? Does it seem to you that almost everyone with whom you've talked about such things is either planning to marry and have children or is already married with children? Or does it seem to be the other way around? Does everyone seem to be planning to avoid marriage, perhaps to live alone, at least for extended periods of his or her life? What's the reality? Studying something systematically, with method, means not relying on your impressions, or on anyone else's, either. It means checking our impressions against the facts. It also means being sure about what one is looking for. One way researchers achieve this level of definiteness is to begin their research with a research question, or a question about one or more topics that can be answered through research. One such focusing question about our initial concern, for instance, might be: Is almost everyone in the country married with children or are they living alone?

research question, a question about one or more topics or concepts that can be answered through research.

The U.S. Census Bureau's *Statistical Abstract* (2009) provides one apparently relevant set of facts. It shows that in 2006, the number of Americans living alone was about 26.6 percent of all households and was greater than the proportion of American households made up of a married couple with children, about 22.7 percent. Until 2000, the proportion of American households made up of married couples with children had always been greater than the proportion of Americans living alone (Hobbs, 2005). This could be the starting point of your investigation.

STOP & THINK   *This would only be a start, however. What about our initial interest in the future of you and people like you is not addressed by the "facts" of the Census report?*

Although the Census report does tell us something that's relevant to our main research—whether young people today are planning to marry and have children or planning not to marry—it requires a little interpreting, doesn't it? At first glance, Census facts seem to almost address this question. They show that, in 2006, more households in the United States were made up of people living alone than of married couples with children and so seem to imply that more people will end up alone than in families with children. Or do they?

We've made up the pie chart in Figure 1.1 to help us think about the question of whether the Census data really settle the issue addressed in our research question. The chart indicates one thing that we already knew: More American households are made up of people living alone than of married couples with children. But, perhaps, looking at the chart will remind you that all households made up of married couples with children had at least *twice as many adults in them* as households made up of people living alone. So, in fact, there actually were more adults living as parts of married couples with children in 2006 than there were adults living alone. It also reminds us that we should consider other kinds of households. One such group is married couples without children, a group that is more numerous than households made up of people living

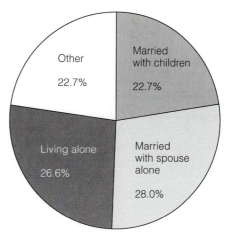

FIGURE **1.1** Total Households in 2006: 114,384,000

Source: Data from U.S. Census Bureau (2009).

alone and, again, would therefore be households that have at least twice as many adults in them as households made up of people living alone. The presence of this large group might make us want to revise our original question to include something about the number of people who plan to live with another adult (or adults) but without children. Moreover, we suspect that more of these people, people in couples without children, "plan" to become couples with children than "plan" to live alone. In effect, then, the chart also leads us to a second research question, a question that, in some ways, might even better embody the concern expressed at the beginning of paragraph one: Are more young people planning to marry and not have children, to marry and have children, or to live alone?

STOP & THINK *Can you think of a better way to find out how young people plan to live in the future than with the Census data of 2006?*

We suspect that you can think of such a way. Maybe you've noticed that Figure 1.1 is misleading in at least two ways. First, it focuses on the way things are, or more precisely, the way things were in 2006. It doesn't help with the question of what people are planning to do in the future. Second, it focuses on households, not people. We've obviously been able to make some inferences about people from the information about households, but the inferences have been somewhat awkward and would be even more so if we focused more attention on the fourth part of Figure 1.1: the part about "other" households. People, not households, are what we're really interested in, so Figure 1.1 ultimately provides information about the wrong kind of thing, the wrong **unit of analysis,** the unit about which information is collected. One of the first decisions you need to make when planning research is, "What should my unit of analysis be?" Moreover, one of the first questions to ask of any research project you read is, "What is this study's unit of analysis?"

unit of analysis, a unit about which information is collected.

See if you can identify the units of analysis for each of the following studies: (1) Southgate and Roscigno's (2009) finding that involvement in music is associated with academic performance for both young children and adolescents; (2) Cooney and Burt's (2008) finding that in American counties where a particular crime occurs frequently, the average punishment for that crime will be less severe than in counties where it occurs rarely; and (3) Soule and King's (2008) finding that social movement organizations (like women's rights organizations) that have to develop specialized goals, rather than more general ones, are less likely to survive than organizations that can maintain generalized goals.

Returning to our question about how you might find out about how young people plan to live in the future, perhaps you've thought of doing some kind of questionnaire survey, the nature of which we talk about more in Chapter 9. Perhaps you've also realized that you'd want to survey some kind of representative sample of young people. We talk about gathering such samples in Chapter 5. Perhaps you've thought a little about the kinds of information you might want. We deal with those issues in Chapter 6. Perhaps you've even thought about examining what others have had to say about the issue. You might, for instance, be interested in Michele Hoffnung's research, the focal piece of Chapter 4. Maybe someone's already collected just the information you need and would be willing to share it with you. We examine this possibility in Chapter 12, on finding and using available data.

Whatever you've thought, you've clearly begun to engage the question, as we (Emily and Roger) have done with questions we've studied, as a mystery to be solved. Learning about social research is a lot like learning to solve mysteries. It's about challenge, frustration, excitement, and exhilaration. We, Emily and Roger, are addicted to social research. Even as we are writing about research, we're also working on a study of how people make the transition to retirement (and will tell you about our findings in Chapter 7). We're passionately interested in this mystery, partly because we're thinking about the process of retirement ourselves.

RESEARCH VERSUS OTHER WAYS OF KNOWING

Knowledge from Authorities

The data about households in America—that slightly more of them are made up of people living alone than of people living with a spouse and children—are fascinating. But perhaps they're not as fascinating as what we, Emily and Roger, learned many years ago, literally at our mothers' knees, that "In fourteen hundred and ninety-two, Columbus sailed the ocean blue," despite nearly everyone's belief that he was sailing in a direction that jeopardized his very existence.

Now, why do we think we "know" these things? Basically, it's because some authority told us so. In the first case, we read it in a Census Bureau report. In the second case, we relied on our moms, who were reporting a commonly accepted version of America's "discovery." **Authorities**, such as the

authorities, socially defined sources of knowledge.

Census Bureau and our moms, are among the most common sources of knowledge for most of us. Authorities are socially defined sources of knowledge.

Many social institutions, such as religion, news media, government agencies, and schools, are authorities, and individuals within them are often seen as having superior access to relevant knowledge. In modern societies like ours, we often attribute authority to what newscasters on television say or what's posted on an online newspaper such as the *Huffington Post*. Sometimes authorities use research as their basis for knowledge, but we usually don't evaluate their sources. For most of us, most of the time, learning from authorities is good enough, and it certainly helps keep life simpler than it would be otherwise. Life would be ridiculously difficult and problematic if, for instance, we who live in the Western world had to reinvent a "proper" way of greeting people each time we met them. At some point, we're told about the customs of shaking hands, bumping fists, or saying "Hi," and we move on from there.

 STOP & THINK *What, do you think, are the major disadvantages of receiving our "knowledge" from authorities?*

Although life is made simpler by "knowledge" from authorities, sometimes such knowledge is inappropriate, misleading, or downright incorrect. Very few people today take seriously the "flat-earth" theories that have been attributed to Columbus's social world. More interesting, perhaps, is an increasingly accepted view that, our mothers' teachings notwithstanding, very few people in Columbus's social world took seriously the flat-earth view either; that, in fact, this view of the world was wrongly attributed to them by late-nineteenth-century historians to demonstrate how misguided people who accepted religious over scientific authority could be (e.g., Gould, 1995: 38–50). In fact, we mean no offense to our moms when we say they represent a whole category of authorities who can mislead: authorities in one area of expertise (in the case of our moms, us) who try to speak authoritatively on subjects in which they are not experts (in the case of our moms, the worldview of people in Columbus's society).

Knowledge from Personal Inquiry

But if we can't always trust authorities, like our moms, or even experts, like those late-nineteenth-century historians (and, by extension, even our teachers or the books they assign) for a completely truthful view of the world, who or what can we trust? For some of us, the answer is that we can trust the evidence of our own senses. Personal inquiry, or inquiry that employs the senses' evidence for arriving at knowledge, is another common way of knowing.

personal inquiry, inquiry that employs the senses' evidence.

 STOP & THINK *Can you think of any disadvantages in trusting personal inquiry alone as a source of "knowledge"?*

The problem with personal inquiry is that it, like the pronouncements of authorities, can lead to misleading, even false, conclusions. As geometricians like to say, "Seeing is deceiving." This caution is actually as appropriate for students of the social world as it is for students of regular polygons because the evidence of our senses can be distorted. Most of us, for instance, developed our early ideas of what a "family" was by observing our own families closely. By the time we were six or seven, each of us (Emily and Roger) had observed that our own families consisted of two biological parents and their children. As a result, we concluded that all families were made up of two parents and their biological children.[1]

There's obviously nothing wrong with personal inquiry ... except that it frequently leads to "knowledge" that's pretty half-baked. There are many reasons for this problem. One of them is obvious from our example: Humans tend to *overgeneralize* from a limited number of cases. Both of us had experienced one type of family and, in a very human way, assumed that what was true of our families was true of all human families. Another barrier to discovering the truth is the human tendency to *perceive selectively* what we've been conditioned to perceive. Thus, even as 10-year-olds, we might have walked into an intentional community such as a commune, with lots of adults and children, and not entertained the possibility that this group considered itself a family. We just hadn't had the kind of experience that made such an observation possible. A third problem with knowledge from personal inquiry is that it often suffers from *premature closure*—our tendency to stop searching once we think we have an answer. At 10, Emily and Roger thought they knew what a family was, so they simply didn't pursue the issue further.

So, neither relying on authorities nor relying on one's own personal inquiry is a foolproof way to the truth. In fact, there might not be such a way. But authors of research methods books (ourselves included) tend to value a way that's made some pretty astounding contributions to the human condition: the scientific method. In the next two subsections, we'll, first, discuss some relative strengths of the scientific method for knowledge acquisition and, second, give you some idea about what this knowledge is intended to do.

The Scientific Method and Its Strengths

scientific method, a way of conducting empirical research following rules that specify objectivity, logic, and communication among a community of knowledge seekers and the connection between research and theory.

positivist view of science, a view that human knowledge must be based on what can be perceived.

Specifying precise procedures that constitute the scientific method is a dicey business at best. Part of what makes it so hard is that there are fundamental philosophical differences in beliefs about what science is. Without getting too bogged down in weighty philosophical matters, we should probably confess that our own philosophical approach to science, and that of most contemporary social scientists, isn't the classical positivist view. A positivist view of science suggests we should stick to those things we can observe and measure directly, such as the time it takes a person to solve a Rubik's cube, whether someone lives alone, or how many times a week an individual goes to church. The goals of positivist science are unchallengeable propositions about the world: like, in the physical sciences, "every action has an equal and opposite reaction." Things that can't be directly observed, such as what people think about President Obama, how they feel about losing a job, or the economic

---

[1] Of course, by the time that Roger, in his early forties, had adopted two children from another country, his ideas of "family" had changed many times.

post-positivist view of science, a view that knowledge is not based on irrefutable observable grounds, that it is always somewhat speculative, but that science can provide relatively solid grounds for that speculation.

objectivity, the ability to see the world as it really is.

intersubjectivity, agreements about reality that result from comparing the observations of more than one observer.

impact of globalization, are irrelevant. Ours is a **post-positivist view of science,** a view that knowledge is not based on irrefutable observable grounds, that it is always somewhat speculative, but that science can provide relatively solid grounds for that speculation. Moreover, most post-positivists would argue that the pursuit of unchallengeable propositions about the world is unrealistic and, perhaps, unscientific. Most physicists today feel that Newton's laws of motion, of which "every action has an equal and opposite reaction" is but one example, are limited in their scope. Good as they are, they simply don't apply when things get very small (say, subatomic) or very fast (say, close to light speed). They're certainly not unchallengeable.

Positivists have often claimed that they strive for **objectivity,** or the ability to see the world as clearly as possible, free from personal feelings, opinions, or prejudices about what it is or what it should be. Post-positivists, on the other hand, suggest that the most even scientists can strive for is **intersubjectivity,** or agreements about reality that come from the practice of comparing one's results with those of others and discovering that the results are consistent with one another. Generally, post-positivists, such as Emily and Roger, believe that scientists may be more careful than everyday people, but their work is nonetheless fallible. The "goal of science," according to Trochian (2006), "is to hold steadfastly to the goal of getting it right about reality, even though we can never achieve that goal!"

But how are we to hold steadfastly to that goal? We'd like to suggest four steps that are often involved in doing science and point to the relative emphasis placed on *care* and *community* that distinguishes science from other modes of knowing, whatever one's philosophical views of science are. An early step is to specify the goals or objectives that distinguish a particular inquiry (here care is paramount). In our first example, we eventually specified the goal of answering the research question, "Are more young people planning to marry and have children or to live alone?" A subsequent step involves reviewing literature or reading what's been published about a topic (here, learning what a relevant community thinks is the goal). We could do worse, in the pursuit of that early research question—about whether young people are planning to marry and have children or to live alone, for instance, than read Michele Hoffnung's work (see Chapter 4) on what an earlier generation of "young people" planned to do. At some point, it becomes important to specify what is actually observed (care again). We'd want to define, for instance, what we mean by "young people" and how we plan to measure what their plans are. A later step is to share one's findings with others in a relevant community so that they can scrutinize what's been done (community again). We might want, for instance, to prepare a paper for a conference of family sociologists. These steps or procedures will come up again throughout this book, but we'd now like to stress some of the strengths that accrue to the scientific method because of their use.

The Promotion of Skepticism and Intersubjectivity

One great strength of the scientific method, over modes that rely on authorities and personal inquiry, is that, ideally, it promotes skepticism about its own knowledge claims. Perhaps you looked at our presentation of the data about today's households (e.g., that fewer are made up of married couples with children than are made up of people living alone) and said, "Hey. Those facts alone don't tell us much about people's intentions. And they don't even indicate that more people are living alone than living in couples today." If so, we applaud your skepticism. One way in which healthy skepticism is generated is through the communities of knowledge seekers. Each member of these communities has a legitimate claim to being a knowledge producer, as long as he or she conforms to other standards of the method (mentioned later). When intersubjective agreement eludes a community of scientists because various members get substantially different results, it's an important clue that knowledge remains elusive and that knowledge claims should be viewed with skepticism. Until the 1980s, for instance, the medical community believed that stomach ulcers were caused by stress. Then a couple of Australian scientists, Barry Marshall and Robin Warren, found, through biopsies, that people with ulcerous stomachs often had the *Helicobacter pylori* bacterium lurking nearby and theorized that the bacterium had caused the ulcers. Marshall and Warren thus became skeptical of the medical community's consensus about the causes of ulcers. Few members of that community took them seriously, however, until Marshall, experimenting on himself, swallowed *H. pylori* and developed pre-ulcerous symptoms. Subsequently, Marshall and Warren found that most stomach ulcers could be successfully treated with antibiotics. For their work in discovering the bacterium that causes stomach inflammation, ulcers, and cancer, Marshall and Warren won the 2005 Nobel Prize for Physiology or Medicine (Altman, 2005: D3; Marshall, 2002).

The Extensive Use of Communication

Another related ideal of the scientific method is *adequate communication* within the community of knowledge seekers, implicit in the scientific procedures of referring to previous published accounts in an area and of sharing findings with others. Unlike insights that come through personal inquiry, scientific insights are supposed to be subjected to the scrutiny of the larger community and, therefore, need to be as broadly publicized as possible. Communication of scientific findings can be done through oral presentations (as at conferences) or written ones (especially through publication of articles and books). For example, Hoffnung has presented her findings about college students' plans for the future at scientific conferences and in scholarly articles. In Chapter 4, she offers you a glimpse of her most recent findings. Increasingly, new technology (discussed in Chapter 12) allows for increased communication about research and the exchange of data. You, for instance, can use a variety of print and online resources that your college library may make available to you to find references to, and even copies of, research articles about your topic. We did and found research by Weer (2006) and Aronson (2008) that supplement Hoffnung's work and our interests in the kinds of family lives that young people envision by examining recent survey data on how young men and women are expecting to deal with work–family conflicts in their future. Once findings are communicated, they then become grist for a critical mill. Others are thereby invited to question (or openly admire) the particular approach that's been reported or to try to reproduce (or *replicate*) the findings using other approaches or other circumstances. Adequate communication thus facilitates the ideal of reaching intersubjective "truths."

Testing Ideas Factually

These communal aspects of the scientific method are complemented by at least three other goals, goals that underscore the care admired by scientists: that "knowledge" be *factually testable*, be *logical*, and be *explicable through theory*. Factual testability means that scientific knowledge, like personal inquiry, must be supported by observation. Unlike positivists, post-positivists don't insist that observations themselves can't be affected by personal biases or theories. For example, people concerned about environmental warming might be most likely to notice that glaciers are melting. All observations are fallible, but most post-positivists would nonetheless accept the notion that if what is being proposed as knowledge doesn't stand up to many people's observation, it's not really knowledge at all. And rather than simply using evidence to support a particular view, as we sometimes do in personal inquiry, scientific observation also includes trying to imagine the kinds of observations that would undermine the view and then pursuing those observations. Confronted with the Census data about, say, the increase in the number of households with people living alone, we questioned whether those data were in fact the best data for answering our question and imagined other possible sources (e.g., a questionnaire survey of our own). Similarly, when confronted with the idea that people of Columbus's day held a "flat-earth" view of the world, historians consulted the writings of scientists of that day and earlier and found evidence of a pretty widespread belief that the earth was spherical—surprisingly similar to our beliefs today (Gould, 1995: 38–50). The pursuit of counterexamples, and the parallel belief that one can never fully prove that something is true (but that one can cast serious doubt on the truth of something), is a key element of the scientific method.

The Use of Logic

Scientists are often thought of being especially logical. Television characters from *Star Trek*'s Spock to *CSI*'s Gil Grissom are meant to embody the desirability of logical reasoning in science. Both Spock and Grissom would recognize the logical fallacy in appealing to authority we've already mentioned. (Just because someone is an authority in field X doesn't necessarily mean that that person's claim, C, in field Y is true.) Spock and Grissom, like most scientists, approved of logical reasoning. You used logical reasoning, too, if when confronted with the Census data about the compositions of households in America in 2006, you wondered if those data were adequately related to the question about young people's plans.

Like many scientists, Spock and Grissom were quick to point out illogical reasoning. If young Roger or Emily had proposed the notion that all families consisted of two biological parents and their offspring and presented the reasoning—"We belong to such families. Therefore everyone must"—both Spock and Grissom might have responded, "That's illogical." One canon of the scientific method is that one must adhere, as closely as possible, to the rigors of logical thinking. Few practicing scientists invoke the scientific standard of logic as frequently as Spock, or even Grissom, did, but fewer still would wish to appear as illogical as Roger and Emily were in their reasoning.

STOP & THINK   *Sam Roberts (2007) broke the news that in 2005, for the first time, more American women were living without a husband than with one. Does it follow logically from this news that more than 50 percent of American women were unmarried?*

Theoretical Explanation

Logically, the fact that more American women are living without a spouse than with one doesn't mean that more than 50 percent of women are unmarried. In fact, in 2005, almost 54 percent of American women *were* married, but almost 5 percent of them were either legally separated or said their spouses weren't living at home for some reason (Roberts, 2007). Nonetheless, by 2005, 51 percent of American women were living without a spouse. Does this single observation, interesting as it is, constitute "knowledge"? Not according to most scientists or, in fact, most other people. In order for it to rise to the level of knowledge, most people (and scientists) would want some kind of explanation of the fact or some kind of **theory**. Roberts (2007) provides such an explanation in terms of two trends: (1) younger women marrying later and (2) older women not remarrying after being widowed or after divorce. The fact that one might logically deduce from these two trends that fewer women would be living with spouses than before really enhances your sense that you know something. Doesn't it?

theory, an explanation about how and why something is as it is.

The relative strengths of the scientific method, then, derive from several attributes of its practice. Ideally, the method involves communities of relatively equal knowledge seekers (scientists) among whom findings are communicated freely for careful scrutiny. Knowledge claims are ideally subjected to factual tests and to tests of logical reasoning. They're also supposed to be explicable in terms of theory.

STOP & THINK   *Suppose I submit a research report to a journal and the journal's editor writes back that the journal won't publish my findings because expert reviewers don't find them persuasive. Which of the strengths of the scientific method is the editor relying on to make his or her judgment?*

THE USES AND PURPOSES OF SOCIAL RESEARCH

By now, you might be saying, "OK, research methods may be useful for finding out about the world, for solving mysteries. But aren't there any more practical reasons for learning them?" We think there are.

We hope, for instance, that quite a few of you will go on and apply your knowledge of research methods as part of your professional lives. We know that many of our students have done so in a variety of ways: as graduate students and professors in the social sciences, as social workers, as police or correctional officers, as analysts in state agencies, as advocates for specific groups or policies, as community organizers, or as family counselors, to name but a few. These students, like us, have tended to direct their research toward two different audiences: toward the scientific community in general, on the one hand, and toward people interested in specific institutions or programs, on the other. When they've engaged the scientific community in general, as Sandra Enos does in the research, reported in Chapter 10, on how women in prison mother their children, they've tended to engage in what is sometimes called basic research. Enos reports to other scientists, for instance, that, while incarcerated, white women tend to place their children in the care of their husbands or the state (e.g., foster care), and African American women tend to place their children in the care of their own mothers. Basic research is designed to add to our knowledge and understanding of the social world for the sake of that knowledge and understanding. Much of the focal research in this book, including both Enos's work on women in prison (Chapter 10) and Hoffnung's work on college students' plans about (and achievements in) their futures (Chapter 4), is basic research.

When our former students have addressed a clientele with interests in particular institutions or programs, like the students quoted in the left-hand column of Box 1.1, they've tended to engage in what is called applied research, which aims to have practical results and produce work that is intended to be useful in the immediate future. Schools, legislatures, government and social service agencies, health care institutions, corporations, and the like all have specific purposes and ways of "doing business." Applied research, including evaluation research and action-oriented research, is designed to provide information that is immediately useful to those participating in institutions or programs. Such research can be done for or with organizations and communities and can include a focus on the action implications of the research. Evaluation research, for example, can be designed to assess the impact of a specific program, policy, or legal change. It often focuses on whether a program or policy has succeeded in effecting intentional or planned changes. Participatory action research, which we'll discuss in Chapter 14, is done jointly by researchers and community members. It often has an emancipatory purpose. Participatory action research focuses on the differentiated consequences of social oppression and "lifts the multiple stories and counter stories dwelling within any complex institution or social arrangement" (Fine and Torre, 2006: 255). You'll find an example of applied research in Chapter 14, where Harrison and Norton-Hawk assess the needs of incarcerated women in Ecuador.

Even if you don't enter a profession in which you'll do research of the sort we discuss in this book, we still think learning something about research methods can be one of the most useful things you do in college. Why? Oddly, perhaps, our answer implicates another apparently esoteric subject: theory.

When we speak of theory, we're referring not only to the kinds of things you study in specialized social theory courses, although we do include those things. We view theories as explanations about how and why things are as they are. In the case of social theories, the explanations are about why people "behave, interact, and organize themselves in certain ways" (Turner, 1991: 1).

basic research, research designed to add to our fundamental understanding and knowledge of the social world regardless of practical or immediate implications.

applied research, research intended to be useful in the immediate future and to suggest action or increase effectiveness in some area.

BOX **1.1**

Examples of Applied Research

Here are just a few examples of applied research some of our graduates have done or are planning to do:

"When I worked for the Department of Children, Youth, and Their families, we conducted a survey of foster parents to see what they thought of foster care and the agency's services. The parents' responses provided us with very useful information about the needs of foster families, their intentions for the future, and the kinds of agency support that they felt would be appropriate."
Graduate employed by a state Department of Children, Youth, and Their Families

"As the Department of Corrections was under a court order because of crowding and prison conditions, it was important that we plan for the future. We needed to project inmate populations and did so using a variety of data sources and existing statistics. In fact, we were accurate in our projections."
Graduate employed by a state Department of Corrections

"I'm working at a literacy program designed to help children in poverty by providing books for the preschoolers and information and support for their parents. I've realized that while the staff all think this is a great program, we've never really determined how effective it is. It would be wonderful if we could see how well the program is working and what we could do to make it even better. I plan on talking to the director about the possibility of doing evaluation research."
Graduate employed by a private pediatric early literacy program

Examples of Basic Research

Here are three more examples of basic research that former students and current colleagues have done:

Paul Khalil Saucier (2008) conducted fieldwork in Greater Boston from 2007 to 2008 to explore ways in which second-generation Cape Verdean youth negotiate their identity as Cape Verdean and as black. He paid particular attention to hip-hop culture to see how it was used as a site where new identities were fashioned and reworked.

Desiree Ciambrone (2001) interviewed women with human immunodeficiency virus (HIV) infection and found that many did not consider HIV to be the most devastating event in their lives. She found, rather, that violence, mother–child separation, and drug use were seen to be more disruptive than HIV infection. You may read more about ethical considerations involved in Ciambrone's work in Chapter 3.

Mikaila Arthur (2007) examined the rise of women's studies, Asian-American studies, and queer studies programs in colleges and universities in the United States. She found, among other things, that while external market forces were weak predictors of curricular change, pressure by students and faculty for such change was the necessary condition for the rise of such programs.

Such explanations are useful, we feel, not only because they affect how we act as citizens—as when, for instance, we inform, or fail to inform, elected representatives of our feelings about matters such as welfare, joblessness, crime, and domestic violence—but also because we believe that Charles Lemert (1993: 1) is right when he argues that "social theory is a basic survival skill." Individuals survive in society to the extent that they can say plausible and coherent things about that society.

Useful social theory, in our view, concerns itself with those things in our everyday lives that can and do affect us profoundly, even if we are not aware of them. We believe that once we can name and create explanations (or create theories) about these things, we have that much more control over them. At the very least, the inability to name and create such explanations leaves us powerless to do anything. These explanations can be about why some people live alone and some don't, why some are homeless and some aren't, why some commit crimes and some don't, why some do housework and some don't, and why some people live to be adults and some don't. These explanations can come from people who are paid to produce them, like social scientists, or from people who are simply trying to make sense of their lives. Lemert (1993) reminds us that the title for Alex Kotlowitz's (1991) *There Are No Children Here* was first uttered by the mother of a 10-year-old boy, Lafeyette, who lived in one of Chicago's most dangerous public housing projects. This mother observed, "But you know, there are no children here. They've seen too much to be children" (Kotlowitz, 1991: 10). Hers is eloquent social theory, with serious survival implications for those living in a social world where nighttime gunfire is commonplace.

We'll have more to say about theory and its connection to research methods in the next chapter. But, for now, we'll simply say that we believe the most significant value of knowledge of research methods is that it permits a critical evaluation of what others tell us when we and others develop social theory. This critical capacity should, among other things, enable us to interpret and use the research findings produced by others. Our simple answer, then, to the question about the value of a research methods course is not that it adds to your stock of knowledge about the world, but that it adds to your knowledge of *how* you know the things you know, *how* others know what they know, and ultimately, *how* this knowledge can be used to construct and evaluate the theories by which we live our lives.

The major purposes of scientific research, in many ways overlapping with the "uses" (of, say, supplying other scientists with basic information, supplying interested persons with information about programs, or developing theories) we've mentioned, include *exploration, description, explanation,* and *evaluation*. Although any research project can have more than one purpose, let's look at the purposes individually.

Exploratory Research

exploratory research,
groundbreaking research on a relatively unstudied topic or in a new area.

In **exploratory research,** the investigator works on a relatively unstudied topic or in a new area, to become familiar with this area, to develop some general ideas about it, and perhaps even to generate some theoretical perspectives on it. Exploratory research is almost always inductive in nature, as the researcher starts with observations about the subject and tries to develop tentative generalizations about it (see Chapter 2).

An example of exploratory research is Enos's study of how women inmates manage to mother their children when they're in prison, presented in Chapter 10. Enos began by observing not only how much the U.S. prison population has grown in recent decades, but also how much the population of women prisoners had grown—between just 1985 and 2000, it had increased by more than 300 percent. Moreover, she noted that the imprisonment of women created a special problem for their children: where and with whom to live. Enos wondered how these women were dealing with an increasingly urgent issue: the "managing of motherhood from prison." Enos wondered whether she'd be able to spot patterns if, first, she observed the interactions of incarcerated women and their children in a prison's weekend "parenting" program and, then, if she interviewed 25 women inmates intensively. What she noticed was that white and African American women had quite distinctive approaches to placing their children while incarcerated (e.g., African American women were much more likely to place their children with their own mothers or other relatives than white women) and that these patterns had a lot to do with the distinctive ways they'd experienced childhood themselves (e.g., African American women were less likely than white women to blame the people in their childhood families for the behavior that led them to prison). Similarly, Traver (2007) interviewed a relatively unstudied population—American parents of children adopted from China. She found that, to a greater or lesser degree, these parents tended to evolve "Chinese" identities whether they themselves were European American, African American, or Asian American, through their acquisition and use of Chinese cultural objects (ranging from Chinese scrolls to posters of the NBA basketball player, Yao Ming). Both Enos and Traver, then, did exploratory research—research into a relatively new subject (mothers in prison and adoptive parents of Chinese-born children), collected data through observations or in-depth interviews of a relatively few cases, and tried to spot themes that emerged from their data. Although exploratory analyses, with their focus on relatively unexplored areas of research, do not always employ this kind of thematic analysis of data on relatively few cases, when they do they undertake what is called **qualitative data analysis**, or analysis that tends to involve the interpretation of actions or the representations of meanings in words (see Chapter 15).

qualitative data analysis, analysis that results in the interpretation of action or representation of meanings in the researcher's own words.

Descriptive Research

In a **descriptive study**, a researcher describes groups, activities, situations, or events, with a focus on structure, attitudes, or behavior. Researchers who do descriptive studies typically know something about the topic under study before they collect their data, so the intended outcome is a relatively accurate and precise picture. Examples of descriptive studies include the kinds of polls done during political election campaigns, which are intended to describe how voters intend to vote, and the U.S. Census, which is designed to describe the U.S. population on a variety of characteristics. The Census Bureau description of the makeup of households in the United States (2006) is just that a description, in this case of the whole U.S. population. As the pie chart in Figure 1.1 indicates, about 26.6 percent of American households in 2006 were made up of people living alone. Unlike exploratory studies, which might help readers become familiar with a new topic, descriptive studies can provide a very detailed and precise idea of the way things are. They can also provide a sense of how things have changed over time. Thus, a study by Clark and Nunes (2008) of the pictures in introductory sociology textbooks found that, while in textbooks of the 1980s only 34 percent of identifiable individuals shown in pictures were women, in textbooks of the 2000s almost 50 percent of those images were of women. And Grady's investigation of a sample of advertising images in *Life* magazine between 1936 and 2000 shows that "while commitment to racial

descriptive study, research designed to describe groups, activities, situations, or events.

integration appears to have taken longer to develop than survey data suggest, this commitment seems to be much firmer than findings based on census data imply" (2007a: 211). Descriptive research, in its search for a picture of how the land lies, can be based upon data from surveys (as the Census report is) or on content analyses (used by Clark and Nunes and by Grady). Often, as in the case of the Census Bureau's study of American households, descriptive research generates data about a large number of cases: there were actually 105,480,101 households enumerated by the census of 2000. To analyze these data meaningfully, descriptive researchers frequently use quantitative data analysis, or analysis that is based on the statistical summary of data (see Chapter 15).

quantitative data analysis, analysis based on the statistical summary of data.

Explanatory Research   WHY things are

Unlike descriptive research, which tends to focus on how things are, the goal of explanatory research is to explain why things are the way they are. Explanatory research looks for causes and reasons. Unlike exploratory research, which tends to be inductive, building theoretical perspectives from data, explanatory research tends to be deductive, moving from more general to less general statements. Thus, for instance, explanatory research often uses preexisting theories to decide what kinds of data should be collected (see Chapter 2). In a study present in Chapter 7, we, Emily and Roger, asked ourselves the explanatory question: Why are some people more likely than others to do work for pay after they retire from their regular lifetime work? We knew that some people were likely to do it because they genuinely needed to augment their post-retirement incomes. But we believed that financial hardship couldn't possibly explain all the variation in retired people's actual participation in post-retirement work. Even some professional workers, people with decent savings and good pensions, returned to work after their official retirements. And so we did a study, primarily of professional people, to see if we could figure out why some of them were more likely than others to do post-retirement work.

explanatory research, research designed to explain why subjects vary in one way or another.

Deductive: General to less general (deduction)

We did what many scientists do: We consulted theory. In fact, we consulted a couple of theories about retirement, one of which is "continuity theory," a theory that suggests that some people are more invested in roles, like work roles, than others.

In Chapter 2, you'll read more about continuity theory, but for now we'll just mention a key element of the theory. It expects retired people, when deciding whether to work for pay or not, to make decisions based upon how much of their self-worth was dependent on work during their working lives. From continuity theory, we derived an expectation about the length of a person's pre-retirement career and his or her likelihood of continuing to work after retirement. After surveying people twice, once before they retired and once afterward, we found, in fact, that people who had had longer careers were considerably more likely to work after retirement than people who had had shorter careers. In our work, we describe something (the length of a professional's career) that "goes with" something else (the likelihood of working after formal retirement). In doing so, we explain this connection with a theory (continuity theory), one of the goals of the scientific method.

Explanatory analyses, with their focus on areas upon which a researcher might be able to shed theoretical light in advance of collecting data, may, like descriptive analyses, generate data about relatively large numbers of cases and employ statistical analyses to make sense of these cases. When they do, they, like many descriptive analyses, involve quantitative data analysis (see Chapter 15).

Evaluation Research

evaluation research, research designed to assess the impacts of programs, policies, or legal changes.

Although evaluation research can be seen as a special kind of explanatory research, it is distinctive enough that we feel it is worth its own place. Evaluation research is research designed to assess the impacts of programs, policies, or legal changes. It often focuses on whether a program or policy has succeeded in effecting intended or planned change, and when such successes are found, the program or policy explains the change. Thus, when Babalola, Folda, and Babayaro (2008) were able to show that exposure to a communication campaign designed to encourage young people to use family planning in Nigeria had, in fact, led to greater contraceptive use, they were providing an explanation of that usage. But this is a different kind of explanation than the theoretical explanation that Roger and Emily used in explaining why some people work after retirement and others do not. We will be discussing evaluation and other applied research at much greater length in Chapter 14.

STOP & THINK   *Suppose you've been asked to learn something about the new kinds of communities that have arisen out of people's use of tweets and twitter. Of the four kinds of research outlined earlier (exploratory, descriptive, explanatory, and evaluation), what kind of study have you been asked to do?*

SUMMARY

We've argued here that, at its best, research is like trying to solve a mystery, using the scientific method as your guide. We've distinguished the scientific approach of social research methods from two other approaches to knowledge about the social world: a reliance on the word of "authorities" and an exclusive dependence on "personal inquiry." We've suggested that the scientific method compensates for the shortcomings of these two other approaches in several ways. First, science emphasizes the value of communities of relatively equal knowledge seekers who are expected to be critical of one another's work. Next, science stresses the simultaneous importance of empirical testing, logical reasoning, and the development or testing of theories that make sense of the social world. Two communities that researchers report to are the community of other scientists (basic research) and communities of people interested in particular institutions or programs (applied research). We've argued that knowing research methods may have both professional and other practical benefits, not the least of which is the creation of usable theories about our social world. We suggest that social research methods can help us explore, describe, and explain aspects of the social world, as well as evaluate whether particular programs or policies actually work.

EXERCISE 1.1

Ways of Knowing about Social Behavior

This exercise compares our recollections of our everyday world with what we find out when we start with a research question and collect data.

Part 1: Our Everyday Ways of Knowing

Pick any *two* of the following questions and answer them based on your recollections of past behavior.

1. What does the "typical" student wear as footwear to class? (Will the majority wear shoes, boots, athletic shoes, sandals, and so on?)
2. While eating in a school cafeteria, do most people sit alone or in groups?
3. Of those sitting in groups in a cafeteria, are most of the groups composed of people of the same gender or are most mixed-gender groups?
4. Of your professors this semester who have regularly scheduled office hours, how many of them will be in their offices and available to meet with you during their next scheduled office hour?

Based on your recollection of prior personal inquiry, describe your expectations of social behavior.

Part 2: Collecting Data Based on a Research Question

Use the two questions you picked for Part 1 as "research questions." With these questions in mind, collect data by carefully making observations that you think are appropriate. Then answer the same two questions, but this time base your answers on the observations you made.

Part 3: Comparing the Ways of Knowing

Write a paragraph comparing the two ways of knowing you used. (For example, was there any difference in accuracy? Was there any difference in ease of collecting data? Which method do you have more confidence in?)

EXERCISE 1.2

Social Science as a Community Endeavor

This exercise is meant to reinforce your appreciation of how important the notion of a community of relatively equal knowledge seekers is to social research. We'd like you to read any one of the "focal research" articles in subsequent chapters and simply analyze that article as a "conversation" between the author and one or two authors who have gone before. In particular, summarize the major point of the article, as you see it, and see how the author uses that point to criticize, support, or amend a point made by some other author in the past.

1. What is the name of the focal research article you chose to read?
2. Who is (are) the author(s) of this article?
3. What is the main point of this article?
4. Does the author seem to be criticizing, supporting, or amending the ideas of any previous authors to whom she or he refers?
5. If so, which author or authors is/are being criticized, supported, or amended in the focal research?
6. Write the title of one of the articles or books with which the author of the focal research seems to be engaged.
7. Describe how you figured out your answer to the previous question.
8. What, according to the author of the focal research, is the central idea of the book or article being discussed?
9. Does the author of the focal research finally take a critical or supportive position in relation to this idea?
10. Explain your previous answer.

EXERCISE 1.3

Ways of Knowing about the Weather

This exercise is designed to compare three ways of knowing about the weather.

Part 1:

Knowledge from Authorities.   The night before, see what the experts have to say about the weather for tomorrow by watching a television report, listening to a radio newscast, or checking online. Write what the experts said about tomorrow's weather (including the temperature, the chances of precipitation, and the amount of wind).

Knowledge from Casual Personal Inquiry.   That day, before you go outside, look through only one window for a few seconds but don't look at a thermometer. After taking a quick glance, turn away, and then write down your perceptions of the weather outside (including the temperature, the amount and kind of precipitation, and the amount of wind).

Knowledge from Research.   Although we're not asking you to approximate the entire scientific method (such as reviewing the literature and sharing your findings with others in a research community), you can use some aspects of the method: specifying the goals of your inquiry and making and recording careful observations.

Your research question is, "What is the weather like outside?" To answer the question, use a method of collecting data (detailed observation of the outside environment) and any tools at your disposal (thermometer, barometer, and so on). Go outside for at least five minutes and make observations. Then come inside and write down your observations of the weather outside (including the temperature, the amount and kind of precipitation, and the amount of wind).

Part 2: Comparing the Methods

Write a paragraph comparing the information you obtained using each of the ways of knowing. (For example, was there any difference in accuracy? Was there any difference in ease of collecting data? Which method do you have the most confidence in?)

CHAPTER **2**

Theory and Research

© Emily Stier Adler

INTRODUCTION

Concepts, Variables, and Hypotheses

We imagine that you've given some thought as to whether abortion is moral or not and therefore whether it should be legal or not. But have you ever considered the ways in which legalized abortion might be associated with legal (or illegal) behaviors of other kinds? In this chapter, on the relationship between theory and research, we will begin with an example that forces us to think about such an association. Let's examine Box 2.1, a summary of a theory by John Donohue and Steven Levitt, that connects the legalization of abortion in the early 1970s and a decrease in crime rates in the United States in the early 1990s. You'll recall from Chapter 1 that a theory is an explanation of the "hows" and "whys" of something.

concepts, words or signs that refer to phenomena that share common characteristics.

Donohue and Levitt's theory about the connection between abortion legalization and crime rates, like all theories, is formulated in terms of **concepts,** which are words or signs that refer to phenomena that share common characteristics. Some, but not all, of these concepts refer to phenomena that are, at least for the purposes of the theory, relatively fixed or invariable in nature. Donohue and Levitt, for example, refer to the "high-crime late adolescent years" and "birth cohort." These concepts require clarification for the purposes of research, or what researchers call **conceptualization,** but once defined, they do not vary. For Donohue and Levitt, the high-crime late adolescent years begin at about 17 years of age, and a birth cohort is the people born in a calendar year. The first birth cohort in the United States that could have been affected by the 1973 *Roe v. Wade* decision would have been born in 1974, therefore Donohue and Levitt expect that there should have been a diminution in this cohort's crime rate in comparison with previous cohorts 17 years later, or about 1991.

conceptualization, the process of clarifying what we mean by a concept.

BOX **2.1**

A Theoretical Statement: Why the Legalization of Abortion Contributed to the Decline of Crime Rates

In 2001, John Donohue and Steven Levitt published a highly controversial research report, "The Impact of Legalized Abortion on Crime," in which they claimed that the introduction of legalized abortion in the United States in the early 1970s led to a drop in crime in the early 1990s. The key theoretical argument in Donohue and Levitt's paper was that abortion "is more frequent among parents who are least willing or able to provide a nurturing home environment" (2001: 386). Moreover, children from less nurturing home environments are more likely than others to engage in criminal activity, once they reach the "high-crime late adolescent years" (386). Therefore, to the extent that the legalization of abortion in the early 1970s enabled parents who would provide the least nurturing home environments to avoid having children, it also led to a lower crime rate when the affected birth cohort reached its late adolescent years.

 **STOP & THINK** *Do you see how the concept of a "high-crime late adolescent years" is relatively fixed? How its definition does not admit much variation? (It begins at about 17 and ends, one guesses, at about 19 years of age.) Can you identify concepts used by Donohue and Levitt that could vary from one person or birth cohort to another?*

In addition to using concepts that identify phenomena that are relatively fixed in nature, which don't vary for the purposes of the research, like "high-crime late adolescent years," Donohue and Levitt use concepts such as "nurturing home environment" and "criminal activity" that might vary from one person to another or from one birth cohort to another. A person might experience a nurturing home or not and might engage in criminal activity later in life or not. A higher percentage of the children in one birth cohort, for instance, might experience a nurturing home environment than those in another birth cohort. A higher percentage of the people in one birth cohort might engage in criminal activity when they are 17 years of age than those in another birth cohort. Whether one focuses on individuals or birth cohorts would, of course, make a big difference for a particular research project. Individuals and birth cohorts are different units of analysis, which are, as you may recall, units about which one collects information. But what do we call the information that is gathered about these units? In the sciences, this information is in the form of categories or values of variables—or categories of characteristics that can vary from one unit of analysis to another or for one unit of analysis over time. One can also think of a **variable** as a concept that varies.

variable, a characteristic that can vary from one unit of analysis to another or for one unit of analysis over time.

 **STOP & THINK** *The lowest possible value for the percentage of a birth cohort that has experienced a nurturing home environment is zero percent. What's the highest possible value? What are the lowest and highest possible values for the percentage of a birth cohort that engages in criminal activity when it is 17 years old?*

 **STOP & THINK AGAIN** *Suppose, in testing Donohue and Levitt's theory, you wanted to make the individual, not the birth cohort, your unit of analysis. What categories could you use to determine whether someone engaged in criminal activity after he or she reached his or her seventeenth birthday?*

To be a variable, a concept must have at least two categories. For the variable "gender," for instance, we can use the categories "female" and "male." Gender is most frequently used to describe individual people. As we've indicated, the variable "criminal activity" used by Donohue and Levitt, however, can be meant, in the theoretical formulation explained earlier, to describe a birth cohort. However, getting information about a birth cohort is very difficult, so Donohue and Levitt use substitutes for birth cohort, focusing instead on the crime rates for the whole U.S. population in different years. They end up measuring the concept of "criminal activity" in a variety of ways. One of these ways is as a characteristic of the whole nation and in terms of the number of crimes committed in a year per 100,000 people (see Table 2.1). This variable could have many more than two, perhaps even an infinite number, of categories. Scientists can use variables to describe the social world, just as you might use the variable "gender" to describe students in your research methods class.

 **STOP & THINK** *Notice that the crime rate is a variable characteristic of the United States because it can vary over time. Notice also that this variable could take on many, rather than just two, values. (Actually, in principle it could vary from zero to infinity. Right?) What pattern in Table 2.1 do you spot in the U.S. crime rate between 1985 and 2006? In what year did a reversal of the*

TABLE **2.1**
What Happened to the Crime Rate in the United States
between 1985 and 2006?

Year	Total Crime Rate (number of crimes per 100,000 people)
1985	5,207.1
1986	5,480.4
1987	5,550.0
1988	5,664.2
1989	5,741.0
1990	5,820.3
1991	5,897.8
1992	5,660.2
1993	5,484.4
1994	5,373.9
1995	5,275.9
1996	5,086.6
1997	4,930.0
1998	4,615.5
1999	4,266.5
2000	4,124.8
2001	4,162.6
2002	4,118.8
2003	4067.0
2004	3977.3
2005	3900.5
2006	3808.1

Source: The Disaster Center (2008).

previous pattern of increases begin? When was the first cohort affected by the 1973 Roe v. Wade decision born? How many years was this before the downturn in the national crime rate? How might this be significant for the thesis advanced by Donohue and Levitt?

When we think of "crime activity" as the crime rate of the country, then, we see that it does decline at about the time that Donohue and Levitt's theory predicts: about 17 years (in 1991) after *Roe v. Wade* would have had its initial effects (in 1974). This illustrates a frequent use of theory for research: to point to variable characteristics of subjects that can be expected to be associated with each other. Thus, Donohue and Levitt are claiming that, among other things, they expect that a downturn in the crime rate should occur in jurisdictions where abortion became legal sometime earlier. Similarly, a research project that we (Emily and Roger) completed—mentioned in Chapter 1—found that people who do volunteer work before they retire are much more likely to do volunteer work after they retire than those who did not do volunteer work before they retire. Using the categories of two variables, "whether one did volunteer work before retirement" and "whether one does volunteer work after retirement," we found that "those who did volunteer work before retirement" were more likely "to do volunteer work after retirement" and "those who did not do volunteer work before retirement" were more likely "to not do volunteer work after retirement."

hypothesis, a testable statement about how two or more variables are expected to relate to one another.

The finding supported an expectation that we had derived from role theory about the way the two variables would be associated. We stated this expectation in the form of a **hypothesis,** which is a statement about how two or more variables are expected to relate to one another. We developed this hypothesis:

> People who have performed volunteer roles before retirement will be much more likely to perform such roles after retirement than people who have not performed such roles. (Chapter 7)

Note that this hypothesis links two variable characteristics of individuals: their participation in volunteer work before retirement and their participation in such work after retirement. You might also note that, like many other social science hypotheses, this one doesn't refer to absolute certainties (such as, "If one does volunteer work before retirement, one will always engage in volunteer work after retirement, and if one does not do volunteer work before retirement, one will never engage in volunteer work after retirement"). Rather, this hypothesis speaks of tendencies.

The business of research, you've probably surmised, can be complex and can involve some fairly sophisticated ideas (theory, hypotheses, concepts, and variables, to name but a few), all of which we'll discuss again later. Meanwhile, we'd like to introduce you to one more fairly sophisticated idea before moving on to a more general consideration of the relationship between research and theory. This is the idea that explanations involve, at minimum, the notion that change in one variable can affect or influence change in another. Not surprisingly, scientists have special names for the "affecting" and "affected" variables. Variables that are affected by change in other variables

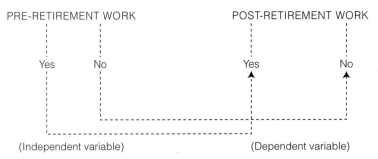

FIGURE **2.1** A Diagram of the Hypothesized Relationship Between Pre-Retirement Volunteer Work and Post-Retirement Volunteer Work

dependent variable, a variable that a researcher sees as being affected or influenced by another variable (contrast with independent variable).

independent variable, a variable that a researcher sees as affecting or influencing another variable (contrast with dependent variable).

are called **dependent variables**—dependent because they depend on change in the first variable. **Independent variables,** on the other hand, are variables that affect change in dependent variables.

Thus, when Donohue and Levitt hypothesize that making abortion legal leads to declines in the crime rate, they are implying that a change in the independent variable (whether abortions are legal) will cause change in the dependent variable (the crime rate). And, when we hypothesize that people who do volunteer work before retirement are more likely to do volunteer work after retirement than those who do not do volunteer work before retirement, we are suggesting that a change in the independent variable (doing or not doing volunteer work before retirement) will cause change in the dependent variable (doing or not doing volunteer work after retirement). Having an arrow stand for something like "is associated with," hypotheses can be depicted using diagrams like the one in Figure 2.1. The arrows in this diagram suggest that having done volunteer work before retirement "is associated with" doing volunteer work after retirement, while not having done volunteer work before retirement "is associated with" not doing volunteer work after retirement. And, in general, when one or more categories of one variable are expected to be associated with one or more categories of another variable, the two variables are expected to be associated with each other.

 **STOP & THINK** *Can you draw a diagram of Donohue and Levitt's hypothesis linking the legalization of abortion to declines in criminal activity? (Hint: It might be easiest to think in terms of individuals, not cohorts or countries. What kind of person, according to Donohue and Levitt's thesis, is more likely to engage in criminal activity: someone born when abortion is legal or some-one who is not born when abortion is legal?)*

 **STOP & THINK AGAIN** *See if you can identify the independent and dependent variable of hypotheses that guided some other recent research: (1) Kravdal and Rindfuss's (2008) reexamination of the hypothesis that countries with higher average levels of education will have lower fertility (birth) rates than countries with lower average levels of education, (2) Lu and Treiman's (2008) reinvestigation,*

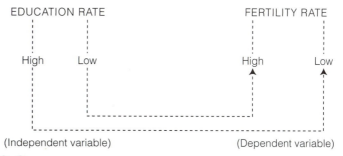

FIGURE **2.2** A Diagram of the Hypothesized Relationship Between Gender and Attribution of Responsibility

---

using Chinese data, of the hypothesis that the larger the number of siblings (brothers and sisters), the lower the educational attainment, and (3) Lundquist's (2008) hypothesis that white females will experience greater job satisfaction while doing military service than white males will experience. You might also see how many of these hypotheses you can draw as a diagram.

We'd like to give you one more example of how to take apart a hypothesis. Let's look at the first hypothesis mentioned in the previous "Stop and Think," the one by Kravdal and Rindfuss's (2008), that countries with higher average levels of education will have lower fertility (birth) rates than countries with lower average levels of education. Here the units of analysis are countries, the independent variable is education rates (which can vary from low to high), and the dependent variable is fertility rates (which can also vary from low to high). We would draw a diagram of this hypothesis as shown in Figure 2.2.

We want to assure you that at this point we don't necessarily expect you to feel altogether comfortable with all the ideas we've introduced so far. They'll all come up again in this book, so you'll gain more familiarity with them. For now, we want to emphasize a major function of theory for research: to help lead us to hypotheses about the relationship between or among variables. We'd like to turn to a more general consideration of the relationship between research and theory. But first let us say a little more, by way of a word of caution, about the difficulty of establishing causality in social science.

Social Science and Causality: A Word of Caution

The fact that two variables are associated with each other doesn't necessarily mean that change in one variable causes change in another variable. Table 2.1 shows, generally, what Donohue and Levitt (2001), in their much more sophisticated analysis, contend: When abortion rates (the independent variable) go up, societies are likely to experience a decline in crime rates (the dependent variable) later, when the more "wanted" generation grows up. But does this mean that making abortion legal causes lower crime rates? Donohue and Levitt suggest that it does. Others argue that it doesn't.

In the social sciences, we have difficulty establishing causality for several reasons. Some reasons are purely technical, and we'll discuss them in greater detail in Chapter 8. One reason establishing causality is difficult is that to show that change in one variable causes change in another; we want to be sure that the "cause" comes before, or at least not after, the "effect." But many of our most cherished methods of collecting information in the social sciences just don't permit us to be sure which variable comes first. Stevic and Ward (2008), for instance, found that students who gave indications, in a questionnaire survey, of having relatively high level of satisfaction with their lives were more likely than others to be involved in changing and developing themselves as individuals. The association of these two variables, however, doesn't prove that life satisfaction causes them to develop as individuals because, given the nature of the questionnaire survey, Stevic and Ward had to collect information about the two variables at the same time. They couldn't tell whether life satisfaction came before their involvement in changing and developing themselves or changing and developing themselves came before their life satisfaction. It would seem that Donohue and Levitt's research at least passes the time-order test: that the "cause" (abortions, subsequent to *Roe v. Wade*) comes before the "effect" (a decline in crime rates in the 1990s). In this case, however, there can remain doubt about whether the "right" time "before" has been specified. Donohue and Levitt, in their sophisticated statistical analysis, make very plausible guesses about appropriate time "lags," or periods over which abortion will have its greatest effects on crime, but even they can't be sure that their "lags" are exactly right.

STOP & THINK *Donohue and Levitt (2001: 382) think there was something pretty significant about the 17 years between when Roe v. Wade would have affected its first birth cohort (1974) and the downturn, beginning in 1991, in the overall crime rate in the United States. Can you see any reasons why 17 years might be a significant length of time? Can you imagine, assuming that abortion has any effect on crime, why one might expect a shorter or longer period of time between the legalization of abortion and the onset of significant declines?*

Another technical problem with establishing causality is that even if we can hypothesize the correct causes of something, we usually can't demonstrate that another factor isn't the reason why the "cause" and "effect" are associated. We might be able to show, for instance, that members of street gangs are more likely to come from single-parent households than nonmembers. However, because we're not likely to want or be able to create street gangs in a laboratory, we can't test whether other things (like family poverty) might create the association between "family structure" (an independent variable) and "street gang membership" (a dependent variable). Similarly, it's possible that some "other thing" (like swings in the economic cycle) might create conditions that make "abortion rates" (people may be more likely to have more abortions during economic downturns) and "subsequent crime rates" (people may be less likely to commit crimes during economic upturns) go together.

Doing experiments (discussed in Chapter 8) is most useful in demonstrating causality, but this strategy frequently doesn't lend itself well (for ethical and practical reasons, discussed in the next two chapters) to social science investigations.

antecedent variable, a variable that comes before both an independent variable and a dependent variable.

The idea of a third variable, sometimes called an **antecedent variable**, which comes before, and is actually responsible for, the association between an independent variable and a dependent variable, is so important that we'd like to give you another example to help you remember it. Firefighters will tell you of the association between the number of firefighters at a fire (an independent variable) and the damage done at the fire (a dependent variable): that the more firefighters at a fire, the more damage occurs.

 **STOP & THINK**   *Can you think of an antecedent variable that explains why fires that draw more firefighters are more likely to do more damage than fires that draw fewer firefighters?*

Perhaps it occurred to you that a characteristic of fires that would account for both the number of firefighters drawn to them and the amount of damage done is the size of the fire. Smaller fires tend to draw fewer firefighters and do less damage than larger fires. In this case, the antecedent variable, the size of the fire, explains why the independent variable and the dependent variable are associated. When this happens, when an antecedent variable provides such an explanation, the original association between the independent variable and the dependent variable is said to be **spurious,** or non-causal. Neither one causes the other; their association is due to the presence of an antecedent variable that creates the association. It is only when the association between an independent variable and a dependent variable is non-spurious (not generated by some third variable acting on the two) that we can conclude that it is causal. But this is hard to demonstrate, because it means taking into account *all possible* antecedent variables, of which there are an infinite number.

spurious, non-causal.

The three conditions that must exist before we can say an independent variable "causes" a dependent variable are, then:

1. that the two variables are associated in fact. In the case of the fires, this would mean showing that fires that drew many firefighters actually had more damage than other fires. This is the condition of *empirical association.*

2. that the independent variable, in fact, precedes, or at least doesn't come after, the dependent variable. In the case of the fires, this would mean showing that fire damage never came before the arrival of the firefighters. This is the condition of *temporal precedence* or *time order.*

3. that there is no third variable, antecedent to the independent variable and the dependent variable, that is responsible for their association. In the case of the fires, this would mean showing that there is no variable, like the size of the fire, that led to the association between the number of firefighters at the fire and the damage done at the fire. This is the condition of *elimination of alternative explanations* (or *demonstrating non-spuriousness*).

Most social science research designs (as you'll see in Chapters 7 and 8) make it difficult to establish the condition of temporal precedence and make it impossible to establish the condition of non-spuriousness, so we can rarely say we're sure that one thing causes another. In fact, only the experimental design is meant to establish causality. We often settle for establishing the first condition, the condition of empirical association, knowing that it is a necessary condition for, but not a completely satisfactory demonstration of, causality.

 **STOP & THINK** *Philip Maymin (reported by Pothier [2009]) reports, after studying the last fifty years of popular music and over 5,000 hit songs, that when the stock market is jumpy, we prefer music with a steady beat (like Kanye West's "Heartless"), but when the stock market is calm, we prefer music with more unpredictable beats (like Sean Paul's "Like Glue"). Assuming Maymin is right, which of the conditions that must exist to show that market behavior affects our musical tastes seems to hold? Which one or ones has not yet been demonstrated?*

Proving causality is, as we've suggested, pretty tough to do, partly because one has to show that no third variable, antecedent to the independent variable and the dependent variable, is responsible for their association. Because there are, in principle, an infinite number of antecedent variables for all combinations of independent and dependent variables, eliminating all of them as possible instigators of the relationship is impossible. While we're thinking about causation, though, we'd like to mention two other kinds of variables—intervening and extraneous variables—whose presence would not challenge the possibility that an independent variable causes variation in a dependent variable. An **intervening variable** is a variable that comes between an independent and a dependent variable. Here, the researcher conceives of the independent variable affecting the intervening variable, which in turn is conceived to affect the dependent variable. Donohue and Levitt, for instance, believe legalizing abortion means that more children experience nurturing homes and that this experience reduces their likelihood of crime. Here, as in the case of all intervening variables, the experience of a nurturing home is posited to be the reason why legalizing abortion reduces the likelihood of crime. An **extraneous variable** is a variable that the researcher sees as having an effect on the dependent variable in addition to the effect of the independent variable. Unlike antecedent variables, whose presence may demonstrate that the relationship between the independent and the dependent variable is non-causal, and intervening variables, whose presence may account for how an independent variable affects a dependent variable, an extraneous variable simply provides a complementary reason for variation in the dependent variable. Social scientists rarely seek a single cause for given effects. They tend, instead, to look for multiple reasons for variation in a dependent variable. Thus, for instance, while Donohue and Levitt were initially interested in the effects of legalizing abortion on rates of crime, Levitt (2004) eventually explained the 1990s downturn in the crime rate in terms of four factors, each extraneous to the other: the legalization of abortion, an increase in the number of police, a rising prison population, and a waning crack epidemic.

intervening variable, a variable that comes between an independent and a dependent variable.

extraneous variable, a variable that has an effect on the dependent variable in addition to the effect of the independent variable.

THE RELATIONSHIP BETWEEN THEORY AND RESEARCH

If theories are explanations of the hows and whys of things, then good *social* theories might be considered to be theories that enable people to articulate and understand something about everyday features of social life that had previously been hidden from their notice. We begin this section with two examples of social theories used by the social-scientist authors of focal research pieces that we use later in this text. Although the complete articles are included in later chapters, here we will excerpt the authors' descriptions of a theory they've employed or developed through their research. The first excerpt is from our (Adler and Clark) essay entitled "Moving On? Continuity and Change after Retirement" (see Chapter 7).

 STOP & THINK *As you read each excerpt, see if you can guess whether it appears before or after the authors' presentation of their own research.*

FOCAL   RESEARCH

Excerpt from "Moving On? Continuity and Change after Retirement": Role Theory

By Emily Stier Adler and Roger Clark

Role theory (e.g., Ashforth, 2001; Wang, 2007) suggests that to the extent people are highly invested in a particular role, their feelings of self-worth are related to the ability to carry out that role. Role theorists also argue that the loss of roles can lead to depression or anxiety. We posit, then, that after retirement, to the extent possible, people will retain previous roles because doing so enables them to maintain feelings of self-worth and avoid feelings of anxiety and depression. Moreover, we posit that pre-retirement expressed desires to perform new roles (like new volunteer roles) after retirement are less likely to be based upon the same emotional calculus of self-worth, depression, and anxiety that informs the desire to continue performing a pre-retirement role. In fact, we expect that all workers have at least some cognitive access to the emotional and psychological needs that work roles have satisfied for them and that their plans to do some kind of work after retirement will therefore be related to whether they actually work for pay after retirement.

References

Ashforth, B. 2001. *Role transitions in organizational life.* Mahwah, NJ: Erlbaum.

Wang, M. 2007. Profiling retirees in the retirement transition and adjustment process: Examining the longitudinal change patterns of retirees' psychological well-being. *Journal of Applied Psychology* 92(2): 455–474.

Our second excerpt is from Jennifer C. Mueller, Danielle Dirks, and Leslie Houts Picca's "Unmasking Racism: Halloween Costuming and Engagement of the Racial Other" (see Chapter 11 for full article).

FOCAL RESEARCH

Excerpt from "Unmasking Racism: Halloween Costuming and Engagement of the Racial Other"

By Jennifer C. Mueller, Danielle Dirks, and Leslie Houts Picca

With respect to theorizing what activates the cross-racial costuming behavior of our respondents, it is useful to draw upon the concept of "rituals of rebellion"—culturally permitted and ritually framed spaces (like New Year's Eve and Mardi Gras) where the free expression of countercultural feelings are tolerated, and protected to some degree by the agents of the official culture (Gluckman, 1963). Interestingly, although the Gluckman framework might predict the use of cross-racial costuming among people of color, it is not immediately apparent that students of color use Halloween as an opportunity to create costume performances that subvert the racial and/or social hierarchy....

In contrast, there does appear to be a unique, ritually rebellious form of performance that occurs among many white students. In the "colorblind" post-Civil Rights era, it has become commonplace for whites to express frustration and resentment toward color-conscious racial remediation programs, such as affirmative action (Feagin, 2000, 2006; Wellman, 1997).

Although in truth white students occupy the dominant racial social identity group, we posit that many may entertain if not a sense of "oppression," at minimum a sense of normative restriction by a social code which prescribes "nonracist" presentations, and for which racialized Halloween "rituals of rebellion" afford some release. One white student praised Halloween as "great" because it eliminates the need to worry about racial offense. For those whites who actively endorse the idea that whites are now victimized by the preferencing of people of color (e.g., in employment, admissions, etc.), Halloween may ironically signify a suspension of this imagined "hierarchy."

References

Feagin, J. 2000. *Racist America: Roots, current realities and future reparations*. New York: Routledge.

Feagin, J. R. 2006. *Systemic racism: A theory of oppression*. New York: Routledge.

Gluckman, M. 1963. *Order and rebellion in tribal Africa: Collected essays with an autobiographical introduction*. London: Cohen and West.

Wellman, D. 1997. Minstrel shows, affirmative action talk and angry white men: Marking racial otherness in the 1990s. In *Displacing whiteness: Essays in social and cultural criticism*, 311–331, edited by R. Frankenberg. Durham, NC: Duke University Press.

Deductive Reasoning

Adler and Clark's description of "role theory" comes at the beginning of their article, before they present their own research; the Mueller, Dirks, and Houts Picca's theory of what might be called the functions of cross-racial costuming by white students during Halloween comes at the end of their article, after the presentation of their findings. Note, however, that both theories attempt to shed light on what might have been a previously hidden aspect of social life. They do so with concepts (such as "self-worth" and "anxiety" or "rituals of rebellion" and "reverse racism") that are linked, explicitly or implicitly, with one another. These concepts are neither so personal as the one espoused by Lafeyette's mother at the beginning of *There Are No Children Here* (remember the boy mentioned in Chapter 1?) nor as grandiose as some theories you might have come across in other social science courses. But they are theories, nonetheless.

What distinguishes these theories, and many others presented by scientists, from those presented by pseudoscientists (astrologers and the like) is that they are offered in a way that invites skepticism.[1] Neither we nor Mueller, Dirks, and Houts Picca are 100 percent sure about the correctness of the explanation for the social world. For both sets of researchers, a major value of research methods is that they offer a way to test or build social theories that fit the observable social world. We'll spend the rest of this chapter discussing how research can act as a kind of "go-between" for theory and the observable world.

Many social scientists engage in research to see how well existing theory stands up against the real world. That is, they want to *test* at least parts of the theory. We did our research because of an interest in explaining why some people work for pay and do volunteer work after retirement and others don't. We started with a general theory of people's behavior in later life: role theory. This theory, like many others, is made up of very general, though not necessarily testable, propositions. One of these propositions is that people's feeling of self-worth in later life is dependent on their ability to continue to play roles they'd played earlier in life. From this theory, we deduced, among others, the testable hypothesis: people who did volunteer work before retirement were more likely to do volunteer work after retirement than those who hadn't.

Having made this deduction, we needed to decide what kind of observations would count as support for the hypothesis and what would count as nonsupport. In doing so, we devised a way of measuring the variables for the purpose of the study. Measurement, as we will suggest in Chapter 6, involves devising strategies for classifying units of analysis (in this case, people around retirement age) by categories (say, those who did volunteer work and those who didn't) to represent variable concepts (say, volunteering). To measure whether respondents did volunteer work before retirement, Emily, who actually did the interviewing, asked questions, during a pre-retirement interview, about work history, followed by a question like, "So thinking about life outside of work, how would you say you spend your free time?" If respondents clearly indicated that part of their free time was spent as volunteers, they were counted doing volunteer work.

measurement, the process of devising strategies for classifying subjects by categories to represent variable concepts.

---

[1] The implicit or explicit invitation to skepticism is what distinguishes scientific theories from pseudoscientific ones, according to Carl Sagan (1995) in *The Demon-Haunted World*.

If respondents didn't mention volunteer work during the response, Emily asked, "Now you've mentioned doing several things with your free time. Do you spend some time doing volunteer work?" If respondents now indicated that they did, they were also counted doing volunteer work. If they indicated that they did not, they were counted as not volunteering. A similar set of questions was used in post-retirement interviews to discover whether respondents were engaged in volunteer work then. (We'll have more to say about measurement strategies used in this research in Chapters 6 and 7.) Responses to questions, then, were the ways in which Emily and Roger observed the social world.[2]

Here we'd like to emphasize that in the process of testing role theory, Emily and Roger did what is typical of most such exercises: They engaged in a largely **deductive reasoning** in which more general statements were used as a basis for deriving less general, more specific statements. Figure 2.3 summarizes this process.

Theory testing rarely ends with the kinds of observational statements suggested in Figure 2.3. These statements are usually summarized with statistics, some of which we'll introduce in Chapter 15. The resulting summaries are often **empirical generalizations,** or statements that summarize a set of individual observations. We report, for example, the empirical generalization that people who did volunteer work before retirement were indeed more likely to do it after retirement as well.

As with our study, research is frequently used as a means of testing theory: to hold it up against the observable world and see if the theory remains plausible or becomes largely unbelievable. When researchers use their work

deductive reasoning, reasoning that moves from more general to less general statements.

empirical generalizations, statements that summarize a set of individual observations.

Type of Statement	Illustrative Statement
Theoretical	After retirement, to the extent possible, people will retain previous roles, including non-work roles, because doing so enables them to maintain feelings of self-worth and avoid feelings of anxiety and depression.
Hypothetical	People who do volunteer work before retirement are more likely to do it after retirement than others.
Observational	Respondent A was a lay leader in his church at the time of his pre-retirement interview. After retirement, he'd continued as a volunteer in his church and found other volunteer opportunities as well, including work at an HIV/AIDS drop-in center at a nearby hospital.

FIGURE **2.3** An Illustration of the Logic of Theory Verification from Adler and Clark.

---

[2] Answers to questions on a questionnaire might seem to be *indirect* ways of discerning things like attitudes. But much scientific observation is indirect. An x-ray of your tooth doesn't provide direct evidence of the presence or absence of a cavity, but, when read by a specialist, it can provide pretty conclusive *indirect* evidence.

to test theory in this way, they typically start by deducing hypothetical statements from more theoretical ones and then make observational statements that, when summarized, support or fail to support their hypotheses. In general, they engage in deductive processes, moving from more general to less general kinds of statements.

Inductive Reasoning

inductive reasoning, reasoning that moves from less general to more general statements.

grounded theory, theory derived from data in the course of a study.

In contrast with the more or less deductive variety of research we've been discussing is a kind that emphasizes moving from more specific kinds of statements (usually about observations) to more general ones and is, therefore, a process called **inductive reasoning**. Many social scientists engage in research to develop or build theories about some aspect of social life that has previously been inadequately understood. Theory that is derived from data in this fashion is sometimes called **grounded theory** (Glaser, 1993, 2005; Glaser and Strauss, 1967).

The process of building grounded theory from data is illustrated by Mueller, Dirks, and Houts Picca's research reported in Chapter 11. These researchers asked college students in their classes to observe, and take notes on, other college students, particularly when they dressed like people of other races. Mueller, Dirks, and Houts Picca did not work deductively but rather focused initially on what individual students had to say about other students' actions and behaviors. They quote, for instance, one young woman who recalled a discussion over costumes before Halloween, "We were all getting dressed up and one person said that they wanted to paint themselves black and wear a diaper and be a black baby."

Having immersed themselves in many such observations, Mueller, Dirks, and Houts Picca then developed statements at a higher level of generality, statements that summarize a much larger set of observations and therefore can be seen as empirical generalizations. One generalization suggests that students who cross-racially dressed as celebrities tended to make relatively inoffensive presentations, while those who dressed in the most essentializing ways (or ways meant to capture a view of what the "typical" person of another race is like) were relatively offensive.

Mueller, Dirks, and Houts Picca might easily have left their study here, providing us with a kind of descriptive knowledge about cross-racial dressing for Halloween. Instead, they offer a theory of such dressing that includes an explanation of why white students do it. They argue that some students engage in such costuming to subvert a system of racial inequality that the students perceive as having turned against them. Mueller, Dirks, and Houts Picca are clear that, in their view, this perception is unfounded in fact, but that it is one that nonetheless exists, engendering a kind of pleasure in those who can act out against it, at least for one day. Mueller, Dirks, and Houts Picca's movement from specific observations to more general theoretical ones is another approach to relating theory and research. Figure 2.4 summarizes this approach.

STOP & THINK *Consider the Donohue and Levitt research on the relationship between abortion and crime. Does that research, as we've described it, seem to rely more on deductive or inductive reasoning? What makes you say so?*

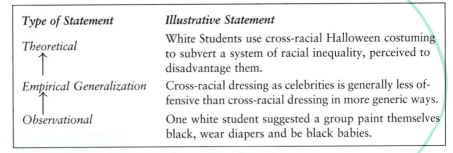

Type of Statement	Illustrative Statement
Theoretical	White Students use cross-racial Halloween costuming to subvert a system of racial inequality, perceived to disadvantage them.
Empirical Generalization	Cross-racial dressing as celebrities is generally less offensive than cross-racial dressing in more generic ways.
Observational	One white student suggested a group paint themselves black, wear diapers and be black babies.

FIGURE **2.4** An Illustration of the Process of Discovery Based on Mueller, Dirks, and Picca (2007).

The Cyclical Model of Science

A division exists among social scientists who prefer to describe the appropriate relationship between theory and research in terms of theory testing and those who prefer theory building. Nonetheless, as Walter Wallace (1971) has suggested, it is probably most useful to think of the real interaction between theory and research as involving a perpetual flow of theory building into theory testing, and back again. Wallace captured this circular or cyclical model of how science works in his now famous representation, shown in Figure 2.5.

The significance of Wallace's representation of science is manifold. First, it implies that if we look carefully at any individual piece of research, we

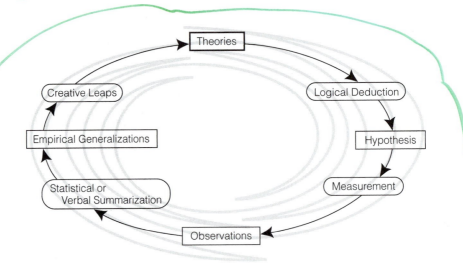

FIGURE **2.5** Wallace's Cyclical Model of Science Theories

Source: Adapted from Walter Wallace (1971). Boxed items represent the kinds of statements generated by researchers during their research. Items in ovals represent the typical ways in which researchers move from one type of statement to the next. Reprinted with permission from: Wallace, Walter L. *The logic of science in sociology* (Chicago: Aldine). Copyright © 1971 by Walter L. Wallace.

might see more than the merely deductive or inductive approach we've outlined. We've already indicated, for instance, that our research on retirement goes well beyond the purely deductive work implied by the "right half" of the Wallace model. We do use statistical procedures that lead to empirical generalizations about the observations. (One such generalization: people who do volunteer work before retirement are more likely than others to do it after retirement.) This led, in turn, to affirmation of "role theory." Moreover, although much of Mueller, Dirks, and Houts Picca's work is explicitly inductive, involving verbal summaries of observations and creative leaps from empirical generalizations to theoretical statements shown on the "left half" of Wallace's model, they actually did quite a bit of hypothesis testing as well. Thus, in arriving at the inference that white students were more likely than others to use cross-racial dressing as a way of subverting the racial hierarchy, they reexamined their data and found that there were no instances of black students cross-racially dressing as an apparent "indictment of whiteness *per se*." This kind of cross-checking of hypotheses against observations is actually an important element in the development of grounded theory (Glaser, 1993, 2005; Glaser and Strauss, 1967).

A second implication of Wallace's model is that, although no individual piece of research need touch on all elements of the model, a scientific community as a whole might very well do so. Thus, for instance, some researchers might focus on the problem of measuring key concepts, whereas others use those measurement techniques in testing critical hypotheses. Joseph Ferrari, in the focal research presented in Chapter 6, tests a hypothesis that students who feel like impostors (or, for the purposes of his research, are high-achieving but have "feelings of intellectual phoniness") are less likely than non-impostors to be academically dishonest. In creating such a test, Ferrari used measurement strategies developed by other researchers, notably Clance's (1985) impostor phenomenon scale and Roig and DeTommaso's (1995) academic practices scale.

Another example of how one might use Wallace's model to analyze the workings of a scientific community comes out of the controversy stirred by Donohue and Levitt's work on abortion and crime. Levitt told how he became interested in the connection (Stille, 2001). It began, Levitt says, with a simple observation: that before *Roe v. Wade* in 1973, about 700,000 abortions were performed each year; the number reached 1.6 million in 1978, remained fairly constant through the 1980s, and had begun to decline during the 1990s. "When I first learned of the magnitude of abortions performed in the late 1970s, I was staggered by the number and thought, 'How could this not have had some dramatic social impact?'" (Stille, 2001). From this simple fact, Levitt reports, and the observation that "the five states that legalized abortion … [before *Roe v. Wade*] saw drops in crime before the other 45 states" (Stille, 2001) (which methodology students will recognize as an empirical generalization), he and Donohue built their theory that increased abortion rates lead, via the reduction of unwanted births, to decreased crime rates (general proposition). They then came up with a series of hypotheses, suggesting, among other things, that various crime rates in different states should drop at particular times after the rise of abortion rates in those states (hypotheses). They collected appropriate data to test these hypotheses (observations) and found that the resulting analyses (giving them empirical generalizations) affirmed those hypotheses (Donohue and Levitt, 2001). But, their research drew a critical firestorm, one that led some of the rivals of Donohue and Levitt to offer alternative theoretical explanations. For instance, Ted Joyce, an economist, suggested that in states where abortion was made legal before 1973, when *Roe v. Wade* was decided, there was not the decline in crime rates that one would have expected, given Donohue and Levitt's thesis, before 1991. Joyce (2003) collected data (observations) and published his own counterstudy (empirical generalizations and theory). This in turn led to Donohue and Levitt's (2003) research, which challenged Joyce's study with the claim that the nationwide crack cocaine epidemic hit the high-abortion early legalizing states harder and earlier (in the late 1980s) than it did other states and that taking this epidemic into account would explain away Joyce's findings (theory and hypotheses). Donohue and Levitt collected the appropriate data (observations) and published their own study (with theory, hypotheses, and empirical generalizations). And, subsequently, Levitt (2004) collected additional data to test a theory that suggests the 1990s downturn in criminality reflected four factors: increase in the numbers of police, a rising prison population, a waning crack epidemic, and the legalization of abortion.

Most important, Wallace's model reinforces the basic point that theory and research are integrally related to each other in the sciences. Whether the scientist uses research for the purpose of theory testing, theory building, or both, the practitioner learns to think of theory and research as necessarily intertwined. Perhaps even more than "the chicken and the egg" of the old riddle or the "love and marriage" of the old song, when it comes to theory and research, "you can't have one without the other."

SUMMARY

In this chapter, we added to our discussion of theory by introducing the terminology of concept, variable, hypothesis, and empirical generalization. Using these terms and several examples of actual research, we illustrated the relationships that most scientists, including social scientists, see existing between research and theory. Many of the nuances of the connections between theory and research have been captured by Walter Wallace's cyclical model of how science works. Wallace's model suggests, on the one hand, how research can be used to test theories or explanations of the way the world really is. On the other, the model suggests how research can be used as a tool to discover new theories. Moreover, Wallace's model suggests that an appropriate model of science encompasses not only both of the previous views of the relationship between theory and research but also the view that research and theory "speak" to each other in a never-ending dialogue.

EXERCISE 2.1

Hypotheses and Variables

1. Name a subfield in the social sciences that holds some interest for you (e.g., marriage and family, crime, aging, stratification, gender studies).
2. Name a topic that you associate with this subfield (e.g., divorce, domestic violence, violent crime, medical care, the cycle of poverty, sharing of domestic work).
3. Form a hypothesis about the topic you've named that involves a guess about the relationship between an independent variable and a dependent variable (e.g., and this is just an example—you have to come up with your own hypothesis—"children whose parents have higher status occupations will begin their work lives as adults in higher status occupations than will children whose parents have low status occupations").
4. What, in your opinion, is the independent variable in your hypothesis (e.g., "parents' occupation")?
5. What is the dependent variable of your hypothesis (e.g., "child's occupation")?
6. Rewrite your hypothesis, this time in the form of a diagram like the one in Figure 2.1, a diagram that connects the categories of your variables.

EXERCISE 2.2

The Relationship between Theory and Research

Using your campus library, find a recent research article in one of the following journals: *Social Science Quarterly, American Sociological Review, American Journal of Sociology, Journal of Marriage and the Family, Criminology, The Gerontologist, Social Forces, Qualitative Sociology,* and *The Journal of Contemporary Ethnography.*

Or find an article in any other social science research journal that interests you. Read the article and determine how the research that is reported is related to theory.

1. What is the article's title?
2. Who is (are) the author(s)?
3. What is the main finding of the research that is reported?
4. Describe how the author(s) relate(s) the research to theory.

   Hint: Do(es) the author(s) begin with a theory, draw hypotheses for testing, and test them? Do(es) the author(s) start with observations and build to a theory or a modification of some theory? Do(es) the author(s) work mostly inductively or deductively? Do(es) the author(s) use theory in any way?

5. Give support for your answer to Question 4 with examples of the use of theory and its connection to observations from the research article.

CHAPTER 3

Ethics and Social Research

© ALEXANDER JOE/AFP/Getty Images

INTRODUCTION

ethical principles in research, the set of values, standards, and principles used to determine appropriate and acceptable conduct at all stages of the research process.

Have you been the victim of a crime—even a relatively small one, such as having had some personal property stolen, or a more serious crime, such as assault? If so, would you be willing to fill out a questionnaire or talk to a researcher who asks about your experiences? Would you have any concerns about participating in such a study?

Researchers have responsibilities to those who participate in their studies. Criminologist Peter Manning (2002: 546) believes "We owe them, I think, succor, aid in times of trouble, protection against personal loss or embarrassment, some distance that permits their growth and perhaps feedback on the results of research if requested or appropriate." Although not every social scientist offers succor and aid in times of trouble, most aim to provide the other items on Manning's list. That is, they strive to act *ethically* as they plan and conduct studies and disseminate their findings. In the next chapters we'll be talking about evaluating studies on practical and methodological grounds, but first we need to judge research using ethical criteria. Here we focus on ethical principles in research, the abstract set of standards and principles used to determine appropriate and acceptable research conduct.

Determining appropriate research behavior is a complex topic and a source of heated debate. Social scientists question if there is a set of universally applicable ethical principles or if it's better to take a case-by-case approach and consider each project's specific circumstances when deciding what is, and what is not, ethical. We also know that all possibilities cannot be anticipated. Unforeseen circumstances and unintended consequences at every stage of research—from planning to publication—can create ethical problems. Finally, there is the issue of whether the ethical *costs* of the study should be measured against the study's *benefits* for the participants or others.

STOP & THINK *Do you know of any research projects in the social sciences, medicine, or biology where researchers did something that was ethically questionable? What was it about the study that raised questions for you?*

A Historical Perspective on Research Ethics

Individual scholars might have been interested in defining appropriate research conduct earlier, but ethics did not become a concern of social science organizations as a whole until the middle of the twentieth century. The first ad hoc committee of the American Psychological Association (APA) established in the late 1930s to discuss the creation of a code of ethics did not support developing one. The APA did not develop and adopt a code of ethics until the 1950s (Hamnett, Porter, Singh, and Kumar, 1984: 18). The members of the American Sociological Association (ASA) were so divided about creating a statement on ethics that it took more than eight years from the proposal of the code to its adoption in 1971 (Hamnett et al., 1984). Today, most professional organizations have a code of ethics for research (see Box 3.1). The most recent version of the ASA Code of Ethics focuses on values for professional and scientific work and presents the rules and procedures of the ASA for investigating complaints of unethical behavior (see Box 3.2).

BOX **3.1**

A Timeline of Milestones in Ethics in Research

1932–1972	The Tuskegee Syphilis Study is conducted.
1948	The Nuremberg Code, a set of ethical directives for human experimentation, is developed.
1953	The APA adopts its first code of ethics.
1961	Stanley Milgram begins experiments on obedience.
1970	Laud Humphreys's book *Tearoom Trade: Impersonal Sex in Public Places* is published.
1967	The American Anthropological Association adopts its first code of ethics.
1971	The ASA adopts its first code of ethics.
1973	The British Sociological Association adopts its first code of ethics.
1974	The National Research Act creates the National Commission for the Protection of Human Subjects of Biomedical and Behavioral Research.
1979	The National Commission drafts the Belmont Report, a statement of basic ethical principles and guidelines in conducting research with human subjects in the United States.
1991	The Federal Policy for the Protection of Human Subjects or the "Common Rule" is adopted by some agencies.
1998	In Canada, the Tri-Council Policy Statement: Ethical Conduct for Research Involving Humans is adopted.
2000	The Academy of Criminal Justices Sciences adopts its first code of ethics.
2006	The Council on Graduate Education receives a grant to expand ethics education to science and engineering students.
2009	Federal Policy or "Common Rule" is now used by 17 federal departments and agencies that sponsor human-subjects research.

Several factors encouraged disciplines to consider ethical issues. One was revulsion at the total disregard for human dignity perpetrated during World War II by researchers in concentration camps controlled by Nazi Germany. After the Nuremberg trials, the United Nations General Assembly adopted a set of principles about research on human beings (Seidman, 1991: 47). The first principle of the Nuremberg Code is that voluntary consent of human subjects is absolutely essential (U.S. Government Printing Office, 1949).

There has also been increasing awareness of and distress among scientists, policy makers, and the general public about problematic biomedical research involving human subjects. For example, government-sponsored research conducted during the 1950s studied the effects of radiation on human subjects. Funded by the Atomic Energy Commission, this research included medical research using radioisotopes to diagnose or cure disease. However, some experiments were gruesome, such as one where people with "relatively short life expectancies," such as semi-comatose cancer patients, were injected with plutonium to determine how much uranium was needed to produce kidney damage (Budiansky, 1994). In cancer research in 1963, physicians injected live cancer cells into 22 elderly patients without informing them that live cells were being used or that the procedure was unrelated to therapy (Katz, 1972).

BOX **3.2**

The ASA's Code of Ethics sets forth the principles and ethical standards that underlie sociologists' professional responsibilities and conduct. These principles and standards should be used as guidelines when examining everyday professional activities. They constitute normative statements for sociologists and provide guidance on issues that sociologists may encounter in their professional work. The ASA's Code of Ethics consists of an Introduction, a Preamble, five General Principles, and specific Ethical Standards.

General Principles

The following General Principles are inspirational and serve as a guide for sociologists in determining ethical courses of action in various contexts. They exemplify the highest ideals of professional conduct.

Principle A: Professional Competence

Sociologists strive to maintain the highest levels of competence in their work; they recognize the limitations of their expertise; and they undertake only those tasks for which they are qualified by education, training, or experience. They recognize the need for ongoing education in order to remain professionally competent; and they utilize the appropriate scientific, professional, technical, and administrative resources needed to ensure competence in their professional activities. They consult with other professionals when necessary for the benefit of their students, research participants, and clients.

Principle B: Integrity

Sociologists are honest, fair, and respectful of others in their professional activities—in research, teaching, practice, and service. Sociologists do not knowingly act in ways that jeopardize either their own or others' professional welfare. Sociologists conduct their affairs in ways that inspire trust and confidence; they do not knowingly make statements that are false, misleading, or deceptive.

Principle C: Professional and Scientific Responsibility

Sociologists adhere to the highest scientific and professional standards and accept responsibility for their work. Sociologists understand that they form a community and show respect for other sociologists even when they disagree on theoretical, methodological, or personal approaches to professional activities. Sociologists value the public trust in sociology and are concerned about their ethical behavior and that of other sociologists that might compromise that trust. While endeavoring always to be collegial, sociologists must never let the desire to be collegial outweigh their shared responsibility for ethical behavior. When appropriate, they consult with colleagues in order to prevent or avoid unethical conduct.

Principle D: Respect for People's Rights, Dignity, and Diversity

Sociologists respect the rights, dignity, and worth of all people. They strive to eliminate bias in their professional activities, and they do not tolerate any forms of discrimination based on age; gender; race; ethnicity; national origin; religion; sexual orientation; disability; health conditions; or marital, domestic, or parental status. They are sensitive to cultural, individual, and role differences in serving, teaching, and studying groups of people with distinctive characteristics. In all of their work-related activities, sociologists acknowledge the rights of others to hold values, attitudes, and opinions that differ from their own.

Principle E: Social Responsibility

Sociologists are aware of their professional and scientific responsibility to the communities and societies in which they live and work. They apply and make public their knowledge in order to contribute to the public good. When undertaking research, they strive to advance the science of sociology and to serve the public good.

The entire code can be found on ASA website at www.asanet.org/galleries/default-file/Code of Ethics. pdf and on the website for this textbook.

One long-lasting research scandal ended as the result of public outrage. The Tuskegee Syphilis Study (labeled as such because the project was conducted in Tuskegee, Alabama), conducted by the U.S. Public Health Service, studied the effects of untreated syphilis to determine the natural history of the disease (Jones, 1981). The 40-year research project began in 1932 with a sample of 399 poor black men with late-stage syphilis. The study ultimately included 600 men, all of whom were offered free medical care in exchange for their medical data, and none of whom were told that they had syphilis. More than 400 of the men were not offered the standard (and often fatal) treatment for syphilis in the 1930s. In later years, they were not provided with penicillin even though it was available by the late 1940s (Jones, 1981). Many of the men died of the disease, and some of them unknowingly transmitted it to their wives and children (*The Economist*, 1997: 27). As late as 1965, an outsider's questioning of the morality of the researchers was dismissed as eccentric (Kirp, 1997: 50). The project ended in 1972, but not until 1997 was an official governmental apology offered by President Clinton to the surviving victims and their families (Mitchell, 1997: 10).

The disgrace of the Tuskegee study was one impetus for the passage of the National Research Act of 1974. This act created a commission to identify principles for conducting biomedical and behavioral research ethically and prompted the development of regulations to protect the rights of research subjects at institutions receiving federal funds (Singer and Levine, 2003). In 1966, Surgeon General William Stewart announced that social research was to be reviewed by local boards at hospitals and universities although defining proper treatment of human subjects was not yet a high priority (Stark, 2007). By the 1970s, federal agencies and professional organizations had developed guidelines for conducting ethical research (see, e.g., the NSF policy on the ethical treatment of subjects in Box 3.3).

BOX **3.3**

Federal Policy on the Ethical Treatment of Research Subjects

Many federal agencies have adopted the "Common Rule," a set of regulations governing human subjects of research with some agencies adding provisions for research with vulnerable populations such as children, prisoners, pregnant women, or handicapped persons.

The policy is designed to ensure minimal standards for the ethical treatment of research subjects. The major goal is to limit harms to participants in research. That means that no one should suffer harm just because they became involved as subjects or respondents in a research project. Institutions engaged in research should foster a culture of ethical research.

ETHICAL RESEARCH rests on three principles:

1. RESPECT for persons' autonomy, meaning the researcher gives adequate and comprehensive information about the research and any risks likely to occur, understandable to the participant, and allows them to voluntarily decide whether to participate.

2. BENEFICENCE, meaning the research is designed to maximize benefits and minimize risks to subjects and society.

3. JUSTICE, meaning that the research is fair to individual subjects and does not exploit or ignore one group (e.g., the poor) to benefit another group (e.g., the wealthy). Research produces benefits valued by society. Regulatory oversight seeks to ensure that any potential harm of the research is balanced by its potential benefits.

Research produces benefits valued by society. Regulatory oversight seeks to ensure that any potential harm of the research is balanced by its potential benefits.

Source: http://www.nsf.gov/bfa/dias/policy/hsfaqs.jsp#relation

institutional review board (IRB), the committee at a college, university, or research center responsible for evaluating the ethics of proposed research.

An **institutional review board** (**IRB** or **REB** in Canada) or committee on human subjects research is the committee charged with evaluating the ethics of research proposals, and almost every college, university, and research center in the United States, Canada, and Europe has one.[1] Before any federal, state, or institutional monies can be spent on research with human subjects, the research plan must be approved by the committee, at least one of whose members is unaffiliated with the institution. For the past 30 years, IRBs have attempted to verify that the benefits of proposed research outweigh the risks and to determine if subjects have been informed of participation risks. Making decisions involves a great deal of interpretation and the decisions of individual IRBs may differ (Stark, 2007: 782). The need to seek approval from an IRB varies depending on the topic, the source of funding, and the specific institutions involved. The federal policy using the Common Rule (outlined in Box 3.3), for example, allows the exemption of surveys and observations when subjects cannot be identified (NSF, 2009). Each researcher needs to consult appropriate ethical codes and be informed about the requirements and procedures for submitting proposals to review committees. Ultimately, the responsibility for ethical research behavior lies with the individual researcher.

There is some controversy about how IRBs deal with social research. The impetus for IRB regulation came from abuses in biomedical research and the members of many boards are mostly individuals in the medical and biological sciences. With few social scientists serving on some IRBs, some argue that social science research is assessed using a biomedical model and that IRB requirements can cause undue hurdles, censorship, and costly delays, creating impediments to a socially relevant and critical sociology (Feeley, 2007: 765–766). However, in general, researchers support the IRB process. A survey of researchers at one university found the overwhelming majority of faculty and graduate students said that IRB approval is an important protection for human subjects; all of the graduate students and 88 percent of the faculty felt that they had been treated fairly by the committee, and very few complained that they could not carry out a research project because of problems with the IRB (Ferraro, Szigeti, Dawes, and Pan, 1999). More recently, a national survey of 40 U.S. health researchers found that they supported the goals of IRB regulation but questioned the effectiveness, efficiency, and fairness of regulation implementation (Burris and Moss, 2006). The IRB at Brown University was very helpful to Desirée Ciambrone, a sociologist whose work we'll read in the following focal research section. While taking a graduate seminar in medical sociology at Brown University, Ciambrone became interested in the topic of women and acquired immunodeficiency syndrome (AIDS). Conversations about AIDS in Africa with a fellow graduate student who had worked on public health and policy issues convinced Ciambrone that focusing on

---

[1] In Chapter 14, we'll read about research conducted by Jill Hume Harrison and Maureen Norton-Hawk in a women's prison in Ecuador. Although their research was approved by an IRB at an American university, they discovered that the prison did not require an IRB and prison administrators were expected to protect the privacy of the inmates.

American women's experiences with human immunodeficiency virus (HIV)/ AIDS was a worthwhile and doable project. As this was her first totally independent research project, she learned a great deal conducting a study from start to finish. She found out about the impact of illness on people's lives and how an HIV diagnosis affects women's social and familial roles. She also learned that submitting a research proposal to an IRB can bring to your attention issues that you might not have considered.

FOCAL RESEARCH

Ethical Concerns and Researcher Responsibilities in Studying Women with HIV/AIDS

By Desirée Ciambrone[2]

Conducting one's first independent research project is a daunting task; eager to immerse oneself in the project, there is a great deal of preparatory work to be done. I chose to study women with HIV infection and found that the anxiety and complexity of conducting social science research are heightened when people facing difficult circumstances are involved.

Studying Women with HIV: Getting Ready
Respondent Recruitment

Finding women with HIV to speak with was no easy task. I sought respondents from health and social service agencies located primarily in the northeastern United States. Some agencies requested copies of my interview guide, my research proposal, and/or the Brown University's IRB approval. I discussed the goal of my study with my contacts and urged them to post/distribute a letter describing the study and ways to contact me with questions about the study or were willing to participate. Better yet, I urged that agency contacts speak with women directly and encourage their participation.

Many organizations I contacted were justifiably protective of their clients and appeared unwilling to assist in respondent recruitment despite their verbal acquiescence. Others were too busy to help me with my study; they posted my letter among a myriad of other brochures and leaflets but did little to personally encourage women to participate. Several key contacts in local hospitals were very helpful and referred most of the 37 HIV-positive women who volunteered to be interviewed for my study.

---

[2] This article was written for this text by Desirée Ciambrone and is published with permission. A complete description of her study and findings can be found in D. Ciambrone, *Women's Experiences with HIV/AIDS: Mending Fractured Selves*, New York: Haworth Press, 2003.

Informed Consent

The protection of human subjects is of utmost importance when conducting any type of research. Thus although my study would require "only an interview" (as opposed to a clinical trial, e.g., where drugs and other treatments may be involved), obtaining *informed consent* from my participants was crucial. The IRB at Brown University provided the guidelines that helped me carry out my research in an ethical way, including providing potential participants with an understanding of the potential risks and benefits of participation. In my initial proposal, I was quick to note the benign nature of my study. I described the benefits of being able to share one's story with an interested listener and the potential impact of my results on social policy and medical care provision. I noted that I wanted to talk to women about their lives and about living with HIV/AIDS. Initially, I did not see any risks that could result from this. Comments from the IRB prompted a more careful reflection of the *potential* risks of participation in my study. Indeed, I realized my questions might elicit a negative emotional response; such uneasiness or distress is, in fact, a risk. Thus, my informed consent was rewritten to include the possibility of emotional discomfort (see Appendix A).

Informed consent means that respondents demonstrate that they understand the study, what is expected of them, and how the information will be used. It also requires that participants *voluntarily* enter into a research project; that they haven't been coerced or duped into participation. Given previous research in which human subjects were misinformed and/or forced to participate in dangerous and life threatening experiments under the guise of science, the apprehensions of vulnerable populations are justified.

Women with HIV, while not unintelligent, are less likely to be highly educated and may not be knowledgeable about research; thus, it was my responsibility to explain the study in a clear, straightforward manner. I explained that I was a graduate student in medical sociology at Brown University who was conducting research in order to learn about the daily lives of women with HIV/AIDS. I told them that I wanted to interview them about their experiences in whatever setting was comfortable for them, their homes or a more neutral setting. I said I would tape-record the interviews in order to have an accurate, complete record of their stories but that, in order to protect their privacy, I would not use their real names on the tapes, on the transcripts, or in the reports and would keep the information confidential. I needed to describe what the study was and, perhaps more importantly, *what it was not*. I had to be clear that I was not a health care provider; I could not offer medical advice or prescribe medications. I was also not a social worker; I could not act as their case manager linking them to needed services. On the other hand, I felt I owed women answers to their questions. So I became educated about local services for persons with HIV/AIDS so that if a woman had a concern I might say, "Well, perhaps X organization can help. I know they deal with those types of issues." I was careful to make certain my respondents understood that while I was knowledgeable about many services, I did not know of *all* that was available nor could I guarantee the organization would be able to help. It was a challenge to be frank, honest, and share whatever I could *and* be sure that women did not perceive me as someone who would solve their problems; even with the best of intensions, that was not in my power.

An IRB follows research from its conception through its completion. Brown University's IRB requires annual reports in which researchers describe progress and explain any changes made to protocols or informed consent forms. This mechanism ensures that changes made are warranted and ethical. As an example, one of the things I reported in my annual progress report was my decision to offer respondents compensation for participation. I felt that this would attract those less accustomed to sharing their stories and offer all the participants some compensation for their time. To this end, I offered each participant a $10.00 gift certificate from a local supermarket chain, a decision which was approved by the IRB.

The IRB protocol outlines the appropriate storage of data. For example, I had to ensure that my respondents' personal identifying information was not attached to the data. Thus, instead of using the real names on the transcripts, pseudonyms were used. In accordance with federal regulations, transcripts, audiotapes, and other materials were stored under lock and key in a limited-access location.

The Interview Experience

Women with HIV/AIDS, particularly those with substance abuse histories, face unique challenges including poverty, caring for young children and other adults, reproductive decisions, access to health care and social services, inadequate housing, negotiating condom use, sexual undesirability, and abandonment. Race and class inequities exacerbate these problems. I knew all of this from my review of the literature but had no *personal experience* with this population.

Being able to talk to the women in person was essential as I was able to listen to them speak, follow up on key issues, observe body language, and see where they lived. It was also important for me to learn that these women are knowledgeable, articulate people—experts on their experiences. I developed a newfound respect for these women's agency; in the face of poverty, drug use, and illness these women managed to live productive, meaningful lives. Several did public speaking and worked to help others with HIV/AIDS in their community.

Connecting with Women with HIV

In graduate school, I had learned about the importance of the re-searcher's role in the research process. Thus, at every stage of my research I made a conscious effort to empower the study participants and present them as experts on living with HIV. Prior to interviewing, I contemplated the worse case interview scenarios, particularly access and rapport problems stemming from differential social positions. I did not encounter skepticism or scrutiny; none of my respondents questioned my interest in HIV/AIDS, nor did they appear uncomfortable sharing their stories with someone unlike themselves. This was probably due to the fact that many of these women were accustomed to talking to "outsiders" about their experiences via their participation in clinical trials of medical treatment at local health care centers. Women who had not participated in such projects may well have been skeptical of my intentions and less frank about their lives.

In addition to social distance and nondisclosure issues, I had also prepared myself for other possible problematic interview situations, such as lack of privacy and dire living circumstances. For most of the interviews, respondents and I had adequate privacy. Despite occasional commonplace interruptions (e.g., phone calls, demands of young children), the interviewees tried to create conducive interview situations.

The living conditions of some women were more troubling than relational issues. Several women lived in poor, traditionally high crime neighborhoods or public housing developments; neighborhoods I was familiar with from a young age as "places not to go." In order to feel as safe as possible, I scheduled most of my interviews, especially those in "tough" neighborhoods, during the day. Entering questionable dwellings was difficult. By and large, the women's homes were adequately furnished and relatively well kept. It was distressing to witness the apparent poverty and the destitute living conditions of some of my respondents. For example, one woman's home was filthy—the floors appeared as if they had not been swept or washed for quite some time, there were dirty dishes and pans overflowing in the sink, and ashtrays were filled to the rim with cigarette butts. I felt somewhat uneasy in homes like this, and less accepting of women's hospitality (e.g., a cup of coffee). Not accepting food and beverages made me feel rude and I never wanted to appear judgmental of women's homes. And although I had no intention of verbalizing my thoughts, I wondered how I could explain that it wasn't the HIV status that made me uncomfortable at times; it was the unsanitary home!

Eliciting Women's Stories and Confidentiality

Given the sensitive nature of my questions (e.g., about sexual relations and illicit drug use), I took great pains to make sure that the women knew the information provided was completely *confidential* and would not interfere with the receipt of medical/social services. I reiterated that they did not have to answer any questions that made them feel uncomfortable. While there were a couple of women who did not want to discuss certain topics, overall, the women I interviewed were incredibly open and often offered more intimate details than I had expected.

Many women were very interested in the purpose of the study. Often, women wanted to know how *their* responses compared to those of other women. Many also asked questions about my academic experience and personal life (e.g., What are you studying? Are you married? Do you have kids?). Recognizing the importance of establishing trust and minimizing power differentials between us, I answered the questions and told them about myself. When asked about the other interviewees, I was concerned that answering questions about the study would lead respondents to feel there was a normative or desired response to my questions that they should confirm. As I was willing to discuss my research, but didn't want women's narratives to be influenced by those of other respondents, I suggested that we talk about the study after the interview so as to be sure we had ample time to discuss their experiences and views. Evading interviewees' questions hinders efforts to establish trust and build rapport with respondents. My self-disclosure put my respondents at ease by reassuring them that the interview was a guided conversation and that their experiences and opinions were what was important, not getting the "right" answer. The open-ended nature of my questions allowed women to highlight the issues that were important to them and explore areas that I may not have inquired about.

Interviewing women who were active drug users posed a challenge. I felt most uncomfortable around these women as their experiences were so far removed from mine. In addition to striving to appear calm and comfortable, I had to remind myself that my role was that of researcher, not social worker or care provider. Thus, I could not offer unsolicited advice about the dangers of drug use on their weakened immune systems and the potential ramifications of non-adherence to their medications or provide solutions for the apparent disrespect the women were shown by their partners or children. Furthermore, when women did ask me for my advice I was careful to make it clear to them that I was a sociologist, not a therapist or a counselor. While I answered their questions about HIV and services available to the best of my ability, I didn't feel it was appropriate to offer unsolicited advice even when I had strong feelings about their choices. For example, one woman was married to a devout Christian who had convinced her to leave her health in God's hands and forgo her treatment regimen. Sheila[3] felt torn between her husband's religious beliefs and her health care providers' recommendations. Despite a dropping T-cell count and increasing viral load, Sheila stopped taking her medications at her husband's request. I found her situation troubling, and while I worried about Sheila's health, I felt it was not my role to encourage medication use or discredit her faith in religion.

Lessons Learned

In designing and conducting this study I learned a great deal about the *process* of conducting research, including the importance of following ethical practices, using informed consent procedures, protecting my respondents' privacy and confidentiality, and preserving the integrity of their narratives. My understanding of how women cope with HIV/AIDS was enhanced as a result.

Appendix A: Informed Consent

Informed Consent for Participation in Women with HIV Infection Study

You are being asked to participate in an important study to learn more about women's experiences with HIV infection. This study is being conducted by Desirée Ciambrone, a graduate student in the Sociology Department at Brown University.

During this interview, I will ask you about your illness, how HIV infection affects your daily life, and about the people who help you if, and/or when, you need assistance.

- This interview will be tape-recorded.
- In order to participate in this project I must be 18 years old or over.
- I may be contacted for further questions regarding this study.
- My participation in this project is completely voluntary; the decision to participate is up to me and no one else.
- I do not have to answer every question; I may choose not to answer certain questions.
- I can ask the interviewer to leave at any time. Even if I sign this form, I can change my mind later and not participate in the study.
- The information I provide will be kept confidential to the extent of the law.
- The only known risk to me is the possibility of some emotional discomfort.
- A direct benefit of participation in this study is the opportunity to talk with an interested listener and to think reflexively about my own experiences. I understand that the information obtained by this study may help other women who have had similar life experiences and may help professionals understand and treat women with HIV infection.

---

[3] Sheila is a pseudonym to protect the respondent's confidentiality.

If I have any questions about this study, I may contact Desirée Ciambrone at (401) 863-1275 at any time. Or, I may contact Alice A. Tangredi-Hannon or Dorinda Williams at the Office of Research Administration at Brown University (401) 863-2777 with questions about my rights as a participant in this study.

I have read the consent form and fully understand it. All my questions have been answered. I agree to take part in this study, and I will receive a copy of this completed form.

Signature of Participant   Date

Signature of Interviewer   Date

PRINCIPLES FOR DOING ETHICAL RESEARCH

Ethics in research is a set of moral and social standards that includes both prohibitions *against* and *prescriptions for* specific kinds of behavior in research. Ethical principles provide a moral compass to help researchers make decisions about their research. Although researchers can scientifically study people's values and their ethical standards, they cannot use scientific methods to determine how to behave ethically in their own work. Instead, they must examine their behavior using other frames of reference. To act ethically, each researcher must consider many factors, including safeguarding the welfare of those who participate in their studies, protecting the rights of colleagues, and serving the larger society. As we consider specific ethical principles, we'll examine these issues.

 **STOP & THINK** *What ethical standards do you think are most important when doing research? For example, if you were asked to observe friends at a party to study social interaction among college students for a class assignment, what would you do? Would you do the observation without first asking your friends for their permission?*

Principles Regarding Participants in Research
Protect Participants from Harm

protecting study participants from harm, the principle that participants in studies are not harmed physically, psychologically, emotionally, legally, socially, or financially as a result of their participation in a study.

First and foremost among ethical standards is the obligation to **protect study participants from harm,** which means that the physical, emotional, social, financial, legal, and psychological well-being of those who participate in a research project must be protected, both during the time that the data are collected and after the conclusion of the study. Overall, there should not be negative consequences as a result of being in a study.

Social science data are frequently collected by asking study participants questions. In the next chapter, we'll read about a research project in which study participants were asked about their career and family goals. Answering such questions did not put people at risk and represented a minimal intrusion on their privacy. In addition, their answers were held in confidence. In contrast, in studies like Ciambrone's, where participants are vulnerable in some way, answering questions is more likely to cause emotional or psychological discomfort. In Ciambrone's case, the IRB requirement to inform potential interviewees about the emotional risk of participation was appropriate.

One project that has been criticized as violating the "no harm" principle and other ethical standards are the Milgram experiments on obedience to authority figures. In a series of famous experiments, psychologist Stanley Milgram (1974) deceived his subjects by telling them that they were participating in a study on the effects of punishment on learning. The subjects, mostly men ages 20 to 50, were told that they would be "teachers," who, after watching a "learner" being strapped into place, were taken into the main experimental room and seated before an impressive shock generator with 30 switches ranging from 15 volts to 450 volts. The subjects were ordered by the experimenter to administer increasingly severe shocks to the "learner" whenever he gave an incorrect answer. They were pressed to do so even when the "learner" complained verbally, demanded release, or screamed in agony. After "receiving" 300 volts, the "learner" would pound on the wall and then no answer would appear on the panel. Subjects typically turned to the experimenter for guidance and were told to treat no response as a wrong answer and continue to shock according to the schedule of increasing volts. While the average subject showed a great deal of tension (including sweating, trembling, groaning, or nervous laughter), the majority continued to shock until 450 volts—two steps above the designation "Danger—Severe Shock." At debriefing, the subjects were told that the "learner," or victim, was part of the research team and had actually received no shocks at all. The intended purpose of the experiment was to see at what point, and under what conditions, ordinary people would refuse to obey an experimenter ordering them to "torture" others (Milgram, 1974).

The results of the experiment were unexpected. Milgram asked groups of psychiatrists, college students, and middle-class adults to predict the results and not one person predicted that any subject would give high-voltage shocks (Milgram, cited in Korn, 1997: 100). Milgram was taken aback to find that despite the anxiety and doubt that many of the hundreds of subjects exhibited while ministering the shocks, with "numbing regularity good people were seen to knuckle under to the demands of authority and perform actions that were callous and severe" (1974: 123).

It might surprise you that Milgram's work was consistent with the ethical principles for psychological research at the time. The APA's 1953 guidelines allowed for deception by withholding information from or giving misinformation to subjects when such deception was, in the researcher's judgment, required by the research problem (Herrera, 1997: 29). Not until the 1992 revision of the APA's statement on ethical principles were additional restrictions placed on the use of deception (Korn, 1997). When his study was published, Milgram found himself in the middle of a controversy over ethics that continued for decades. Detractors blamed him for deceiving his subjects about the purpose of the study and for causing his subjects a great deal of emotional stress both during and after the study (Baumrind, 1964), but supporters cited the importance of knowing the dark side of conforming to authority (Sokolow, 1999).

Milgram's study raises the issue of whether the end can justify the means. Thinking about Milgram's work, you might feel that the study was acceptable because it made an important contribution to our understanding of the dangers of obedience. Or, you might agree with those who have vilified Milgram's work because it violated the ethical principle of protecting subjects from harm.

Whatever your perspective on Milgram's study, his research points out that in many projects, the line between ethical and unethical methods is not always obvious. It can be difficult to design a study in which there is absolutely no risk of psychological or emotional distress to participants. For example, for at least a few people, being interviewed about their marriages or their relationships with siblings can dredge up unhappy memories. Similarly, for others, talking about experiences with sexual assault, divorce, or the death of a loved one can produce adverse effects or create difficulties for them in groups to which they belong. Sometimes methodological decisions can be based on the needs of subjects—that is, to protect them from harm that might occur. In a study of families with terminally ill children, for example, the researchers added a mentor to the research team to act as a resource in the event that the interviews became emotionally charged and the researchers did not ask questions about the diagnosis, prognosis, or death (Connolly and Reid, 2007: 1045).

It is impossible to anticipate each and every consequence of participation for every research participant. But, even foreseeing possible negative reactions by subjects does *not* mean a study should be abandoned. Instead, a realistic and achievable ethical goal is for studies to create, *on balance*, more positive than negative results and to give participants the right to make informed decisions about that balance. Researchers should try to anticipate needs for help or support. In a study of children's experiences of violence, for example, the participants were informed of avenues of support, such as known and trusted adults, and given toll-free phone numbers to talk to counselors, which were, in fact, used by some children to talk about upsetting events (Carroll-Lind, Chapman, Gregory, and Maxwell, 2006: 985). In a study about partner violence (Logan, Walker, Shannon, and Cole, 2008: 1232), the researchers learned that victims wanted referrals with phone numbers and information about the agencies. The research staff networked with the agencies, learned about them, visited them, and periodically checked the agency phone numbers to provide the women with accurate information on community resources.

Ciambrone told her participants about the possible emotional consequences of participating and occasionally offered referrals to agencies. She found that most seemed to enjoy the interview and liked the idea of telling their stories to help others with HIV. They said they appreciated her interest and in having the opportunity to share their views. A main theme was that the interview made them feel "normal," accepted, and valued rather than being defined exclusively by their HIV status.

STOP & THINK   *If you were asked to be in a study what would you first like to know? One review board administrator, Robert L. Johnson of Appalachian State University, asks students about their concerns and says that the usual responses are "What are you doing? Will you use my name? Who's responsible if I get hurt? What's in it for me?" (Cohen, 2007). Do you agree with these students or is there other information you'd want before making a decision about participating?*

Voluntary Participation and Informed Consent

voluntary participation, the principle that study participants choose to participate of their own free will.

The ethical strengths of Ciambrone's study include the fact that all the interviewees volunteered to participate after being informed of the researcher's identity and affiliation and the study's purpose, risks, and topics. The ethical principles being followed were **voluntary participation** and **informed consent.** These principles require that, *before participating*, all potential participants be told that they are not obliged to participate, that they can discontinue their participation at any time, that they be given accurate information about the study's purpose, the method of data collection, whether the information obtained will be collected anonymously or kept confidential, and the availability of results. Violations of ethical standards include not asking for consent, presenting inadequate information or misinformation, and using pressure or threats to obtain participation.

informed consent, the principle that potential participants are given adequate and accurate information about a study before they are asked to agree to participate.

One way to provide potential respondents with information about the study is through a written **informed consent form,** like the ones included in the Appendix of Ciambrone's article and in Chapter 10. Anonymous questionnaires that pose few risks for participants may be exempted from the requirement for informed consent, and some researchers rely on verbal rather than written permission as illustrated in Box 3.4. In recent years, it is more common to use written statements to provide potential interviewees with

informed consent form, a statement that describes the study and the researcher and formally requests participation.

BOX **3.4**

An Example of Verbal Parental Consent

A research project by the National Alliance for Caregiving (2005: 45) was conducted to determine the impact of children's participation in caregiving within households. First an adult in the household was interviewed by phone and then asked for permission to speak directly to his/her child. A verbal consent statement was read that began: "Before we proceed, I would like to tell you about the interview so you feel comfortable having me talk to your child. I will be asking him/her about the kinds of things s/he typically does, including his/her responsibilities and free time activities. We will also talk about the kinds of things s/he does for the person she/he helps care for and how the responsibilities affect him/her. Please be assured that your responses and your child's will be kept strictly confidential. Individual results will not be released to anyone.... If you'd like, you may listen to the interview."

Source: http://www.uhfnyc.org/usr_doc/Young_Caregivers_Study_083105.pdf

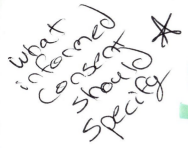

sufficient information to decide about participation. Informed consent should specify the purpose of the study, the benefits to participants and to others, the identity and affiliations of the researcher, the nature and likelihood of risks for participants, a statement of how results will be disseminated, an account of who will have access to the study's records, and a statement about the sponsor or study's funding source.

STOP & THINK *Who should give informed consent in a study involving children under 18? The children? Their parents? Both?*

There is some debate about the issue of who can appropriately give informed consent. Of particular concern are children, adolescents, and individuals unable to act in an informed manner. Most agree that, for minors, it is necessary to obtain both the child's and a parent's permission. However, there is debate about the age at which someone can decide for himself or herself to grant or withhold consent and who should be authorized to give consent for those judged unable to decide for themselves (Ringheim, 1995). Studying anorexic teenagers, Halse and Honey (2005: 2152) note that difficult family relationships or histories could make parental consent a "double-edged sword, protecting some girls and erasing other girls' potential for agency by increasing the opportunity for parental coercion." One PhD candidate interested in interviewing gay teenagers he had contacted through clubs decided that his research was unfeasible when his university's ethics committee required parental consent before the interview (Crow, Wiles, Heath, and Charles, 2006: 89).

Some researchers are granted IRB approval to waive parental permission. Diviak, Curry, Emery, and Mermelstein (2004), noting that teen smokers might not be willing to admit their smoking habit to parents, surveyed 21 researchers engaged in smoking cessation research with adolescents and found that 13 had asked for parental permission to be waived. Four of the researchers' projects were authorized to collect teen permission only.

Methods of obtaining permission vary. In school settings, for example, active consent requires someone, such as a parent or guardian, to sign and return a form to give permission for their child to participate in a study, whereas passive consent requires them to sign and return a form if they do *not* want their child to participate. Getting consent forms returned can be quite difficult. In a study of high school students in Australia, de Meyrick (2005) decided on active consent and mailed 2,500 consent letters to parents, but only 200, or less than 8 percent, were returned. In comparison, researchers used passive consent at one U.S. high school; only 4 percent of parents notified by mail of a study on parent–child interaction contacted the school to refuse permission for their child's participation (Bogenschneider and Pallock, 2008). Passive consent may work particularly well in giving children the opportunity to describe experiences like child abuse when their parents might want to prevent discussion (Carroll-Lind et al., 2006). The issue of getting participants may be why Morris and Jacobs (2000) found in a survey of researchers that only 69 percent of them agreed that passive consent was definitely or probably ethically problematic.

passive consent, when no response is considered an affirmative consent to participate in research; also called "opt out informed consent," this is sometimes used for parental consent for children's participation in school-based research.

 **STOP & THINK** *Have you ever been asked by a faculty member while you were his or her student to participate in a study that he or she was conducting? Were you ever asked while in class to complete a questionnaire or participate in an interview? Did you feel any pressure or did you feel you could decline easily? Do you think that students in this kind of situation who agree to participate are true volunteers?*

As a way of ascertaining if participants are really *volunteers*, researchers should ask the following questions: Do potential participants have enough information to make a judgment about participating? Do they feel pressured by payments, gifts, or authority figures? Are they competent to give consent? Are they confused or mentally ill? Do they believe that they have the right to stop participating at any stage of process? Those who are dependent on others, such as those in shelters or in nursing homes, might be especially vulnerable to pressure to cooperate. For example, in her discussion of research with homeless women, Emily Paradis (2000) notes the possibility of a dependent or coercive relationship when the institution in which the research takes place is the major source of subsistence, advocacy, or emotional support for the participants and when the participants are used to making accommodations to survive. Such relationships might make it harder for those approached to decline participation in a study.

Other studies raise different questions about consent. For a study on police interrogations, for example, researcher Richard Leo (1995) was required by his university's IRB to secure informed consent from the police officers, but not from the criminal suspects. In another research project, Julia O'Connell Davidson obtained informed consent from a prostitute and her receptionist but did not ask for permission from the prostitute's clients who did not know she was listening to their conversations in the hall or watching them leave the house and did not consent to having the prostitute share information about their sexual preferences or physical and psychological defects (Davidson and Layder, 1994). Davidson was not troubled by the uninvited intrusion into the clients' world, arguing that the prostitute had willingly provided knowledge that belonged to her.

New technologies can lead to ethical dilemmas. For example, what about using information from Facebook or similar sites? Some professional organizations have turned their attention to this issue (see Box 3.5). Although some consider Internet postings to be public and informed consent unnecessary, others disagree. Similar differences of opinion can be found about other sources of data such as field research, which is where the researcher observes daily life in groups, communities, or organizations.

 **STOP & THINK** *How about observing people in public, such as watching crowd behavior at a sporting event? What research procedures would be ethical? Would you or could you ask for permission? What would you do if you saw illegal behavior? How about if law enforcement personnel asked you for the notes you took while watching the crowd?*

Most researchers would argue that it's unnecessary, impractical, or impossible to get informed consent when observing in public places. Sometimes

BOX **3.5**

New Media, Old Concerns

In 2000, international scholars came together to found the Association of Internet Research (AoIR) to promote research into the social, cultural, political, economic, and aesthetic aspects of the Internet. In 2002, the organization's membership approved a set of recommendations for ethical decision making in Internet research. That same year, the British Sociological Association (BSA) included in its statement of ethical practices a warning to members to "take special care when carrying out research via the Internet" and to "consider caution in making judgments affecting the well-being of online research participants" (BSA, 2002).

While the kinds of media researchers study may be new, the ethical concerns are not. Whether using blogs, Google searches, personal e-mail, and the like as data, researchers must be aware of the obligation to protect the rights of subjects, including the right to privacy. AoIR (2002) reminds researchers that "the greater the vulnerability of the author/subject, the greater the obligation of the researcher to protect the author/subject."

there is no simple distinction between "public" and "private" settings. For the study described in *My Freshman Year: What a Professor Learned by Becoming a Student*, 52-year-old anthropologist Rebekah Nathan obtained approval from her university's IRB and followed the ethical protocol of informed consent when conducting formal interviews with students (Hoover, 2005: 36). However, in day-to-day interactions with students, Nathan let students assume she was one of them, which some might argue is unethical. How about attending a meeting of Alcoholics Anonymous? Maurice Punch (1994: 92), a qualitative researcher, asks "Can we assume that alcoholics are too distressed to worry about someone observing their predicament (or that their appearance at A.A. meetings signal their willingness to be open about their problem in the company of others)?" He argues that observation in many public and semi-public places is acceptable even if subjects are unaware of being observed. The Code of Ethics developed by the ASA (1999: 14) supports observing without permission in public places but suggests consultation with an IRB if there is any doubt about the need for informed consent.

Anonymity and Confidentiality

Even if informed consent is not obtained, the ethical principles of anonymity and confidentiality do apply. Anonymity is when it is impossible for anyone, including the researcher, to connect specific data to any particular member of the sample. To collect data anonymously, data must be collected without names, personal identification numbers, or information that could identify subjects.

If it is not possible or practical to collect data anonymously, research ethics dictate keeping them confidential. Confidentiality, or privacy, is keeping the information disclosed by study participants, including their identities, from *all* other parties, including parents, teachers, and school administrators, and others

anonymity, when no one, including the researcher, knows the identities of research participants.

confidentiality, also called privacy, is when no third party knows the identities of the research participants.

who have given permission or aided participation in the research. To keep material confidential, the data collected from all respondents must be kept secure. Identifying respondents by code numbers or pseudonyms are useful ways to keep identities and data separate. When results are made public, confidentiality can be achieved by not using real names or by grouping all the data together and reporting summary statistics for the whole sample. In addition, researchers can change identifiers like specific occupations or industries and names of cities, counties, states, and organizations when presenting study results.

The issues of anonymity and confidentiality are at the very heart of ethical research. One way to ensure the security of participant data is to destroy all identifying information as soon as possible. Such a step, however, eliminates the possibility of doing future research about those participants because collecting additional data from the same participants means being able to identify them. If a researcher wants to keep in touch with study participants over a number of years, then records of names, addresses, and phone numbers must be kept. However, keeping identifying information means that confidentiality could be at risk because it is possible through subpoena to link a given data record to a specific person (Laumann, Gagnon, Michael, and Michaels, 1994: 72).

One very controversial study, *Tearoom Trade*, by Laud Humphreys (1975), illustrates several ethical principles in doing research with human subjects—perhaps more by violation of than by compliance with the principles. This well-known study gives an account of brief, impersonal, homosexual acts in public restrooms (known as "tearooms"). Humphreys was convinced that the only way to understand what he called "highly discreditable behavior" without distortion was to observe while pretending to "be in the same boat as those engaging in it" (Humphreys, 1975: 25). For that reason, he took the role of a lookout (called the "watchqueen"), a man who watches the interaction and is situated so as to warn participants of the approach of anyone else.

Humphreys observed in restrooms for more than 120 hours. Once, when picked up by the police at a restroom, he allowed himself to be jailed rather than alert anyone to the nature of his research. He also interviewed 12 men to whom he disclosed the nature of his research and then decided to obtain information about a larger, more representative sample. For this purpose, he followed men from the restrooms and recorded the license plate numbers of the cars of more than 100 men who had been involved in homosexual encounters. Humphreys obtained the names and addresses of the men from state officials who were willing to give him access to license registers when he told them he was doing "market research." Waiting a year, and using a social health survey interview that he had developed for another study, Humphreys and a graduate student approached and interviewed 50 men in their homes as if they were part of a regular survey. Using this deception, Humphreys collected information about the men and their backgrounds.

Humphreys viewed his primary ethical concern as the safeguarding of respondents from public exposure. For this reason, he did not identify his subjects by name and presented his published data in two ways: as aggregated, statistical analyses and as vignettes serving as case studies. The publication of

Humphreys's study received front-page publicity, and because homosexual acts were against state law at the time, he worried that a grand jury investigation would result. Humphreys (1975: 229) spent weeks burning tapes, deleting passages from transcripts, and feeding material into a shredder and resolved to go to prison rather than betray his research subjects. Much to his relief, no criminal investigation took place.

 **STOP & THINK** *Review the principles of ethical research that have been discussed in this chapter. Before we tell you what we think, decide which of the principles you believe Humphreys violated. Make some guesses about why he behaved the way he did.*

Two principles that Humphreys violated are voluntary participation and informed consent. By watching the men covertly, tracking them through their license plates, asking them to participate in a survey without telling them its true purpose, and lying to the men about how they were selected for the study, Humphreys violated these principles several times over. After the study's publication, critics took him to task for these actions.

Humphreys's work provided insights about sexual behavior that had received little scholarly attention and told about groups that were often targets of hostile acts. Because of the usefulness of the work and his observational setting, Humphreys maintained that he suffered minimal doubt or hesitation. He argued that these were *public* restrooms and that his role, a natural one in the setting, actually provided extra protection for the participants (Humphreys, 1975: 227). But, reevaluating his work in the years after the study's publication, Humphreys came to agree with critics about the use of license plate numbers to locate respondents and approach them for interviews at their homes. He came to feel that he had put the men in danger of public exposure and legal repercussions. We will never know how many sleepless nights and worry-filled days the uninformed and unwilling research subjects had after the publication of and publicity surrounding Humphreys's book. However, no subpoena was ever issued; the data remained confidential, and no legal action was taken against the subjects.

Decades later, the Humphreys study continues to be a source of debate about where to draw the line between ethical and unethical behavior. Although many of its substantive findings have been forgotten, Humphreys's work has become a classic because of the intense debate about ethics that it generated. Discussing the choices that researchers like Humphreys made can help others to understand more clearly the legal and ethical issues implicit in doing social research. Debate about the costs and benefits of deception continues. For a less common point of view, see Box 3.6.

Ethical Issues Concerning Colleagues, the Community, and the General Public

 **STOP & THINK** *Assume that you have designed a study and you feel confident that the participants will face minimal risks. Can you think of any other ethical responsibilities that should concern you?*

BOX **3.6**

Another Point of View

In "The Ethics of Deception in Social Research: A Case Study," Erich Goode argues a different point of view:

> [P]rotecting our respondents from harm is the researcher's *primary* interest ... [but] I intend to argue that it *is* ethical to engage in *certain kinds* of deception ... the ethics of disguised observation [should be] evaluated on a situational, case-by-case basis.... In *specific* social settings, some kinds of deception should be seen as entirely consistent with good ethics....
>
> I strongly believe every society *deserves* a skeptical, inquiring, tough-minded, challenging sociological community. It is one of the *obligations* of such a society to nurture such a community and to tolerate occasional intrusions by its members into the private lives of citizens.... I have no problem with the *ethics* of the occasional deceptive prying and intruding that some sociologists do. I approve of it as long as no one's safety is threatened. (1996: 14 and 32)

In addition to treating research participants ethically, researchers have responsibilities to colleagues, the community, and the public at large. Soliciting input from groups involved in the research topic, such as local advocacy groups or community leaders, is an important part of ethical behavior (Dickert and Sugarman, 2005). Researchers can act ethically by partnering with community groups and using the research findings to promote better conditions. For example, after studying mental health services in Canadian immigrant communities, researchers presented the concerns of the participating communities about mental health services to community service providers, policy makers, and other decision makers (Malter, Simich, Jacobson, and Wise, 2009: 320). Similarly, in Chapter 14, you'll read Harrison and Norton-Hawk's assessment of a prison in Ecuador. After completing their research, they met with prison officials, quasi-governmental organizations, and medical personnel to help improve the lives of the inmates.

honest reporting, the ethical responsibility to produce and report accurate data.

A central ethical principle is **honest reporting,** which is the responsibility to produce accurate data, report honestly, and disseminate it in both professional and community forums. Recent scandals in the biological sciences and a survey of scientists that found a significant number anonymously admitting to having committed some form of scientific misbehavior (Singer, 2005) indicate that dishonest reporting might be more widespread than previously thought. We want to emphasize that important responsibilities to colleagues, the community, and the general public include the presentation of data without plagiarism, fraud, falsification, distortion, or omission; the disclosure of the research's limitations; and a willingness to admit when hypotheses are not supported by the data.

As the actions of an unethical researcher can affect the reputations of other social scientists and the willingness of potential research participants in the future, acting ethically is part of the responsibility to colleagues. Research funding can be a source of concern as funding agencies and sponsors might influence research agendas and methodologies by making awards on the basis of certain criteria. In addition, they can limit the freedom of researchers by imposing controls on the release of research reports and limit public disclosure of findings. A researcher hired and paid by an organization can face expectations to publish findings that support its goals. Criminologist Charles Thomas, for example, studied the impact of privatizing prisons while employed by the Florida Correctional Privatization Commission. After reporting results exaggerating the positive impact and reduced recidivism rates of juveniles leaving private versus public facilities, he was criticized by other criminologists for acting unethically and producing tainted results because of his funding source (Fogel, 2007: 114).

Sometimes a conflict of interest exists between the aims of the study's sponsor and those of the larger community, especially if the sponsor is interested in influencing or manipulating people, such as consumers, workers, taxpayers, or clients (Gorden, 1975: 142). For example, in one city, a team of researchers representing a city planning commission, a community welfare council, a college, and several cooperating organizations, including local Boy Scout officials, studied the metropolitan area's social problem rate. At the conclusion of the study, the planning commission, the welfare council, and a local council of churches attempted to suppress the research findings that did not fit their groups' expectations or images. In each case, "it was one of the front-line field workers, viewed as 'hired hands' by the sponsoring agencies, who had to press for full dissemination of the findings to the public at large" (Gorden, 1975: 143–144).

ETHICAL DILEMMAS DURING AND AFTER DATA COLLECTION

> Who owns the data? ... Whose interpretation counts? Who has veto power? ... What are the researcher's obligations after the data are collected? Can the data be used against the participants? Will the data be used on their behalf? Do researchers have an obligation to protect the communities and the social groups they study or just to guard the rights of individuals? (Marecek, Fine, and Kidder, 1997: 639)

Most examples of unethical behavior in the social and behavioral sciences research literature are rooted in ethical dilemmas rather than in the machinations of unethical people. Many of the dilemmas involve *conflicts*—conflict between ethical principles, between ethical principles and legal pressures, or between ethical principles and research interests.

Conflict between Ethical Principles

In some research situations, researchers encounter behavior or attitudes that are problematic or morally repugnant. Although ethical responsibilities to subjects might prohibit us from reporting or negatively evaluating such behavior, other ethical principles hold us responsible for the welfare of others and the larger society. James Ptacek encountered an example of this ethical dilemma in his study of wife batterers. He recruited his sample from a counseling organization devoted to helping such men, paid each participant a nominal fee, and told the men that participation would help the organization and that the interview was not a formal counseling session (Ptacek, 1988: 135). He used a typical sociological interviewing style—staying dispassionately composed and nonjudgmental even when the men described bloody assaults on women. Even though Ptacek's goal was to facilitate their talking rather than to continually challenge the men, he did worry about the moral dimension of his impartial approach. In the three interviews where men reported ongoing if sporadic violence, Ptacek switched from an interviewer role to a confrontational counselor role after the formal questions had been completed. He noted, "At the very least, such confrontation ensures that among these batterers' contacts with various professionals, there is at least one place where the violence, in and of itself, is made a serious matter" (Ptacek, 1988: 138). Facing a conflict between ethical principles, Ptacek gave a higher priority to the elimination of abuse than to the psychological comfort of study participants but others may have made different choices. In contrast, Yassour-Borochowitz (2004) interviewed a group of men who had abused their partners but she worried about the ethics of the pain and hurt that she caused her subjects just by asking them to talk about the subject with her.

 **STOP & THINK** *What would you do if someone you were interviewing told you about illegal behavior they'd engaged in? What about if they were planning self-destructive acts or to hurt others? Would you keep such information confidential or report it to someone?*

Conflict between Ethical Principles and Legal Concerns

Sometimes during the course of research, observing others or interviewing study participants, the researcher finds out about illegal behavior. In his study of the nighttime economy in Manchester, England, David Calvey (2008) took the role of a "bouncer" in a club and hid his role as a researcher. He observed drug taking, violence, and the withholding of information from the police without comment. He kept the data confidential but knew that his situation might be problematic and that the encounters might lead to a difficult legal position. He felt fortunate because he felt he was doing "fingers crossed" research where his luck could have run out (Calvey, 2008: 913).

Counting on luck isn't wise as researchers can face pressure to violate confidentiality. Although ethical responsibilities to subjects prohibit the reporting of what study participants tell us, other ethical principles hold us responsible for the welfare of others and to the larger society. If we were told about someone planning to hurt themselves or others, we need to consider our ethical and legal responsibilities. For example, in 22 states, studying a topic such as child abuse or neglect means a researcher will face the legal requirement that anyone having reason to suspect maltreatment must report it to the authorities (Socolar, Runyan, and Amaya-Jackson, 1995: 580). This has significant implications for the informed consent process, as there are consequences of participating, including legal risks for subjects and legal liabilities for investigators (Socolar et al.,

1995: 580). Some researchers have tried to avoid a legal obligation to report suspected child maltreatment by collecting data anonymously, but others would argue that ethical obligations remain. The ASA's Committee on Professional Ethics has considered the choice that must be made if, for example, a researcher who gets permission from a family to observe their interaction discovers child abuse. Should the child abuse be reported and the confidentiality that might have been promised to the parent be breached, or should the confidentiality be maintained and risk future harm to the child? The ASA (1999: 12) and other organizations conclude that in these cases, sociologists must balance guarantees of confidentiality with other principles in the Code of Ethics, standards of conduct, and applicable law. One approach is to make to disclose the limits of the guarantee of confidentiality. For example, in a study of parents and children in 70 households, the researchers made the legal limits to confidentiality clear in the informed consent form and in discussion with participants. During the course of the study, because of serious concerns about the welfare of the children in two households, the researchers made two referrals to Children's Services (Gorin, Hooper, Dyson, and Cabral, 2008).

But, even if the researcher decides to keep confidentiality, there may be legal consequences. Russel Ogden, studying assisted suicide, used a common law claim of privilege and became the only researcher in Canada to win privilege and avoid contempt of court after being subpoenaed and asked to provide information about the identity of a research subject (Palys and Lowman, 2006). In the United States, under current rules of evidence, social scientists have not been awarded the legal right to refuse to turn over materials by claiming privileged relationships with participants (Leo, 1995: 124). For example, following the 1989 environmental disaster created when an oil tanker, the Exxon *Valdez*, ran aground off the coast of Alaska, research projects were conducted to assess the impact of the spill, including the social and psychological consequences. After the data were collected, some of the researchers and their associates were subpoenaed and deposed by Exxon, and one researcher, J. Steven Picou, was ordered to turn over all the information he had gathered (McNabb, 1995: 331). Although the case was settled before going to trial, the researchers learned that anything (files, agreements, tapes, diaries, etc.) can be subpoenaed and that criminal prosecution can result if the material is not released (McNabb, 1995: 332).

Researcher Richard Leo learned the same lesson on a much smaller scale. Leo had spent more than 500 hours "hanging out" inside the criminal investigation division of a large urban police department, observing police interrogation practices with departmental permission. Leo was subpoenaed by a suspect's public defender as a witness in a felony trial. Facing the threat of incarceration, Leo (1995: 128–130) complied and testified at the preliminary hearing but ultimately felt that he had betrayed confidentiality and spoiled the field for future police researchers because, as a result of his testimony, the suspect's confession was excluded from the jury.

In a case with a different outcome, sociologist Rik Scarce (1995) went to jail to protect the confidentiality of participants in his research. After completing a study of radical environmental activists, Scarce remained in contact with some of them. At one point, one activist, Rod Coronado, stayed at Scarce's house for several weeks while Scarce was away. The time in question included the evening that the Animal Liberation Front took an action resulting in the release of 23 animals and $150,000 in damages at a local research facility. Scarce was subpoenaed to testify about Coronado before a grand jury. Even after being granted personal immunity, Scarce (2005: 16) refused to answer some questions because they called for information he had by virtue of confidential disclosure given in the course of his research. The district court judge held Scarce in contempt of court, ruling that if the government has a legitimate reason for demanding cooperation with a grand jury, a social scientist cannot claim academic privilege; the appeals court upheld the ruling. After spending 159 days in jail, the judge decided that further incarceration was not likely to pressure Scarce into giving grand jury testimony and released him (Scarce, 2005: 207). The Supreme Court never heard Scarce's case and the precedent his case set stands. Sociologists Patricia Adler and Peter Adler (2005: xii–xiii) believe that "Rik Scarce has laid it on the line for the rest of the community of scholars to ponder. He has not merely paid lip service to the ideals he cherishes, but he has served his time to honor them.... We can only hope that we are moving toward a time when the ethical becomes both respectable and legal."

Situations involving subpoenas are rare but it is important to be aware of their potential beforehand and IRBs may ask researchers to specify that confidentiality is limited by law. At the request of the IRB at Brown University, Ciambrone included in her informed consent form the statement that information provided by interviewees would be kept confidential to the extent of the law. She tells us that she was fortunate that no interviewee told her anything of concern, so that she did not have to consider breaking confidentiality.

Conflict between Ethical Principles and Research Interests

 **STOP & THINK**   *You're thinking of a research project that involves studying people's behavior by observing them. You wonder what the effect of asking for permission will be. Could there be situations where it might be better for the outcome of the study if researchers **don't** ask for informed consent?*

What if you want to study powerful organizations? And what if you want to study something that organizations do not want you to study? Would you be tempted to be deceptive to open closed doors? Robert Jackall (1988) was turned down by 36 corporations when he asked for permission to study how bureaucracy shapes moral consciousness and managerial decisions. His analysis of the occupational ethics of corporate managers was made possible only when he used his personal contacts to gain access and recast his proposal using "bland, euphemistic language" (Jackall, 1988: 14). He was able to do field work and interviews in a chemical company, a defense contractor, and a public relations firm because the companies thought his interest was in issues of public relations. Jackall completed an analysis of the "moral mazes" navigated by executives only after he made a few twists and turns of his own. His work answers some questions and raises others: Can the knowledge gained by a study outweigh the ethical costs? What about choosing between using ethically compromised methods and not doing the research? We must ask at what point the researcher should refrain from doing a study because of the ethical problems. Did Milgram's work contribute so much to our understanding of human nature that the violations of ethics are worthwhile? How important were Humphreys's contributions? At what point should the Tuskegee study have been stopped?

In future chapters we'll discuss issues of study design and methods of data collection and see how even the most benign of studies can run into ethical challenges. At this point, we'll present a few examples of potential conflicts between methodological and ethical considerations.

Dilemmas can occur when recruiting a sample of participants for a study if when disclosing the specific focus of a project will lead to many more refusals to participate than a more general description of a study. Ruth Frankenberg (1993), for example, originally told people, accurately, that she was "doing research on white women and race" and was greeted with suspicion and hostility. When she reformulated her approach and told white women that she was interested in their interactions with people of different racial and cultural groups, she was more successful in obtaining interviews.

In observing groups or organizations who meet in private settings, it is ethical to ask for permission from those you'll be observing. If, however, those being observed are likely to change their behavior if they know about the observation, then you will need to make a choice. Consider, for example, studies that have used "pseudo-patients," research confederates who go into mental hospitals feigning symptoms. Some argue that such research creates high-quality data and yields new knowledge by observing true-to-life situations that would be distorted if informed consent were requested (Bulmer, 1982: 629–630). Judith Rollins (1985) faced this issue and decided to not disclose her identity as a graduate student researcher when she also worked as a domestic worker and collected data about her interactions with her employers. Rollins acknowledged feeling deceitful and guilty, especially when her employers were nice. However, she feels that her behavior was acceptable because there was no risk of harm or trauma to the subjects.

Matthew Lauder also feels that deception can be ethically defensible. In a study of a Canadian neo-National Socialist organization that had a history of racism and using violence, Lauder (2003) first tried collecting information by interviewing people using informed consent. But many group members refused to sign informed consent forms and his formal interviews produced little useful information about the anti-Semitic and religious character of the organization's ideology. Rather than abandoning the project, Lauder decided to change tactics. Using deception, he moved from an overt to a covert role by pretending to have converted to the group's worldview (Lauder, 2003: 190–191) and then by talking to members informally. He collected a great deal of information covertly and protected confidentiality when using the data. Lauder feels that he balanced the needs of society against the needs of individuals and concluded that he was morally obligated to conduct the study.

Ward and Henderson (2003) came down on the side of ethics over research considerations as they conducted a long-term study of young people leaving the care of state service agencies. Noting that study participants had faced painful childhood experiences and emotional upset, they decided that some were too vulnerable to participate in the follow-up interview and others needed it postponed. They gave up research data because they concluded that it was their responsibility to avoid placing participants in situations where it was very difficult for them to discuss sensitive issues (Ward and Henderson, 2003: 258).

ETHICS AND PUBLISHING RESULTS

When a study is completed and the results are made public, if there is a possibility that the participants can be identified, the researcher faces another dilemma. Published material can have an impact on the study's participants. Clark and Sharf (2007), for example, cite several researchers who made decisions to omit material from their publications because they felt they might be harmful to their research participants. Other researchers make different decisions. When Carolyn Ellis studied isolated fishing communities, she didn't worry about how the people would respond. She thought, "They're illiterate; they'll never read this work," but found out later that it was painful for many members of the community when they learned what she had written (Ellis et al., 2008: 271). Conflict between reporting results and maintaining confidentiality also confronted the authors of *Small Town in Mass Society: Class, Power and Religion in a Rural Community*. Vidich and Bensman (1964: 314) described their study of a town in upstate New York as unintentional and unplanned, a by-product of a separate organized and formal research project. One of the major problems they faced occurred because, during the data collection stage of the research, the townspeople of "Springdale" had been promised confidentiality and purely statistical analyses of the data. The researchers ultimately came to focus on the town's political power structure and decided on a case study as the most appropriate method of analysis. Upon publication, identity concealment of townspeople was impossible despite the use of pseudonyms and the changing of some personal and social characteristics to describe the members of the power hierarchy. The identities of those in the town's "invisible government" were recognizable to the residents, and community leaders were distressed to be identified as part of such a government. Townspeople felt humiliated and felt that their trust in the researchers had been betrayed (Vidich and Bensman, 1964: 338). Vidich and Bensman recognized that by publishing their study, they gave greater priority to the ethic of scientific inquiry than to the ethic of confidentiality. They came down on the side of honest reporting to colleagues and the public, but others might have chosen differently.

It is better for researcher to anticipate ethical problems than to face difficulties at publication. Sandra Enos and a group of researchers studying an orphanage thought ahead about the way they wanted to present their results and thereby saved themselves from breaking confidentiality (see Box 3.7).

BOX **3.7**

Interviewing and Confidentiality by Sandra Enos*

In 1979, Rhode Island closed the state orphanage that had provided institutional care for thousands of dependent and neglected children since 1885. In 2002, an interdisciplinary team began the first study of the state's experiment to provide care for children who otherwise would have been placed in local and state almshouses or in private orphanages. After public meetings and tours attracted former residents and workers to the site, the State Home and School Oral History Project was established to create a social history of the institution.

Typically in sociological research, data are aggregated and presented statistically or pseudonyms are used. Because we wanted to use historical documents and excerpts from some of the audio- and video-taped interviews in publications and public programming, we faced particular challenges concerning confidentiality. Our solution, approved by a local IRB, was to create a declaration of participants' rights and a disposition form with options. We allowed the former residents in our sample to review and edit transcripts, to decide whether to have some or all of their information closed to public

view and to have it released at a later date, such as after their deaths.

We were able to balance our research goals with participant rights. Participation in our study was voluntary and all members of the sample controlled the privacy of their data. As we expected, the process of being interviewed about this time in their lives was hard for some participants. Some had never before told the story of their years at the State Home.

We found that residents who came to the State Home in the 1920s through the 1940s thought it provided them a good and safe place to live; those who were residents in the 1960s and later were critical of the care they received. If you would like to learn about the study, hear excerpts from the CD, *Let Us Build a Home for Such Children: Stories from the State Home and School*, view video segments, or read the informed consent documents, you can do so at the project's website http://www.ric.edu/statehomeandschool/cdProject.htm#section02

*Sandra Enos, PhD, was a member of the State Home and School Oral History Project and is an Associate Professor of Sociology at Bryant University.

MAKING DECISIONS TO MAXIMIZE BENEFIT AND MINIMIZE RISK

Although social scientific research does not usually place subjects in situations that jeopardize their health and well-being, most social research involves some risk. Discomfort, anxiety, reduced self-esteem, and revelation of intimate secrets are all possible costs to subjects who become involved in a research project. No investigator should think it possible to design a risk-free study, nor is this expected. Rather, the ethics of human subject research require that investigators calculate the risk–benefit equation and seek to balance the risks of a subject's involvement in the research against the possible benefits of a study for the individual and for the larger society.

Many colleges and universities have research review boards and most professional organizations have codes of ethics. The guidelines they provide can make us aware of the ethical issues involved in conducting a study and offer advice for appropriate behavior. Ultimately, however, the responsibility for ethical research lies with each individual researcher.

To make research decisions, it's wise to use separate but interconnected criteria focusing on the ethics, practicality, and methodological appropriateness of the proposed research. In selecting a research strategy, we should ask these questions: Is this strategy practical? Is it methodologically appropriate? Is it ethically acceptable? When we answer "yes" to all three questions, there is no dilemma. But, if we feel that it will be necessary to sacrifice one goal to meet another, we must determine our priorities.

Each researcher must think about the consequences of doing a given study as opposed to *not* doing the study and must consider *all* options and methods to find a research strategy that balances being ethical and being practical with the likelihood of obtaining good quality data. In doing research, each of us must recognize that we are balancing the rights of study participants against our desire for research conclusions in which we have confidence and that we can share with the public.

SUMMARY

Even though researchers often face unanticipated situations, most aspire to conduct their studies ethically and to apply the values and standards of ethical behavior to all stages of the research process. Recognizing the need to use feasible and methodologically acceptable strategies, we should strive for appropriate and acceptable conduct during the planning, data collection, data analysis, and publication stages of research, as summarized in Box 3.8.

Being ethical in the treatment of study participants means collecting data anonymously or keeping identities confidential, providing adequate information about the study to potential participants, and, most important, working toward keeping negative consequences of participation to a minimum. Many ethical dilemmas can be anticipated and considered when planning a research strategy. IRBs and the codes of ethics of professional associations provide guidelines for individual researchers. When researchers face unusual ethical and legal dilemmas, they need to take special care in planning and conducting their studies.

In addition to the treatment of participants, ethical principles require researchers to conduct themselves in ways that demonstrate that they are

BOX **3.8**

Summing Up Ethical Principles

- Protect research participants from harm.
- Get informed consent.
- Be sure that study participants have not been pressured into volunteering.
- Collect data anonymously or keep data confidential.
- Submit the research proposal to a review board.
- Provide accurate research findings.
- Consider responsibilities to research participants, colleagues, and the general public at all stages of research including after project's completion and publication.
- Maximize benefits and minimize risks.

---

responsible to other scholars and to society as a whole. Scholars contribute an important service to their fields and the general public by providing and disseminating accurate information and honest analyses. Researchers will confront difficult situations if ethical principles conflict with each other, when legal responsibilities and ethical standards clash, or when collecting high-quality data means violating ethical principles. These situations must involve careful decision making. The best way to prepare for ethical dilemmas is to try to anticipate possible ethical difficulties and to engage in open debate of the issues.

EXERCISE 3.1

Knowing Your Responsibilities

All institutions that receive federal funds and most that conduct research have a committee or official group to review the ethical issues involved in studies done on campus or by members of the campus community. Called an IRB, a Human Subjects Committee, or something similar, the group must approve research involving human subjects before such research begins. Your college or university almost certainly has such a committee.

Find out if there is a committee on your campus that evaluates and approves proposed research. If there is no such group, check with the chairperson of the department that is offering this course and find out if there are any plans to develop such a committee and describe those plans. If there is an IRB, contact the person chairing this committee and get a copy of the policies for research with human subjects. Answer the following questions based on your school's policies. Attach a copy of the policies and guidelines to your exercise.

1. What is the name of the review committee? Who is the committee's chairperson?
2. Based on either written material or information obtained from a member of the committee, specify the rules that apply to research conducted by *students* as part of course requirements. For example, does each project need to be reviewed in advance by the committee or is it sufficient for student research for courses to be reviewed by the faculty member teaching the course? If the committee needs to review all student projects, how much time would you need in advance of the project's start date to allot to the review process?
3. What are some advantages and disadvantages on your ability to collect data of your institution's procedures for reviewing the ethics of student research?
4. Based on the information you've received from the committee, describe at least three of the ethical principles that are included in the standards at your school.

EXERCISE 3.2

Ethical Concerns: What Would You Do?

In the following examples, the researcher has made or will need to make a choice about ethical principles. For each situation, identify the ethical principle(s), speculate about why the researchers made the decisions they did, then comment whether you agree with their decisions.

A. Professor Ludwig research question is whether there is a connection between experiencing family violence and academic achievement among students. He contacts the principal of a local high school who agrees to the school's participation. At a school assembly, the principal hands out Professor Ludwig's questionnaire, which includes questions about family behavior including violence between family members. The principal requests that the students fill out the survey and put their names on it. The following day, Professor Ludwig uses school records to obtain information about each student's level of academic achievement so that he can answer his research question. He keeps all information that he obtains confidential and only presents a statistical analysis of the data.

1. What ethical principle(s) do you think was (were) violated in this research?
2. Describe the probable reason(s) Professor Ludwig violated the principle(s) you noted in your answer to Question 1.
3. Describe your reactions to the way this study was done.

B. Professor Hawkins receives funding from both a state agency and a national feminist organization to study the connection between marital power, spouses' financial resources, and marital violence. Deciding to interview both spouses from at least 100 married couples, she contacts potential participants by mail. Because she is a university professor, she writes to each couple on university letterhead stationery and asks each couple to participate in her study. She specifies that she will be interviewing them about their marriage but does not specify the specific topics that the interview will cover or the source of her funding. She specifies that everything the couple tells her will be kept confidential.

1. What ethical principle(s) do you think was (were) violated in this research?

2. Describe the probable reason(s) Professor Hawkins violated the principle(s) you noted in your answer in Question 1.
3. Describe your reactions to the way this study was done.

C. To study a radical "right-to-life" group, Professor Jenkins decides to join the group without telling the members that he is secretly studying it. After each meeting, he returns home and takes field notes about the members and their discussion. After several months of attending meetings, he is invited to become a member of the group's board of directors and he accepts. One evening at the group's general membership meeting, one member of the board becomes enraged during the discussion about abortion information and procedures available at a local women's health clinic. He shouts that the clinic staff deserves to die. Later that week, someone fires a gun through several windows of the clinic. Although no one is killed, several clinic staff and one patient are injured. The police find out that Professor Jenkins has been studying the local group and subpoena his field notes. Professor Jenkins complies with the subpoena, and the board member is arrested.

1. What ethical principle(s) do you think was (were) violated in this research?
2. Describe the probable reason(s) Professor Jenkins violated the principles you noted in your answer to Question 1.
3. Describe your reactions to the way this study was done and the researcher's response to the subpoena.

© Image copyright Adam Gregor, 2009. Used under license from Shutterstock.com

CHAPTER **4**

Selecting Researchable Topics and Questions

INTRODUCTION

Do you know people who lived through a natural disaster such as a hurricane, a tornado, a tsunami, or the like? In what ways have their lives changed? Do you think their experiences are typical or unusual? Could you create a research question about the social impact of living through a natural disaster?

Creating research questions by noticing patterns of behavior is a common way to begin the research process. In 1886, for example, after widespread rioting in London, sociologist Charles Booth published a paper reporting the results of his survey of poverty in London's East End (Bales, 1999). The press at the time was warning of revolution in the streets, but Booth argued that the working class was not well organized and did not pose a significant threat to the social order. He argued for social reform to address the poverty he documented. Booth's analyses were widely reported around the world in many newspapers, making him perhaps the first sociological "household name" (Bales, 1999). In Booth's work, we find an early example of how the events can shape research and, conversely, how research can shape the social and political responses to events and underlying social problems.

In 2006, four months after Hurricane Katrina, sociologist Ronald Kessler, director of the Hurricane Katrina Advisory Group Initiative, announced the start of two-year study of survivors of the hurricane. Kessler told the media, "The situation is worse here than in a lot of other disasters … [survivors] are spread out all over God's creation. There are people in Los Angeles and Oshkosh and in Florida and in New Jersey, so … we have to beat the bush all over the country" (Allen, 2006: A4). Kessler and his colleagues (2008) are continuing their study and finding high rates of hurricane-related mental illnesses years after the hurricane. Their work illustrates a connection between social problems and social research.

research questions, questions about one or more topics or concepts that can be answered through research.

In Chapter 1 we introduced the topic of **research questions**, questions about topics that can be answered through research. You might have asked yourself, "Research questions *about what?* Where does the idea for a research project come from?" Now that you're acquainted with some aspects of research, including ethical issues involved, we'll consider research questions in more depth.

The public reaction to the 1896 riots that gripped London was the impetus for Booth's study of poverty (Bales, 1999). He ultimately asked and answered three questions: How many people were living in poverty? What brought them to poverty and kept them impoverished? What could be done to alleviate the poverty?

In 2005, the suffering caused by Hurricane Katrina and the national attention to the survivors motivated Kessler and his colleagues to find out about the adequacy of their medical care, the long-term mental health consequences of the hurricane, and the extent to which the survivors will rebuild their cities and their lives (Allen, 2006). As we'll read in this chapter, developmental psychologist Michele Hoffnung had personal experiences and professional interests that set the stage for her study of women's lives. Hoffnung and Kessler are doing work that will yield important research findings: *basic* information about the social world and understandings that can be *applied* to creating social policy.

Research questions can be about local or global phenomena; can be about individuals, organizations, or entire societies; can be very specific or more general; can focus on the past, the present, or the future; and can ask basic questions about social reality or seek solutions to social problems. Research questions are similar to hypotheses, except that a **hypothesis** presents an *expectation* about the way two or more variables are related, but a research question does not. Both research questions and hypotheses can be "cutting edge" and explore new areas of study, can seek to fill gaps in existing knowledge, or can involve rechecking things that we already have evidence of. Whereas research projects that have explanatory or evaluation purposes typically begin with one or more hypotheses, most exploratory and some descriptive projects start with research questions.

hypothesis, a testable statement about how two or more variables are expected to be related to one another.

The American Sociological Association (ASA, 2009), founded in 1905, is a nonprofit organization dedicated to helping sociologists professionally and serving the public good. As we discussed in Chapter 3, the ASA provides a code of ethics. With over 14,000 members, 20 percent of whom work in government, business, or nonprofit organizations, the ASA hosts 44 special interest sections that allow members to interact with others specializing in a wide range of topics, including aging and the life course, animals and society, education, family, gender, and health. The annual meeting attracts some 6,000 participants who attend sessions focusing on a wide variety of topics. A few of the research questions covered in papers presented at a recent conference are found in Box 4.1.

BOX **4.1**

Examples of Research Questions Asked in Papers Presented at Meeting of the ASA

- Malcolm Fairbrother (2008) asked about national trade policies. If economic and political elites in countries are strongly committed to free markets, what explains the substantial international conflict? Why don't countries quickly agree and act upon a neoliberal course of action in which they open up their economies?

- Peggy Giordano and Stephen Cernkovich (2008) asked about the impact of education on men's likelihood of committing property crimes. To what extent does attending college encourage, rather than deter, social deviance and risk-taking among men? What is the effect of alcohol consumption and unstructured socializing patterns on this relationship?

- Jessica Halliday Hardie and Amy Lucas (2008) asked about the impact of economic factors on young cohabitating couples. Do earnings affect the quality of the relationship and is the impact the same for both women and men?

- Kyrsi Mossakowski (2008) asked what the effects of socioeconomic status, race, and ethnicity are on rates of depression among adolescents. Are racial and ethnic differences in mental health explained by family background, wealth, and the duration of poverty during the transition to adulthood?

- Ana Villalobos (2008) asked what factors contribute to "intensive mothering." More specifically, do women compensate by spending a greater proportion of their time with their children when there is insecurity in their adult partnerships?

- Irina Voloshin (2008) asked what the effect of family is on student work patterns. Do household resources and structure have an impact on the types of jobs held by students and the numbers of hours worked per week?

STOP & THINK   *We're willing to bet that by this point in your college career you've read dozens of articles and most likely some books that report the results of research. You can probably identify the research questions or hypotheses that the studies started with. But, think about some articles you've read in the recent past, and try to recall if there was any information about the actual genesis of each project. That is, do you know why those particular social or behavioral scientists did those particular research projects at that point in time?*

If the ones you've read are like most journal articles or papers, including the ones reported in Box 4.1, they probably included a review of the literature on the research topic and made connections between the research question and at least some of the previous work in the field. Research questions and hypotheses are usually presented as if they've emerged dispassionately simply as the result of a careful and thorough review of the prior work on a given topic.

As a result of the formal and impersonal writing style that is often used in academic writing, we often don't learn much about the beginnings and the middles of many projects. Although real research is "messy"—full of false starts, dead ends, and circuitous routes—we frequently get sanitized versions of the process. There are some wonderful exceptions,[1] but the "inside story" of the research process is rarely told publicly. More typically, written accounts of research projects are presented in formalized ways. Articles often include clearly articulated hypotheses, summaries of the work in the field that the study builds on, descriptions of carefully planned and flawlessly executed methods of collecting data, and analyses that support at least some of the hypotheses. Researchers will frequently comment on the limitations of their work, but only sometimes on problems encountered, mistakes made, and pits fallen into. Most articles published in scholarly journals present the research process as a seamless whole. There is little sense that, to paraphrase John Lennon, real research is what happens to you when you're busy making other plans. Nor is there a sense that researchers are multidimensional people with passions, concerns, and lives beyond that of "researcher."

Because our primary goal is to explain research techniques and principles, we've tried to do more than reproduce conventional social science writing in the excerpts of research that we've included. To give you a realistic sense of the research process, we've mostly focused on small, manageable research projects of the kind being done by many social scientists. In each case, we wanted to get "the story behind the story." We have been most fortunate to have found a wonderful group of researchers, each of whom was willing to give us a candid account. From them, we learned about the interests, insights, or experiences that were the seeds from which their projects grew, the kinds

---

[1] See, for example, some of the classics in social research including *Street Corner Society* (Whyte, 1955), *Sociologists at Work* (Hammond, 1964), and *Tell Them Who I Am* (Liebow, 1993) as well as the more recent *Moving Up and Out: Poverty, Education and the Single Parent Family* (Holyfield, 2002).

of practical and ethical concerns they had at the start, and how "real life" shaped their data collections and analyses. As a result, in each chapter, we are able to share at least a little of "the inside scoop."

As you read this, researcher Hoffnung is likely still working on her study of women's lives. Here she tells us about the impetus for and some current data from her research project.

FOCAL RESEARCH

Studying Women's Lives: Family Focus in the Thirties

By Michele Hoffnung[2]

The Research Questions

In 1992, I began a long-term study of women's work and family choices. Several factors led me to undertake this research project. First and foremost, my personal experience as a woman committed to both career and motherhood gave me an understanding of what such a commitment entails. Second, my concern about the unequal status of women in the labor force led me to focus on the costs of the conflicting demands of employment and family for women. Third, the scarcity of theories and longitudinal research about women's adult development presented a challenge to me as a feminist researcher.

Even as a girl, I always knew I wanted children. My mother was one of eleven children; my father one of eight. While I was one of three children, I wanted to have *at least* four. I also expected to go to college. During my college years, I became very interested in psychology. My academic success led my professors to suggest graduate school. I went and loved it. At the University of Michigan I committed myself to an academic career and I also got married. My desire for motherhood remained unchanged; with my course work completed, I had my first child at the end of my third year of graduate school.

The first year of motherhood, I worked on my dissertation. Since my son was a good sleeper, my husband was willing and able to share infant care, and we were able to hire a student to help us for a couple of hours a day, I managed. But, the next year was far harder. Starting as a new assistant professor was more taxing and less self-scheduled than researching and writing a dissertation. I found balancing the new career and caring for the baby extremely difficult. I spent three days at school and four days at home with my son,

---

[2]This article was written for this text by Michele Hoffnung, PhD, Quinnipiac University, and is published with permission.

writing lectures during nap time and late into the night. By mid-year I was exhausted. Luckily, I had an intersession in which to partially recover, so I survived.

What, I wondered, happened to women with less flexible jobs, husbands with less time or willingness to share family responsibilities, less money for outside help, and no intersessions? The social science literature I read focused on what women should be doing or feeling rather than describing what women *actually* did and felt. When my second child was out of infancy, I found the time to study mother's lives. I conducted in-depth interviews with forty mothers of at least one preschool child and discovered that the mothers who were pursuing careers were more likely to plan family life with rigorous use of birth control, systematic preparation for childbirth, and careful selection of child care. They formed marital relationships with a less gender-stereotyped division of labor than the non-career mothers; their husbands shared parenting. I also found that all of the women were grappling with the issue of employment; even at-home mothers were concerned with when and whether they would be able to work again (Hoffnung, 1992).

The women I interviewed were products of the fifties and sixties, when women with career aspirations were not common. What, I wondered, were the realities for women growing up in more recent decades? Studies done during the 1980s and 1990s indicated that a vast majority of college women wanted it all: marriage, motherhood, and career. While we know that most college women say that they plan to have a career as well as a family, we have little data on the kinds of families they want or how they are going to balance career and family. When Kristine Baber and Patricia Monaghan (1988) questioned 250 college women, they found that all planned careers, 99 percent planned to marry as well, and 98 percent also planned to be mothers. When asked how they would accomplish their goals, 71 percent said they planned to establish their careers before having their first child and the more career-oriented the women, the later they projected their first birth. The most common plan was to return to work part time after the child was at least a year old. Such plans for "combining," if implemented, can have a negative impact on a woman's career development. We also know that, regardless of their intentions, most women are employed these days, even mothers of preschool children. Many women have jobs rather than careers; others are underemployed given their talents and training, often because of their family commitments.

My experiences, previous research, and current interests led me to plan a long-term study of a group of young women. I thought that starting at a time when they had plans to "do it all," and continuing on as they made life choices, would reveal patterns in their lives and the efficacy of various choices. For example, does the level of career commitment affect family choices? Does it affect work satisfaction? My research goal was to see how women's ideas and expectations change as they become more educated, face the world of graduate and professional school and employment, and make the choices of whether and when to marry, whether and when to have children, and whether and how to combine career and motherhood. These are important questions for our time, as most mothers work outside the home, even though the family continues to have sole responsibility for infant and preschool care.

The Study Design

My earlier study was retrospective, with adult respondents recalling their childhood hopes and dreams; this was prospective, first asking young women still in the relatively sheltered environment of college to look toward their futures and then following them over time. My strategy was designed to avoid the problem of having respondents' memories altered by the choices they've actually made, but it has required a long-term commitment with methodological challenges.

In 1992–1993, I conducted in-person interviews with 200 seniors at five New England colleges about their expectations and plans for the future regarding employment, marriage, and motherhood. Since then, I have followed their progress, most years with a brief mailed questionnaire. During the seventh and fourteenth years of the study, I interviewed the participants by telephone. This regular contact has helped me maintain a high response rate over the years. In 2008, 151 women, 75.5 percent of the sample, returned their questionnaires.

Collecting Data

My study requires that respondents participate over a period of many years so I have had to interest the women enough in my project to commit to responding annually. Because American young adults move frequently, I have devised ways to contact them as they travel and relocate. In this ongoing process, I see my relationships with the women as more ongoing, personal, and collaborative than is typical of more traditional and impersonal survey procedures. At the same time, I strive to maintain a value-free, open-ended environment in which they can honestly tell about their individual lives (Hoffnung, 2001).

A major methodological concern was to select a sample that included white women and racial/ethnic women from a wide variety of backgrounds. This was important because until recently, most social science research, even research on women, has focused on members of the white middle class. I wanted to recruit a sample that would be varied and accessible to me. I decided to select a stratified random sample of female college seniors from five different colleges and universities. While my sample excludes the lower ranges of the socioeconomic spectrum, by including a small private college of mostly first-generation college students, a large state university, two women's colleges, and an elite private university, I reached women with a wide range of family, educational, and economic backgrounds. My proposal was approved by my university's IRB; each college required me to follow its own procedures to gain access to the senior class lists from which I picked random samples of both white women and racial/ethnic women. University approval gave me credibility when I called to ask the women to participate and allowed alumni associations to help me locate students in the years after graduation.

Since for the initial in-person interview I could expect students to come to me only if I made it convenient, I needed an office to do my interviews in private and a phone for people to call if they couldn't come. One college was my home institution, where I already had an office. I applied for and received a fellowship and office space at a research institute housed on another's campus. For the remaining campuses, I used strategies based on contacts I had. At one school, I called three feminist psychologists whom I had met professionally; one offered me her own (but phoneless) research space. At the second, I called the chairman of the psychology department and he provided me with unused space reserved for student projects. A campus phone in the hall enabled me to call interviewees, although I couldn't receive calls. In the third, I called the chairman of the history department, who knew my husband, and asked for help. His secretary provided me with the office and phone of a faculty member on sabbatical.

With the office problem solved, I called students, recruited, scheduled, reminded, and, if they did not show, rescheduled. Each interview took about an hour, but many more hours went into phone calls and waiting for those who forgot or skipped their appointments. I quickly learned to make reminder phone calls, and I always called and rescheduled interviews for those who missed an appointment. As a result, most of the random sample I contacted completed the interview. I suspect a couple of women finally showed up to stop my phone calls!

In the initial interview, I used a set of questions that was consistent from respondent to respondent, yet was flexible enough to allow respondents to explain themselves. I asked questions about their families of origin, attitudes about work, marriage, motherhood, and short- and long-term plans for the future. In the mailed surveys I've used in subsequent years, I've asked about relationships, work, school, and a few open-ended questions about aspirations, feelings, and concerns. Notes and comments from the women have helped to shape the next year's questionnaire or interview. In the recent phone interviews, I focused on work experience and aspirations, personal histories and expectations, and experiences and expectations concerning motherhood. Because not all of the women are married and only some are mothers, this set of interviews took different forms with different participants.

Findings

What have I found so far? As I anticipated, in their senior year, while only one woman was married and another one was pregnant, virtually all planned to have a career (96 percent) and to have children (99 percent). Only 87 percent expected to marry—less than in earlier studies of young women. Since white, non-Hispanic women were significantly more likely to expect to marry than other women, the smaller percentage may be due to my more culturally diverse sample of women (118 white, 25 African American, 20 Hispanic, and 36 Asian American). This appears to be a realistic appraisal, since other data indicate that African American women have the lowest marriage rates; Chinese and Japanese American women the next lowest rates; and European Americans have the highest marriage rates (Ferguson, 2000).

For these 1993 graduates, career was their primary focus in their twenties. Fifteen years after college graduation, more than half had obtained graduate or professional degrees. One third earned a master's degree, and 21 percent a doctorate (including MD, JD, and PhD), and some are still students. In contrast, their spouses are not as well educated: 21 percent have less than a bachelor's degree, 16 percent a bachelor's degree, 21 percent a master's degree, and 28 percent a doctorate.

By 2008, about 85 percent have been married at least once (4 percent are separated, divorced, or widowed; 5 percent are remarried). Another 7 percent report being engaged or in a committed relationship and some others say they expect to marry when the right person comes along.

As seniors in college virtually all of these women wanted children. By year fifteen, 70 percent are mothers. Of those, 24 percent have one child; 59 percent have two; 11 percent have three; 6 percent have four or five. About 40 percent of those who are not yet mothers expect to have children. In their thirties, family has become a focus of their attention.

Motherhood has a significant impact on women's employment. In 2008, 62 percent of the women were employed full time, 19 percent were employed part-time, and 19 percent were unemployed. Of the mothers, 49 percent worked full time and 51 percent had reduced or relinquished employment. Only 11 percent of the childfree women are not employed full time. About a fifth (19%) of the mothers were at home full time caring for children, whereas only 1 percent of their spouses were.

I started this research project to chart their paths and to find out if these women would "have it all." I've found that at average age 37, while others may join their numbers, 51 percent of the women "have it all": marriage, at least one child, and full or close to full-time work. Most interesting, when asked if they were happy with the work–family balance in their lives, I found no significant differences between stay-at-home moms and employed moms (about 78 percent report being happy: 81 percent stay-at-home and 78 percent employed); between moms and non-moms (about 77 percent report being happy: 78 percent moms and 76 percent non-moms); or between full-time employed moms and full-time employed non-moms (75 percent report being happy: 75 percent moms and 76 percent non-moms).

References

Baber, K. M., and Monaghan, P. 1988. College women's career and mother-hood expectations: New options, old dilemmas. *Sex Roles* 19: 189–203.

Ferguson, S. J. 2000. Challenging traditional marriage: Never married Chinese American and Japanese American women. *Gender & Society* 14: 136–159.

Hoffnung, M. 1992. *What's a mother to do? Conversations on work and family*. Pasadena, CA: Trilogy.

Hoffnung, M. 2001. *Maintaining participation in feminist longitudinal research: Studying young women's lives over time*. Paper presented as part of the Feminist Research Methods Symposium at the National Conference of the Association for Women in Psychology, Los Angeles, CA.

THINKING ABOUT ETHICS   *Before we consider the other issues in this focal research, let's note the ethical issues and how they were handled. Hoffnung submitted her proposal to her university's Institutional Review Board (IRB) where it was approved; participation in her study was and continues to be voluntary; she does not believe that sharing information about their lives has harmed any participants; she assured her participants of confidentiality and has kept all materials confidential; when she writes about her research, she will use pseudonyms rather than the real names of participants.*

SOURCES OF RESEARCH QUESTIONS

STOP & THINK   *Think about your life, the issues you're facing, and the concerns you have. Are there any questions you'd like to have answers to? This is one way to create a research question.*

In the focal research, Hoffnung tells us the genesis of her study: her personal experiences and values, her choice of a professional specialization, and the lack of psychological research and theory on women's lives when she began her work. The selection of a research question is often the result of many factors, including personal interests, experiences, values, and passions; the desire to satisfy scientific curiosity; previous work—or the lack of it—on a topic; the current political, economic, and social climates; being able to get access to data; and having a way to fund a study.

Values and Science

During the nineteenth century and for a good part of the twentieth, it was commonly thought that all science was "value free," but today it is more widely believed that values—social, moral, and personal—are part of all human endeavors, including science. Science and even the most objective scientific products are socially situated (McCorkel and Myers, 2003: 201). Some argue that dominant groups (e.g., privileged, white males) are especially poorly equipped to detect values and interests in their own work (Harding and Norberg, 2005: 2010). Group interests and values can subtly influence scientific investigations, and they are especially influential during the creation and evaluation of research questions and hypotheses since, "At any given time, several competing hypotheses may explain the facts equally well, and each may suggest an alternate route for further research" (National Academy of Sciences, 1993: 342).

Social and personal values do not necessarily harm science but researchers should be aware of their values and their influence on their work. Hoffnung, for example, is clear that her values include a positive view of combining motherhood and career. Rose Weitz (1991) tells us that she began a study of people living with human immunodeficiency virus (HIV) or acquired immunodeficiency syndrome (AIDS) in the 1980s after learning about a governmental policy mandating the reporting of people with HIV which she opposed.

The desire to produce knowledge is a social value. So is the desire to do accurate work and the belief that such work can ultimately benefit rather than harm humankind. We know that values contribute to the motivations and conceptual outlook of scientists. The danger comes when scientists allow their values to introduce biases into their work that distort the results of scientific investigation (National Academy of Sciences, 1993: 343). Articulating values can help others evaluate a researcher's work, including the choice of research questions and methodological decisions, and may make the researcher more aware and perhaps more objective as a result.

Personal Factors

 **STOP & THINK** *Compare your life today to your parents' lives when they were your age. How did they keep in touch with their friends? What are some of the differences in the technologies that you use that they didn't (such as cell phones, e-mail, Twitter, texting, blogging, and Facebook)? Think about the social and psychological impacts of the widespread use of the new technologies, and create a research question about one or more of these technologies.*

Researchers often undertake studies to satisfy intellectual curiosity and to contribute to society and to social science. However, personal interests and experiences can influence the selection of a specific research topic or question. It shouldn't surprise us that researchers often study topics with special significance for them. After all, research projects can take months or years to complete; having a strong personal interest can lead to the willingness to make the necessary investment of time and energy.

Hoffnung tells us how her personal experiences in juggling career and motherhood influenced her work. Jane Bock writes of conducting a study of single mothers by choice as a "midlife single woman who had long considered single motherhood but had not yet taken any action" (2000: 66). Randolph Haluza-DeLay (2008) tells us how his personal involvement with environmental concerns, including teaching in environmental programs and taking wilderness leadership positions, along with his personal activities in Christian spirituality and theology, came together to contribute to his research agenda on environmental justice.

One important study came about through very difficult personal circumstances. In 1984, anthropologist Elliot Liebow learned that he had cancer and a limited life expectancy. Not wanting to spend what he thought would be the last months of his life working at his government job, Liebow (1993: vii) retired on disability but found himself with a great deal of time on his hands. He became a volunteer at a soup kitchen and later at an emergency shelter for homeless women and found he enjoyed meeting the residents. Liebow (1993: viii) was struck by the enormous efforts that most of the women made to secure even elementary necessities and he appreciated their humor and lack of self-pity. Liebow volunteered for two years, but then, out of either habit or training, during the last three years of his work at the shelter, he collected data as a participant observer—talking to, observing, and taking notes about the women and their lives. Before his death, Liebow completed an important contribution to helping others, the thoughtful and sympathetic analysis, *Tell Them Who I Am: The Lives of Homeless Women* (1993).

Other less dramatic examples of personal interests and experiences can be found in many research projects. Jonah Berger, co-author of a study on when losing leads to winning, said the idea for the project came to him when he was coaching youth soccer and "always felt like the kids worked harder when we were slightly behind at halftime" (Cammorata, 2009: C2). Emily's first major study was a study of the division of labor and the division of power in marriage. She began the study in the first few years of her own marriage, as half of a two-career couple. And, perhaps because both of us, Emily and Roger, are parents who have read hundreds of books to children over the years, we've produced, separately and together, more than a dozen studies on children's literature. In Box 4.2, Jessica Holden Sherwood recounts her search for a research topic and how she came to study members of exclusive country clubs, a social world that she had some connection to.

BOX **4.2**

Reflections on "Studying Up" by Jessica Holden Sherwood*

I wasn't one of those people who had a plan right from the beginning. Approaching my doctoral dissertation, I was like any student in need of a paper topic. At my university, graduate students must choose specialty areas within sociology. I picked social inequality, as in race, class, and gender. Often, studying inequality means studying minorities, the poor, and women. Missing from this picture is the other side of each equation: the white, the rich, and men. Sometimes I would read or hear a social scientist point out the importance of "studying up" as well as "down" regarding inequality. While these groups are not what we usually think of when we think of race, class, and gender, that's part of the point. If they *don't* get studied, we'll never have a complete understanding of inequalities—who maintains them, why, and how.

The call to "study up" or study the "unmarked" (the categories that seem standard and go unnoticed) resonated with me. My graduate experience at a large state university has been eye-opening in terms of my own position on the socioeconomic scale. Coming from an upper-middle-class background, my socioeconomic position had been unmarked for me. Now I recognize its significance: first, that as a member of the unmarked group (like "white"), one of my privileges has been never to think about that group membership**; and second, that my life was organized in such a way (such as attending an expensive private college) as to preserve that obliviousness well into my twenties.

One reason that "studying up" is rare is that it's considered difficult. Especially when it comes to class, advantaged group members are thought of as elusive subjects of study. I thought that my own social location—being white, knowing some people who are rich and/or powerful, and having the cultural capital*** to interact with them effectively and comfortably— might enable me to overcome that elusiveness.

Another facet of my biography is that I have one Jewish parent and one WASPy or white Anglo-Saxon Protestant parent (WASP is a common label in the Northeast). So, I have a tenuous relationship with the mythical "old-WASPy-Yankee" peoples. It is not hopelessly foreign and distant, but it's also not a circle I wholly belong to. So I am faced with the option of making efforts to belong to that circle, and also asking

myself why I even want to. This tenuous relationship is probably one reason that I care about, and have chosen to study, this particular slice of social life.

Next, I needed a place to find these people, a context in which to study them. I attended an elite boarding school, and I considered doing research at that school or a similar one. But, I was more interested in the doings of adults than teenagers. I'm familiar with a small island nearby that's a destination for wealthy vacationers, and I considered studying its social life. But, anybody can visit that island, and the wealthy vacationers don't all know each other, so it's not exactly a coherent community. I've settled on something that's convenient and concrete: a study of country clubs and other exclusive social clubs in one section of the Northeast. These private social clubs provide a specific context for the social lives of people who are "unmarked" and advantaged rather than disadvantaged.

I'm finding a unique blend of pros and cons as I do this research. On the one hand, I'm having a perfectly smooth time securing interviews with the people who are members of the clubs. On the other hand, though I expect my dissertation to be critical, I'll hesitate to condemn the people I've studied, because some of them are neighbors and friends. But in any case, I'm pleased with the research topic I've settled on and feel it's enriched by my consideration of its connection to my biography.

*Jessica Holden Sherwood wrote this while she was a graduate student in the Department of Sociology, North Carolina State University. Her dissertation, *Talk about Country Clubs: Ideology and the Reproduction of Privilege*, can be found at http://www.lib.ncsu.edu/theses/available/etd-04062004-083555. This article is published with permission.

**Peggy McIntosh, 1992, "White Privilege and Male Privilege: A Personal Account of Coming to See Correspondences through Work in Women's Studies," in *Race, Class and Gender: An Anthology*, ed. Margaret Anderson and Patricia Hill Collins, Wadsworth, Belmont, CA.

***Cultural capital is an acquired cluster of habits, tastes, manners, and skills, whose social display often sends a message about one's class standing. From Pierre Bourdieu, 1977, "Cultural Reproduction and Social Reproduction," in *Power and Ideology in Education*, eds. Jerome Karabel and A. H. Halsey, Oxford University Press, New York.

STOP & THINK *Newspapers and newscasts have stories about social phenomena all the time. Today there might be a story about how a family is dealing with parental unemployment, a graph showing changes in the crime rate, an editorial about laws allowing gay and lesbian couples to marry, or an article on the job market for new college graduates. Do you have a personal interest in any of these topics? If you needed to design a research project for a class assignment, which of these or other topics might you select to study?*

Research and the Social, Political, and Economic World

Personal interests alone rarely account for the selection of a research topic. Knowledge is socially constructed and socially situated. Social relations produce individual research projects and the enterprise of research itself. In fact, all knowledge claims "bear the fingerprints of the communities that produce them" (Harding, 1993: 57). As social, political, and economic climates change, the kinds of research questions that social scientists pose and the kinds of questions that are acceptable and of interest to the research community and the larger society change with them.

 **STOP & THINK** *Some topics have been the focus of considerable research in the social sciences for a century or more, but others have received attention only within the past decade or two. Can you think of examples of topics that have been studied only recently?*

Some topics have been largely invisible to the general public and generally ignored by social scientists until recently. In Chapter 12, we'll read about an interesting study that compares homicide rates of different groups. While in graduate school, one of the study's authors, Ramiro Martinez, was told that homicide was the leading cause of death among young African American males. But when he asked about the Latino-specific rate, he was told that this information was not available. This omission became one of the reasons he began his work (Martinez, personal communication).

Before the 1970s, few studies focused on women and, when considered at all, data about women were analyzed using male frames of reference (Glazer, 1977; Westkott, 1979).[3] However, as a result of the women's movement, which first developed outside of the university in the late 1960s and early 1970s, scholars began to study women and their lives. Hoffnung, for example, noticed that few of the long-term studies in her specialty, adult development, included women as subjects. She found herself perplexed that the theories that she had studied as a student focused on only men and their lives. In her professional work, she sought to redress the imbalance and began using new theories being developed by feminist thinkers such as Nancy Chodorow, Carol Gilligan, and Jean Baker Miller (Hoffnung, personal communication).

---

[3] The natural, no less than the social, sciences offer many interesting examples of such male bias. In one federally funded study examining the effects of diet on breast cancer, only men were used as sample subjects (Tavris, 1996).

Critiques of gender imbalance are found frequently in social science. For example, in the mid-1990s, when three out of five American women were employed, Weaver (1994) noted that the number of studies focusing on women's retirement were quite small. Studies of the homeless have followed a similar pattern, with early efforts ignoring women's experiences (Liebow, 1993; Lindsey, 1997). Given the recent changes in the social climate, current research is much more likely to focus on women and men.

Change in society also has influenced the amount of research on other topics. In 1995, Allen and Demo argued that sexual orientation was understudied, especially in research on families. Analyzing nine journals that published more than 8,000 articles on family research between 1980 and 1993, they found that only 27 articles focused explicitly on gay men, lesbian women, or their families (Allen and Demo, 1995). The data led them to conclude that sexist and heterosexist assumptions continued to underlie most of the research on families, with the research focusing mainly on heterosexual partnerships and parenthood. Allen and Demo (1995: 112) felt that studying lesbians and gay men as individuals, but not as family members, reflected a society-wide belief that "gayness" and family are mutually exclusive concepts. To see if this criticism was still valid, Emily did a search of *Sociological Abstracts*, a database of scholarly sources that we'll talk about later in this chapter. Her search found citations for only 20 articles about gay men and lesbians as family members published between 1996 and 2000, more than 200 published between 2001 and 2005, and over 800 published between 2006 and 2009. If this particular omission has been rectified, there will always be topics that need more research. Chito Childs (2005), for example, noticing that black women's voices were overlooked in research on interracial relationships, designed a study focusing on them; Hagestad and Uhlenberg (2005) point out that while ageism has been studied extensively, age segregation is a neglected topic; Szinovacz and Davey (2006) note that research on grandchild care has ignored the impact of other simultaneous life transitions, such as retirement.

When researchers decide to break new ground, they may find a lack of support for their projects. Helena Lopata, a well-established sociologist, encountered negative reactions from many in her social and professional circles when she decided to study the social role of widows. She pursued her interest despite the view that the subject was judged "depressing and as unworthy of serious sociological research while 'more important' scientific problems remained unstudied" (Lopata, 1980: 69).

Topics and issues become the focus of research when they become visible, more common, and/or a source of concern to the public or the scholarly community. We've seen, for example, more studies on the elderly as the proportion of the population older than 70 has increased. Similarly, as the number of children living in single-parent and in two-worker households has risen, the impacts on children of family structure and workplace demands have received increasing scholarly attention. Finally, consider the annual global costs of public expenditures on criminal justice—which in the late 1990s was $360 billion (Farrell and Clark, 2004)—and then note how many hundreds of articles on crime and criminal behavior are published annually.

Social changes and current events can lead to specific research questions. In a study that is the focal research of the next chapter, Scott Keeter and his colleagues at the Pew Research Center for the People & the Press describe the effect of cell phone use on public opinion polling. Because of the dramatic increase in the use of cell phones in recent years, they found reporters and political observers constantly asking about the impact and decided to study the topic (Keeter, personal communication).

Although the social and psychological aspects of natural disasters have been studied for decades, recent events are probably why research on the social dimensions of disasters has increased. A quick search of articles indexed in *Sociological Abstracts* finds the number of articles presenting research on the social impact of hurricanes, earthquakes, and tsunamis to have exploded in recent years: 12 articles between 1996 and 2000; 36 articles between 2001 and 2005; 82 articles between 2006 and 2009. A similar trend is likely to occur concerning the severe economic downtown that was underway by 2008. For example, there were about 100 articles published between 2000 and 2008 indexed in *Sociological Abstracts* about the effects of unemployment on families, but coming years are likely to see many more. As social problems come to the fore, social scientists are more likely to select them as research topics and to consider possible solutions.

Research Funding

As one set of researchers has noted, "It is always difficult to raise money for social science research; after all, it neither directly saves thousands of lives nor enables one to kill thousands of people" (Fischman, Solomon, Greenspan, and Gardner, 2004: x). Even a small study can be expensive and the cost of doing research can run into hundreds, thousands, or millions of dollars, so funding is an important consideration. Many research projects, including those in the social sciences, are funded by federal agencies, private foundations, local and state institutions, or corporate sponsors.

National funding for behavioral and social science research has been significant since the middle of the twentieth century. In the United Kingdom, the government-supported Economic and Social Research Council, the country's leading social and economic research and training agency, has an annual budget of around £105 million that supports research centers, groups, resources, and programs (ESRC, 2009).

In the United States, social and behavioral research is funded by the National Institutes of Health (NIH), the National Science Foundation (NSF), the National Institute of Justice, the Department of Education, the National Institute on Aging, and other agencies. Although the amount of funding decreased in the early 1990s, with the share of total federal research funding for the social and behavioral sciences falling from 8 to 4.5 percent (Smith and Torrey, 1996: 611), government support remains important. In 2004, the NSF awarded $5.5 million for research in sociology with awards that ranged from $12,000 for a project entitled "Collaborative Research: The Extent and Underlying Cause of Corporate Malfeasance in US Corporations" to $425,000 for "Family Resource Allocation in Urban and Rural Communities" (ASA, 2005). The NSF

Division of Social and Economic Sciences that covers the other social sciences as well received $92 million in funding in fiscal year 2005 (COSSA, 2005: 45). Within months of Hurricane Katrina, a project on the mental health of the survivors of the hurricane was funded by a $1 million grant from the National Institute of Mental Health (Allen, 2006).

Many foundations offer support for the social sciences. In an analysis of a sample of over 150,000 grants from large foundations, the Foundation Center found that funding for the social sciences increased from $259 million in 2006 to $295 million in 2007. However, they found the social sciences continued to receive the smallest share of the grant money—just 1.4 percent of these grants—with an average grant of $50,000 (Lawrence, Mukal, and Atlenza, 2009: 24). Some foundations support specific kinds of social research. In 2005, the Robert Wood Johnson Foundation gave over $100 million in grant awards to support 329 projects involving research and evaluation in the areas of health and heath care with a focus on finding answers relevant to policy decisions (Knickman, 2005).

Funding research expresses a value choice—the judgment not just that knowledge is preferable to ignorance, but that it is worth allocating resources to research as opposed to other priorities (Hammersley, 1995: 110). The particular values associated with specific projects affect funding as the appropriations process for research is part of a larger political process. Each year, for example, after the president submits the budget, constituent institutes appear before appropriations committees to defend it. Interest groups, such as the Consortium of Social Science Associations, the umbrella organization of the professional societies in the behavioral and social sciences, testify as well. Research that is perceived to be threatening to those making decisions might have a more difficult time getting financial support (Auerbach, 2000).

Funding can be in the form of grants or in the form of a contract for an agency or sponsor. Grants typically allow the researcher more control of the topic, methods of data collection, and analysis techniques, whereas contracts usually commit the researcher to a specific topic, timetable, and methods. Although some sources are more restrictive than others, the availability of funding and economic support can influence a study, specifically the questions asked, the amount and kinds of data collected, and the availability of the resulting research report.

A fascinating example of the effect of funding on research is the story behind one of the largest study of sexual practices and beliefs in the United States (Laumann, Gagnon, Michael, and Michaels, 1994). The project was influenced by a combination of factors including a major health crisis, public values, politics, and funding. The initial motivation for the project was a growing AIDS epidemic and a public health community with little data on sexual behavior because of the controversial nature of doing such research in our society (Laumann et al., 1994: xxvii). In 1987, the National Institute of Child Health and Human Development requested proposals to design studies—one on adult sexual behavior and the other on adolescent sexuality (Laumann et al., 1994: 39). Researchers at the National Opinion Research Center in Chicago won the contract and designed the adult study, but political appointees at the highest administrative levels of the Department of Health and Human Services refused to allow approval of even a narrowly focused survey of sexual practices. When public funding was not forthcoming, the researchers found financial support from a consortium of private foundations. With the funding in place, the researchers were able to complete the first large, nationally representative study of human sexuality in decades. It is ironic that a *refusal* to sponsor research ultimately led to a much more comprehensive study than had been proposed originally (Laumann et al., 1994: 41). Another example of the political nature of research can be found in the 2003 action of U.S. Representative Patrick Toomey, who sponsored an amendment to a bill that, had it succeeded, would have rescinded the National Institute of Health funding for five already approved grants on sexual behavior and health with goals such as preventing the spread of AIDS (Silver, 2006: 5). This action was followed by intensive lobbying by the Traditional Values Coalition, a self-described conservative lobbying group, against such research (Kempner, 2008). A recent survey of NIH researchers working on these grants found the political controversy led to a "chilling effect" with many researchers engaging in self-censorship practices, such as eliminating controversial language from grant applications, reframing studies, and removing research topics from their agendas (Kempner, 2008).

DEVELOPING A RESEARCHABLE QUESTION

Regardless of the source of an idea for a project—scientific curiosity, personal experiences, social or political climate, the priorities of funding agencies, or even chance factors—once a researcher has a topic, there's still work ahead.

 STOP & THINK   *Let's say you were interested in the topic mentioned earlier and created a research question about the social and psychological impacts of the use of technologies such as cell phones or Facebook. Where would you go from here?*

researchable question, a question that is feasible to answer through research.

Several steps are needed to turn a research question into a **researchable question,** a question that is feasible to answer through research. The first step is to narrow down the broad area of interest into something that's manageable. You can't study *everything* connected to cell phones, for example, but you could study the effect of these phones on family relationships. You can't study *all* age groups, but you could study a few. You might not be able to study people in *many* communities, but you could study one or two. You might not be able to study *dozens* of behaviors and attitudes and how they change over time, but you could study some current attitudes and behaviors. While there are many research questions that *could* be asked, one possible researchable question is: In the community in which I live, how does cell phone use affect parent–child relationships; more specifically, how does the use of cell phones affect parents' and adolescents' attempts to maintain and resist parental authority?

Reviewing the Literature

 **STOP & THINK**   *Let's assume that we've selected the impact of cell phones on family relationships as our research topic, and we want to begin a review of the literature. Where should we start?*

The accumulation of scientific wisdom is a slow, gradual process. Each researcher builds on the work others have done and offers his or her findings as a starting point for new research. Reviewing previous work helps us to figure out what has already been studied, the conclusions that were reached, and the remaining unanswered questions. Although a thorough review is time consuming, it can generate a more cumulative social science, with each research project becoming a building block in the construction of an overall understanding of social reality. Reading, summarizing, and synthesizing at least a significant portion of the work that has preceded our own is an important step in every project. We can look for warnings of possible pitfalls and reap the benefits of others' insights.

literature review, the process of searching for, reading, summarizing, and synthesizing existing work on a topic or the resulting written summary of a search.

Because both the *process* of reviewing previous work and the resulting *written summary* of the work that has been reviewed are called a **literature review,** we'll discuss both usages of that term in the sections that follow.

Academic Sources

To start the *process* of reviewing the literature means figuring out what literature or sources you want to search. Books, articles, and government documents are among the most commonly used sources for materials that describe research. Popular literature, including newspapers and magazines, might be good sources of research ideas, but academic journals will be more useful for your literature review.

A good place to start looking for books and published government documents is in your university's library. Many libraries allow others in addition to their patrons to search their catalogs online. If you find a book, article, or government publication that you want to use at a library other than your own, you can often request it through an interlibrary loan system. In addition, most government agencies have a great deal of material that can be downloaded from their websites. In addition to the sites of specific agencies, www.fedstats. gov links to information from more than 100 government agencies.

Hundreds of print and online journals publish research in the social sciences and related fields. Many of them are peer-reviewed journals, meaning that the articles have been evaluated by several experts in the field before publication. Some well-known sociology journals are *American Journal of Sociology, American Sociological Review, Crime and Delinquency, Gender & Society, Gerontology, Journal of Health and Social Behavior, Marriage and Family, Social Psychology Quarterly,* and *Sociology of Education.*

Almost every university library has electronic databases that allow you to search many journals simultaneously. Some databases list citations and abstracts and others provide the full text of the articles. *Sociological Abstracts* is a database that indexes the international literature in sociology and related disciplines in the social and behavioral sciences. It provides abstracts of journal articles drawn from over 1,800 publications, as well as abstracts of books, book chapters, dissertations, and conference papers. Other useful databases that many libraries have include *Academic Search, Education Resources Information Center (ERIC), LexisNexis Academic, Criminal Justice Abstracts, PsycArticles, PsycInfo,* and *Worldwide Political Science Abstracts.* You might also want to check out Google Scholar at http://scholar.google.com/, a free search engine that covers peer-reviewed papers, books, and articles from academic publishers, professional societies, and universities.

keywords, the terms used to search for sources in a literature review.

Although the databases vary, most allow you to search titles and abstracts only or to search the entire article using one or more **keywords**, which are the terms and concepts of interest. You'll generate numerous citations (title of the article, name of the author, name of the journal, volume, date of publication, and pages) with commonly used keyword(s). With common keywords, narrow the focus by using multiple keywords—such as by including "and" between the terms. For more unusual topics, think of broader terms or synonyms that could be used as keywords in additional searches. If you want articles that focus on research findings, use the word "research" and/or "study" as a key term in the search.

STOP & THINK *If you were going to use a database to look for scholarly articles on the impact of the cell phone on family relationships, what keywords might you use?*

Using the electronic *Sociological Abstracts* as our database, we searched for articles published between 2005 and 2009 by using the following terms: "mobile" or "cell," "phone" or "phones," and "research" or "study" searching the entire article record. We obtained 170 peer-reviewed articles as results, but when we added other search words, such as parent, family, or adolescent, the number of results dropped significantly.

Once a list of citations is generated, the titles and abstracts can help you decide if the book or article is relevant to your work. Most university libraries have many thousands of books, documents, and academic journals (in print, in microforms, or online). If a library doesn't subscribe to a specific journal or own the book, you can usually request a copy through interlibrary loan. When reading a journal article, you'll find that it will usually include a literature review with references to additional sources that you might want to check out. Box 4.3 describes the sections in a typical journal article, including the review of the literature.

A thorough literature search is a process of discovery. In addition to providing useful information, each article, book, or other item can direct you to new ideas and sources. Just as when you're on a hike in the woods with only a general idea of the direction that you want to travel, you'll need a good compass and some common sense to find your way.

Using the Literature in a Study

When you read the focal research sections in the chapters to come, you'll find that most of the researchers have used the existing literature to develop their own research questions or hypotheses. Some researchers, however, postpone a complete literature review until after data collection so they can develop new concepts and theories without being influenced by prior work in the field. In both approaches, prior work in the field is used at *some point* in the research process.

Existing research helps put new work into context. What comes before determines if a study is breaking new ground, using an existing concept in a

BOX **4.3**

Parts of a Research Article

The title tells what the research is about. It usually includes the names of at least a few of the major concepts.

The abstract is a short account of what the research is about including a statement of the paper's major argument and its methods.

Introduction will introduce the topic, discuss previous work and, sometimes, present relevant theories. Part of the introduction is a *literature review* as the author discusses how this research will fit into the work and gives credit to those who have contributed ideas, theories, conceptual definitions, or measurement strategies.

Data and Methods sections describe and explain the research strategies and usually cover the population, sample, sampling technique, study design,

method of data collection and measurement techniques that were used, and any practical and ethical considerations that influenced the choices.

Results or Findings are usually the point of the paper. In findings sections, researchers incorporate data or summaries of data.

Discussion section highlights the major findings and typically connects the findings to the literature review.

Conclusions usually sum up the current study, point out the directions for future work, and may also caution the reader about the study's limitations.

References list the books and articles referred to in the article.

new way, expanding a theory, or replicating a study in a new setting or with a new sample. When writing your own literature review, you will want to be up to date and thorough, but you will have to choose which sources to use and *give credit to,* using an appropriate method of citing sources. In writing a literature review for a research proposal or project, you should pick the materials that focus on the most relevant concepts, theories, and findings to provide a context for your own work.

You can use the literature review sections of published articles and research monographs as guides for writing your own literature reviews. A good literature review provides a perspective for the study you are proposing, in much the same way that a frame helps to focus attention on a painting. They can serve many purposes, among them a discussion of the significance of a research question, providing an overview of the thinking about a research question as it has developed over time, and presenting one or more theoretical perspectives on the topic. One endpoint of a literature review could be one or more research questions to be answered with data. In addition, a literature review can include definitions of useful concepts and variables, identify testable hypotheses deduced from prior research, and suggest some useful strategies and methodologies. Desirée Ciambrone, in her study of women with HIV/AIDS described in Chapter 3, found that doing a thorough literature review was one of the most important steps in the research process, as it honed her thinking on the topics she selected for her own interviews. For a discussion of writing a research report including some tips on writing the literature review, see Appendix C of this text.

STOP & THINK   *Let's assume that we've picked a topic and developed a research question about technologies, like cell phones, on family relationships. Let's also assume that as a result of the literature review, we have some ideas for a research strategy. Now, we need to see if our study is "doable."*

Practical Matters

Even with a general idea of a research strategy, it's important to make sure that the strategy is feasible before beginning the actual data collection. No matter how important or interesting a question is, we don't have a researchable question unless it is feasible or practical to answer the question through research. Feasibility or practicality means considering three separate but related concerns: access, time, and money.

feasibility, whether it is practical to complete a study in terms of access, time, and money.

Access

access, the ability to obtain the information needed to answer a research question.

Access is the ability to obtain the information needed to answer a research question. Although it's very easy to get some data, other kinds are much more difficult to secure. In her study of college seniors, Hoffnung had relatively easy access. Once her research proposal was approved by an IRB, Hoffnung was permitted to obtain a random sample of names and phone numbers of seniors. When contacting students on her list to request participation, Hoffnung could mention the university's "stamp of approval." Relatively easy access, IRB approval, and Hoffnung's personal and professional skills all contributed to the success of the project, as most of the women selected agreed to participate in the initial and later phases of the study.

 STOP & THINK   *College students are the subjects of many studies in the social and behavioral sciences, partly because of convenient access. Are there other groups or situations that you think are also easily accessible? Any that you think are particularly difficult to study?*

Most public information is easy to obtain and therefore practical to analyze. Typically, materials produced by government or public agencies and most media, such as newspapers, children's books, and television commercials, are quite easy to access. For example, if you were interested in comparing states by the percentage of residents of Hispanic or Latino origin, you would find this information very easy to obtain (see, for example, http://quickfacts. census.gov/qfd/states/).

People and information that are connected in some way to organizations and institutions may be fairly accessible, although the researcher will often have to go through a gatekeeper to gain access or get permission to do a study. Some gatekeepers will deny access because they see the research as disruptive of organizational or institutional activities or feel that participation would be risky for some reason. Examples of individuals who may be easily accessed through organizations once the gatekeeper agrees include employees of businesses, members of professional unions or associations, patients in a hospital, residents of a nursing home, college students, pet owners whose pets get care from veterinarians, and families whose children attend a private school. In each case, there

BOX **4.4**

Access Denied by Rachel Filinson*

The impact of Hurricane Katrina demonstrated the lack of effective emergency preparedness by governments and institutions. The negative consequences were particularly pronounced for vulnerable populations such as the poor, minorities, disabled, and older adults. Because of the hurricane and its aftermath, students enrolled in a course entitled "Community" at Rhode Island College decided to study the level of disaster preparedness in nursing homes in their state.

Federal law requires Medicare and Medicaid certified facilities (which are virtually all of the long-term care facilities) to have written plans and procedures to meet all potential emergencies and to provide emergency training for employees. The student researchers wanted to compare the content of the emergency plans with the model recommended by the Office of the Inspector General after the devastating Gulf coast hurricanes. They obtained the list of the 86 facilities in Rhode Island from the Center for Medicare and Medicaid Services Nursing Home Compare website that provides a complete listing of Medicare/Medicaid certified facilities for each state. The students

sent a letter to each facility director explaining the study's purpose, assuring confidentiality, and offering to send a copy of the study's aggregated results. The letter further indicated that a student would be in contact with the facility to arrange retrieval of its emergency plan. Despite repeated efforts to procure the plans—and many promises by the facilities to provide them—only 26 of 86 facilities ultimately did so.

Perhaps the difficulty in obtaining the data was because the law does not require the plan to be publicly accessible even though it (the plan) is required by law. It may be that administrators had more pressing responsibilities or they worried about disclosing deficient plans. While our response rate was disappointing, we were able to compare the respondents with the non-respondents on a variety of characteristics documented on the Nursing Home Compare website. On these variables, we found no significant differences between the nursing homes that shared their disaster plans and those that didn't.

*Rachel Filinson, PhD, is the Director of the Gerontology Program at Rhode Island College. This article is published with permission.

is a place that allows us to easily meet or contact people to ask for their participation and an organizational structure that could facilitate the request. For Ciambrone, finding women with HIV/AIDS to interview was the most difficult part of her study. She found enough study participants only when the gatekeepers, her contacts at a local hospital, started referring women who were participating in an ongoing clinical trial of a medical treatment.

Sometimes gaining access to data can be quite challenging. In a study of institutional emergency plans and procedures, Rachel Filinson had easy access to a list of long-term care facilities but had great difficulty in getting the information she wanted from the facility directors (see Box 4.4). In a study of commercial fishing families, researchers mailed out 2,000 surveys but had only 43 returned. Phone calls inquiring about reasons for refusal prompted fishermen to reply in anger that the study was a waste of taxpayers' money (McGraw, Zvonkovic, and Walker, 2000: 69). Interested in the health of women moving from welfare to employment, Kneipp, Castleman, and Gailor (2004) were granted access to the Florida Department of Children and Family's database. They randomly selected the names of 150 women from one county who had received welfare at some time in the past year and had left welfare for employment, but found that with the phone numbers provided, they could not

BOX **4.5**

Groups or Records with Difficult Access

Group or Record	Examples
People without an institutional connection	Children 2 years of age
	Homeless people
	Those not registered to vote
	Atheists
People who are not easily identifiable	Undocumented immigrants
	Italian Americans
	HIV-positive individuals
People who don't want attention from outsiders	Members of gangs
	Anonymous philanthropists
	The Amish
	Child abusers
People who are very busy or who guard their privacy	U.S. Senators
	Celebrities
	Chief executive officers
Records that are private or closed	Diaries
	Tax records
	Census data for individual households

locate two thirds of the women. In their work on attitudes about DNA testing, Turney and Pocknee (2005) found it difficult to recruit men involved in father's rights groups and, as a result, employed a variety of strategies, including e-mail, telephone, and asking participants for the names of other fathers. (See Box 4.5 for other examples of difficult access groups and records.)

In some situations, gaining access is an ongoing process as people "check out" researchers, find out if they are to be trusted, and then gradually let them into various parts of their lives and organizations. Jessica Holden Sherwood, in the research described earlier in this chapter, was interested in an elite group: members of exclusive country clubs. Beginning with a few personal acquaintances who were club members, she asked them to identify others and then had no trouble securing interviews, most likely because she always introduced herself using the names of others who were participating (Sherwood, 2004). In Chapter 12, you'll learn about a study of homicide by Ramiro Martinez and Matthew Lee. Martinez was introduced to the chief of police for Miami through a colleague of a colleague and eventually received permission to do the study (Martinez, personal communication). By spending countless hours accompanying detectives on calls and demonstrating that he was trustworthy, Martinez was able to gain their confidence and that of the administrators and, as a result, access to the data (Lee, personal communication).

Time and Money

Research can be time consuming and expensive. In a study with human subjects, approval from an IRB must be obtained. In a study to examine the circumstances of hospitalized children with complex health needs (Stalker, Carpenter, Connors, and Phillips, 2004), the researchers obtained funding from a foundation and then ran into access problems. With a variety of health authorities involved, there was a complex hierarchy of gatekeepers and getting the applications approved took six months, longer than the researchers had anticipated.

If study participants are interviewed, the budget must include salaries for interviewers and transcribers. If the interviewees are to be given small gifts as compensation for their time and encouragement for participation, like the $10 gift certificates that Ciambrone gave the women in her study, the costs increase. If phone surveys are done, the researchers need to pay for interviewers and phone lines. As Scott Keeter and his colleagues at the Pew Research Center for the People & the Press found, calling cell phones involves additional cost, including the $10 that they gave to each respondent to defray costs of receiving their calls. For the study you'll read about in the next chapter, they did 20-minute phone interviews during the 2008 presidential election. For each of their polls, they contacted approximately 1,500 Americans, doing around 1,125 landline interviews and 375 cell phone interviews for a cost of between $75,000 and $80,000 (Keeter, personal communication).

If mailed questionnaires are planned, costs include paper, printing, and postage. If historical or archival documents are used, there could be copying costs or time expenditures. In the study using homicide records that you'll read in Chapter 12, the researchers needed to fly to where the records were, rent a car, and stay in hotels in that city. They even ended up buying their own copy machine because they found the one in the police department frequently out of order or unavailable (Martinez, personal communication).

Some projects can be completed quickly and very inexpensively. For example, the data collection time and costs might be negligible when using easily obtained public documents or books that can be borrowed from libraries. In the study of long-term facilities and their emergency plans, Filinson and her students did not get paid for their research time and so had only the costs of mailing letters to the facilities and following them up with phone calls. Doing surveys where data is collected using self-administered questionnaires, especially if they can be handed out in groups, can be relatively inexpensive. Even less expensive than doing a survey is using survey data that were collected by others and are archived on a database such as the one maintained by the Inter-university Consortium for Political and Social Research (ICPSR).

Typically, most web-based information is easily accessible and inexpensive. For example, the Statue of Liberty-Ellis Island Foundation recently made accessible through their site (www.ellisislandrecords.org) a great deal of information, including ship manifests and personal stories, about many of the millions of immigrants and passengers that came through Ellis Island from 1892 to 1924. Although there is no charge for the data, it can take many hours to find information about specific individuals.

Projects that collect data over time can be much more costly than those with only one data collection. Similarly, projects that study large numbers of people or groups are more expensive than those that study smaller samples. Researchers can sometimes rely on donated materials and volunteer workers to make a study feasible. Other times, unanticipated costs and events can lead to a scaled-down project.

research costs, all monetary expenditures needed for planning, executing, and reporting research.

Although each method of data collection has unique costs and time expenditures, we can identify some general categories of expense. **Research costs** include the salaries needed for planning the study; any costs involved in pretesting the data collection instruments; the actual costs of data collection and data analysis such as salaries of staff, payments to subjects for their time, rent for office space or other facilities; the cost of equipment (such as recording devices and computers) and supplies (such as paper or postage); and other operating expenses (such as paying for cell phone calls). **Time expenditures** include the time to plan the study, obtain approval, complete the data collection process, and organize and analyze the data.

time expenditures, the time it takes to complete all activities of a research project from the planning stage to the final report.

The focal research in this chapter is an example of an ambitious and a "doable" study. Hoffnung has invested a great deal of her own time in the project for many years and expects to do so for some years to come. By doing the interviewing herself, she is able to control the quality of the interviews and avoid the expense of hiring interviewers. She has obtained small research grants to cover costs of postage, tapes, recorders, and, the largest expense, interview transcription. She has kept costs to a minimum by employing work-study students as assistants for the follow-up questionnaires, by using her existing and donated office space, and by working on her personal computer.

SUMMARY

Research questions can vary in scope and purpose. Some projects are designed to address very broad and basic questions whose answers will help us describe and understand our social world more completely. Other projects work toward solving social problems or creating information that can be used directly by members of communities and participants in programs or organizations. Questions can focus on one or more concepts, can be narrowly or broadly defined, and concentrate on the present or changes over time. Questions can be generated from a number of sources, such as values, interests, and experiences; the social, political, economic, and intellectual climates; previous research; or the ability to get funding. The availability of funding can determine not only the size and scope of a project, but also the specific question, the timetable, and the methodology.

A review of the literature is essential at some point in the research process. By reviewing the work that other researchers have done, each new study builds upon that work and provides a foundation for the next wave of data collection and analyses.

Planning a study means considering the practical matters of time, money, and access to data. It's necessary to plan a feasible study. No matter how important or interesting a research question is, unless the research plan is practical, the study will not be completed. Hoffnung gained access to a sample of women to interview and she has been able to keep in contact with most of them since 1992. She has had sufficient resources of time and funding to be able to continue to collect and analyze data so that she will be able to answer the research questions she set for herself many years ago.

EXERCISE 4.1

Selecting a Topic

1. Get a copy of a local or regional newspaper or go to a newspaper's website. Find three articles that deal with social behavior or characteristics. List the name of the paper, the date of publication, and the headline for each of these articles.
2. Select one of the stories and make a copy to turn in with this exercise.
3. Create one research question on the topic of the article that could be answered by social research and state it.

4. Assuming you could get enough funding and had the time to do research to answer the question you stated, would you do the research? Describe why you would or would not do the research by focusing on your personal interests, today's social and political climate, and the contribution you think the research would make to social science or society.

EXERCISE 4.2

Starting a Literature Review

Select a topic you are interested in or one of the topics from Hoffnung's study, such as career aspirations, self-concept, two-career family, family planning attitudes, or role conflicts. Using your library's resources, find five citations of original social research from social science journals on this topic that are no more than five years old. (Be careful not to get an article that summarizes other research or is solely a discussion of theory.) Complete the following exercise:

1. Write down the topic you selected and the database(s) you used for your search.
2. List the five citations giving full bibliographic information for each including name(s) of the author(s), title of the article, name of the journal, date of publication, and pages of the article.

3. Select one of the articles from your five citations that is available to you either online or in print and read it. Put an asterisk (*) next to the name of the article that you've read.
4. Describe the review of the literature that the researcher(s) presented in the article. For example, is the literature review extensive? To what extent does it focus on theories and theoretical perspectives? To what extent does it present the research findings from previous research? How adequate is the literature review in covering previous work on the topic?
5. Does the researcher(s) start with a research question or hypothesis? If so, list one research question or hypothesis. Otherwise, list the topic of the study. Note the page number of the article where you found the research question, hypothesis, or the topic.

EXERCISE 4.3

Using the Internet and the World Wide Web

1. Select one of the following topics: two-career family, death penalty legislation, or unemployment. Write down the name of your topic.
2. Using a search engine on the web, such as Google, Yahoo, or Altavista, do a search for research on this topic by searching for an exact phrase that's of interest (such as "death penalty legislation") and the word *research*.
3. How many results did you obtain in your search? If you ended up with more than a few hundred items, narrow down your topic in some way. Describe what you did to narrow the search. Describe the number and quality of the results you obtained.
4. Check out at least three of the sites that you found in your search. Identify the website, describe the information on the site, and evaluate the usefulness of this information for you and/or other researchers.

EXERCISE 4.4

Summing Up Practical Matters

Think about the following research questions and the suggested ways that a researcher might collect data. For each, make an estimate of the cost and time and your evaluation of whether access will be easy or hard.

Research question	Will access be easy or hard?	Time consuming or not?	What are some expenses?
1. Do teachers' characteristics and teaching styles have an impact on the amount of student participation in class? (Suggested method: Observe 50 classes for several hours each.)			
2. What kinds of countries have the highest infant mortality rates? (Suggested method: Use government documents for a large number of countries.)			
3. Is the psychological impact of unemployment greater for people in their twenties or fifties? (Suggested method: Do a survey using a questionnaire e-mailed to 300 people who have contacted the unemployment office in one state.)			
4. Which regions of the United States have experienced the greatest population growth over the past 20 years? (Suggested method: Use data from the U.S. Bureau of the Census.)			
5. Are people who are more religiously observant more satisfied with their lives than those who are less observant? (Suggested method: Interview 100 people for approximately one hour each.)			

CHAPTER 5

Sampling

© Ellen McKnight / Alamy

INTRODUCTION

sampling, the process of drawing a number of individual cases from a larger population.

Have you ever asked yourself, "How do pollsters figure out what Americans feel about things like the economy, the war in Afghanistan, gay marriages, presidential candidates, and all those other things they tell us about on the news? They surely don't ask *all* Americans about their feelings, do they?" An important part of the answer to this question involves the topic of this chapter: sampling.

Sampling is the process of drawing a number of individual cases from a larger population. Researchers sample in the hopes that they can gain insight into a larger population without studying each member of the population. Presidential polls are based upon samples of the population that might vote in an election. We'd like to begin this chapter on sampling with three stories about presidential election polls (see Box 5.1): two that are examples of how misleading sampling can be and the other of how informative it can be.

 **STOP & THINK** *How much better were the 2008 presidential polls at predicting the winner of the presidential election than the 1936 and 1948 polls?*

Why did the *Literary Digest* poll, with two million respondents, do so poorly in 1936? Why did Gallup, with his apparently superior technique, miss the mark in 1948? And how was it possible that the RealClearPolitics

---

BOX 5.1

Three Different Sets of Pre-Presidential Election Polling Results

1936 *Literary Digest* Poll Predicts That Alf Landon Will Beat Franklin Delano Roosevelt, 57 Percent to 43 Percent

Based on a survey of over two million people, a survey that had predicted with "uncanny accuracy" (its own words), the outcome of the 1924, 1928, and 1932 presidential elections, the *Literary Digest* predicted that Alf Landon would win the 1936 election by a margin of 14 percentage points over the then President Franklin Delano Roosevelt: 57 percent to 43 percent. Imagine the *Digest*'s surprise when Roosevelt actually won and won big: 61 percent to 39 percent.

1948 Gallup Poll Predicts That Thomas Dewey Will Beat Harry Truman

Using a technique called quota sampling, George Gallup correctly predicted that Roosevelt would win the 1936 election, even if his prediction that Roosevelt would receive 54 percent of the vote was somewhat wide of the 61 percent Roosevelt actually received. Gallup got it completely wrong in 1948, however,

when he predicted that Thomas Dewey would beat Harry Truman in that year's presidential election. Truman actually won. It was a very good day for Truman, but a very bad day for Gallup.

2008 Pre-Election Polls Say Obama Will Win Election

Based on an average of the 15 national surveys taken between November 1 and November 3, 2008, the RealClearPolitics (2008) webpage predicted, the day before the election, Barack Obama would win the 2008 Presidential election over John McCain by a 52.1 percent to 44.5 percent margin. The predicted spread, of 7.6 percent, missed the real mark by only 0.3 percent, as Obama actually won the general election on November 4 by 52.9 percent to 45.6 percent, a 7.3 percent spread. How could the RealClearPolitics prediction have been so much better than the *Literary Digest* prediction in 1936 and the Gallup Poll prediction in 1948? The answer is improvements in the science of sampling.

average of the 15 pre-election polls could come with 0.3 percent of the actual spread in the race between Obama and McCain in 2008? The main reason has to do with the topic of this chapter: sampling. The science of sampling has progressed dramatically since the 1930s, to the point where even a basic introduction must touch on some relatively technical matters. But the logic of sampling can be easily grasped, and we think you'll actually enjoy learning something about it. Let's start at the beginning by addressing a question that has probably already occurred to you: why sample?

Why Sample?

Sampling is a means to an end: to learn something about a large group without having to study every member of that group. The *Literary Digest*, George Gallup, and the authors of the 2008 polls mentioned in Box 5.1 wanted to know how Americans were going to vote for president. Why didn't the *Literary Digest*, Gallup, and the 2008 poll takers simply survey the whole of the American voting-age population? Put this way, one answer to the question, "Why sample?" is obvious: We frequently sample because studying every single instance of a thing is impractical or too expensive. The *Literary Digest*, Gallup, and 2008 poll takers couldn't afford to contact every American of voting age.

We sample because studying every single instance is beyond our means. Suppose, for example, you wanted to know how all Americans felt about something? Why not ask them all? The U.S. Census Bureau is, in fact, obliged by law to learn something about every American every 10 years. But it's very expensive. The 2000 census, during which 281,421,906 Americans were enumerated, cost in excess of $4.5 billion (Gauthier, 2002), maybe 60 percent of which was spent to employ the half-million temporary workers used to follow up on the (34 percent of) households that failed to return their forms by mail (U.S. Bureau of the Census, 2000). The projected cost of the 2010 census is estimated at between $13.7 and $14.5 billion (Rogers, 2008). Such costs, in terms of money and people, are beyond the means of even the wealthiest research organizations.

So we sample to reduce costs. But we also sample because sampling can improve data quality. Sandra Enos, whose study of how imprisoned women handle mothering we present in Chapter 10, wanted to understand women's approaches to mothering in an in-depth fashion, with a real appreciation of each individual woman's social context. Enos intensively interviewed 25 incarcerated women, not only because studying the whole population of such women would have been expensive, but also because she wanted to spend her time gathering in-depth information about each of her subjects, rather than more superficial data on all of the women in that prison.

In short, we sample because we want to minimize the number (or quantity) of things we examine or maximize the quality of our examination of those things we do examine.

 **STOP & THINK** *Using the logic of this section, when do you think sampling is unnecessary?*

We don't need to sample, and probably shouldn't, when the number of things we want to examine is small, when data are easily accessible, and when

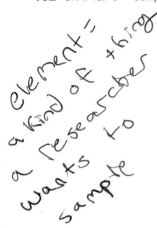

element, a kind of thing a researcher wants to sample. *(handwritten margin note)*

data quality is unaffected by the number of things we look at. For instance, suppose you were interested in the relationship between team batting average and winning percentage of major league baseball teams last year. Because there are only 30 major league teams, and because data on team batting averages and winning percentages are readily available (see any of this year's almanacs), you'd simply be jeopardizing the credibility of your study by using, say, 10 of the teams. So we would examine all 30 of them or, in other words, use the whole population.

We need to remind ourselves of some terms, and define a few others, before proceeding much further. First, we need a term to describe the kind of thing about which we want information—for example, a major league baseball team, an adult American of voting age, all Americans. As you learned in Chapter 1, social scientists usually refer to such things as units of analysis, or the things about which information is collected. Sometimes, however, they call these things elements, especially at the stage of sampling. Social scientists normally use the term *element* when they're referring to members of a **population** (the entire group of interest). In the case of presidential elections in the United States, the entire group of interest, or the population, is the voting population. Elements of this population are voters. Lastly, one can define a **sample** as a subset of a population. As you'll see later, there are a number of ways of picking a sample of elements from a population. When selecting members of a population to include in the sample, those selected members are referred to as elements of the population. Once one collects information about those members, and analyzes it, they become units of analysis. See Box 5.2 for an example of how elements became units of analysis in Hoffnung's research, reported in Chapter 4.

element, a kind of thing a researcher wants to sample.

population, the group of elements from which a researcher samples and to which she or he might like to generalize.

sample, a number of individual cases drawn from a larger population.

BOX **5.2**

The Story of Some Elements That Became Units of Analysis

You may recall that, for the research reported in Chapter 4, Michele Hoffnung wanted to study the plans and experiences of women who graduated from college in 1993. Her population of interest was women graduating from college in 1993. So far, she was still thinking of this population in terms of elements: all graduating women. She knew she wouldn't be able to collect information from all women graduating in 1993 with the intensive interviews she wanted to use, so she decided on a number she thought she could interview: 200. As her sampling frame, or a list of elements from which she'd actually sample (more about this later), she chose those women graduating

from five New England colleges. So far, those women were still elements, from Hoffnung's points of view. Hoffnung then used a sampling procedure called stratified random sampling (more about this later) to select the 200 women she would actually interview. So far, from her point of view, they were still elements. Finally, though, Hoffnung reached the point when she was actually interviewing individual members of her sample—when she was collecting information from them. At that point, the students, who had formerly been elements of a larger population, started to yield data for analysis … and so became units of analysis.

 **STOP & THINK** *Suppose you're interested in describing the nationality of Nobel prize-winning scientists. What would an element in your study be? What would the population be?*

Sampling Frames, Probability versus Nonprobability Samples

We can now begin to address the question of why the *Literary Digest* poll did so poorly in 1936, why Gallup did badly in 1948, and why the 2008 final pre-election polls did so spectacularly well at predicting the outcome of the 2008 election. All of these polls sought information about the same population: people who would vote on Election Day. But the concept of "people who would vote on Election Day" is elusive, just as many target populations are. A **target population** is the population of theoretical interest. In the case of "people who would vote on Election Day," the target is notoriously difficult to hit, especially *before* an election, when even the most well-intentioned citizens can't be sure what might keep them from voting on Election Day (e.g., weather, natural or social disaster, death). One thing most researchers must do, then, is come up with a list or description of elements that approximates the elements of the population of interest. This list is referred to as a **sampling frame** or **study population.** A large part of the reason why today's pre-election polls are so much better that the ones of the past has to do with refinements in sampling frames. The list or sampling frame that the *Literary Digest* used for its 1936 poll consisted of names from the list of *Literary Digest* readers, automobile registrations, and telephone directories. But, in 1936, at the height of the Great Depression, substantial numbers of people voted who weren't readers of the *Literary Digest* and owned neither a car nor a telephone. So, even though the *Literary Digest* received 2 million responses to the (10 million) ballots it sent out, the best it could hope to do was get responses from people either who were readers (of the *Digest*) or who owned cars or telephones—people who, because of their wealth, were more likely to vote for the Republican Landon than the Democrat Roosevelt. All of the 2008 polls were of people who had telephones but, because the percentage of U.S. households with telephones had reached 97.8 percent by 2004, and held steady after that (Blumberg and Luke, 2008), this sampling frame provided a much greater chance of reaching the desired population than the one used by the *Literary Digest* in 1936 had. Modern surveys conducted over the Internet, when intended to provide insights about the total American population, are a more recent example of surveys that are subject to (admittedly decreasing) inaccuracy because of a discrepancy between the target population and the sampling frame. As of 2008, only about 72 percent of the U.S. population used the Internet (Internet World Stats, 2008).

So, the main reason the *Literary Digest* poll did so poorly is that it employed a relatively poor sampling frame. A major problem with Gallup's poll in 1948 was that he, like the *Literary Digest* pollsters, employed a nonprobability, as opposed to a probability, sample. **Nonprobability samples** are

---

target population, the population of theoretical interest.

sampling frame or **study population,** the group of elements from which a sample is actually selected.

nonprobability samples, samples that have been drawn in a way that doesn't give every member of the population a known chance of being selected.

those in which members of the population have an unknown (or perhaps no) chance of being included. But Gallup used a fairly sophisticated type of non-probability sampling called quota sampling. Quota sampling begins with a description of the population: its proportion of males and females, of people in different age groups, of people in different income levels, of people in different racial groups, and so on. Based on this description, which in Gallup's case was from the U.S. Census, the researcher draws a sample that proportionately represents the sampling frame in all ways that are deemed important (e.g., gender, age, income levels, race). The main problem in 1948 was that the description of the American population that Gallup used was one based upon the 1940 census, a census that, because of all that had gone on between 1940 and 1948 (World War II, etc.), didn't provide a particularly good profile of the target population—adult Americans who would vote.

probability samples, samples drawn in a way to give every member of the population a known (nonzero) chance of inclusion.

A better approach for predicting elections is to use a probability rather than nonprobability sample of Americans. **Probability samples** are selected in a way that gives every element of a population a known (and nonzero) chance, or probability, of being included. The most recent presidential election polls used probability samples that were designed to give every likely voter of the United States a (known) chance of being included in the samples.

biased samples, samples that are unrepresentative of the population from which they've been drawn.

The advantage of probability over nonprobability samples is not, as you might hope, that they ensure that a sample is representative or typical of the population from which it is drawn. Still, probability samples are *usually* more representative than nonprobability samples of the populations from which they are drawn. To use the jargon of statisticians, probability samples are less likely to be **biased samples** than are nonprobability samples. Biased means that the cases that are chosen are unrepresentative of the population from which they've been drawn. And because probability samples increase the chances that samples are representative of the populations from which they are drawn, they tend to maximize the **generalizability** of results. Gener-

generalizability, the ability to apply the results of a study to groups or situations beyond those actually studied.

alizability here refers to the ability to apply the results of a study to groups or situations beyond those actually studied.

But the greatest problems afflicting the *Literary Digest* and Gallup polls were likely that their sampling frames didn't match their target populations very well. Finding the right sampling frame is a problem that continues to plague polling organizations. This chapter's focal research piece, by Scott Keeter, Michael Dimock, and Leah Christian, describes how the Pew Research Center, like other polling organizations, dealt with another sampling frame issue: that associated with the growing percentage of the American population that uses cell phones and doesn't own landline phones.

 **STOP & THINK** *Can you think why researchers haven't used cell phone numbers in polling until recently? What problem may result from using only landline numbers?*

FOCAL RESEARCH

Calling Cell Phones in '08 Pre-Election Polls

By Scott Keeter, Michael Dimock, and Leah Christian*

Public opinion polling faced many challenges during the 2008 presidential election. None was more daunting than the rising number of "cell phone only" voters who could not be reached over landline telephones. In previous elections, polling was generally limited to people with landline telephones, partly because cell phone owners in the U.S. are charged for calls and partly because relatively few adults with cell phones didn't have landlines. Cell phone only adults constituted only two percent of those with telephones in 2001 (Weisberg, 2005: 210). The latest estimates from the National Health Interview Survey (Blumberg and Luke, 2008)—the most comprehensive measure available—suggest that in the fall of 2008 nearly 18% of households were wireless only; in the National Exit Poll (Mokrzycki, 2008) conducted on Nov. 4, 20% of Election Day voters identified themselves as "cell-only" voters. What made this growth so important in the recent election is that many news commentators (e.g., Wihbey, 2008) suggested that Barack Obama would prove more popular among cell phone users, because of their youth and education, than his opponent in the presidential election, John McCain. This paper, in fact, examines the hypothesis than Barack Obama fared better in probability samples including landline-and cell phone-users than in samples including landline users alone.

Method

Results for this report are based on combined data from six surveys conducted from September 9 through November 1, 2008. The data are from telephone interviews conducted under the direction of Princeton Survey Research Associates International and Abt SRBI, Inc., using a nationwide probability sample of adults, 18 years of age or older. In total, 10,430 adults were interviewed on a landline telephone and 3,460 on a cell phone, including 1,160 who had no landline telephone. This sample composition is based on a ratio of approximately three landline interviews to each cell phone interview. This ratio is based on an analysis that attempted to balance cost and fieldwork considerations in addition to resulting demographic composition of the sample and the need for a minimum number of cell-only respondents in each survey.

---

* This is an abridged and revised version of a report that appeared as a release from the Pew Research Center for the People & the Press on December 18, 2008. It is reprinted with permission from the Pew Research Center.

Pew's cell phone and landline contact and cooperation rates for election surveys conducted between September and November were essentially the same. (Contact rates for landline respondents were 77%, while for cell phone respondents, they were 76%; cooperation rates for landlines were 31% and 30% for cell phones. Thus, response rates for landlines were 22%, while for cell phones, they were 21%.) The response rate in Pew's cell phone surveys may benefit from the decision to offer cell respondents a modest $10 reimbursement for potential phone charges they might incur. In contrast, landline respondents were not offered any reimbursement.

Results

An analysis of six Pew surveys conducted from September through the weekend before the election shows that estimates based only on landline interviews were likely to have a pro-McCain tilt compared with estimates that included cell phone interviews (see Table 5.1). But the difference, while statistically significant, was small in absolute terms. Obama's average lead across the six surveys was 9.9 points among registered voters when cell phone and landline interviews were combined. If estimates had been based only on the landline samples, Obama's average lead would have been 7.6 points, indicating an average bias of 2.3 percentage points. Limiting the analysis to likely voters rather than all voters produced similar results. Obama's average lead among likely voters was 8.2 points across all six surveys versus 5.8 points in the landline sample.

TABLE **5.1**

Samples with Cell Phones Showed a Larger Obama Advantage

	Landline/Cell phone Sample	Landline Sample
Registered Voters	%	%
Obama	49.9	48.5
McCain	40.0	40.9
Other/DK	10.1	10.6
Obama Adv.	+9.9	+7.6
Sample size	(11,964)	(9,228)
Likely Voters		
Obama	49.8	48.5
McCain	41.6	42.7
Other/DK	8.6	8.8
Obama Adv.	+8.2	+5.8
Sample size	(10,819)	(8,143)

Obama enjoyed a large advantage over McCain among younger people, but this advantage would have been underestimated by polls using landline samples only. Obama led McCain by 33 points (63% – 30%) in the full-frame sample of registered voters, compared with his 29-point advantage in the landline sample. The sample difference among likely voters under 30 was even larger, where the samples involving both cell phone- and landline-owning respondents set the difference at 33 percentage points, but the landline-only sample set it at 26 points. According to the national exit poll, Obama won this age group by 34 points, 66%–32%.

Discussion

The wireless-only population has been growing steadily as many young people enter adulthood without ever getting a landline and others drop their landline telephone service. Our projection for the fall of 2008, based on previous data from the National Health Interview Survey, is that 18% of adults are wireless only, with a rate of increase of roughly 3% annually.

Although there is little evidence of sizeable non-coverage bias in landline surveys of the general public, the potential for bias exists and has been documented in studies of certain sub-populations including young and low income people (Blumberg and Luke, 2007). Pew's research found that differences in presidential preferences expressed by samples of landline respondents alone and those involving both landline and cell phone respondents were small but could have been important in the 2008 election cycle had the election been closer. Still, it is not possible at this stage to offer a blanket recommendation to researchers regarding the inclusion of cell phones. There are considerable extra costs involved in cell phone interviewing; our cell phone interviews cost approximately two to two-and-one-half times as much as a landline interview. If present trends in the growth of the cell-only population continue, it is likely that the bias involved in contacting only landline users in the future will make these costs a necessary part of doing polling.

References

Blumberg, Stephen J. and Julian V. Luke. 2007. Coverage Bias in Traditional Telephone Surveys of Low-Income and Young Adults. *Public Opinion Quarterly* 71 734–749. http://poq.oxfordjournals.org/cgi/reprint/71/5/734

Blumberg, Stephen J. and Julian V. Luke. 2008. Wireless Substitution: Early Release of Estimates from the National Health Interview Survey, January–June 2008. *National Center for Health Statistics*. Retrieved February 6, 2009. http://www.cdc.gov/nchs/data/nhis/earlyrelease/wireless200812.htm

Mokrzycki, Mike. 2008. Exit Poll: 1 in 5 Voters Only Have Cell Phones. *Newsday*. November 7. http://www.newsday.com/news/local/politics/ny-bc-polls-cellphones1107nov07,0,7248380.story

Weisberg, H. F. 2005. *The Total Survey Error Approach.* Chicago: The University of Chicago Press.

Wihbey, John. 2008. The Cell Phone Polling Gap: An Artificially "Close" Race? *The Huffington Post.* October 1. http://www.huffingtonpost.com/john-wihbey/the-cell-phone-polling-ga_b_131021.html

THINKING ABOUT ETHICS   *Because of the sampling technique employed, the Pew pollsters never knew the identity of their respondents, so respondent anonymity was never in danger. Moreover, participation in the survey was voluntary.*

SOURCES OF ERROR ASSOCIATED WITH SAMPLING

Keeter, Dimock, and Christian report that Obama's advantage over McCain was slightly, and significantly, greater in samples that included both landline and cell phones than in samples that included landline phones alone. And then they go on to suggest that part of the reason that this might be the case is that young people are more likely to be cell phone users only and that young people were more likely to be Obama supporters.

STOP & THINK   *If Keeter, Dimock, and Christian are right about age, what kind of variable would it (age) be in relation to their independent variable (type of phone contacted) and their dependent variable (the candidate a respondent preferred)?*

Keeter, Dimock, and Christian propose that a respondent's age comes before and affects both the respondent's likelihood of using a cell phone and the candidate who is preferred. It is therefore seen as an antecedent variable to both, a variable that, at least in part, accounts for their association. (You may recall, from Chapter 2, that this would mean the association between cell phone usage and presidential preference is at least, in part, non-causal. But who would have thought that owning a particular kind of phone would cause a presidential preference anyway?)

Keeter, Dimock, and Christian's larger point, however, is this: that one does not get exactly the same results in polls from samples using owners of landline phones as a sampling frame and samples using owners of landline phones and cell phones as a sampling frame.

STOP & THINK   *Which, in your opinion, is the better sampling frame if the target population is all American voters: owners of landline phones alone or owners of landline and cell phones?*

If it is true that increasing percentages of Americans will be depending on cell phones in the future, it makes sense that using sampling frames that exclude cell phones users will increasingly "miss the mark." But this raises the

coverage error, an error that results from differences between the sampling frame and the target population.

nonresponse error, an error that results from differences between non-responders and responders in a survey.

sampling error, any difference between sample characteristics and the equivalent characteristics in the sampling frame, when this difference is not due to nonresponse error.

parameter, a summary of a variable characteristic in a population.

statistic, a summary of a variable in a sample.

larger question of why there might be errors involved in surveys using samples, and it is to this question that we'd now like to turn.

The overarching reason why samples fail to represent accurately the populations we're interested in is that populations themselves aren't uniform or *homogeneous* in their characteristics. If, for instance, all human beings were identical in every way, any sample of them, however small (say, even one person), would give us an adequate picture of the whole. The problem is that real populations tend to be heterogeneous or varied in their composition—so any given sample might not be representative of the whole. Obviously, not all eligible voters in the United States vote for the same candidate in every election. If they did, and did so reliably, any sample, however selected, would perfectly predict the election's outcome. Weisberg (2005) lists three sources of survey error that are associated with sampling for populations that are not homogeneous, however: **coverage error, nonresponse error,** and **sampling error** (see Figure 5.1).

Coverage Errors

Coverage errors are attributable to the difference between a sampling frame and a target population. It's useful to introduce a little more terminology here. A **parameter** is a summary of a variable *in a population*. In the 1936 election, about 61 percent of the voting public favored Roosevelt over Landon for president. This percentage is a parameter of the population that voted for president in 1936. A **statistic,** on the other hand, is a summary of a variable *in a sample*. The best the *Literary Digest* could do was come up with a statistic, that Roosevelt had received 43 percent of the ballots cast in their poll. But the largest problem was that the sample with which that statistic was calculated was drawn from a sampling frame, lists of *Digest* readers, telephone owners, and car owners, which bore only a remote resemblance to the target population. Hence the coverage error, defined precisely as the difference between the parameter for the sampling frame, or study population, and the parameter for the target population, was necessarily large.

 **STOP & THINK** *Can you imagine any sources of coverage error in the 2008 election-eve polls?*

The 2008 polls also involved some coverage error. After all, they were based upon sampling frames that included only people with telephones and then only people who said they were "likely voters" in those households. When we recall that since 2004 the percentage of U.S. households with phones has hovered around 98 percent, not 100 percent, and that even people

Coverage errors	Nonresponse errors	Sampling error

FIGURE **5.1** Sources of Error Associated with Sampling

Source: Adapted from Weisberg (2005).

who thought they were "likely" voters couldn't be sure they'd vote (and those who thought they weren't likely voters couldn't be sure they wouldn't), we can see that the sampling frames for these polls were imperfect, if pretty good, matches for their target populations (people who did vote in the election).

 **STOP & THINK** *What do you think survey organizations do about the problem of unlisted telephone numbers?*

random-digit dialing, a method for selecting participants in a telephone survey that involves randomly generating telephone numbers.

One of the great potential sources of coverage errors is that many telephone numbers are not listed in telephone books. Survey organizations have gotten around this problem through a process called **random-digit dialing.** Incidentally, the Pew Research Center drew probability samples using random-digit dialing for both landline and cell phones. In doing so, they used random telephone numbers to select the people they talked to. We'll be saying more about random-digit dialing a little later. For now, notice that one of the great advantages of random digit-dialing is that it avoids the problem of unlisted telephone numbers, even if it brings into play numbers, like those of businesses, that are best left out of surveys of households.

Most of the time, when we think of coverage errors, we think of the eligible elements of a target population that are excluded in the sampling frame. The eligibles who were excluded by the *Literary Digest* poll, for instance, were voters who weren't *Digest* readers or didn't own a car or didn't own a telephone. But coverage errors also might occur when elements that are ineligible for the target population are included in the sampling frame. Thus, a voting survey of households with telephones could contact noncitizens—ineligibles in U.S. elections. Similarly, a household survey, especially one based, as many large-scale surveys are today, on random-digit dialing, could contact some people at non-residential (perhaps business) numbers. The problem of ineligibles, however, can, in most cases, be dealt with by the use of screening questions ("Are you an eligible voter?" "Is this your personal phone?"). The problem of excluded eligibles, such as those without a personal phone, is more challenging, but not always insurmountable.

The issue of excluded eligibles had become particularly problematic with the increased popularity of cell phones. Increasingly, as pointed out by Keeter, Dimock, and Christian, cell phone–only users possess only a cell phone, with no landline. The figure was only 2 percent in 2001, but had reached nearly 18 percent by 2008, and may have been even higher among those who actually voted, according to exit polls on Election Day. Because most cell phone plans charge users for calls they receive, survey organizations used to avoid calling numbers assigned to cell phones. But several polling organizations, including the Pew Research Center, offered cell phone users compensation for receiving their calls. The fact that the RealClearPolitics prediction that the spread between Obama and McCain (7.6 percent—a statistic) was so close to the actual spread in the election (7.3 percent—a parameter) indicates, among other things, that coverage error for this election was quite small.

The list of technological innovations that currently frustrate survey research does not stop at cell phones alone, however. Answering machines and caller ID, their functional equivalents on cell phones, and other devices used to screen calls are increasingly used by "eligibles" to shield themselves from participation in surveys. And one can imagine that the increasing use of Internet technology as a substitute for traditional landline telephone connections will, in future, create a new coverage problem for telephone surveys. But now our discussion is bleeding into the next source of survey error: nonresponse error.

Nonresponse Errors

Another source of error in both nonprobability and probability samples is nonresponse error. Nonresponse refers to the observations that cannot be made because some potential respondents refuse to answer, weren't "there" when contacted, and so forth. Nonresponse error is the error that results from differences between those who participate in a survey and those who don't. Nonresponse can involve large amounts of bias, as it was suspected to do in the *Literary Digest*'s 1936 presidential poll, when about 8 million of the *Digest*'s 10 million contacts simply didn't return their ballots. In this case, responding took a good deal of work (filling out forms, affixing stamps, which cost money, and mailing) and probably reflected unusual degrees of commitment to the candidates that respondents favored. The response rates for the Pew Research polls on which Keeter, Dimock, and Christian report (22 percent for landline phones and 21 percent for cell phones) weren't much higher than that of the *Literary Digest* and seem to be in line with long-term trends toward nonresponse. Curtin, Presser, and Singer (2005) report, for instance, that response rates to the University of Michigan's telephone Survey of Consumer Attitudes dropped from about 72 percent in 1979 to about 48 percent in 2003. But recent pre-election polls have been done on the (evidently plausible) assumption that there really isn't much difference between those who respond and those who don't. In fact, several scholars, including Keeter, Miller, Kohut, Groves, and Presser (2000), have found that, with particular topics, there aren't necessarily very large differences in the findings of probability samples with substantially different levels of response. Moreover, Groves (2006) and Groves and Peytcheva (2008) demonstrated that nonresponse errors are only indirectly related to nonresponse rates and are particularly likely to exist when the survey's variable of interest is strongly associated with the likelihood of responding. Thus, if people are simply more averse than they once were to *surveys of all kinds*, lower response rates, generally, don't necessarily mean greater levels of nonresponse bias or error. As Keeter says, with respect to political polls:

> Basically we think that non-response is largely random and not correlated with the things we want to study—e.g., non-respondents are not disproportionately conservatives who hate the media, or more affluent people who are too busy to talk on the phone, etc. Instead, people have varying propensities to participate, and this propensity probably varies over time—you might be a willing respondent one day but too busy the next. (personal communication)

But when one has reason to believe that the likelihood of responding is affected by the variable of interest, all bets are off. Thus, Travis Hirschi (1969) aptly argued that nonresponse bias is enormous in studies of delinquency

among school-aged adolescents when the questionnaire is given in school because delinquents are much more likely than their nondelinquent counterparts
to have dropped out of school or to skip sessions when questionnaires are
administered.

 **STOP & THINK** *What kinds of people might not be home to pick up the phone in the early*
evening when most survey organizations make their calls? What kinds of
people might refuse to respond to telephone polls, even if they were
contacted?

If sampling is done from lists, as is frequently done in probability sampling,
it is sometimes possible to study the nature of nonresponse bias, especially if the
lists include some information about potential sampling units. Thus, in a study
of married people, Benjamin Karney and his colleagues (1995) were able to determine that people who did not respond to their questionnaire survey generally
had less education and were employed in lower-status jobs than those who did
respond. Karney and his colleagues used information from the marriage certifi-
cates of those who constituted their sampling frame. Weisberg (2005) points
out that nonresponse can come at the unit level (when there is no response at
all from the respondent) or the item level (when a respondent fails to answer a
particular question) and that the former can be due to an inability to contact
the respondent, an incapacity on the part of the respondent to be interviewed,
or noncooperation on the part of the respondent. The latter, noncooperation,
is most likely to be associated with error, especially insofar as respondents'
lack of cooperation is systematically related to their attitudes toward the subject
of the survey, as, for instance, when people with conservative social values
refuse to answer a survey on sexual behavior (Weisberg, 2005: 160). Keeter,
Dimock, and Christian point out that some nonresponse in the Pew 2008 pre-
election surveys was due to noncontact (the contact rates were 77 percent for
landline respondents and 76 percent for cell phones) but that cooperation rates
for those contacted (31 percent for landlines; 30 percent for cell phones) were
lower. A key issue is whether cooperation rates had more to do with people's
attitudes toward presidential candidates or with a general unwillingness to be
bothered. The evident accuracy of the Pew, and many other pre-election, polls
suggests it had more to do with the latter than the former.

Sampling Errors

Sampling error is the error that arises when sample characteristics are different from the characteristics of the sampling frame, assuming this difference is
not due to nonresponse error. Here the question isn't whether there's a perfect match between the sampling frame and the target population or whether
non-responders are different from responders. Theoretically, if not practically,
one can imagine no coverage errors (due to differences between the sampling
frame and the target population) or nonresponse errors (due to differences between responders and non-responders) and still imagine that a sample does
not perfectly represent the population from which it is drawn. In fact, because

the 15 election-eve polls of likely voters conducted by various nonpartisan polling agencies and used by RealClearPolitics to compose its final average showed "spreads" between candidates Obama and McCain of between 2 and 11 percent (see Box 5.3), it is clear that at least some of the samples on which the polls were based did not give a perfectly accurate picture of the spread among all likely voters (the sampling frame for these polls). If they all did, their results would all be the same.

Some of the differences in the election-eve polls were probably due to differences in the sampling frames (e.g., some of the polls, like Pew, polled both landline and cell phone users; some, just landline users; two sampled just "battleground states," while the rest, the whole country). And some were probably due to different nonresponse errors. But much of the difference was probably due to another source of error: **sampling variability,** or the variability in sample statistics that can occur when different samples are drawn from the same study population or sampling frame.

sampling variability, the variability in sample statistics that can occur when different samples are drawn from the same population.

BOX **5.3**

The Final Presidential Poll Results by Non-Partisan Polling Agencies, 2008 (RealClearPolitics, 2008)

Poll	Obama	McCain	Spread
Marist	52	43	Obama +9
Battleground (Lake)	52	47	Obama +5
Battleground (Tarrance)	50	48	Obama +2
Rasmussen Reports	52	46	Obama +6
Reuters/C-Span/Zogby	54	43	Obama +11
IBD/TIPP	52	44	Obama +8
FOX News	50	43	Obama +7
NBC News/Wall St. Jrnl	51	43	Obama +8
Gallup	55	44	Obama +11
Diageo/Hotline	50	45	Obama +5
CBS News	51	42	Obama +9
ABC News/Wash Post	53	44	Obama +9
Ipsos/McClatchy	53	46	Obama +7
CNN/Opinion Research	53	46	Obama +7
Pew Research	52	46	Obama +6
RealClearPolitics Ave.	52.1	44.5	Obama +7.6
Actual Election Results	52.9	45.6	Obama +7.3

Even probability samples entail sampling variability and, therefore, frequently involve sampling error. To demonstrate, we'd like you to imagine drawing random samples of two elements from a population of four unicorns: a population with four members named, let us say, Leopold, Frances, Germaine, and Quiggers. Let us also say that these unicorns were one, two, three, and four years of age, respectively.

 STOP & THINK *How could you draw a sample of two unicorns, giving each unicorn an equal chance of appearing in the sample?*

simple random sample, a probability sample in which every member of a study population has been given an equal chance of selection.

A sample in which every member of the population has an equal chance of being selected is called a **simple random sample.** Simple random samples, because they give every member of the population an equal chance of being selected, also give every member a known chance and are therefore probability samples. Perhaps we should emphasize that the word *random* in the phrase *simple random sample* has a special meaning. It does not mean, as it can in other contexts, haphazard, erratic, arbitrary, funny, or crazy; here it connotes a "blind" choice, almost lottery-like, of equally likely elements from a larger population. One way of drawing (selecting) a simple random sample of two of our unicorns would be to write the four unicorns' names on a piece of paper, cut the paper into equal-sized slips, each with one unicorn's name on it, drop the slips into a hat, shake well, and, being careful not to peek, pick two slips out of the hat. Let's say we have just done so and drawn the names of Frances and Quiggers (ages two and four). We thereby achieved our goal of a sample of unicorns whose average age ([2 years + 4 years]/2 = 3 years) was *not* the same as that of the population (2.5), hence proving that it is possible to draw a probability sample that misrepresents the larger population and therefore entails sampling error.

Of course, if we had selected a sample of Frances and Germaine, the average age of sample members ([2 years + 3 years]/2 = 2.5 years) would have been the same as the average age for the whole population of unicorns. But we would also have proven that it is possible not only, as we did in our first attempt, to draw a sample that misrepresents the larger population, but also to draw probability samples from the same population that have different statistics (and, so, show sampling variability).

 STOP & THINK *How many possible unique samples of two unicorns could be drawn from our population?*

You could randomly select six possible unique samples of two unicorns from our population. These would be: Leopold and Frances (average age = 1.5), Leopold and Germaine (2.0), Leopold and Quiggers (2.5), Frances and Germaine (2.5), Frances and Quiggers (3.0), and Germaine and Quiggers (3.5).

We could make a visual display of these samples, called a **sampling distribution,** by placing a dot, representing each of the sample averages, on a graph big enough to accommodate all possible sample averages. A sampling distribution is a distribution of a sample statistic (like the average) computed from more than one sample. The sampling distribution of any statistic, according to

sampling distribution, the distribution of a sample statistic (such as the average) computed from many samples.

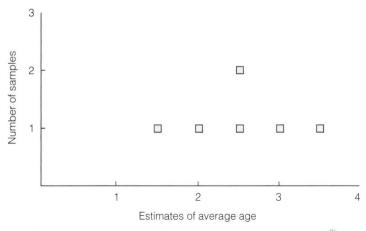

FIGURE **5.2** The Sampling Distribution of Unique Samples of Two Unicorns

probability theory, tends to hover around the population parameter, especially as the number of samples gets very large. Figure 5.2, a sampling distribution for the possible unique samples of two unicorns, suggests this property by showing that the sampling distribution of the average age for each possible pair of unicorns is centered around the average age for unicorns in the population (2.5 years).

 **STOP & THINK** *You're obviously pretty imaginative if you've followed us up to here. But, can you imagine the sampling distributions of all possible unique samples of one unicorn? Of four unicorns? What, if anything, do you notice about the possible error in your samples as the number of cases in the samples grows?*

The sampling distribution of the average ages of samples of one unicorn would have four distinct points, representing one, two, three, and four years, all entailing sampling error. The sampling distribution of the average age for a sample of four unicorns would have one point, representing 2.5 years, involving no sampling error at all. Figure 5.2 suggests that the sampling distribution of samples with two cases generally has less average sampling error than that for samples with one case, but generally more average sampling error than the sampling distribution with four cases. This illustrates a very important point about probability sampling: Increases in sample size, referred to as *n*, improve the likelihood that sampling statistics will accurately estimate population parameters.

The 2008 pre-election polls shown in Box 5.3 were based on samples of between 800 likely voters (the two Battleground polls) and 3,000 likely voters (the Rasmussen Reports poll) and all came reasonably close to predicting the actual spread in the election of 7.3 percentage points. One, the Battleground (Tarrance) poll was off by 5.3 percentage points, but most, like the Pew

Research poll (predicting an Obama win by 6 points), hovered within 1 or 2 percentage points of the final outcome.

A major point of this section has been to demonstrate that even when one employs probability sampling, sampling error can and frequently does occur. This means, unfortunately, that no sample statistic can be relied on to give a perfect estimate of a population parameter. Another point of this section has been to suggest something about the nature of sampling distributions. Notably, they tend to be centered around the population parameter when the number of samples is large. Mathematicians have given extensive thought to the nature of sampling distributions. A practical offshoot of this consideration is a branch of mathematics known as probability theory.

margin of error, a suggestion of how far away the *actual* population parameter is likely to be from the statistic.

A tremendous advantage of probability samples is that they enable their users to use probability theory to estimate population parameters with a certain degree of confidence. All of the final polls reported by RealClearPolitics (2008), for instance, were associated with an estimate of **margin of error,** a suggestion of how far away the *actual* population parameter (say, the spread among all likely voters) was likely to be from their statistic (of the spread within its sample). Thus, when the Pew Research poll told us on the eve of the 2008 election that the margin of error of its survey, suggesting that Obama would win 52 percent of the vote, was 2 percent, it meant they were saying that there is a 95 percent chance that the actual percentage was somewhere between 50 and 54 percent of the votes. (And, in fact, Obama won 52.9 percent of the votes.)

We haven't shown you how to calculate margins of error here, but they are not difficult to do and perhaps you'll learn how to do them in a statistics class. As our unicorn example suggests, though, the accuracy with which sample statistics estimate population parameters increases with the size of the sample. Probability theory tells us that if we want to achieve a margin of error of 10, we need a sample of about 100—and it doesn't matter what the size of the population is. If you'd like to reduce the margin of error from 10 percent to 5 percent, you'd need a sample of 400. To achieve its 2 percent margin of error, Pew Research talked with 2,587 likely voters. In general, *one needs to increase the sample size by about four times to achieve a halving of the margin of error.* Because of the costs involved in sampling more people, Weisberg (2005: 229) sees little reason for trying to achieve, say, a reduction in the margin of error from 3 percent to 1.5 percent by surveying 4,800 respondents rather than 1,200. It's far better, once sampling error has been reduced to, say, 3 percent or even 5 percent, to spend time and energy trying to reduce other sources of error, such as coverage, nonresponse, and measurement error, a kind of error we'll be focusing on in Chapters 6 (on measurement) and 9 and 10 (on questionnaires and interviews). These kinds of error are normally far greater contributors to what Weisberg calls "total survey error" than can possibly be offset by increases in sample size beyond, say, 1,200.

TYPES OF PROBABILITY SAMPLING

In the previous section, we introduced you to simple random sampling, a type of probability sampling that is appropriate for many statistical procedures used by social scientists. Here, we'd like to introduce you to other ways of doing simple random sampling and to other kinds of probability sampling and their relative advantages.

Simple Random Sampling

Simple random samples are selected so that each member of a study population has an equal probability of selection. To select a simple random sample, you need a list of all members of your population. Then, you randomly select elements until you have the number you want—perhaps using the blind drawing technique we mentioned earlier.

To show you another way of drawing a simple random sample, we'd like you to consider the list (or sampling frame) of Pulitzer Prize winners in Journalism for Commentary until 1993. The list of such award winners, given only since 1970, is relatively short and so is useful for illustrative purposes (though if, in fact, your study population included so few elements, you would probably want to use the entire population). From 1970 to 1993, there were 24 members of this population:

01. Marquis Childs	09. William Safire	17. Jimmy Breslin
02. William Caldwell	10. Russell Baker	18. Charles Krauthammer
03. Mike Royko	11. **Ellen Goodman**	19. Dave Barry
04. David Broder	12. Dave Anderson	20. Clarence Page
05. Edwin Roberts	13. Art Buchwald	21. Jim Murray
06. **Mary McGrory**	14. Claude Sitton	22. Jim Hoagland
07. Red Smith	15. Vermont Royster	23. **Anna Quindlen**
08. George Will	16. Murray Kempton	24. **Liz Baalmaseda**

 **STOP & THINK** *We have shown the female members of the population in boldface print. Because we've presented the population in this way, it is possible to calculate one of its parameters: the fraction of award winners who have been female. What is that fraction?*

Let us draw a simple random sample of six members from the list of award winners (i.e., we want an *n* of 6). If we don't want to use the "blind" drawing technique, we could use a random number table. To do so, we would, first, need to locate a random number table, such as the one we've provided in Figure 5.3, a larger version of which exists in Appendix E of this book. (Such tables, incidentally, are generally produced by computers in a way that gives every number an equal chance of appearing in a given spot in the table.) Second, we'd need to assign an identification number to each member of our population, as we have already done by assigning the numbers 1 through 24 to the list of award winners. Now we need to randomly select a starting number whose number of digits is equal to the digits in your highest identification number. The highest identification number in our list was 24 (with two digits), so we need to select (randomly) a number with two digits in it. Let's say you did this by closing your eyes, pointing toward the page, and (after opening

5334	5795	2896	3019	7747	0140	7607	8145	7090	0454	4140
8626	7905	3735	9620	8714	0562	9496	3640	5249	7671	0535
5925	4687	2982	6227	6478	2638	2793	8298	8246	5892	9861
9110	2269	3789	2897	9194	6317	6276	4285	0980	5610	6945
9137	8348	0226	5434	9162	4303	6779	5025	5137	4630	3535
4048	2697	0556	2438	9791	0609	3903	3650	4899	1557	4745
2573	6288	5421	1563	9385	6545	5061	3905	1074	7840	4596
7537	5961	8327	0188	2104	0740	1055	3317	1282	0002	5368
6571	5440	8274	0819	1919	6789	4542	3570	1500	7044	9288
5302	0896	7577	4018	4619	4922	3297	0954	5898	1699	9276

FIGURE **5.3** Random Number Table Excerpt

your eyes again) underlining the closest two-digit number—perhaps, the number 96 (see the "96") underlined in Figure 5.3.

Then, proceeding in any direction in the table, you could select numbers, discarding any random number that does not have a corresponding number in the population, until you'd chosen the number of random numbers you desire. Thus, for instance, you could start with 96 and continue across its row, picking off two-digit numbers, then down to the next row, and so forth, until you'd located six of them (remember our desired *n* numbers).

 **STOP & THINK** *What numbers would you select if you followed the procedure described in the previous sentence?*

The numbers you'd select, in this instance, would be the following:

96, 30, **19**, 77, 47, **01**, 40, 76, **07**, 81, 45, 70, 90, **04**, 54, 41, 40, 86, 26, 79, **05**, 37, 35, 96, **20**

Of course, you'd keep only six of these: those in bold print in the previous array—19, 01, 07, 04, 05, 20. Incidentally, you'd be working with a definition of simple random selection that assumes that once an element (corresponding to, say, the number 19) is selected, it is removed from the pool eligible for future selection. So if, say, during your selection number "07" had appeared more than once, you would have discarded the second "07." We'll also be using this definition later on in our discussions.

Your resulting sample of six journalists would be: Marquis Childs (case 01), David Broder (04), Edwin Roberts (05), Red Smith (07), Dave Barry (19), and Clarence Page (20).

 **STOP & THINK** *What fraction of this sample is female? Does this sample contain sampling error relative to its gender makeup?*

Pick a different simple random sample of six journalists. (Hint: Select another starting number in the random number table and go from there.) Who appears in your new sample? Does this sample contain sampling error relative to gender makeup?

A quicker way of drawing a random sample is to use Urbaniak and Plous's (2009) website, the Research Randomizer (http://randomizer.org). To random sample the same Pulitzer Prize winners, you'd simply tell the *Research Randomizer* you wanted one set of six numbers from the numbers 1 through 24 and, presto, you'd have your set of random numbers.

Systematic Sampling

systematic sampling, a probability sampling procedure that involves selecting every *k*th element from a list of population elements, after the first element has been randomly selected.

Even with a list of population elements, simple random sampling, done manually, can be tedious. It is rarely used in practice. Only five articles abstracted in *Sociological Abstracts* between 2006 and 2008 inclusive, for instance, employed it as a sampling technique. Botha and Pienaar (2006) used it, for instance, in a study of occupational stress among corrections officers in South Africa, drawing a simple random sample of 157 from a total list of 605 officers. If one does have a list of sampling frame elements, a much less tedious probability sampling technique is **systematic sampling.** In systematic sampling, every *k*th element on a list is (systematically) selected for inclusion. If, as in our previous example, the population list has 24 elements in it, and we wanted six elements in our sample, we'd select every fourth element (because 24/6 = 4) on the list. (If the division doesn't yield a whole number, you should round to the nearest whole number.) To make sure that human bias wasn't involved in the selection and that every element actually has a chance of appearing, we'd select the first element randomly from elements 1 through *k*. Thus, in the example of the Pulitzer Prize winners, you'd randomly select a number between one and four, then pick the case associated with that number and with every fourth case after that one on the list.

Try to draw a sample of six award winners using systematic sampling. What members appear in your sample?

In general, we call the distance between elements selected in a sample the selection interval (4 in the award-winner example but this is generally referred to by the letter "k"), which is calculated using the formula:

$$\text{selection interval (k)} = \frac{\text{population size}}{\text{sample size}}$$

We've mentioned the relative ease of selecting a systematic sample as one of the procedure's advantages. There is at least one other advantage worth mentioning: If your population is physically present (say, file folders in an organization's filing cabinets), you don't need to make a list. You need only determine the population size (i.e., the number of folders) and the desired sample size. Then, you'd just need to look at every *k*th element (or folder) after you'd randomly selected the first folder. Ian Rockett, a colleague in epidemiology, tells us that he has used a modified form of systematic sampling in a hospital emergency use study. A potential interviewee was the first eligible patient entering the emergency room after the hour. Eligibility criteria included being 18 years of age or older and able to give informed consent to participate in the study. In each hospital, interviewing lasted three weeks and covered each of three daily eight-hour shifts twice during the three-week period.

The only real disadvantage of systematic sampling over simple random sampling, in fact, is that systematic sampling can introduce an error if the sampling frame is cyclical in nature. If, for instance, you want to sample months of several years (say, 1900 through 2006) and your months were listed, as they frequently are, January, February, March, and so on for each year, you'd get a very flawed sample if your selection interval turned out to be 12 and all the months in your sample turned out to be Januarys. Imagine the error in your estimates of average annual snowfalls in North Dakota if you just collected data about Januarys and multiplied by 12! A word to the wise: When you're tempted to use systematic sampling, check your population lists for cyclicality. Perhaps for fear of unsuspected cyclicality, and despite its relative ease, systematic sampling is used only a little more in social science than simple random sampling. In fact only six articles mentioned in *Sociological Abstracts* between 2006 and 2008 employed systematic sampling. One, by Lin, Chen, Wang, and Cheng (2007), was based on a questionnaire survey of 1 out of 10 people entering a fitness center in Taiwan to produce 424 respondents in an examination of the relationship between extroversion and leisure motivation, concluding that extroverts are more highly motivated to attend fitness centers than non-extroverts.

Stratified Sampling

stratified random sampling, a probability sampling procedure that involves dividing the population in groups or strata defined by the presence of certain characteristics and then random sampling from each stratum.

One disadvantage of simple random sampling is that it can yield samples that are flawed in ways you want to avoid and can do something about. Take the earlier example of the Pulitzer Prize winners. Suppose you knew that approximately one sixth of that population was female and you wanted a sample that was one sixth female (unlike the one that we drew in the example). We could use a procedure called **stratified random sampling.** To draw a stratified random sample, we group the study population into strata (groups that share a given characteristic) and randomly sample within each stratum. Thus, to draw such a sample of the prizewinners, stratified by gender, we'd need to separate out the 4 female winners and the 20 male winners, enumerate each group separately, and random sample, say, a quarter of the members from each group.

 **STOP & THINK** *See if you can draw a stratified random sample of the prizewinners, such that one female and five males appear in the sample.*

More complexity can be added to the basic stratifying technique. If, for instance, you were drawing a sample of students at your college, using a roster provided to you by the registrar, and the roster provided information not only about each student's gender but also about his or her year in school, you could stratify by both gender and year in school to ensure that your sample wasn't biased by gender or year in school.

Stratified sampling can be done proportionately and disproportionately. The Pulitzer Prize sample you drew for the "Stop and Think" exercise was a proportionately stratified random sample because its proportion of females and males is the same as the proportion of females and males in the population. However, if you wanted to have more female prizewinners to analyze (maybe all of them), you could divide your population by gender and make sure your sample included a number of females that was disproportionate to their number in the population.

 **STOP & THINK** *How would you draw a disproportionate stratified random sample of eight prizewinners, with four females and four males in it?*

Disproportionate stratified sampling makes sense when the given strata, with relatively small representation in the population, are likely to display characteristics that are crucial to a study and when proportionate sampling might yield uncharacteristic elements from those strata because too few elements were selected. Suppose, for instance, you had a population of 100, 10 of whom were women. You probably wouldn't want to take a 10 percent proportionate sample from men and women because you'd then be expecting 1 woman to "represent" all 10 in the sample. You could oversample the women and then make an adjustment for their over-representation in the sample during your analysis. This is what Hoffnung did in her study of student views of their futures focused on in Chapter 4: she oversampled African American, Hispanic, and Asian American women to have large numbers of women of color in the study (personal communication).

It's also what Bostock and Daley (2007) did when they wanted to estimate lifetime and current sexual assault and harassment victimization rates of active-duty women in the Air Force. They drew a disproportionate random sample of women with senior ranks to make up for their relatively small numbers. Among their findings was that the prevalence of lifetime rape among Air Force women (28 percent) was more than twice as high as it is in national samples of women (13 percent).

Stratified random sampling has major advantages over simple random and systematic sampling, including its capacity to deal with otherwise underrepresented groups, and thereby reduce sampling error with respect to those groups. For this reason, it is used more frequently by researchers. In contrast with the five articles that *Sociological Abstracts* identifies has having simple random samples, and the six with systematic samples, 37 articles published between 2006 and 2008 were identified as having stratified random samples.

Cluster Sampling

Simple random, systematic, and stratified sampling all assume that you have a list of all population elements and that the potential distances between potential respondents aren't likely to result in unreasonable expense. If, however,

cluster sampling, a probability sampling procedure that involves randomly selecting clusters of elements from a population and subsequently selecting every element in each selected cluster for inclusion in the sample.

data collection involves visits to sites that are at some distance from one another, you might consider cluster sampling. **Cluster sampling** involves the random selection of groupings, known as clusters, from which all members are chosen for study. If, for instance, your study population is high school students in a state, you might start with a list of high schools, random sample schools from the list, ask for student rosters from each of the selected schools, and then contact each of those students.

 **STOP & THINK** *How might you cluster sample students who are involved in student organizations that are supported by student activity fees at your college?*

Multistage Sampling

multistage sampling, a probability sampling procedure that involves several stages, such as randomly selecting clusters from a population, then randomly selecting elements from each of the clusters.

Cluster sampling is used much less frequently than **multistage sampling,** which frequently involves a cluster sampling stage. Forty articles published between 2006 and 2008 are displayed when *Sociological Abstracts* is searched for multistage sampling, though the actual number would be far greater if one included all articles based on available data (see Chapter 12) offered by survey organizations to researchers. Two hundred and fifty-one articles using General Social Survey data alone are displayed when one queries *Sociological Abstracts* about research from 2006 to 2008. (And the General Social Survey, as indicated in Chapter 12, is but one of many kinds of "available" data sets used by researchers these days.) Two-stage sampling, a simple version of the multistage design, usually involves randomly sampling clusters and then randomly sampling members of the selected clusters to produce a final sample. Ozbay and Ozcan (2008), for instance, took a random sample of high schools in Ankara, Turkey, and then a random sample of students from each of those high schools. With their sample of 1,710 students, they found that Travis Hirschi's social bonding theory worked better for explaining the delinquent behavior of boys than of girls.

Random-digit dialing is generally a form of multistage sampling, especially when nationwide surveys are being conducted. The Pew Research Center for People & the Press (2009) uses slightly different approaches to random-digit dialing of landlines and cell phones, but both involve multistage sampling. When it comes to landlines, for instance, the Pew procedure is to random sample telephone numbers within a given county in proportion to the country's share of telephone numbers in the United States. (This process is imaginative. You might have a look at http://people-press.org/methodology/sampling/ for more details.) Then, once a number is contacted, the interviewer asks to speak with "the youngest male, 18 years or older, who is now at home." If there is no eligible male at home, the interviewer asks to speak with "the youngest female, 18 years or older, who is now at home." Pew has found that this procedure, once a number is contacted, is a good way of assuring that all adult members of a household has something like an equal chance of appearing in the sample.

 **STOP & THINK** *What coverage error might result from questioning the first adult to answer the phone?*

Some people who are more likely than others to answer phones are the able-bodied, those who have friends they like to chat with on the phone, those who aren't afraid that creditors might be calling, and so forth. Any number of biases might be engendered if you queried only those who answered the phone. But if you, say, asked to speak to the adult who had the most recent birthday (as Bates and Harmon, 1993, did), you'd be much more likely to get a representative sample of the adults living in the residences you contacted. Keeter admits that, theoretically, the Pew procedure of asking, first, for the youngest adult male and, then, for the youngest adult female, has no clear basis for yielding probability samples at the household level, but argues that, in fact, it's better at doing so than asking for the person with the most recent birthday:

> We get a lot of criticism for this method (though it is used by several other national polling organizations). Critics rightly charge that there is no statistical theory on which to base the selection, and thus we throw a wrench into the probability nature of the sample. We agree at some level, but the household enumeration that is required to do a true probability selection is very onerous and likely to result in lower response rates. "Last birthday" methods that have at least some claim to randomness don't work very well in practice—you get too many women, indicating that many respondents are not complying with the request. (personal communication)

The Pew Research Center multistage design, like all probability sampling designs, is much better at ensuring representativeness (and generalizability) than any nonprobability sampling design of a comparable population is. There are times, however, when nonprobability sampling is necessary or desirable, so we now turn our attention to some nonprobability sampling designs.

TYPES OF NONPROBABILITY SAMPLING

We've already suggested that, although nonprobability samples are less likely than probability samples to represent the populations they're meant to represent, they are useful when the researcher has limited resources or an inability to identify members of the population (say, underworld or upperworld criminals) or when one is doing exploratory research (say, about whether a problem exists or about the nature of a problem, assuming it exists). Let us describe each of these advantages as we describe some ways of doing nonprobability sampling.

Purposive Sampling

purposive sampling, a nonprobability sampling procedure that involves selecting elements based on the researcher's judgment about which elements will facilitate his or her investigation.

For many exploratory studies and much qualitative research, **purposive sampling,** in any of a number of forms, is desirable. In purposive sampling, the researcher selects elements based on his or her judgment of what elements will facilitate an investigation. Ghose, Swendeman, Sheba, and Chowdhury (2008), for instance, drew upon sex workers who had been mobilized by an organization for sex workers in Kolkata, India, to see why there was such a high-rate condom use among sex workers in that city. Because it can be used to study elements of particular interest to the researcher, even when comprehensive lists of or reliable access to such elements is nonexistent, purposive sampling is used with great frequency by social scientists. A query of *Sociological Abstracts* about articles based on purposive sampling or purposive samples generated a list of 58 articles published between 2006 and 2008.

quota sampling, a non-probability sampling procedure that involves describing the target population in terms of what are thought to be relevant criteria and then selecting sample elements to represent the "relevant" subgroups in proportion to their presence in the target population.

Quota Sampling

Quota sampling was the type of sampling used by Gallup, and others, so well in presidential preference polls until 1948. (You might remember that Gallup's projection, picking Roosevelt over Landon, was better for the 1936 election than was the *Literary Digest*'s, which depended on surveying owners of cars and telephones.) Quota sampling begins with a description of the target population: its proportion of males and females, of people in different age groups, of people in different income levels, of people in different racial groups, and so on.

 **STOP & THINK** *Where do you think Gallup and other pollsters might have gotten such descriptions of the American public?*

Until 1948, Gallup relied on the decennial censuses of the American population for his descriptions of the American voting public. Based on these descriptions, a researcher who used quota sampling drew a sample that "represented" the target population in all the relevant descriptors (e.g., gender, age, income levels, race).

Quota sampling at least attempts to deal with the issue of representativeness (in a way that neither convenience nor snowball sampling, discussed later, do). In addition, quota samples are frequently cheaper than probability samples are. Finally, when it yields a sample that consists of groups whose members are distinguished from one another in a theoretically significant way, it can be used as a way of making useful tests. Thus, Moller, Theuns, Erstad, and Bernheim (2008) took a quota sample of two groups in South Africa—regular community residents and patients at a tuberculosis (TB) hospital—to examine the usefulness of a new measure of subjective well-being. They found that the new measure showed that patients in the TB hospital had, as one might expect theoretically, lower subjective well-being than community residents, and that it did so more sensitively than better-established measures of life satisfaction and happiness. Quota sampling is, however, subject to some major drawbacks. First is the difficulty of guaranteeing that the description of the target population is an accurate one. One explanation given for the inability of Gallup and other pollsters to accurately predict Truman's victory in the 1948 presidential election is that they were using a population description that was eight years old (the decennial census having been taken in 1940), and much had occurred during the forties to invalidate it as a description of the American population in 1948. Another drawback of quota sampling is how the final case selection is made. Thus, the researcher might know that she or he needs, say, 18 white males, aged 20 to 30 years, and select these final 18 by how pleasant they look (in the process, avoiding the less pleasant-looking ones). Pleasantness of appearance is likely to be one source of bias in the final sample. Quota sampling is considerably less popular today than it was in the 1930s and 1940s when it was very popular among large public polling companies (Berinsky, 2006). In contrast with the 58 sociological articles that seem to have been based upon purposive sampling, a *Sociological Abstracts* search for articles using quota sampling (or quota samples) yielded only seven articles published between 2006 and 2008.

Snowball Sampling

snowball sampling, a nonprobability sampling procedure that involves using members of the group of interest to identify other members of the group.

When population listings are unavailable (as in the case of "community elites," gay couples, upperworld or underworld criminals, and so forth), **snowball sampling** can be very useful. Snowball sampling involves using some members of the group of interest to identify other members. One of the earliest and most influential field studies in American sociology, William Foote Whyte's (1955) *Street Corner Society: The Social Structure of an Italian Slum*, was based on a snowball sample: Whyte talked with a social worker, who introduced him to Doc, his main informant, who then introduced him to other "street corner boys," and so on. More recently, Jacinto, Duterte, Sales, and Murphy (2008) used a snowball sample of 80 women and men who sold Ecstasy in private settings to demonstrate how such sellers denied the identity of being "dealers" on three grounds: Ecstacy's benign reputation, the fact that they sold it in private places, and the fact that they sold it mainly to friends. Snowball sampling's capacity for making such discoveries about deviant subcultures makes it particularly popular in the study of crime and deviance: most of the 59 articles based on snowball sampling and published between 2006 and 2008, identified through a *Sociological Abstracts* search, focused on crime or deviance.

Convenience Sampling

convenience sample, a group of elements that are readily accessible to the researcher.

A **convenience sample** (sometimes called an available-subjects sample) is a group of elements (often people) that are readily accessible to, and therefore convenient for, the researcher. Psychology and sociology professors sometimes use their students as participants in their studies. Landra and Sutton (2008) used a convenience sample of 160 black, white, and racial minority students in two introductory sociology classes and asked them to take a "Black IQ Test," consisting of questions to test facility with African American slang and cultural references. They report that blacks outperformed nonblacks on all questions, effectively demonstrating the importance of racial bias in standardized testing.

Chase, Cornille, and English (2000) made very creative use of convenience sampling in their study of life satisfaction among people with spinal cord injuries. They sent letters requesting participation in their study to five listservs related to disability, to a website bulletin board, and to organizations that focused on spinal cord injuries and disabilities. They also sent the letter to various organizations that were likely to deal with people with spinal cord injuries. Eighty-nine percent of their 158 respondents chose to use a web-based form. Chase and colleagues argue that their own convenience in using the web was matched by the convenience of using the web for their specific population, people with spinal cord injuries, many of whom would have had much more difficulty responding through other media. The Internet might be a pretty good medium for convenience samples since about 72 percent of the U.S. population uses the Internet (Internet World Stats, 2008).

Convenience samples are relatively inexpensive and they can yield results that are wonderfully provocative and plausible. A *Sociological Abstracts* search yielded 97 articles published between 2006 and 2008, with research based on convenience sampling or convenience samples—suggesting that convenience sampling is probably the most popular of the nonprobability sampling techniques in the social sciences. Our sampling of articles (through the search mechanism provided by *Sociological Abstracts*) is, nevertheless, itself a kind of convenience sampling and, like other forms of convenience sampling, should entail a realization that generalizing beyond the samples involved is dangerous and should be tempered with considerable caution.

 **STOP & THINK** *A student who hands out 100 questionnaires in a campus café might be tempted to say he or she collected a "random" sample. Why is his or her really not a random sample?*

CHOOSING A SAMPLING TECHNIQUE

Choosing an appropriate sampling technique, like making any number of other decisions in the research process, involves a balancing act, often among methodological, practical, theoretical, and ethical considerations. First, one needs to consider whether it's desirable to sample at all, or if the whole population can be used. Then, if it is methodologically important to be able to generalize to some larger population, as in political preference polls, choosing some sort of probability sample is of overwhelming importance. The methodological importance of being able to generalize from such a poll can clash with the practical difficulty of finding no available list of the total population and, so, dictate the use of some sort of multistage cluster sampling, rather than some other form of probability sampling (e.g., simple random, systematic, or stratified) that depends on the presence of such a list. Or, however desirable it might be to find a probability sample of corporate embezzlers (a theoretical concern), your lack of access to the information you'd need to perform probability sampling (a practical concern) could force you to use a nonprobability sampling technique—and perhaps finally settle for those you can identify as corporate embezzlers and who volunteer to come clean (ethical).

You might find that your study of international espionage would be enhanced by a probability sample of the list of Central Intelligence Agency (CIA) agents your roommate has downloaded after illegally tapping into a government computer (methodological and practical considerations), but that your concern for going to prison (a practical consideration) and for jeopardizing the cover of these agents (practical and ethical) leads you to sample passages in the autobiographies of ex-spies (see Chapter 13 on content analysis) rather than to survey the spies themselves. In another study, your exploratory (theoretical) interest in the psychological consequences of adoption for nonadopted adult siblings could dovetail perfectly with your recent introduction to three such siblings (practical) and their obvious desire to talk to you about their experiences (practical and ethical).

Method, theory, practicality, and ethics—the touchstones of any research project—can become important, then, as you begin to think about an appropriate sampling strategy.

SUMMARY

Sampling is a means to an end. We sample because studying every element in our population is frequently beyond our means or would jeopardize the quality of our data—by, for instance, preventing our close scrutiny of individual cases. On the other hand, we don't need to sample when studying every member of our population is feasible.

Probability samples give every member of a population a chance of being selected; nonprobability samples do not. Probability sampling is generally believed to yield more representative samples than nonprobability sampling. Both probability and nonprobability sampling can be associated with coverage errors, nonresponse errors, and sampling error itself. Coverage errors are caused by differences between the sampling frame, or study population, and the target population. Nonresponse errors are caused by differences between nonresponders and responders to a survey. Sampling error is caused by differences between the sample and the study population, or sampling frame, from which it is drawn. A great advantage of probability samples, however, is that they allow the estimation of sampling error and thereby generate estimates of things like margins of error. Keeter, Dimock, and Christian's article reminds us that new sources of coverage errors may be entailed by technological change, as it demonstrates that the increased reliance on cell phones by the American public means that sampling frames that ignore cell phone–only respondents may become increasingly inadequate ways of estimating public attitudes.

We looked at five strategies for selecting probability samples: simple random sampling, systematic random sampling, stratified random sampling, cluster sampling, and multistage sampling. We then looked at four common strategies for selecting nonprobability samples: purposive sampling, quota sampling, snowball sampling, and convenience sampling. We emphasized that the appropriate choice of a sampling strategy always involves balancing theoretical, methodological, practical, and ethical concerns.

EXERCISE 5.1

Practice Sampling Problems

This exercise gives you an opportunity to practice some of the sampling techniques described in the chapter.

The following is a list of the states in the United States (abbreviated) and their populations in 2005 to the nearest tenth of a million.

State	Population	State	Population
AL	4.5	CO	4.5
AK	0.6	CT	3.5
AZ	5.5	DE	0.8
AR	2.7	FL	17.0
CA	35.5	GA	8.6

State	Population	State	Population
HI	1.3	NM	1.9
ID	1.4	NY	19.2
IL	12.7	NC	8.4
IN	6.2	ND	0.6
IA	2.9	OH	11.4
KS	2.7	OK	3.5
KY	4.1	OR	3.6
LA	4.5	PA	12.4
ME	1.3	RI	1.1
MD	5.5	SC	4.1
MA	6.4	SD	0.8
MI	10.0	TN	5.8
MN	5.1	TX	22.1
MS	2.9	UT	2.4
MO	5.7	VT	0.6
MT	0.9	VA	7.4
NE	1.7	WA	6.1
NV	2.2	WV	1.8
NH	1.3	WI	5.4
NJ	8.6	WY	0.5

1. Number the states from 01 to 50, entering the numbers next to the abbreviated name on the list.
2. Use the random number table in Appendix E and select enough two-digit numbers to provide a sample of 12 states. Write all the numbers and cross out the ones you don't use.
3. List the 12 states that make it into your random sample.
4. Now, if you have easy access to the Internet, locate the *Research Randomizer* (http://randomizer.org/) and draw one set of 12 numbers from 01 to 50 from it. Then list the 12 states that would make it into your random sample this time.
5. This time, take a stratified random sample of 10 states, one of which has a population of 10 million or more and 9 of which have populations of less than 10 million. List the states you chose.
6. How might you draw a quota sample of 10 states, one of which has a population of 10 million or more and 9 of which have populations of less than 10 million?
   a. Describe one way of doing this.
   b. Describe, in your own words, the most important differences between the sampling procedures used in Questions 5 and 6a.

EXERCISE 5.2

Examining Sampling in a Research Article

Using your campus library (or some other source), pick a research article from any professional journal that presents the results of a research project that involves some sort of sampling. Answer the following questions:

1. Name of author(s):
   Title of article:
   Journal name:
   Date of publication:
   Pages of article:

2. What kind of sampling unit or element was used?

3. Was the population from which these elements were selected described? If so, describe the population.
4. How was the sample drawn?
5. What do you think of this sampling strategy as a way of creating a representative sample?
6. Do(es) the author(s) attempt to generalize from the sample to a population?
7. Did the author(s) have any other purposes for sampling besides trying to enable generalization to a population? If so, describe this (these) purpose(s).
8. What other comments do you have about the use of sampling in this article?

CHAPTER 6

Measurement

© David Young-Wolff/PhotoEdit Inc.

INTRODUCTION

measurement, the process of devising strategies for classifying subjects by categories to represent variable concepts.

quantitative research, research focused on variables, including their description and relationships.

qualitative research, research focused on the interpretation of the action of, or representation of meaning created by, individual cases.

measure, a specific way of sorting units of analysis into categories

If you see a good friend today and he or she asks how you are, what will you say? What will you think your friend means by his or her question? What kinds of things might come to mind as you think about the answer? What would you mean by your answer? These kinds of questions are like the ones researchers ask themselves when they try to decide how to measure concepts.

Measurement generally means classifying units of analysis (say, individuals) into categories (say, satisfied with life or not satisfied with life) to represent variable concepts (say, life satisfaction). Measurement in quantitative research and qualitative research is often different, relating to differences in the purposes of the two enterprises. While distinctions between quantitative and qualitative research can be overdrawn, one thing is quite clear: a defining characteristic of **quantitative research** is its focus on variables. The Pew polling of prospective voters before the 2008 elections (see Chapter 5) is an example of quantitative research inasmuch as its primary function was to determine who was going to vote for Barack Obama and who was not—a variable characteristic of the American adult population. Consequently, measurement, with its focus on sorting cases into categories of variables, is central to quantitative research. **Qualitative research**, on the other hand, tends to be less focused variables per se and more concerned with interpreting action and sorting out meaning—and is typically more focused on individual cases than on variable characteristics of many cases. Ciambrone was primarily interested in eliciting the stories of individual women with HIV/AIDS (see Chapter 3) to learn what it was like to deal with the disease. Eventually, however, Ciambrone did become interested in variable characteristics that distinguished the women—and it was at that point that she became interested in measurement.

Often, in quantitative research, the researcher begins the measurement process by identifying a variable of concern, works out categories of that variable that make sense, and then classifies, through observation, various units of analysis. In contrast, in qualitative research, the process typically begins with close observation of individual units. The researcher may move to composing categories that help make sense of the units and sometimes goes on to making decisions about variables that might link the categories. We devote more of the current chapter to quantitative measurement because measurement has received more attention in quantitative than in qualitative research. But, by the end, you'll know quite a bit about both quantitative and qualitative measurement.

A **measure** of a variable is a specific way of sorting units of analysis into categories. For example, we could decide that our measure of gender would be people's answer to the question, "What is your gender?," on a questionnaire. But measuring a variable can, and sometimes should, be considerably more complicated. And since measurement can be an important source of both insight and error in research, it is a process that requires considerable care during the preparation of research, and attention when we read a research report. The building blocks of research are concepts or ideas (like "impostor," "fatherhood," "satisfaction with life," "suicide," and "power") about which we've developed words or symbols. Such concepts are almost always abstractions that we make from experience. Concepts cannot often be observed directly. What quantitative researchers do during the measurement process is to define the concept, for the purposes of the research, in progressively precise ways so that they can take indirect measurements of it.

Suppose we want to determine (or measure) whether a death was a suicide. We might, for instance, begin with a dictionary definition or something we will call a theoretical definition of suicide as "the act of killing oneself intentionally." Then, we might decide to take any clear evidence of a deceased person's intention to take his or her life (e.g., a note or a tape recording) as an indicator of suicide. We might look for more than one such indicator and would need to figure out what to do if some of the evidence was contradictory. Whatever decisions we make along the way would be part of our measurement strategy—one that usually starts with the broader issues of how to conceptualize something and then goes to the more specific issue of how to operationalize it.

CONCEPTUALIZATION

Researchers who start by creating a research question the way Michele Hoffnung did in the study discussed in the focal research section of Chapter 4 are working with the building blocks of all scientific, indeed of all thinking, concepts—words or signs that refer to phenomena that share common characteristics. Hoffnung asked questions about what would happen to young women who had graduated from college in 1993: Would they "have it all"? Would they establish long-term relationships, become mothers, share family responsibilities, and have careers?

Whenever we refer to ideas such as "sharing family responsibilities," "long-term relationships," or "careers," we are using concepts. But let's notice something

---

BOX 6.1

What Do You Think about Life?

Take a moment to answer the following questions suggested by Diener, Emmons, Larsen, and Griffin (1985). Later in the chapter, we'll talk about what these questions are intended to measure.

Following are five statements with which you may agree or disagree. Using the 1–7 scale shown, indicate your agreement with each item by placing the appropriate number on the line preceding that item. Please be open and honest in responding.

The 7-point scale is as follows:

    7—Strongly agree

    6—Agree

    5—Slightly agree

    4—Neither agree nor disagree

    3—Slightly disagree

    2—Disagree

    1—Strongly disagree

_____ In most ways my life is close to my ideal.

_____ The conditions of my life are excellent.

_____ I am satisfied with my life.

_____ So far I have gotten the important things I want in life.

_____ If I could live my life over, I would change almost nothing.

important: We each develop our basic understandings of concepts through our own idiosyncratic experiences. You initially might have learned what an "accident" is when, as a child, you spilled some milk, and a parent assured you that the spilling was an accident. Or you might have heard of a friend's bicycle "accident." Someone else might have had a baby brother who had an "accident" in his diaper. We might all have reasonably similar ideas of what an accident is, but we come to these ideas in ways that permit a fair amount of variation in just what we mean by the term.

So beware: Concepts are *not* phenomena; they are only symbols we use to refer to (sometimes quite different) phenomena. Concepts are abstractions that permit us to communicate with one another, but they do not, in and of themselves, guarantee precise and lucid communication—a goal of science. In fact, in nonscientific settings, concepts are frequently used to evoke multiple meanings and allusions, as when poets write or politicians speak. If, however, concepts are to be useful for social science purposes, we must use them with clarity and precision. The two related processes that help us with clarity and precision are called conceptualization and operationalization.

conceptualization, the process of clarifying just what we mean by a concept.

Conceptualization is the process of clarifying just what we mean by a concept. This process usually involves providing a theoretical or **conceptual definition** of the concept. Such a definition describes the concept through other concepts.

conceptual definition, a definition of a concept through other concepts.

STOP & THINK *What concepts are part of your definition of suicide?*

If we use *Webster's* dictionary definition (or conceptualization) of "suicide" as "the act of killing oneself intentionally," this definition employs the concepts of "killing" and "intention." Or, if we define "satisfaction with life" using the definition of Diener, Emmons, Larsen, and Griffin (1985) as "a cognitive judgment of one's life as a whole," we'd have to understand what "cognitive judgment" and "life as a whole" mean. We could define "power," using Max Weber's classic formulation, as the capacity to make others do what we want them to do despite their opposition—a definition that refers, among other things, to the concepts of "making others do" and "opposition." In each case, we would be defining a concept through other concepts: ideas (like "intention," "opposition," and "cognitive judgment") that themselves might deserve further clarification, but which nonetheless substantially narrow the possible range of meanings a concept might entail. Thus, for instance, in Diener's conceptual definition, he is implicitly suggesting that life satisfaction involves the conscious overall evaluative judgment of life using the person's own criteria. He implies that the concept doesn't involve the degree of frequency that the person feels joy, sadness, or anxiety—things that others might associate with life satisfaction. In doing so, he narrows and clarifies the meaning of his concept. Being specific and clear about a concept are the goals of conceptualization.

As part of creating and considering a conceptual definition, we should ask if the concept is so broad and inclusive that it could profitably be divided up into aspects or dimensions to further clarify it. A concept that refers to a phenomenon that is expected to have different ways of showing up or manifesting

dimensions, aspects or parts of a larger concept.

multidimensionality, the degree to which a concept has more than one discernible aspect.

itself is said to be **multidimensional**. For example, we often hear about the "fear of crime," which can be conceptually defined as the fear of victimization of criminal activity. However, we could follow Mark Warr and Christopher Ellison's (2000) lead and divide it up into two dimensions or two kinds of fear of crime: a "personal fear" of crime, which is when people worry about their own personal safety, and an "altruistic fear" of crime, when they worry about the safety of others, such as spouses, children, and friends.

OPERATIONALIZATION

As we've said, we often can't perceive the real-world phenomena to which our concepts refer, even after we've provided relatively unambiguous conceptual definitions of what we mean by the concepts. "Satisfaction with life" and "fear of crime" and "religion" are very hard to observe directly, and even such a "concrete" concept as "suicide" can elude our direct perception. Someone found dead in a body of water might have been drowned (died by homicide), or accidentally drowned (died by accident), or died as result of his or her own effort (died by suicide)—or even died of natural causes. When the phenomena to which concepts refer can't be observed directly, we might make inferences about their occurrence (or lack or degree thereof).

operationalization, the process of specifying what particular indicator(s) one will use for a variable.

The process of specifying what particular indicator(s) one will use for a variable is called **operationalization**. Indicators are observations that we think reflect the presence or absence of the phenomenon to which a concept refers. Indicators may be answers to questions or comments, things we observe, or material from documents and other existing sources. For an example, let's take a concept that we're all familiar with—age. Sounds simple enough, right? However, reading the gerontology literature (see Atchley, 2003: 6–8), we could notice that there are several ways to conceptualize age: as chronological age (time since birth), functional age (the way people look or the things they can do), and life stages (adolescence, young adulthood, and so on).

indicators, observations that we think reflect the presence or absence of the phenomenon to which a concept refers.

How we conceptualize age will make a huge difference in what we do when we operationalize the concept. Let's say we conceptualize age in a particular way—as chronological age; we'll still have to decide on an operational definition. We could, for example, operationalize chronological age by using as our indicator the answer to a variety of questions, some of which follow:

1. Are you old?
2. How old are you? Circle the letter of your age category:
   a. 25 or younger   b. 26 to 64   c. 65 or older
3. How old are you?

And many others. Each question is a possible indicator of age. Nevertheless, some of these might be much more useful than others are.

An Example of Conceptualization and Operationalization

Stevanovic and Rupert (2009) were interested in how personal feelings of accomplishment at work affected feelings of family-related stress and, in turn, life satisfaction among professional psychologists. To study the topic, they sent out a questionnaire to practicing psychologists, receiving completed questionnaires from 485 of them. To examine their three variables of interest—personal feelings of accomplishment at work, family-related stress, and life satisfaction, they used well-established conceptualizations and operationalizations available in the social-psychological literature. *Importantly, like all other researchers, they conceptualized and operationalized each of these variables separately.* When it came to measuring "life satisfaction," Stevanovic and Rupert employed the tools that Diener et al. (1985) had provided. Like Diener et al., they conceptualized life satisfaction as "a cognitive judgment of one's life as a whole." Note again that this definition of "life satisfaction" not only provides the reader with an idea of what Stevanovic and Rupert mean by "life satisfaction," it also eliminates other possible meanings—like the degree to which one feels (or lacks) anxiety on a daily basis. Furthermore, when they wanted to operationalize the "life satisfaction," Stevanovic and Rupert used the five questions that Diener et al. developed and that we've reproduced in Box 6.1. The index—or composite measure that is constructed by adding scores of individual items—for life satisfaction is created by adding the scores of the five items so that the totals range from 5 (low satisfaction) to 35 (high satisfaction). In Figure 6.1 we summarize the process researchers go through to measure each variable using the Stevanovic and Rupert measurement strategy for life satisfaction as our example.

index, a composite measure that is constructed by adding scores from several indicators.

STOP & THINK *Just for fun, why don't you use your answers to the five statements in Box 6.1 to calculate a "life satisfaction" score for yourself.*

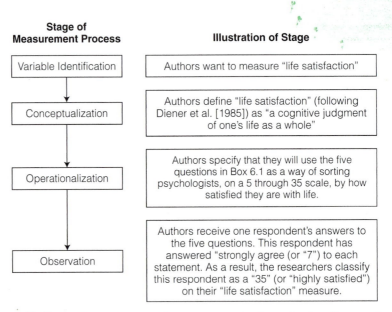

FIGURE **6.1** An Illustration of Measurement in Quantitative Research Based on Stevanovic and Rupert (2009)

Stevanovic and Rupert used their measure of life satisfaction to show that feelings of accomplishment at work are associated with reduced feelings of family-related stress that, in turn, correlate with increased life satisfaction among professional psychologists. Other researchers have used this life satisfaction measure to determine that perceived social support among bisexual college-aged young adults is an important predictor of life satisfaction (Sheets and Mohr, 2009) and that social support from significant others is predictive of life satisfaction among men in the nursing profession (Rochlan, Good, and Carver, 2009). They've also used it in a variety of countries, including the United States, Japan, Australia, Northern Ireland, Portugal, Cape Verde, Korea, and England (Kim and Hatfield, 2004; Lewis, Dorahy, and Schumaker, 1999; Neto and Barros, 2000; Schumaker and Shea, 1993; Smith, Sim, Scharf, and Phillipson, 2004).

 STOP & THINK *Hamarat et al. (2001: 189) studied life satisfaction and found that in their sample of 192 people, the older adults experienced the highest satisfaction and the younger adults the lowest. The average scores were 21.9 for those 18 to 40, 24.2 for those 41 to 65, and 26.4 for those 66 and over. How did your score compare to those in your age group?*

Composite Measures

composite measures, measures with more than one indicator.

Indexes like the one of life satisfaction, and other **composite measures,** abound in the social sciences. All composite measures, or measures composed of more than one indicator, are designed to solve the problem of ambiguity that is sometimes associated with single indicators by including several indicators of a variable in one measure. The effort to confirm the presence (or absence or the degree of presence) of a phenomenon by using multiple indicators occurs in both qualitative and quantitative research.

In qualitative research (see especially Chapters 10 and 11), a researcher might or might not need to prepare measurement strategies in advance of, say, an interview or observation. The careful qualitative researcher is likely to use multiple indicators of key concepts. A researcher can judge a study participant's "emotional tone" and body language and pay attention to the actual answer to a question like "How are you today?" Also, a question can be followed up with other questions. In Box 6.2, Hoffnung's coding of interviewees' answers to questions for two of her variables is described. A little later in this chapter, in the section on "Measurement in Visual Analysis and Qualitative Research," we'll give you an idea of how qualitative researchers frequently do measurement, not in advance of, but after making their observations or collecting their data.

In quantitative research, there has been much sophisticated thinking about the preparation of composite measures in advance of data collection. A full discussion of this thinking is beyond the scope of this chapter, but we would like to introduce you to it briefly. All strategies for making composite measures of concepts using questions involve "pooling" the information gleaned from more than one item or question. Two such strategies, the index and the scale, can be distinguished by the rigor of their construction. An

BOX **6.2**

Coding Qualitative Data*

To illustrate how social researchers code qualitative data, let's return to Michele Hoffnung's work on women's career and family choices.

In 2000, seven years after they graduated, Michele Hoffnung re-interviewed the young women in her study, asking them a series of questions about their current lives and future plans. (See Chapter 4 for Hoffnung's focal research article.) After the taped interviews were transcribed, Hoffnung and an associate read each interview, coded the data, and then identified one category of each variable in which to locate each participant. In most cases, the two coders agreed on the category each participant's answers seemed to best fit. (Comparing coders' decisions is called checking inter-coder reliability, a topic we'll cover later in the chapter.) In the section that follows, Hoffnung generously contributed her questions, coding decisions, and examples from interviews about the variable: Number of Children Wanted.

Variable: Number of Children Wanted

The indicators were different for women who were mothers and those who were not. Women without children were asked, "Would you like to have children? Why or why not?" If they said they wanted children, they were later asked, "How many children would you like to have?" Women with children were asked, "Do you plan to have any more children? Under what circumstances?"

The categories were the actual numbers of children that the women specified, including none. If the woman listed a range of numbers, then the midpoint of the range was used as the category. Below are examples from three study participants.

Participant 1
Q. Would you like to have children?
A. Yes.
Q. Why?
A. In my mind they've always been a part of a family. Um, you know I grew up in a family with six kids. And have always been used to having, you know, kids around at least when I was growing up. I see myself having between two and four kids.

Participant 2
Q. How many children would you like to have?
A. We've said four, but we are going to go one at a time.

Participant 3
Q. How many children would you like to have?
A. Three.

*The data were collected by Michele Hoffnung, Professor of Psychology, Quinnipiac University, as part of her study of women's lives, and is published with permission.

---

index is constructed by adding scores given to individual items, as occurs in the computation of the life satisfaction score. The items don't have to "go together" in any systematic way because indexes (sometimes called indices) are usually constructed with the explicit understanding that the given phenomenon, say "life satisfaction," can be indicated in different ways in different elements of the population. (Incidentally, we define a **scale** as an index in which some items can be, or are, weighed more than others and will illustrate scales a bit later in the chapter.)

scale, an index in which some items are given more weight than others in determining the final measure of a concept.

If a concept is multidimensional—that is, has different ways of showing up or manifesting itself—then an index can sum up all the dimensions. The Consumer Price Index, for instance, summarizes the price of a variety of goods and services (e.g., food, housing, clothing, transportation, recreation) in the United States as a whole in different times and for different cities at a given time. It is an index (of a multidimensional concept) because one wouldn't necessarily expect the price of individual items (say, the price of food and the price of recreation) to go up and down together. The Consumer Price Index is, however, a good way of gauging changes in the cost of living over time or space.

Other examples of indexes are evident in this chapter's focal research on impostor tendencies and academic dishonesty, by Joseph Ferrari. Ferrari became interested in the concept of impostor tendencies when a colleague in clinical psychology introduced him to the impostor phenomenon and he began to wonder whether many students might "shortchange" themselves by thinking they can't be as bright or capable as their accomplishments make them appear to be. He hoped that by examining the impostor phenomenon, he might shed light on it and help students see their own merits. While you read Ferrari's essay, see if you can answer the questions: How does Ferrari conceptualize and operationalize the concepts of "impostor tendencies" and "academic dishonesty?"

FOCAL   RESEARCH

Impostor Tendencies and Academic Dishonesty

By Joseph R. Ferrari*

The impostor phenomenon refers to feelings of intellectual phoniness experienced by high achieving individuals (Clance & Imes, 1978), despite objective evidence to the contrary. Impostors believe others will discover that they are not truly intelligent, but are in fact "impostors" (Clance, 1985). Repeat successes fail to weaken impostors' feelings of fraudulence or to strengthen a belief in their ability. Impostors react either by extreme overpreparation, or by initial procrastination followed by frenzied preparations (Chrisman, Pieper, Clance, Holland, & Glickauf-Hughes, 1995).

Cowman and Ferrari (2002) found that shame-proneness and self-handicapping best predicted impostor fears among students. Consistent with this outcome, Ferrari and Thompson (2005) found that impostor scores by students correlated positively with favorable impression management strategies, and that impostors were more likely than were nonimpostors to engage in self-handicapping acts (like not practicing before an upcoming task) in order to reduce the likelihood of unexpected performance success. Together, these studies suggest that impostors may sabotage their task performances, perhaps to demonstrate that they do not deserve their obtained successes—to show they are, in fact, frauds. Within a college setting, the opportunity

*This is an abridged version of an article that appeared in the *Social Behavior and Personality*. It is reprinted by permission from the Social Behavior and Personality, 2005, Vol. 33, No. 1, 11–18, published by the Society for Personality Research (Inc.).

to engage in destructive behaviors may include cheating on assignments and examinations, or plagiarizing written works. Because impostors state that they self-handicap, and that they believe they do not deserve their successes (Cowman & Ferrari), it is possible that impostors would engage in academic dishonesty in order to "prove" to others that they do not deserve success.

On the other hand, Whitley and Keith-Spiegel (2002) report that students who engage in academic dishonesty have moderate expectations for success, anticipate high rewards for their success, and are competitive about obtaining good grades. Such a profile seems opposite to the characteristics of an impostor, who does not believe success is warranted. Consequently, and as an alternative to the previous hypothesis, one might expect impostors would not cheat as an academic self-handicapping strategy: that, in fact, impostors, compared to nonimpostors, would not use cheating and plagiarism to obtain academic success. The present study explores whether increased or decreased reports of academic dishonesty were more likely among impostors or nonimpostors.

Method

Participants

A total of 124 students (92 women, 32 men; mean age = 20.9 years old) enrolled in a medium-size, private, urban Midwestern teaching university participated in the present study as part of course credit for an introductory psychology class.

Psychometric Measures

Participants completed Clance's (1985) 20-item self-reported unidimensional Impostor Phenomenon Scale (CIPS) to assess individuals' experience of impostor fears. Sample items are "I can give the impression that I'm more competent than I really am" and "At times I feel my success is due to some kind of luck." Respondents indicate their endorsement of items on five-point scales with end-point designations *not at all true* (1) and *very true* (5). With the present sample, the coefficient alpha was 0.91 (mean CIPS score = 54.24).

All participants also completed Roig and DeTommaso's (1995) Academic Practices Survey, a 24-item scale that assesses students' understanding and practices associated with (1) *plagiarism* when asked to perform a written assignment (16 items: "I've added sources not read to the reference section of a paper." "I've taken one or two sentences from someone else's written work, changed them moderately, and inserted this information into my paper."); and (2) *cheating* while completing an examination (eight items: "Copied answers from another student during an exam." "Used hidden notes, books, or calculators during an exam even though such use was prohibited."). All items are rated on a five-point scale (1 = *never*; 5 = *very frequently*). Within the present sample, coefficient alpha for the *plagiarism* subscale was 0.89 and 0.76 for the *cheating* subscale. In addition, all participants completed both sections of Paulhus' (1991) Balanced Inventory of Desirable Responding (BIDR), a well-known, reliable and valid 40-item measure of social desirability responding (see Endler & Parker, 1998, for details). This measure permits a control for the degree to which respondents engage in *self-deception enhancement* (that is, the degree to which they respond honestly but with an inflated sense of confidence, overclaiming their successes, and self-inflation of their skills) and in *impression management* (or with "faking" or "lying" on self-report measures).

Procedure

During a large testing session, a research assistant distributed to students the psychometric scales, along with forms to collect background information, after each student had returned a signed, confidential consent form. After scales had been collected, the assistant explained the purpose of this study.

Results and Discussion

Scores on the impostor scale were divided into extreme low or extreme high categories, to ensure the creation of two independent samples based on the personality style, using the bottom 25 percent and top 25 percent of scores, respectively. Persons who scored lower than or equal to 45 were labeled *nonimpostors* (n = 31: 20 women, 11 men: M score 37.58), whereas persons who scored higher than, or equal to, 63 were labeled *impostors* (n = 32: 22 women, 10 men: M score 71.00). There were no significant differences between men and women in their tendency to be impostors or nonimpostors. There were also no significant differences between impostors and nonimpostors in their likelihood of engaging in *self-deception enhancement* or *impression management*, as measured by the BIDR indexes.

Nonimpostors were significantly more likely than impostors to cheat on examinations (mean cheating score for impostors = 11.38; mean score for nonimpostors = 14.04) and to plagiarize in written assignments (mean plagiarism score for impostors = 24.38; mean score for nonimpostors = 31.38), and they were likely to do so, even when self-deception enhancement and impression management were controlled using the BIDR measures. These results indicate that perhaps among college students, nonimpostors engage in dishonest behaviors to succeed more often than do impostors. Alternatively, it is possible that impostors simply do not report academic dishonesty.

Still, the present study suggests that the academic success of impostors is unlikely to be due to unusually high levels of academic dishonesty. Some scholars (e.g., Cozzarelli and Major, 1990) have claimed that impostors have strong cognitive intelligence despite their beliefs to the contrary. The present study suggests that impostors do not handicap their academic performance through dishonest practices, even if they do engage in self-sabotaging behaviors at other times and in other ways (Cowman & Ferrari, 2002; Ferrari & Thompson, 2005). It should be noted that psychometric self-report scales were administered together in the same testing session to convenience samples and that the results might not reflect a causal relationship between the impostor phenomenon and academic honesty. The results, however, merit further investigation in other contexts. In summary, though, it seems that impostors are persons who are not likely to "cheat their way to the top." Nonimpostors are actually more likely to practice academic dishonesty to succeed.

References

Chrisman, S. M., Pieper, W. A., Clance, P. R., Holland, C. L., & Glickauf-Hughes, C. 1995. Validation of the Clance Impostor Phenomenon scale. *Journal of Personality Assessment*, 65, 456–467.

Clance, P. R. 1985. *The impostor phenomenon: Overcoming the fear that haunts your success*. Atlanta: Peachtree Publishers.

Clance, P. R., & Imes, S. A. 1978. The impostor phenomenon in high achieving women: Dynamics and therapeutic intervention. *Psychotherapy: Therapy, Research and Practice*, 15, 241–247.

Cowman, S., & Ferrari, J. R. 2002. "Am I for real?" Predicting impostor tendencies from self-handicapping and affective components. *Social Behavior and Personality*, 30, 119–126.

Cozzarelli, C., & Major, B. 1990. Exploring the validity of the impostor phenomenon. *Journal of Social and Clinical Psychology*, 9, 410–417.

Endler, N. S., & Parker, J. D. A. 1998. *Paulhus deceptions scales*. (Manual available from Mental Health Services, Inc.)

Ferrari, J. R., & Thompson, T. 2005 or 2006. Impostor fears: Links with self-perfection concerns and self-handicapping. Forthcoming in *Personality and Individual Differences*.

Paulhus, D. L. 1991. Measurement and control of response bias. In J. P. Robinson, P. R. Shaver, & L. S. Wrightman (Eds.), *Measures of personality and social psychological attitudes* (pp. 17–59). New York: Academic Press.

Roig, M., & DeTommaso, L. 1995. Are college cheating and plagiarism related to academic procrastination? *Psychological Reports*, 77(2), 691–698.

Whitley, B. E., & Keith-Spiegel, P. 2002. *Academic dishonesty: An educator's guide*. Mahwah, NJ: Lawrence Erlbaum, Publishers.

 **THINKING ABOUT ETHICS** *Students were given informed consent forms to sign before they were asked to fill out Ferrari's questionnaire. By not putting their names on the questionnaires, the anonymity of respondents is guaranteed.*

 **STOP & THINK** *How is the impostor phenomenon conceptualized by Ferrari and by other social psychologists? Are you surprised by this way of thinking about "impostors"?*

Ferrari conceptualizes the impostor phenomenon, as other social psychologists, and particularly Clance, have as "feelings of intellectual phoniness experienced by high achieving individuals." Like all scientific conceptualizations, this one winnows away many possible connotations of the word "impostor" in favor of one with relative precision. You may notice that this conceptualization stresses an attitude some people might have (that their successes are undeserved), rather than an action (like misrepresenting themselves to others) or an attribute (like actually being a fraud). Ferrari further clarifies what he means by the "impostor phenomenon" in the way he operationalizes the concept, in terms of the series of questions called the Clance Impostor Phenomenon Scale (CIPS) that he asks his student respondents to answer. The 20 questions that constitute CIPS are presented in Box 6.3.

BOX **6.3**

The CIPS

Instrument scoring: 1 = not at all true, 2 = rarely, 3 = sometimes, 4 = often, 5 = very true

Question

- I have often succeeded on a test or task even though I was afraid I would not do well before I undertook the task.
- I tend to remember the incidents in which I have not done my best more than those times I have done my best.
- If I'm going to receive a promotion or gain recognition of some kind, I hesitate to tell others until it is an accomplished fact.
- I can give the impression that I'm more competent than I really am.
- I often compare my ability to those around me and think they may be more intelligent than I am.
- If I receive a great deal of praise and recognition for something I've accomplished, I tend to discount the importance of what I have done.
- I often worry about not succeeding with a project or on an examination, even though others around me have considerable confidence that I will do well.
- It's hard for me to accept compliments or praise about my intelligence or accomplishments.
- I feel bad and discouraged if I'm not "the best" or at least "very special" in situations that involve achievement.

- I'm often afraid I may fail at a new assignment or undertaking, even though I generally do well at what I attempt.
- I rarely do a project or task as well as I'd like to do it.
- Sometimes I'm afraid others will discover how much knowledge or ability I really lack.
- When people praise me for something I've accomplished, I'm afraid I won't be able to live up to their expectations of me in the future.
- I avoid evaluations if possible and have a dread of others evaluating me.
- I'm disappointed at times in my present accomplishments and think I should have accomplished much more.
- I'm afraid people important to me may find out that I'm not as capable as they think I am.
- When I've succeeded at something and received recognition for my accomplishments, I have doubts that I can keep repeating that success.
- I sometimes think I obtained my present position or gained my present success because I happened to be in the right place at the right time.
- At times, I feel my success has been due to some kind of luck.
- Sometimes I feel or believe that my success in life or in my job has been the result of some kind of error.

Source: Oriel, Plane, and Mundt, 2004.

Ferrari's students, we learn from his article, scored, on average, 54.24 on CIPS, and that the top 25 percent, whom Ferrari calls "impostors," scored over 63 and that the bottom 25 percent, whom he call "nonimpostors," scored under 45.

STOP & THINK   *Answer the CIPS questions in Box 6.3 and then calculate your score by adding all the scores to the individual items. How does your score compare to the mean score among Ferrari's students? Would you have qualified as an impostor or a nonimpostor?*

Ferrari uses two other indexes: the Academic Practices Survey (APS) and the Balanced Inventory of Desirable Responding (BIDR). He uses them to find out which of two alternative hypotheses has more support: impostors are unusually likely to plagiarize and cheat or that they're *not* unusually likely to plagiarize and cheat. He finds that imposters are actually less likely to plagiarize and cheat than nonimpostors.

But Ferrari wondered whether part of this difference might be due to the fact that impostors, more than nonimpostors, might be interested in giving a good impression of themselves. It is possible, for instance, that, when asked whether they'd ever "copied answers from another student during an exam," impostors might be more likely than nonimpostors to lie. Ferrari was aware of this issue and has asked us to remind you that sometimes people falsely state information in surveys to look good (Ferrari, personal communication). Weisberg (2005: 86–89) suggests that lying is a major source of **measurement error** in all surveys. Measurement error, unlike the kinds of error we discussed in Chapter 5—coverage error, nonresponse error, and sampling error—refers to the kind of error that occurs when the measurement we obtain is not an accurate portrayal of what we tried to measure. We'll have more to say about measurement error associated with surveys in Chapter 9. To deal with the measurement problem associated with people trying to give a good impression of themselves on surveys, Ferrari employed the BIDR, which not only measures how much respondents are likely to deceive themselves (by asking 20 questions like "I always know why I do things" [Bonanno, Rennicke, and Dekel, 2005]) but also measures how much they are likely to try to give a good impression (by asking 20 questions about, for instance, whether they sometimes tell lies and have some pretty awful habits [von Hippel, Schooler, Preacher, and Radvansky, 2005]). Using a statistical procedure (analysis of covariance), he is able to determine that impostors are less likely to plagiarize and cheat than nonimpostors, even when you hold constant any differences that might exist between impostors and nonimpostors in their chances of lying to give a good impression.

Indexes can be used to measure multidimensional concepts, then, and we often compute an index score, as you did for CIPS in the previous "Stop and Think" by adding the scores on individual items. Methodologists often call indexes in which some items can be, or are, weighed more than others as scales. One kind of scale, the Bogardus social distance scale (designed by Emory Bogardus), was created to ascertain how much people were willing to

measurement error the kind of error that occurs when the measurement we obtain is not an accurate portrayal of what we tried to measure.

interact with other kinds of people. Suppose you wanted to find out how accepting people are of Martians. You could ask them questions like:

1. Would you be inclined to let Martians live in the world?
2. Would you be inclined to let Martians live in your state?
3. Would you be inclined to let Martians live in your neighborhood?
4. Would you be inclined to let a Martian be your next door neighbor?
5. Would you be inclined to let a Martian marry your child?

Because you might expect that a person who answered yes to Question 5 would also answer yes to all the other questions, you could reasonably give a higher score to a yes for Question 5 than the others. Similarly, you could reasonably give a higher score to a yes for Question 4 than for yeses to Questions 3, 2, and 1.

STOP & THINK   *Sometimes indexes that are referred to as scales don't use items that might be sensibly weighted the way our Bogardus scale does but sometimes they do. Have a look at the items in the CIPS (in Box 6.3). When used by Ferrari, all items were given equal weight.*

Indexes and scales are not hard to make up, though assuring yourself of their validity and reliability can be challenging. (See the discussion of these topics that follows.) You're frequently well advised to find an index whose validity and reliability has already been tested if one exists for the variable you're trying to measure. (See Touliatos, Perlmutter, and Straus, 2001, or Miller and Salkind, 2002, and see also the journal, *Social Indicators Research*, for a list of indexes that are often used in the social sciences.) Keep your eyes peeled during your literature review and you might see ways (or references to ways) that others have operationalized the concepts you're interested in. If you think the same approach might work for you, try it out (and remember to cite your source when you write up your results). There's no sense in reinventing the wheel.

Measurement in Visual Analysis and Qualitative Research

Before turning to some finer detail associated with quantitative research measures, we'd like to introduce you briefly to qualitative measures and, particularly, to a specialized version of such measures: those used in visual analysis, an exciting new field in the social sciences. **Visual analysis** refers to a set of techniques used to analyze images. As John Grady (2001: 83) has observed, "visual media and images increasingly dominate mass communications in contemporary society" and not to study how such images are produced and what they tell us is a serious oversight.

visual analysis, a set of techniques used to analyze images.

Images should be treated as data—about the image-maker's view of the world, if nothing else. We encourage you, for instance, to take a critical look at the cover image of our book. Consider, for instance, what it tells you about what we want you to think about social research methods. Here you couldn't begin, as Ferrari did in his study, with predetermined variables and predetermined measurement strategies. You'd have to see what themes emerged as you examined the picture. In the case of the cover picture, you

coding, assigning observations to categories.

might notice, for instance, that it is actually made up of multiple images, each having something to convey about our message about research. Then, using what qualitative researchers call **coding**, or the assigning of your observations to categories, you could begin to create your own variables. You might notice, for instance, how many of the images include women and how many include men. In doing so, you'd not only be "measuring" our presentation of social researchers by "gender," but you might also begin to suspect that we (Emily and Roger) are "suggesting," through the images we've chosen, that social research can be and is done by both women and men.

STOP & THINK *Can you code the images on our cover for what they seem to say about researchers' ethnic backgrounds?*

Grady (2007a) used a very similar process for measuring the messages of 590 advertisements containing images of African American in *Life* magazine from 1936 to 2000. His approach, called "grounded theory" coding, involved successive observations of the advertisements, seeing what codes occurred to him, and then assigning each advertisement to categories of emerging variables. One thing he discovered was a "steadily growing white commitment to racial integration" (Grady, 2007a: 211) expressed through the magazine editors' choice of published advertisements. Notice that the creation of variables for Grady, as in our example of analyzing the cover of this book, came after the collection of data (the images) and is comparatively ad hoc and idiosyncratic to the individual researcher. Another researcher might have chosen different codes and different variables.

We have more to tell you about measurement in qualitative research in Chapter 15. For now, however, we'd like to direct you to Figure 6.2, an image meant to suggest that, in qualitative research, unlike quantitative research (see Figure 6.1), measurement often begins after observations have been made and involves the coding of those observations until one begins to imagine "variables" that make meaning of the codings.

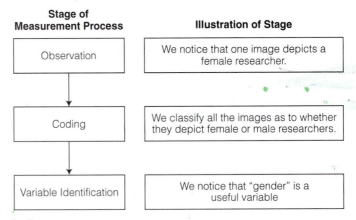

FIGURE **6.2** An Illustration of Measurement in Qualitative Research from the Examination of this Book's Cover Image

The kind of measurement we've attributed to versions of visual analysis is actually fairly typical of qualitative research generally. Another example of qualitative coding comes from the author, Sandra Enos, of the focal research in Chapter 10. She was curious about how incarcerated women "mothered" from prison. Early on, she took copious notes on her observations at a parenting program in a prison in the Northeast. She then coded the notes, finding that some of the women's children were under the care of their fathers, some were under the care of another relative, and some were in foster care. She eventually noticed that she made a variable of the various residences of the children and that this variable included the following categories: grandparent, husband, other relative, and nonrelative.

Exhaustive and Mutually Exclusive Categories

Measuring a concept (e.g., life satisfaction or age) means providing categories for classifying units of analysis as well as identifying indicators. For "life satisfaction," for example, Diener (2001) suggests creating seven categories by taking the answers to the five questions, adding the individual scores together (for a total of between 5 and 35), and classifying the score as follows:

31 to 35	extremely satisfied
26 to 30	satisfied
21 to 25	slightly satisfied
20	neutral
15 to 19	slightly dissatisfied
10 to 14	dissatisfied
5 to 9	extremely dissatisfied

But what of the nature of these categories? In this section, we'd like to emphasize two rules that should stand as self-checks any time you engage in measuring a variable concept.

The first rule for constructing variable categories is that they should be exhaustive. That is, there should be a sufficient number of categories to enable you to classify every subject in your study. The aim of creating exhaustive categories doesn't require any particular number of categories, although the minimum number of categories for a variable concept is two. As an example, consider measuring political party affiliation with the question, "What is your political party affiliation?" followed by answer categories: (1) Republican, (2) Democrat, and (3) other.

exhaustive, the capacity of a variable's categories to permit the classification of every unit of analysis.

STOP & THINK *Can you think of any likely instance that couldn't be "captured" by the net offered by this question's answer categories? If you can't, the categories are exhaustive.*

mutually exclusive, the capacity of a variable's categories to permit the classification of each unit of analysis into one and only one category.

The second rule for variable construction is that the categories should be mutually exclusive. This means you should be able to classify every unit of analysis into one and only one category—that is, your categories should have no overlap. A woman can give birth as the result of a planned pregnancy, but then it won't be unplanned. A man can fall into the category "65 or older," but if he does, he can't also fall into the "26 to 64" category.

mutually exclusive categories
should not overlap

STOP & THINK *Suppose you asked the question "How old are you?" and then told the person to circle the number associated with his or her age category, giving them the answer categories of (1) 25 or less, (2) 25 to 65, and (3) 65 and older. Which of the rules for variable construction would you be offending? Why?*

The example in the "Stop and Think" is of a variable whose categories are not mutually exclusive because they overlap. A person who was 25 years of age, for instance, could put himself or herself in both categories (1) and (2) and a person who was 65 could fall into both categories (2) and (3). But if we asked the question "How old are you?" and told the respondents to circle the number associated with his or her age category, giving them the answer categories of (1) 25 or less and (2) 26 to 65, then we would be violating the rule of exhaustiveness for categories of variables because there are people, 66 and older, who wouldn't fit into a category.

QUALITY OF MEASUREMENT

reliability, the degree to which a measure yields consistent results.

validity, the degree to which a measure taps what we think it's measuring.

We'd now like to introduce two other important qualities of measurement: **reliability** and **validity**. A measurement strategy is said to be reliable if its application yields consistent results time after time. If you were to use a scale to measure how much a rock weighs, the scale would provide you with reasonably reliable measurements of its weight if it tells you that the rock weighs the same today, tomorrow, and next week. A measurement strategy is valid if it measures what you think it is measuring. If it's the weight of a rock you want to measure, a scale works pretty well. But, however reliable (and valid) your scale is when measuring the rock's weight, it cannot validly measure its shape or texture—it's just not supposed to.

STOP & THINK *Suppose you attempted to measure how prejudiced people are with the question "What is your sex?" Which quality of measurement—reliability or validity—is most apt to be offended by this measurement strategy? Why?*

Checking Reliability

test–retest method, a method of checking the reliability of a test that involves comparing its results at one time with results, using the same subjects, at a later time.

The reliability of a measurement strategy is its capacity to produce consistent results when applied time after time. A yardstick can consistently tell you that your desk is 30 inches wide, so it's pretty reliable (even if it's possibly inaccurate—we'd have to check whether the yardstick's first inch or so has been sawed off). There are many ways to check the reliability of a measurement strategy. One is to measure the same object(s)—like your desk—two or more times in close succession. This **test–retest method** has been used by the researchers who created the life satisfaction measure. Diener et al. (1985) gave the life satisfaction questions to 76 students twice, two months apart. They found that students who scored high on the index the first time around were much more likely to do so the second time than students who had not scored high the first time.

The major problems with test–retest methods of reliability assessment are twofold. First, the phenomenon under investigation might actually change between the test and the retest. Your desk won't grow, but a person's life satisfaction might actually change in two months. So if people's categories change *because* there really have been changes in what's being measured, this does not mean that the measurement strategy is unreliable. Second, earlier test results can influence the results of the second. Do you think that if you completed a life satisfaction questionnaire today, it might influence your answers on the same questionnaire next month?

The **internal consistency method** is one way to deal with the problems of the test–retest method. It relies on making more than one measure of a phenomenon at essentially the same time. That's in fact what Diener et al.'s five life satisfaction questions (in their life satisfaction index) do. And some researchers, including Vera et al. (2008) in their study of subjective well-being among urban adolescents of color, have checked the degree to which the answers to these questions are correlated with one another. Among their sample, Vera et al. found that individual indicators of the life satisfaction measure were highly correlated with one another. In other words, they all seem to give consistent reports about adolescents' satisfaction. The CIPS used by Ferrari consisted of the 20 items shown in Box 6.3. Ferrari reports, for instance, that he used the scores of the items and calculated a statistic, coefficient alpha, finding it to be 0.91. Coefficient alpha is actually a way of estimating the internal consistency of items. An alpha of 0.70 or greater suggests that items in a scale "go together" very well—that is, that they are internally consistent, so an alpha of 0.91 indicates high internal consistency.

> **internal consistency method,** a method that relies on making more than one measure of a phenomenon at essentially the same time.

 **STOP & THINK** *Go back to the focal research piece and look at the alphas for the "plagiarism" and "cheating" subscales of the APS among Ferrari's respondents. Do these subscales have high internal consistency?*

Another common way to check the reliability of a measurement strategy is to compare results obtained by one observer with results obtained by another using exactly the same method. We'll see this **interobserver reliability or inter-rater reliability method** figure extensively in later chapters. In Chapter 13, on content analysis, Roger and three colleagues report that two raters calculated the percentage of pages devoted to women's works, to African Americans' work, to Native Americans' works, to Hispanics' works, and to Asian Americans' works in a large sample of American literature anthologies. The fact that the correlation of each of these percentages for the two raters was never less than 0.90 means that the two raters used the same measurement strategy to come up with virtually the same results all the time. In other words, the measurement scheme yielded consistent results when different raters used it.

> **interobserver (or inter-rater) reliability method,** a way of checking the reliability of a measurement strategy by comparing results obtained by one observer with results obtained by another using exactly the same method.

STOP & THINK *Suppose that you found a way of measuring homophobia with 10 questionnaire items and you wanted to use those questions in a questionnaire to see if there's a connection between homophobia and knowledge of AIDS. Describe a way in which you might check the reliability of the homophobia measure you were using.*

Checking Validity

Validity refers to how well a measurement strategy taps what it intends to measure. You might wonder how to tell whether a strategy is valid. After all, most measurement in the social sciences is indirect. Asking people "All things considered, how satisfied are you with your life as a whole these days?" isn't exactly the most direct way of determining personal well-being, even though this what is used in the World Values Survey in 81 countries (Kahneman and Krueger, 2006).

Ultimately the validity of a measure cannot be proven, but we can lend credence to it. Perhaps the most intuitively satisfying check on validity is this question: Does this measurement strategy *feel* as if it's getting at what it's supposed to? If a measure seems as if it measures what you and I (and maybe a whole lot of others) believe it's supposed to be measuring, we say it has **face validity**, and we can feel pretty comfortable with it. The answer to the question, "How old are you?" seems like a pretty good measure of age and, therefore, has high face validity. On the other hand, the question, "What, in your opinion, is the proper role of the military in foreign affairs?" would yield a poor measure of age and, therefore, has low face validity.

face validity, a test for validity that involves the judgment of everyday people, like you and me.

Content validity is similar to face validity, except that it generally involves the assessment of experts in a field, rather than just regular people, like you and me. Oransky and Fisher (2009), for instance, developed a 34-item Meanings of Adolescent Masculinity Scale (MAMS) by first doing a qualitative analysis of boys' own conceptions of what it means to be masculine in today's society. Oransky and Fisher used this analysis to come up with a tentative list of 60 items for their index and then asked a series of focus groups, made up of diverse adolescent males (the real experts, in the view of Oransky and Fisher, in what it takes to possess adolescent masculinity), to pick the items *they* believed most descriptive of what they viewed as "being a man"—and then to add some others. The resulting 34-item masculinity scale, then, had been tested and approved by a group of experts—in this case, adolescent males.

content validity, a test for validity that involves the judgment of experts in a field.

Content valid

Establishing **predictive criterion validity** is different from establishing face validity and content validity, inasmuch as it doesn't directly involve the judgment of people about the measure. Predictive criterion validity involves establishing how well the measure predicts future behaviors you'd expect it to be associated with. Srivastava, Tamir, McGonigal, John, and Gross (2009), for instance, measured the suppression of emotion (the inhibiting of the outward expression of emotion) in a sample of students entering college. They predicted that such suppression would lead to both lower social support for students and less closeness to others within a short period after the start of college. When they found that this was true, they inferred that their measure of emotional suppression had predictive criterion validity, because it did, in fact, go together with the predicted behavioral outcomes. Importantly, the reason a predictive criterion validity test involves the adjective criterion is that, implicitly, it's the behavior against which the new measure is being tested that is the variable of greatest interest to the researchers. In this case, Srivastava et al. were crucially interested in a condition they felt would predict

predictive criterion validity, a method that involves establishing how well the measure predicts future behaviors you'd expect it to be associated with.

or help them understand the capacity of students to develop satisfying relationships, early on, in college.

A similar test for validity is embodied in **concurrent criterion validity** tests. Here the concern isn't so much with behaviors a measure should predict, but to things with which it should happen at the same time. Salmela-Aro, Kiuru, Leskinen, and Nurmi (2009), for instance, devised a measure of school burnout, consisting of dimensions related to exhaustion with school, cynicism toward the purpose of school, and a sense of inadequacy in relation to school. The researchers figured that something that might go along with school burnout would be poor academic achievement. When they found, in fact, that their measure of burnout was related to poor school performance, they concluded that their measure was valid because it was associated with a behavior it should be associated with.

Again, the distinguishing characteristic of predictive and concurrent validity checks is that the criterion (e.g., how one interacts with fellow students at school or school performance) is the researcher's focus, or concern, rather than the variable that's being measured (e.g., emotional suppression or school burnout). The variable that's being measured (e.g., burnout) can been seen as being designed as a tool for predicting or understanding the behavior (e.g., school performance) of interest. Construct validity, on the other hand, is used when the trait or quality being measured itself is of central importance, rather than the measures used to check its validity (Cronbach and Meehl, 1955). **Construct validity** refers to how much a measure of one concept is associated with a measure of another concept that some theory says it should be associated with. Howe, Cate, Brown, and Hadwin (2008), for instance, developed a test of empathy in preschool children. They checked its construct validity by seeing how it was correlated with something their theory suggested it should be associated with: pro-social behavior (as evaluated by their teachers). Now, one might say that the main reason for being concerned with preschool children's capacity for empathy would be its association with pro-social behavior (and that Howe et al. were therefore checking their measures through concurrent criterion validity). But that's not the way Howe et al. saw it. They believed such empathy was crucial in and of itself and so were measuring its construct validity when they correlated it with teacher assessment of pro-social behavior.

Remembering all these specific ways of checking validity is perhaps not as important as remembering the two basic strategies for doing so: assessing or having someone else assess how well a measure seems to be doing its job (involved in both face and content validity checks) and seeing how well a measure correlates with things it should be correlating with if it's measuring what it's supposed to be measuring (involved in predictive criterion, concurrent criterion, and construct validity checks).

⭐ **STOP & THINK** *What are two ways by which you might check the validity of the measure of life satisfaction that Diener created?*

There are, of course, many sources of measurement error associated with the wording of questions (Weisberg, 2005) other than that they might engender self-enhancing responses or might not have mutually exclusive and exhaustive answer categories. Some other considerations are that questions ask respondents about topics that are relevant to them or about which they've given some thought and that they (the questions) not be misleading. We'll have more to say about such sources of measurement error in Chapters 9 and 10.

One last quality of measurement is its precision or exactness. Some measurement strategies allow greater precision than others do. If we ask people how old they were at their last birthday, we can be more precise in our categorization than if we ask them to tell us their ages using broad categories such as "26 to 64." If we use the seven categories of life satisfaction (from extremely satisfied to extremely dissatisfied) that Diener (2001) suggests, we can be more precise than if we use only two categories (satisfied and dissatisfied), though less precise than using the 5- to 35-point scale on which the categories are based.

LEVEL OF MEASUREMENT

Consideration of precision leads us to consider another way that variable categories can be related to one another—through their level of measurement or their type of categorization. Let's look at four levels of measurement: nominal, ordinal, interval, and ratio, and the implications of these levels.

Nominal Level Variables

nominal level variables, variables whose categories have names.

Nominal (from the Latin, *nomen*, or name) **level variables** are variables whose categories have names or labels. Any variable worth the name is nominal. If your variable is "cause of death" and your categories are "suicide," "homicide," "natural causes," "accident," and "undetermined," your variable is a nominal level variable. If your variable is "sex" and your categories are "male" and "female," your variable is nominally scaled. Nominal level variables permit us to sort our data into categories that are mutually exclusive, as the variable "religion" (with categories of "Christian," "Jewish," "Muslim," "none," and "other") would. Another example of a nominal level variable is marital status, if its categories are "never married," "married," "divorced," "widowed," and "other."

A nominal level of measurement is said to be the lowest level of measurement, not because nominal level variables are themselves low or base, but because it's virtually impossible to create a variable that isn't nominally scaled.

Ordinal Level Variables

ordinal level variables, variables whose categories have names and whose categories can be rank-ordered in some sensible way.

Ordinal (from the Latin, *ordinalis*, or order or place in a series) **level variables** are variables whose categories have names or labels and whose categories can be rank-ordered in some sensible fashion. The different categories of an ordinal level variable represent more or less of a variable. If racism were measured using categories like "very racist," "moderately racist," and "not racist," the "very racist" category could be said to indicate more racism than either of the other two. Similarly, if one measured social class using categories like "upper class," "middle class," and "lower class," one category (say "upper class") could be seen to be "higher" than the others.

All ordinal level variables can be treated as nominal level variables, but not all nominal level variables are ordinal. If your variable is "sex" and your categories are "male" and "female," you'd have to be pretty sexist to say that one category could be ranked ahead of the other. Sex is a nominal level variable, but not an ordinal level variable. On the other hand, if your categories of age are "less than 25 years old," "between 25 and 64," and "65 or older," you can say that one category is higher or lower than another. Age, in this case, is ordinal, though it can also be treated as nominal. Ordinal level variables guarantee that categories are mutually exclusive and that the categories can be ranked, as the variable "team standing" (with the categories of "1st place," "2nd place," and so on) does in many sports.

Interval Level Variables

interval level variables, variables whose categories have names, whose categories can be rank-ordered in some sensible way, and whose adjacent categories are a standard distance from one another.

Interval level variables are those whose categories have names, whose categories can be rank-ordered in some sensible way, and whose adjacent categories are a standard distance from one another. Because of the constant distance criterion, the categories of interval scale variables can be meaningfully added and subtracted. SAT scores constitute an interval scale variable, because the difference between 550 and 600 on the math aptitude test can be seen as the same as the difference between 500 and 550. As a result, it is meaningful to add a score of 600 to a score of 500 and say that the sum of the scores is 1100. Similarly, the Fahrenheit temperature scale is an interval level variable because the difference between 45 and 46 degrees can be seen to be the same as the difference between 46 and 47 degrees. All interval level variables can be treated as ordinal and nominal level variables, even though not all ordinal and nominal level variables are interval level.

Ratio Level Variables

ratio level variables, variables whose categories have names, whose categories may be rank-ordered in some sensible way, whose adjacent categories are a standard distance from one another, and one of whose categories is an absolute zero point—a point at which there is a complete absence of the phenomenon in question.

Ratio level variables are those whose categories have names, whose categories can be rank-ordered in some sensible way, whose adjacent categories are a standard distance from one another, and one of whose categories is an absolute zero point—a point at which there is a complete absence of the phenomenon in question. Age, in years since birth, is often measured as a ratio level variable (with categories like one, two, three, four years, and so on) because one of its possible categories is zero. Other variables such as income, weight, length, and area can also have an absolute zero point and are therefore ratio level variables. They are also interval, ordinal, and nominal level variables. For practical purposes, however, interval and ratio level variables are similar, and we will refer to them later as either interval ratio or simply interval level variables. This raises the question of what practical purposes there are for learning about levels of measurement in the first place, however, and that's the topic of our next section.

 **STOP & THINK** *When Ferrari compares impostors with nonimpostors, what level of measurement is he using? What makes you think so? When he reports that the mean score of his student respondents on the CIPS was 54.24, what level of measurement is he ascribing to CIPS? What makes you think so?*

The Practical Significance of Level of Measurement

Perhaps the most practical reason for being concerned with levels of measurement is that frequently researchers find themselves in the position of wanting to summarize the information they've collected about their subjects, and to do that they have to use statistics (see Chapter 15). An interesting thing about statistics, however, is that they've all been designed with a particular level of measurement in mind. Thus, for instance, statisticians have given us three ways of describing the average or central tendency of a variable: the mean, the median, and the mode. The mean is defined as the sum of a set of values divided by the number of values. The mean of the ages 3 years, 4 years, and 5 years is $(3 + 4 + 5)/3 = 4$ years. Thus, the mean has been designed to summarize interval level variables. If age were measured only by categories "under 25 years old," "25 to 64," and "65 and older," you couldn't hope to calculate a mean age for your subjects. The median, however, has been designed to describe ordinal level variables. It's defined as the middle value, when all values are arranged in order. You could calculate a median age if your (five) respondents reported ages of, say, 3, 4, 5, 4, and 4 years. (It would be 4, because the middle value would be 4, when these values were arranged in order—3, 4, 4, 4, 5). You could calculate a median because all interval level variables (as age would be in this case) are ordinal level variables. But you could also calculate a median age even if your categories were "under 25 years old," "25 to 64," and "65 and older."

STOP & THINK *Suppose you had five respondents and three reported their ages to be "under 25 years old," one reported it to be "25 to 64," and one reported it to be "65 and older." What would the median age for this group be?*

A third measure of central tendency or average, the mode, has been designed for nominal-level variables. The mode is simply the category that occurs most frequently. The modal age in the previous "Stop and Think" problem would be "under 25 years old" because more of the five respondents fell into this category (three) than into either of the other categories. The modal age for a group with ages 3, 4, 5, 4, 4, would be 4, for the same reason. You can also calculate the modal sex, if of five participants, three reported they were female and two reported they were male. (It would be "female.") And, you wouldn't have been able to calculate either a median or a mean sex.

The upshot is that the scale of measurement used to collect information can have important implications for what we can do with the data once they've been collected. In general, the higher the level of measurement, the greater the variety of statistical procedures that can be performed. As it turns out, all statistical procedures that scientists use are applicable to interval (and, of course, ratio) data; a smaller subset of such procedures is applicable to ordinal data; a smaller subset still to nominal data.

SUMMARY

Measurement means classifying units of analysis by categories to represent a variable concept. Measuring concepts in quantitative research involves conceptualizing, or clarifying, what we mean by them, and then operationalizing them, or defining the specific ways we will make observations about the concepts in the real world. Measurement in qualitative research often involves making observations, coding those observations into categories, and then looking for relationships among the categories that might constitute usable variables.

The first step of quantitative measurement is to construct a conceptual definition of a concept. The second step is to operationalize the concept by identifying indicators with which to classify units of analysis into two or more categories. The categories should be exhaustive, permitting you to classify every subject in the study, and mutually exclusive, permitting you to classify every subject into one and only one category. Depending on the complexity of the concept involved, you can use simple or complex measures of it and be more or less precise in your measurement. Indexes, like those used by Ferrari in his study of impostors, frequently involve the summation of scores on a variety of items used to measure a concept.

The best measurement strategies are reliable and valid. Reliability refers to whether a measurement strategy yields consistent results. We can examine the reliability of a measurement strategy with a test–retest, internal consistency, or interrater reliability check. The calculation of coefficient alphas, like those used by Ferrari in his study, involve a kind of internal consistency check in which all items in an index are used as checks for consistency against all other items. Validity refers to whether a measurement strategy measures what you think you are measuring. We can examine the validity of a measurement by considering its face validity (by seeing how we and others feel about its validity), by considering its content validity (by seeing how experts evaluate its various elements), by testing its predictive or concurrent validity (by seeing how it's correlated with future or current behaviors of interest), or by testing its construct validity (by seeing how it's correlated with some other variable that theory predicts it will be associated with).

There are three significant levels of measurement (nominal, ordinal, and interval-ratio) in the social sciences. In general, the data analyses that can be used with a given set of variables depend on the level of measurement of the variables.

EXERCISE 6.1

Putting Measurement to Work

This exercise gives you the opportunity to develop a measurement strategy of your own.

1. Identify an area of the social sciences you know something about (aging, marriage and the family, crime and criminal justice, social problems, social class, and so on).
2. Identify a concept (e.g., attitudes toward the elderly, marital satisfaction, delinquency, social class) that receives attention in this area and that interests you.
3. Describe how you might conceptualize this concept.
4. Describe one or more measurement strategies you might use to operationalize this concept. List at least one or more indicators and two or more categories.
5. Identify the level of measurement of this concept given the operationalization you suggested. Give support for your answer.
6. Discuss the reliability of your measurement strategy. How might you check the reliability of this strategy?
7. Discuss the validity of your measurement strategy. How might you check the validity of this strategy?

EXERCISE 6.2

Comparing Indicators of Life Satisfaction: Global Evaluation versus Areas of Life

Some researchers have asked people to evaluate how satisfied they are in various domains of their lives rather than asking for an overall evaluation. The Gallup poll, for example, asks a series of questions about different areas of life as one of the ways it measures life satisfaction. In this exercise, you're asked to compare two alternative sets of indicators for life satisfaction—the one suggested by Diener and his associates and one used by the Gallup poll.

1. Make two copies of the five questions on life satisfaction suggested by Diener et al. at the beginning of the chapter and find two people to answer the questionnaires.
2. After the people complete the questionnaire, interview them and ask the following questions adapted from a Gallup poll. Read the questions to them and write in the answers they give. You could also answer the questions yourself to see how you feel about the issues.

Using the answers "very satisfied," "somewhat satisfied," "somewhat dissatisfied," or "very dissatisfied," how satisfied are you with each of the following aspects of your life?

a. Your community as a place to live in   ———
b. Your current housing   ———
c. Your education   ———
d. Your family life   ———
e. Your financial situation   ———
f. Your personal health   ———
g. Your safety from physical harm or violence   ———
h. The opportunities you have had to succeed in life   ———
i. Your job, or the work you do   ———

3. Americans responded to these questions in a survey conducted on June 11–17, 2001. Review Table 6.1 from Gallup.

TABLE **6.1**
Percentage Answering "Very Satisfied" to Questions

Adults	Men	Women	Non-Hispanic	Whites	Blacks	Hispanics
Your family life	69%	67%	72%	72%	52%	73%
Your current housing	63	63	63	66	41	45
Your community as a place to live in	58	59	57	61	42	53
Your education	45	48	44	48	38	38
Your financial situation	26	28	25	29	15	23
Your personal health	45	57	52	54	54	57
Your safety from physical harm or violence	55	61	50	59	33	47
The opportunities you have had to succeed in life	48	53	43	49	40	44
Your job, or the work you do	50	50	50	51	36	51

Source: Saad, L. "Blacks Less Satisfied Than Hispanics with Their Quality of Life," retrieved from http://www.gallup.com/poll/releases/pr010622.asp

4. Discuss the measurement of "life satisfaction" by answering the following questions:
   a. Which set of questions—Diener's or Gallup's—do you think results in a more valid measure of life satisfaction? Why?
   b. Which set of questions do you think results in a more reliable measure of life satisfaction? Why?
   c. What would be an advantage of using both the Gallup and the Diener measures of satisfaction in a study?
   d. Can you suggest an index that could be constructed based on the nine questions asked by the Gallup organization?
   e. Suggest one or more ways of measuring "life satisfaction" in addition to the two sets of questions presented here. Describe one or more additional questions that could be asked, one or more observations that could be made, or one or more pieces of information that could be obtained from records or documents.

CHAPTER **7**

Cross-Sectional, Longitudinal, and Case Study Designs

INTRODUCTION

You are probably at the beginning of planning your career. How sure are you about your plans? Have you changed your mind in recent years? What factors do you think have affected your career choices? Can you think about being at the end of your career—in 30 or 40 years—and imagine what you'd do after leaving the labor force? If you were to create a hypothesis about work and careers based on your own experiences or those of friends or relatives, what would it be? And, what kind of a study could you do to test it?

Let's say you created a hypothesis about the connection between people's social class backgrounds and their career aspirations. You would need a design to figure out how to get from "here" to "there," where "here" is the research question and "there" is the conclusion or answers to your question (Yin, 2009: 26). Based on the first part of this text, you could now plan most aspects of a study to test the hypothesis, including making some connections between theory and research, being ethical in conducting your study, deciding on a sample, figuring out a methodology that is feasible in terms of time, money, and access, and considering validity and reliability in measuring the concepts you're interested in. We now turn to the remaining research decisions you'd need to make: the selection of a study design, which we'll cover in this chapter and the next, and the choice of one or more methods of data collection, which we'll discuss in Chapters 9 through 13.

Study Design

study design, a research strategy specifying the number of cases to be studied, the number of times data will be collected, the number of samples that will be used, and whether the researcher will try to control or manipulate the independent variable in some way.

A **study design** is a research plan that results from several interconnected decisions. These decisions are made on ethical, practical, and methodological bases. For all research, obviously data must be collected at least once from at least one unit of analysis, but that is the minimum. The study design choices that must be made are

1. Is it appropriate to study only one or a few cases or is a larger sample needed?
2. Is it appropriate to collect only once or more than once?
3. Is it appropriate to collect data about only one sample or more than one sample?
4. Is it appropriate to measure the independent variable or should it be manipulated or "controlled" in some way?

causal hypothesis, a testable expectation about an independent variable's effect on a dependent variable.

For a study with a **causal hypothesis,** where a researcher speculates that one variable, the independent variable, is the *cause* of another variable, the dependent variable, there is an additional decision. As we discussed in Chapter 2, it is difficult to establish causality in part because to show that change in the independent causes the change in the dependent variable, we must know that the independent variable comes first. One way to determine the time order would be to *control or manipulate the independent variable* in some way. In a study you'll read about in Chapter 8, for instance, Chris Caldeira wanted to know if using film clips in class is an effective way to help students comprehend core concepts in sociology. The independent variable (whether or not students saw film clips) could be manipulated so that some students saw film clips and some did not. The study is an experiment—a design in which the manipulation of an independent variable is integral.

Researchers must balance methodological concerns against practical matters and ethical responsibilities when designing a study. They'll consider if the research is practical—that is, "doable" with the available resources. Because it is almost always easier, less time consuming, and less expensive to collect data once, from one sample, and to measure rather than control or manipulate independent variable(s), making decisions that will cost more in time or money must be justified. In some cases, it is just not possible to control the independent variable; in other cases, it might be unethical to do so. Researchers know that controlling or manipulating something is an important responsibility because of the potential effect on people's lives. In addition, studying a sample more than once means that the practical issue of keeping the names and addresses of study participants raises the ethical concern of the risk of a loss of confidentiality.

Connections between the Uses of Research, Theory, and Study Design

Decisions about the design are based partly on the purposes of the research, the researcher's interest in testing causal hypotheses, and the use of theory. You might remember that in earlier chapters we talked about using research for purposes of exploration (breaking new ground), description (describing groups, situations, activities, events, and the like), explanation (explaining why things are the way they are), and evaluation (assessing the impacts of programs, policies, or legal changes). We described theory as an explanation about how and why something is as it is and compared the deductive and inductive approaches to theory—seeing that theory can be used to deduce a testable hypothesis or can be generated after data are collected and analyzed. Our point in recalling the purposes and the deductive and inductive approaches here is that these purposes and approaches have implications for research design decisions.

STUDY DESIGN CHOICES

 **STOP & THINK** *Let's say you were interested in studying what people do with their time after they retire from their career jobs. How might you go about planning this study? Would you decide to talk to people who have already retired or those who are planning to retire? Would you observe people at Senior Centers? Ask for information from local organizations that serve older members of the community? How might you pick your population and sample? Would you need information about many of them or just a small group? Would you want to find about their plans or their actual behavior? Would you need to collect data more than once?*

Cross-Sectional Study Design

Some researchers were interested in the employment statuses and plans of older physicians because of a projected shortage of primary care doctors. Phillip Sloane and his colleagues wanted to know whether the shortage could be partially alleviated by physicians working in part-time or volunteer capacities after retirement. They mailed a four-page questionnaire to a random sample of North Carolina physicians in direct care–related specialties who were 55 or older and received 910 responses (a 59 percent response rate) (Sloane et al., 2008). They asked about current work status, age, health, hours working for pay, retirement plans, current volunteer activities, and interest in volunteerism in retirement. Fifty-seven percent of the respondents were working full-time, 22 percent were working part-time, and 21 percent were currently not working for pay. Forty percent were currently doing some kind of volunteer work within medicine (such as working in a free medical clinic, disaster relief services, or volunteer medical teaching), with those employed having the highest rate (54 percent) and those not working for pay having the lowest (37 percent). Interest in volunteering in retirement was high: 43 percent of those working full-time, 38 percent of those not working for pay, and 33 percent of those working part-time expressed this interest (Sloane et al., 2008: 320). Based on their data, Sloane and his colleagues can *describe* the current behavior and future plans of the sample and some of the relationships among the variables. Of course, they can't know whether the doctors will actually follow through with their plans after retirement.

cross-sectional study, a study design in which data are collected for all the variables of interest using one sample at one time.

The research by Sloane and his colleagues is an example of the **cross-sectional study**, the most frequently used design in social science research. In the cross-sectional design, data are collected about one sample at one point in time, even if that "one time" lasts for hours, days, months, or years. It might have taken the research team weeks or even months to obtain all of the completed questionnaires, but each physician answered the questions only once. Other examples of the cross-sectional design are Desiree Ciambrone's study of women living with human immunodeficiency virus (HIV)/acquired immunodeficiency syndrome (AIDS) (the focal research in Chapter 3) and Joseph Ferrari's study of the impostor phenomenon (the focal research in Chapter 6), where a sample of college students completed a questionnaire that included the "impostor scale" and indicators of other variables only one time.

Cross-sectional designs are widely used because they enable the description of samples or populations and because it is least expensive to collect data once from one sample. Cross-sectional studies often have large samples and usually have data that lend themselves to statistical analyses. In looking for patterns of relationships among variables, the sample can be divided into two or more categories of an independent variable to identify differences and similarities among the groups on some dependent variable. Sloane et al. (2008), for example, can analyze the connection between age and employment status by comparing the percentages of each age group that are employed full-time and part-time. Cross-sectional studies like Sloane et al.'s that have at least one independent and one dependent variable (age and employment status, e.g.) can be drawn as a diagram[1] as shown in Figure 7.1.

---

[1] This diagram and the others used in this chapter and Chapter 8 have been influenced by those presented by Samuel Stouffer (1950) in his classic article "Some Observations on Study Design."

	time 1	
one sample (divided up into categories of the independent variable during analysis)	category 1 of independent variable *ages 55 to 59*	measure of dependent variable *83% are employed full-time 13% are employed part-time*
	category 2 of independent variable *ages 60 to 64*	measure of dependent variable *61% are employed full-time 28% are employed part-time*
	category 3 of independent variable *ages 65 to 69*	measure of dependent variable *35% are employed full-time 28% are employed part-time*

FIGURE **7.1** The Cross-Sectional Study Design

 STOP & THINK *Here are four variables that might "go together" in one or more patterns:*

1. student's gender
2. parental annual income
3. student's perceptions of peers' alcohol use
4. student's alcohol use

Use at least two of these variables and construct one or more hypotheses. Think about the time order and speculate about which of these variables might "go before" others. State your hypothesis and think about how to do a cross-sectional study to test your hypothesis.

Cross-Sectional Studies and Causal Relationships

Cross-sectional designs are sometimes used to examine causal hypotheses. You might remember from our discussion in Chapter 2 that just because two variables "go together" doesn't mean that one variable *causes* change in another. Let's take a simple example of a causal hypothesis and hypothesize that the sight of a bone makes your dog salivate.

Our example would need to meet three conditions before we could say a **causal relationship** exists. In a causal relationship, there must be an *empirical association* between the independent and dependent variables. In the case of the dog and the bone, first, we'd need to see the dog generally salivating in the presence of a bone and generally failing to salivate in its absence. Second, we'd need to make sure that the *temporal precedence* or *time order* was correct—that is, we'd need to make sure that the bone appeared before, rather than after, salivation. (This is to make sure that the bone causes salivation, rather than salivation "causing," in some sense, the bone.)

causal relationship, a nonspurious relationship between an independent and a dependent variable with the independent variable occurring before the dependent variable.

spurious relationship,
a non-causal relationship
between two variables.

Finally, we'd want to make sure that it is not a **spurious relationship**, or caused by the action of some third variable, sometimes called *antecedent variable*, that comes before, and is actually responsible for, making the independent variable and the dependent variable vary together. We'll illustrate spuriousness with a different example. Consider, for instance, that in some regions there is a relationship between two autumnal events: leaves falling from deciduous trees and the shortening of days. One might be tempted, after years of observing leaves falling while days shorten (or, put another way, days shortening while leaves fall), to hypothesize that one of these changes causes the other. But, the relationship between the two events is really spurious, or non-causal. Both are caused by an antecedent factor: the change of seasons from summer to winter. It is perfectly true, that in some regions, leaves fall while days shorten, but it is also true that both of these events are due to the changing of the seasons, and so neither event causes the other. Therefore, the two events are spuriously related.

requirements for supporting causality,
the requirements needed to support a causal relationship include a pattern or relationship between the independent and dependent variables, determination that the independent variable occurs first, and support for the conclusion that the apparent relationship is not caused by the effect of one or more third variables.

The reason we mention the criteria for determining causation in the context of research designs is simple: Some research designs are better than others for examining the **requirements for supporting causality**. Thus, for instance, although it is possible to use a cross-sectional design to test causal hypotheses, this design might be less than ideal for such tests in some circumstances. Collecting data one time for two or more variables, we can determine the relationships between them and sometimes figure out the time order as well. Vivian Tseng (2004) did a cross-sectional analysis of family interdependence and academic adjustment. Comparing young adults of different ethnicities and foreign-born to non-immigrants, Tseng found that Asian youth and those who were foreign-born put greater emphasis on family obligations than those of European backgrounds and those born in the United States. While the effects of third variables still need to be considered, the time order is clear: ethnicity and place of birth come before attitudes toward family obligations. However, it is often difficult to disentangle the time order. It is always possible to ask questions about past behavior, events, or attitudes, but such retrospective data might involve distortions and inaccuracies of memory.

 **STOP & THINK** *In the focal research in Chapter 6, Ferrari reports an association or connection between imposter tendencies and academic dishonesty. Can you think of a problem with concluding that being an imposter **causes** a student to behave with less academic honesty?*

While Ferrari finds evidence of a relationship between the two variables, he doesn't know whether imposter tendencies or academic dishonesty comes first. It seems to make as much sense to argue that a person is more likely to be an imposter as an effect of academic dishonesty as arguing that being an imposter leads to academic dishonesty. Since Ferrari measured both variables at the same time, he can't distinguish the cause and the effect. Nor can he determine if a third, antecedent, variable came first—perhaps the amount of parental pressure to succeed—and contributed to both of the variables. Time order can be established by using an alternative study design, one that enables the collection of data more than once. That way the researchers can determine whether one variable occurs first, thereby establishing the temporal sequencing of the variables.

So, be cautious in interpreting the results of studies that use a cross-sectional design for explanatory purposes. The cross-sectional study design is very appropriate for describing a sample on one or more variables and for seeing connections between the variables. It can also be useful for studying causal hypotheses when the time order between the variables is easy to determine and when sophisticated statistical analyses can be used to control for possible antecedent variables.

Longitudinal Designs: Panel, Trend, and Cohort Designs

Emily and Roger studied the attitudes and experiences of workers, including feelings about current jobs, plans for retirement, and actual experiences in retirement but, unlike Sloane and his colleagues, did not do a cross-sectional study. In the focal research that follows you'll learn about the study we designed and some of our findings.

FOCAL RESEARCH

Moving On? Continuity and Change after Retirement

By Emily Stier Adler and Roger Clark*

Retirement is an important life transition. In addition to the issues of income and identity, retirement means changes in daily routines and typically provides the retiree with more discretionary time. Before a planned and voluntary retirement, workers think about post-retirement activities. Often their plans include a list of things they won't be doing—such as not setting an alarm clock, not eating lunch at the desk, and not having a schedule for every hour during the week—as well as a list of added activities—such as volunteering, getting exercise, doing some work for pay, spending time with grandchildren, reading novels, and the like. With the first of the 76 million "Baby Boomers"—people born between 1946 and 1964—starting to retire (U.S. Census Bureau, 2006), it's especially important to know how people will spend their time after retirement.

Labor Force Participation, Retirement, and Volunteerism

The National Health and Retirement Study (HRS), a biannual panel study of a nationally representative sample of more than 20,000 Americans over the age of 50, finds that the majority of men and women in their fifties work for pay (70% and 60%, respectively), most on a full-time basis, but

---

*This article was written for this text and is published with permission.

that labor force activity declines after age 60 (HRS, 2007, p. 41). By age 62, approximately half of men and more than 40% of women are working for pay, at least two thirds of them full-time; by age 65, half of those employed are working part-time and the labor force participation of both women and men is close to half of the participation rate of people in their fifties (HRS, 2007, p. 41). Analysis of HRS data also finds increasing proportions of those in their early to mid-fifties who *expect* to work full-time after age 65. For example, among college educated men, the proportion increased from just over 30 percent in 1992 to almost 50 percent in 2004, while for college educated women it went from under 30 percent to almost 40 percent (HRS, 2007, p. 49). A Pew Research Center (2006) survey finds that 77% of workers expect to do some kind of work for pay after retirement. The same survey, however, found that only 12 of those retired were currently working for pay either full-time or part-time (Pew Research Center, 2006, p. 2).

Retirement may be a gradual process. One analysis of the HRS data finds that at least half of those moving out of full-time career jobs don't leave the labor force, but at some point take short-duration, part-time jobs or transitional jobs (often called bridge jobs) after retirement (Cahill, Giandrea, and Quinn, 2006). In their analysis of the HRS data, Gustman and Steinmeier (2000, p. 63) found that from 1992 to 1998, 71% of the sample reported no change in status, 17% moved from greater to less labor force participation, and 6% moved from less to more participation.

There are other ways to be productive beyond paid employment, including volunteering for community organizations. Such activities may replace paid work and can provide the individual with an identity, a social role, friendship networks, a sense of being productive, and daily or weekly structure for activities. In 2008, 26% of the American population and 28% of those 55 to 64 years of age did volunteer work, spending about an hour a week on these activities (U.S. Department of Labor, 2009). A survey found that 62% of Baby Boomers who currently volunteer expect to devote more time to service in retirement while 36% who are currently not actively engaged in volunteering expect to do so (AARP, 2004). While it might be expected that those employed in the labor force would have less time to volunteer, those who are employed part-time have the highest rate of volunteering (34%), followed by those employed full-time (28%) and those who are unemployed or not in the labor force have the lowest rate of volunteering (22%) (U.S. Department of Labor, 2009).

As workers retire in the next few decades, how will they spend their time? Role theory and continuity theory can help us predict role constellations in the immediate post-retirement period. *Role theory* (e.g., Ashforth, 2001; Wang, 2007) suggests that to the extent people are highly invested in a particular role, their feelings of self-worth are related to the ability to carry out that role. Role theorists also argue that the loss of roles can lead to depression or anxiety. We posit, then, that after retirement, to the extent possible, people will retain previous roles because doing so enables them to maintain feelings of self-worth and avoid feelings of anxiety and depression. Moreover, we posit that pre-retirement expressed *desires* to perform new roles (like new volunteer roles) after retirement are less likely to be based upon the same emotional calculus of self-worth, depression, and anxiety that informs the desire to continue performing a pre-retirement role. In fact, we expect that all workers have at least some cognitive access to the emotional and psychological needs that work roles have satisfied for them and that their plans to do some kind of work after retirement will therefore be related to whether they actually work for pay after retirement. More specifically, our research is guided by three hypotheses, generated from role theory:

> **Hypothesis 1:** People who express a desire for work after retirement are more likely to actually engage in post-retirement work than others.

> **Hypothesis 2:** People who have performed volunteer roles before retirement will be much more likely to perform such roles after retirement than people who have not performed such roles.

> **Hypothesis 3:** People who before retirement express a desire to perform roles (such as volunteer roles) that they have not previously performed are not more likely to actually perform those roles after retirement than people who do not express this preretirement desire.

Role theory alone cannot explain which people will be attracted to working for pay after retirement because, by definition, everyone who retires is retiring from a work role. *Continuity theory*, on the other hand, suggests reasons why there may be a *selective investment* in roles (including occupational roles), such as differential feedback on the performance of roles (Atchley, 1994). Within a given occupational or professional level one may expect that those who have worked for more years to derive greater satisfaction from the work role. The longer people participate in a role, we posit, the longer they will have had to organize their ideas and feelings about themselves in relationship to those roles and the more likely they will be to seek the structure of organized roles after retirement. More specifically, our research is guided by one additional hypothesis derived from continuity theory:

> **Hypothesis 4:** The longer a person has been in the workforce, the more likely she/he is to pursue a workplace role after retirement.

Methodology

In order to study the *transition* to retirement, it is necessary to use a longitudinal design. We used a panel study with two data collections, one before and one after retirement. Because there are no lists of workers who plan to retire from a variety of occupations, we selected a convenience sample of middle- and upper-middle-class workers who had definite plans to retire. A call for participants, describing the project, was distributed. We requested participants from those who were college educated or working as professionals, currently working at least 25 hours a week and had specific plans to retire within the next 18 months. Using e-mail, letters, Internet discussion group notices, announcements in a college bulletin, and personal contacts, we asked that the call for participants be passed along to others that might fit the sample and estimate that approximately 1,000 people received information about the study.

The final sample included 44 North American college-educated and/or professional workers, currently working for pay 25 hours or more per week with plans to retire from full-time career employment. At time 1, the participants lived in 15 different states and worked in a variety of occupations: six college professors, four managers or administrators for large organizations, five public school teachers, four administrative assistants, three engineers, three computer information technology managers, three nurses working in hospital administration, three state or city department heads, two physicians, two school administrators, a psychologist, a dentist, an accountant, a podiatrist, a clergyman, a small business owner, an elementary school staff member, a social researcher, and a social worker. All of them identified their retirements as voluntary. They'd worked for pay between 16 and 51 years with a mean of 35.5 years in the labor force. Twenty-seven (61%) were female and 17 (39%) were male; all were white; 77% were married; and the rest were never married, divorced, widowed, or cohabitating with a same or opposite sex partner. Their anticipated ages at retirement ranged between 52 and 75, with an average age of 61 at planned retirement. The actual mean age at retirement of the 44 respondents was 61.4 years.

All of the study participants signed informed consent forms and were interviewed by the same member of the research team twice. The pre-retirement interviews lasted between one and three hours and were conducted between November 2001 and July 2002 using a semi-structured format. The interviews were tape-recorded and transcribed. As the period immediately after retirement is often seen as a vacation or "honeymoon" and is frequently a transitional period before settling into new patterns, the second interview was conducted between 18 and 20 months after retirement. The participants were contacted within weeks of their expected retirement to see if they had in fact retired as planned. If not, they were re-contacted every few months until retirement. There was no panel attrition; all 44 post-retirement interviews were completed by 2007.

Findings
Post-Retirement Paid Work
Three quarters of the study participants did retire within a few weeks of their planned dates. The 25% who did not retire as anticipated postponed retirement from two months to four years with a median of

one year. The major reason for postponing retirement was financial—either incentives to continue to work or worries about financial readiness for retirement—but a few retirees waited because of worries about how they would spend their time once retired.

Before retirement, participants were asked about their plans to work for pay after retirement, and if they planned to work, their expectations about the kind of work and the hours/days per week they'd thought they'd work. Ten (23%) of the sample did not expect to work after retirement, 14 (32%) were unsure of their plans to work, and 20 (46%) thought they probably would work after retiring. None expected to work full-time after retirement. In reality, at the time of the second interview, 15 (34%) were doing no paid work, 11 (25%) were working eight hours or fewer per week for pay, 13 (30%) were working more than 8 hours but less than 24 hours a week for pay, and 5 (11%) were working at least 24 hours a week for pay.

Our first hypothesis received support, as expectations of working after retirement were positively associated with both whether respondents were actually working after retirement and the number of hours they were employed. The Spearman's rho for the association between expectation of work and work after retirement is .46, while that between expectation of work and hours of work after retirement is .47 (Table 7.1). In fact, 90% of those planning to work after retirement were actually employed in the labor force at the time of the second interview, while 40% of those with no plans to work and 50% who had been unsure were employed.

When we looked at the number of years in the labor force, we found support for hypothesis 4 as years of paid work was correlated with pursuing a role in the workplace after retirement (rho is .37) and with number of hours employed (rho is .39) (see Table 7.1).

TABLE **7.1**

Associations of Work after Retirement and Hours of Work after Retirement with Expectation of Work, Age at Retirement, Years in the Labor Force, and Gender using Spearman's Rho (N = 44)

	Work after Retirement (1)	Hours of Work after Retirement
Expectation of Work (2)	.46**	.47***
Age at Retirement	−.08	.04
Years in Labor Force	.37*	.39**
Gender (3)	−.18	−.14

*Indicates significant at .01 level; **indicates significant at .001 level; ***indicates significant at .05 level.
(1) Work after retirement coded Yes = 1, No = 0; (2) Expectation of work coded No = 1, Maybe = 2, Yes = 3; (3) Gender coded female = 1, Male = 0.

The correlation between years in the labor force and expectations of work at retirement was positive but weak (rho is .13). Tests not shown here suggest that both years in the labor force and expectations of work at retirement have independent effects on both whether people will work after retirement and how much they will work.

Table 7.1 also shows that a person's age at retirement has almost no association with his or her chances of working after retirement (rho is −.08) and that gender is not much associated with either of these two variables, even though males are slightly more likely than females to have been engaged in work for pay after they retired (rho is −.18).

Some people seem to be able to predict their post-retirement needs for work very well before retirement. Joe, a high school math teacher planning to retire after 36 years, believed that he would have too much time on his hands if he did not work. He thought he would teach math and coach in some capacity after retirement because of his need for some structure and the advantages of extra income. And Joe was, in fact, teaching part-time at a local college and coaching track and field at a nearby high school two years after retirement. Joy, a high school biology and chemistry teacher for 30 years, on the other hand, had no plans to work after her retirement and, in fact, did not work afterward. Although she had been offered several jobs, including part-time teaching, Joy turned them down. She had suffered and recovered from a major health crisis, felt financially secure, and enjoyed being away from the stresses associated with work.

But some people do not foresee the extent to which work has become central to their lives or the extent to which they'll enjoy their lives without paid employment. Michael, an engineer for 42 years before he retired, did not expect to work afterward. In the years before retirement, he disliked the substantial amounts of travel his job required. He was able to separate his pleasure in his work from his dislike of the travel, asserting that "if you took away the travel, I would say I love the job," but he could not foresee wanting to work again. He spent the month after retirement hunting, hiking in the woods, and painting and repairing his house. "And then it became boredom time," Michael said. He found he did not like the lack of structure and the lack of contact with other people in retirement. By the time of the second interview, he had returned to his old job, minus the travel, on a contract basis, full-time for about half the year.

On the other hand, some people found that, despite their pre-retirement intention to do some kind of remunerative work after retirement, they changed their minds. Before retirement, Anne, a department head in a state agency, had planned on post-retirement work as a consultant one or two days a week. But after a very few short-term consulting projects, Anne was a bit surprised to find she found excuses to turn down jobs until she realized, as she noted, "Wait a minute. I've done all this. There are other things I want to do with my life."

We wondered whether our data could help us determine when a career became long enough to have the relationship outlined above. We found that 30 years of labor force participation seemed to maximize the differences. Thus while 9 (64%) of 14 respondents with 30 or fewer years on the job were not engaged in work for pay two years after retiring, only 6 (20%) of the 30 respondents with 31 or more years of work were not so engaged. Moreover, all five of our respondents who were working at least 24 hours per week two years after retirement had worked 31 or more years before retiring.

Volunteer Work

Based on role theory, we hypothesized that those who had taken a volunteer role in the community before retirement were more likely to be volunteering after retirement. Before retirement, 19 (43%) reported no formal volunteer work, 21 (48%) reported up to four hours of formal volunteer work each week, and 4 (9%) reported five or more hours of volunteer work a week. After retirement, 43% still reported no formal volunteer work while 39% percent reported up to four hours of formal volunteer work and 18% reported five or more hours a week of formal volunteer work. Pre-retirement formal volunteer work is very strongly associated with post-retirement formal volunteer work (rho = .63).

Before retirement many participants told us that they wanted to spend time volunteering but we find no association between the expressed hope to do more volunteer work and the reported level of post-retirement volunteer work (rho = .02). But what about the hope to do more volunteer work after retirement with the level of pre-retirement volunteer work controlled? (After all, one might reasonably claim that the hope for doing more volunteer work after retirement should only be associated with *change* in levels of volunteer work, not with absolute levels of volunteer work after retirement.) Additional tests, not shown here, suggest that the pre-retirement desire to do more volunteer work after retirement still has no association with reported levels of post-retirement volunteer work, even after pre-retirement levels are controlled. Moreover, we find the number of hours participants work for pay after retirement, age at retirement, or years in the labor force are not significantly associated with post-retirement volunteer work.

Discussion

People on the cusp of retirement have expectations or hopes about their lives after retirement. They may plan to work for pay part-time or hope to become more engaged in their communities. In this paper, we have considered hypotheses generated from role and continuity theory to test whether people's expectations about these activities or their past behaviors are better predictors of post-retirement activity and found support for a continuation of activities. Among our participants, retirement seems a time of continuity and only moderate change. Before retirement, for instance, all of our participants expressed their expectations about how much they would work in the early years after retirement. They realized these expectations to a significant, but imperfect degree. However, an equally powerful predictor of their work after retirement was the number of years they had spent in the labor force. The longer people had channeled their energies into work, the more likely they were to be employed and employed for more hours two years after retirement, no matter what their plans were before retirement. For volunteer work, the major predictor of post-retirement behavior was not expectation but previous degree of engagement in volunteer work.

Interestingly, post-retirement working for pay and involvement in volunteer work was not substantially related in our sample. It is not true that people who are engaged in substantial amounts of post-retirement paid work are less likely than others to be involved in their communities.

Ours is not a representative sample and our causal conclusions are speculative. Our data are supportive of a causal connection between the length of one's career and the likelihood of working after retirement. We cannot be sure that the continuities we find in pre- and post-retirement activities are not spurious—perhaps the result of some personality need for external structure and community involvement pre-dating both work years and retirement. This said, among our participants, retirement does not seem to be a time in which people reinvent themselves. When it comes to both paid and volunteer work, the habits of a lifetime seem to be good predictors of post-retirement activity.

There are several implications of our study for members of a Baby Boomer generation about to retire. If people hope to engage in activities like volunteer work after retirement, they should begin these activities in the years prior to retirement and increase their time commitments once their paid work responsibilities have diminished. If they haven't had the chance to join organizations or volunteer prior to retirement, they may need to make special efforts to begin. On the other hand, if they have put in a full lifetime of work, they may be surprised how hard it is to withdraw from the workforce cold turkey or, conversely, how necessary it may feel to continue working for pay, at least to some degree. Given the enormous size of this generation, as well as its work experience, one macro-level implication is that it may be a supply of productive energy for years to come.

References

AARP. 2004. Baby Boomers Envision Retirement II: Survey of Baby Boomers' expectations for retirement. Retrieved February 15, 2009, from http://assets.aarp.org/rgcenter/econ/boomers_envision.pdf

Ashforth, B. 2001. Role transitions in organizational life. Mahwah, NJ: Erlbaum.

Atchley, R. C. 1994. *Social forces and aging*. Belmont, CA: Wadsworth Publishing Company.

Cahill, Kevin E., Michael D. Giandrea, and Joseph F. Quinn. 2006. Retirement patterns from career employment. *The Gerontologist* 46: 514–523.

Gustman, A. L., and T. L. Steinmeier. 2000. Retirement outcomes in the health and retirement study. *Social Security Bulletin* 63(4): 57–71.

HRS. 2007. *Growing older in America: The health and retirement study*. Retrieved February 15, 2009, from http://hrsonline.isr.umich.edu/docs/databook/HRS_Text_WEB_Ch2.pdf

Pew Research Center. 2006. Working after Retirement: The Gap between Expectations and Reality. Retrieved February 15, 2009, from http://pewresearch.org/pubs/320/working-after-retirement-the-gap-between-expectations-and-reality

U.S. Census Bureau. 2006. Facts for Features Special Edition Oldest Baby Boomers Turn 60! Retrieved February 15, 2009, from http://www.census.gov/Press-Release/www/2006/cb06ffse01-2.pdf

U.S. Department of Labor. 2009. Volunteering in the United States, 2008. Retrieved February 15, 2009, from http://www.bls.gov/news.release/pdf/ volun.pdf

Wang, M. 2007. Profiling retirees in the retirement transition and adjustment process: Examining the longitudinal change patterns of retirees' psychological well-being. *Journal of Applied Psychology* 92(2): 455–474.

 **THINKING ABOUT ETHICS**   *We submitted our research to our university's Institutional Review Board (IRB) for approval. The participants volunteered to be part of the study by responding to the request for participation; they also voluntarily agreed to the second interview and each time gave informed consent before being interviewed. Furthermore, we believe that no harm came to any of the participants as a result of being part of the study, we have kept all names and information confidential, and, in all reports, we have identified none of the participants by their real names.*

 **STOP & THINK**   *In our study we collected data twice from the same sample and therefore went beyond the cross-sectional design. Why do you think we didn't use a cross-sectional study design?*

internal validity, agreement between a study's conclusions about causal connections and what is actually true.

We could have talked to a group of retired people and, using a cross-sectional design, asked these participants questions about what they used to feel about their jobs, what their activities outside of work had been, what they thought they might do in retirement, and what they were actually doing in retirement. However, using a cross-sectional design with retrospective questions would have been problematic for the **internal validity** or accuracy of the study's conclusions. Validity concerns would result, in part, because of reliance

longitudinal research, a research design in which data are collected at least two different times, such as a panel, trends, or cohort study.

on people's answers about what they felt or did years before. Memories fade, and current realities affect recollections and interpretations of past experiences and attitudes. We instead chose an over-time or a **longitudinal** approach by collecting data twice—once before and once after retirement from career jobs.

 STOP & THINK

*We found no relationship between people's expectation to do volunteer work after retirement and the degree to which they actually did volunteer work after retirement. Do you see how people who, say, weren't doing volunteer work after retirement might have (falsely) remembered that they **didn't** expect to do such work before retirement if we'd waited until then to ask them about those expectations?*

The Panel Study

panel study, a study design in which data are collected about one sample at least two times where the independent variable is not controlled by the researcher.

The design that we, Emily and Roger, selected is a **panel study,** a design that follows one sample over time with two or more data collections. (See Figure 7.2 for a representation of the minimum panel design using the Adler and Clark study as an example.) The study by Michele Hoffnung, described in Chapter 4, is also a panel study—one in which 200 women who graduated from college in 1992 were followed until 2009 with many, repeated data collections.

A cross-sectional study would not have worked for us as we needed more than a snapshot of people's lives at one time. The longitudinal or over-time nature of a panel study allowed us to document patterns of change and to establish time order sequences. We wanted to find out about people's expectations of retirement and learn what they did in their time away from work while they were still employed. We also needed post-retirement information about what they actually were doing and their views of retirement. Remembering accurately what one did or felt years before is problematic. Sometimes people make the past "fit" the present which leads to misremembering actual behavior and previous attitudes.

	time 1 *2001–2002*	time 2 *2003–2006*
one sample of professional or college educated workers	All variables are measured *Expectations of retirement; Current activities, including number of hours working for pay and doing volunteer work*	All variables are measured *Views of retirement; Current activities, including number of hours working for pay and doing volunteer work*

FIGURE **7.2** The Minimum Panel Study Design

BOX **7.1**

A Question of Timing

Studying children's development calls for longitudinal research to see changes over time. One important study has answered many questions about development while raising methodological questions about the *timing* of data collection. The National Institute of Child Health and Human Development (NICHD) Study of Early Child Care (SECC) is a comprehensive longitudinal study on the connection between child care experiences and children's developmental outcomes. In 1991, researchers enrolled 1,364 infants in a study to follow them through adolescence. The results about the effect of child care have changed over time (Barry, 2002). One of the first findings presented was that children's attachments to their mothers were not connected to kind of care: At 15 months, children in day care were as strongly attached to their mothers as children reared at home. But at age two, day care seemed to have negative effects: Those spending many hours in day care showed less

social competence and more behavior problems. At age three, the news was different: Those in day care at least 10 hours a week had better cognitive and language skills. But a year later, the researchers announced that children who spent more than 30 hours a week in day care displayed higher incidence of disobedience, aggression, and conflict with caregivers. In 2008, the researchers reported that *size* of the groups counted: Children who had spent more time in small-sized peer groups in child care when younger were seen as more sociable and co-operative in third grade but were rated by teachers as more aggressive; children who had spent more hours in medium-sized groups also received higher ratings for peer aggression by their teachers (NICHD ECCRN, 2008). The study's findings are important to parents, educators, and policy makers. Methodologists will note that **timing matters** when doing repeated data collections. To read more, see http://secc.rti.org/summary.cfm

One key issue is selecting the time frame for the second and later data collections. In the retirement study, we were interested in the initial transition so our second data collection was approximately 18 months after retirement. Results might be different if we'd waited five or more years after retirement for the follow-up. As the research described in Box 7.1 demonstrates, selecting the time periods affects the findings.

 **STOP & THINK** *Although contacting each person in the sample two or more times has advantages, it also has disadvantages. What do you think are some of the difficulties facing a researcher who collects data more than once from the same sample?*

The ability to see change over time is very important, but there are offsetting costs and difficulties in doing a panel study. The most obvious concerns are time and money. Assuming identical samples and no inflation, a study with three data collections takes much longer and costs three times as much as a one-time study. With the same resources and three data collections, a researcher could select a sample only one third as large as for a study using a cross-sectional design. Therefore, it's important to consider the need for a longitudinal approach to establish time order or improve data validity before incurring extra expenses and time or decreasing sample size.

Funding is always a concern for research, and in a longitudinal study, this is especially true. The expenses connected to research—in this case, travel and the costs of transcription—were two of the reasons for our small sample size. We felt fortunate that some of the research costs were covered by a small faculty research grant from our institution.

Keeping track of study participants and encouraging them to continue participation are very important for panel studies. Data must be kept confidential as anonymity cannot be promised to participants because identifying information, including names and addresses, is necessary to re-contact them. In our highly mobile society, keeping track of respondents is difficult, especially in long-term studies, even when the participants are highly motivated. **Panel attrition,** the loss of subjects from the study because of disinterest, death, illness, or moving without leaving a forwarding address, can become a significant issue. In one study of the career and family goals of academically talented individuals, for example, the researchers began with a group of 1,724 high school seniors who completed an initial survey in 1989 (Perrone, Civiletto, Webb, and Fitch, 2004). The participants were mailed annual questionnaires with self-addressed, stamped response envelopes, but many were returned unopened and marked "cannot forward." For the thirteenth survey, Perrone and his colleagues (2004) report that they received only 113 responses—a response rate of about 6 percent! In contrast, Michele Hoffnung has done a much better job in keeping in touch with the participants in her long-term panel study. In the focal research in Chapter 4, she notes that the response rate for the fourteenth data collection was 76 percent.

Some have speculated that material incentives can decrease panel attrition. One project found that small gifts and being entered into a cash lottery does increase initial participation in a panel study, but that the effect of such incentives faded by later data collections (Goritz, 2008).

In our research on retirement, we had no panel attrition. Our sample was motivated to participate; we had phone numbers, e-mail addresses, and/or home addresses for all of the participants and information about a contact person, such as a son or daughter; and typically less than two years passed between contacts with them.

Another concern for panel studies is **panel conditioning**, the effect of repeated measurement of variables. Participants tend to become more conscious of their attitudes, emotions, and behavior with repeated data collections. This awareness can be at least partially responsible for differences in reports of attitudes, emotions, or behavior over time (Menard, 1991: 39). At the second interview, some of our participants mentioned that talking about their feelings about work and their expectation about retirement during the first interview was useful to them and helped them sort out some of their feelings and think about their post-retirement plans. This is an example of panel conditioning.

Finally, the issue of changes due to time period needs to be considered as "…while each generation's path shares many milestones and processes in common, each generation's journey is unique" (Smith, 2005: 177). The events of September 11, 2001, occurred right before we began our study; the economic crisis of 2008 and beyond occurred afterward. Those events and many others can affect how people see retirement. Historical events of each time period are intertwined with the experiences of individual transition, and unless

panel attrition, the loss of subjects from a study because of disinterest, death, illness, or inability to locate them.

panel conditioning, the effect of repeatedly measuring variables on members of a panel study.

multiple samples of individuals who are experiencing the same transition in different decades are followed over time, a panel study like ours cannot sort out the effects of historical era from other variables.

 **STOP & THINK** *Can you think of any situations where you'd want to study changes over time and therefore would want to do a longitudinal study, but would prefer to select a new sample each time rather than contact the original sample multiple times?*

The Trend Study

Selecting new samples for longitudinal studies can be very useful when researchers are interested in identifying changes over time in large populations, such as registered voters in one country or college students around the world. It can be very time consuming and expensive to re-locate the same individuals to track changes in attitudes, opinions, and behaviors of a very large sample over a number of months or years. It is much more practical to select a new probability sample each time. In addition, selecting new samples allows for anonymous data collection, which can be a more valid way to find out about embarrassing, illegal, or deviant behavior.

trend study, a study design in which data are collected at least two times with a new sample selected from a population each time.

The longitudinal design that calls for the selection of a new probability sample from a population for at least two data collections is called a **trend study.**

 **STOP & THINK** *As an example, let's think about investigating changes in beliefs about gender equality or gay rights. Suppose we did a trend study by studying the attitudes of a population two or more times, selecting a new sample each time. What would be the advantages and disadvantages of this strategy?*

Trend studies avoid panel attrition and panel conditioning, save the expense of finding the original participants, and enable the researcher to collect data anonymously. Furthermore, trend studies are useful for describing changes in a population over time. For these reasons, studies of changes in public opinion and election behavior often use the trend study design. Such studies can tell us when social phenomena emerge in a society. The American National Election Studies (ANES) provide interesting data in this regard. Each data collection in this series is based on 1,000 to 2,000 interviews with a new, multistage probability sample of citizens of voting age that live in private households (Miller and Traugott, 1989). Jeff Manza and Clem Brooks (1998) analyzed the ANES results from 1952 to 1992 to see if the "gender gap" in voting began with the 1980 presidential election. They found the difference between men's and women's voting patterns had emerged much earlier, with women disproportionately supporting the Democratic candidates since 1952. They conclude that women's rising labor force participation rates over time best explains the gender gap (Manza and Brooks, 1998: 1261).

Another widely used source of information is the General Social Survey (GSS) that is conducted periodically in a wide variety of countries, including Japan, German, Great Britain, and Australia. In the American GSSs, probability samples of more than 1,000 adults living in households in the United States are interviewed with some topics rotating in and out and others continuing (Smith, Kim, Koch, and Park, 2005). Some uses of the GSS as trend data include studies of the acceptance of gays and lesbians (Keleher and Smith, 2008), support for funding for public schools (Plutzer and Berkman, 2005), and attitudes toward childless couples (Koropeckyj-Cox and Pendell, 2007).

Trend studies can also include a cross-sectional aspect, when, for example, comparisons are made between groups within each sample. Josephine Olson and her colleagues (2007) were interested in the trends in beliefs in equality for women and men among university students in the United States and in Central and Eastern Europe. Using 22 items of the Attitudes Toward Women Scale, (with five answer categories of strongly agree to strongly disagree) such as "Husbands and wives should be equal partners in planning the family budget" and "Vocational and professional schools should admit the best qualified students," Olson et al. (2007: 301) compared data from new samples selected between 1991 and 2004 to see whether there were any changes over time. In Figure 7.3, a diagram of the trend study design using the American data presented in Olson et al. (2007), we can see that women are more supportive of gender equality than men and that this difference has stayed constant over time.

An advantage of the trend study is its use in studying the impact of time periods. Using a trend design, Celia Lo (2000) compared data collected annually from high school students from 1977 to 1997. Using a multistage sampling

	time 1 *1991*	time 2 *1997*	time 3 *2004*
first sample *US College* *students*	Measure all variables *Men's mean gender role* *equity score: 3.9* *Women's mean gender role* *equity score: 4.3*		
Second sample *US College* *students*		Measure all variables *Men's mean gender role* *equity score: 3.8* *Women's mean gender* *role equity score: 4.3*	
Third sample *US College* *students*			Measure all variables *Men's mean gender role* *equity score: 3.9* *Women's mean gender* *role equity score: 4.3*

FIGURE **7.3** The Trend Study Design

procedure, each year approximately 130 schools were selected, and all students who attended classes on the day of data collection were asked to complete a questionnaire that included questions on drinking and drug use. A one-time, cross-sectional study would be appropriate for an analysis of the connection between the onset age of drinking and other drug-using behaviors, but by using a trend study, Lo (2000) was able to determine if the variables and their relationships had been consistent and stable over two decades. Using calendar year as a unit of analysis, she found that delaying the onset of drinking reduced illicit drug-using behaviors and that, starting in 1993, an increase in the onset drinking age was associated with drops in alcohol-use measures.

Trend studies have limitations. Because a trend study does not compare a specific sample's responses at two or more times, the trend study can only identify *aggregated* changes, not changes in individuals. People move in and out of populations, such as the population of college students, so identified changes in attitudes, opinions, or characteristics over time could reflect changes in the composition of the population or in the sample rather than changes in the individuals. If a trend study finds a change over time, such as in beliefs in equality for the genders, we don't know if there would have been no change, the same change, or a more extreme change if the *same* people had been used at two or more times (i.e., if a panel study had been done). In addition, it is possible for "no change" to result for the aggregate even if individual members of the population change in opposite directions between data collections. True to its name, the trend study design is best for describing trends in a population; it cannot identify how individuals have changed, nor can it pinpoint the causes of change.

The Cohort Study

cohort study, a study that follows a cohort over time.

A **cohort study** is a longitudinal study that follows a cohort over time. In social science research, a **cohort** is a group of people who have experienced the same significant life event at the same time. The most frequently selected cohorts are birth cohorts, that is, people born in the same year. Cohorts can also be groups that experience a life event at the same time, such as graduating from college, retiring from the labor force, or getting divorced. Although people can "exit" from a cohort (e.g., by dying), no one can join after its formation.

cohort, a group of people born within a given time frame or experiencing a life event, such as marriage or graduation from high school, in the same time period.

In a cohort study, the same population is followed at least two times using either a panel or a trend study design. In the panel study approach, a sample that is a cohort is followed with at least two data collections. Since differences in ages, life course events, and time periods can all influence behavior and attitudes, cohort studies can help to sort out these factors.

A cohort study can examine a birth cohort or an entire generation, such as the "Baby Boomers" or "Generation X." One extraordinary study, the National Longitudinal Survey of Mature Women Cohort, a long-term project of the Department of Labor, started with a nationally representative sample of more than 5,000 women in 1967 when the sample was between 30 and 44 years old and followed this sample until 2003 (Bureau of Labor Statistics [BLS], 2001). The response rate remained high: more than three decades after the initial interview, 46 percent of the initial sample (or 55 percent of those who were still alive) responded to the annual survey in 1999 (BLS, 2001: 20).

The periodic data collection originally focused on the cohort's labor force experiences and marital and fertility histories and then later on retirement, pensions, and economic status. Elizabeth Hill (1995), for example, analyzed the women's experiences between 1967 and 1984. She wanted to know whether there were any advantages in receiving at least one form of training—formal schooling, on-the-job training, or some other kind of training—after the usual schooling age; she found that women who had obtained training any time between 1967 and 1984 had received larger wage increases during those years and were more likely to be working in 1984 than women who had not received it. The data have also been analyzed to determine the ages at which significant physical functional limitations arose for the members of the cohort (Long and Pavalko, 2004), the long-term effects of mid-life work history on economic well-being (McNamara, 2007), and the changes in women's retirement expectations over time and the connections between the expectations and race, marital status, and health (Wong and Hardy, 2009).

The focal research study on the transition to retirement is a cohort study in addition to being a panel study because it followed a cohort of people who retired within a few years of each other. Michele Hoffnung's study of 200 women who graduated from college in 1992 is also both a panel and a cohort study.

A trend study approach can also be used in cohort studies. In a regular trend study, if we study the same population over time (e.g., all married men in Rhode Island, all university students in France, all registered voters in Canada), the population itself changes over time as people move in and out of the population. But in a cohort study using the trend design, the same population is followed over time. If, for example, we were studying the birth cohort of Americans born in 1995 over time, we could ask probability samples of this cohort questions about their attitudes toward cohabitation, marriage, and divorce every other year. We would then have the ability to describe changes in attitudes in this cohort as they aged. We might find the cohort becoming more liberal, more conservative or remaining constant in their attitudes as they moved through their teens and into their twenties.

Norval Glenn's (1998) cohort study used a trend study design to investigate a hypothesis derived from cross-sectional research that marital satisfaction is "U shaped," that is higher in the beginning and later stages of marriage and lower in the middle with data from the American GSS from 1973 to 1994. Using the almost annual survey data from these different nationally representative samples of Americans 18 and older, Glenn computed the year of first marriage and the age at first marriage to compare marital satisfaction over time among five marriage cohorts. He concluded that, *in the aggregate*, among all cohorts, the average levels of marital satisfaction decline steeply in the first decade of marriage and continued to decline moderately over the next two decades (Glenn, 1998: 574).

The Case Study

case study, a research strategy that focuses on one case (an individual, a group, an organization, and so on) within its social context at one point in time, even if that one time spans months or years.

The other non-experimental study design is the **case study**, a design with a long and respected history. In a case study, the researcher purposively selects one or a few individuals, groups, organizations, communities, events, or the like and analyzes the selected case(s) within their social context(s). Each case might be studied over a brief period of time or for months or years. The distinguishing feature of this study design is its *holistic* approach in understanding the case. A case study is distinguished from a cross-sectional study that uses a small sample because in a case study design, the researcher does *not* analyze relationships between variables but rather makes sense of the case or cases as a whole. The case study design is especially useful when the research has a "how" or "why" question and the researcher is studying contemporary events that he or she cannot control (Yin, 2009: 13).

Case study analysis is somewhat paradoxical. On the one hand, case studies "frankly imply particularity—cases are situationally grounded, limited views of social life. On the other hand they are something more … they invest the study of a particular social setting with some sense of generality" (Walton, 1992: 121). While case studies describe specific people and places, the best of them help us understand more general categories of the social world. Typically, the case study provides an in-depth understanding because it relies on several data sources, such as observations, interviews, photographs, and other materials.

Many classic studies have used the case study approach. For example, "Goffman (1961) tells us what goes on in mental institutions, Sykes (1958) explains the operation of prisons, Whyte (1943) and Leibow (1967) reveal the attractions of street corner gangs, and Thompson (1971) makes food riots sensible" (Walton, 1992: 125).

Robert Stake (2003: 136–138) identifies three types of case studies: the *intrinsic*, undertaken because the particular case itself, with its particularity and ordinariness, is of interest, the *instrumental*, where a typical case is selected to provide insight into and understanding of, or a generalization about, an issue that it illustrates, and the *collective*, where the instrumental approach is extended to several cases selected to lead to a better understanding or theory about a larger collection of cases.

In Box 7.2, Susan Chase describes her instrumental case study of one university. Using a variety of data sources, her work focuses on how undergraduates in one university engage with each other across social differences in an environment that supports this behavior.

Other projects demonstrate the range of case studies as they use different units of analyses, kinds of data, and approaches. Leigh Culver (2004) wanted to investigate the impact of new immigration patterns on police–Latino community relations in rural Midwestern communities. She selected three neighboring communities in central Missouri that had recently experienced substantial increases in the Latino population and collected a variety of data—including observations of uniformed patrol officers and formal and informal interviews with officers, community, and government leaders. Noting the language barriers and the amount of fear and distrust in interactions, Culver (2004) was able to describe the difficulties in developing cooperative, working relationships between the Hispanic community and the police.

Jo Reger (2004) studied the New York City chapter of the National Organization for Women. Using documents and in-depth interviews with chapter members, Reger (2004: 220) concludes that this organization illustrates

BOX **7.2**

A Case Study of City University by Susan E. Chase*

My current research uses a case study approach to understanding undergraduates' engagement with diversity issues. I selected City University (CU, a pseudonym), a predominantly white institution, because I knew from an earlier comparative study that something was working well on this campus. I studied CU to see how students speak and listen to each other across social differences and to discover how the campus environment can encourage that kind of interaction.

I collected a variety of data. I did individual and group interviews: with faculty, staff, and administrators to get their perspectives on CU as an institution with a commitment to diversity; with a variety of students of different backgrounds and interests; and with leaders of student organizations to get their perspectives on how diversity issues are addressed on campus. With the help of Jessie Finch and Misti Sterling, two sociology majors at my university, I also studied documents. We analyzed every issue of the student newspaper for two years, the minutes of every student government meeting for three years, the university's monthly calendar of events for two years, and the undergraduate course schedule for three years. We counted how often specific diversity issues (race, class, gender, sexual orientation, ability, and global/international) were addressed in these documents. I found that in all the documents (newspaper, minutes, events, and curriculum) there was considerable

discourse about diversity with race the most prominent diversity category in three contexts (newspaper, minutes, and events).

I also did a qualitative analysis of CU's student newspaper and student government minutes for 2004–2005, the year I did the interviews. Focusing on which diversity issues got play and how they were addressed, I found that specific racial issues were not only more prominent but also more contentious. For example, issues related to sexual orientation were usually presented in educational or supportive tones in both the newspaper and the student government minutes, while race-related issues evoked heated exchange in both contexts. This analysis provided strong corroboration of the understandings I had gained from the interviews.

Using a case study design with multiple data sources allows me to understand the "how" and "why" of CU's environment where diversity issues of all types are frequently on the table and in which race is the most prominent and contentious diversity issue, especially in comparison to sexual orientation. I have insight into the reflexive relationship between CU's narrative environment and students' narratives about how they have learned to speak and listen across social differences.

*Susan E. Chase is an associate professor of Sociology at the University of Tulsa. Her current project is *Learning to Speak, Learning to Listen: How Diversity Works on Campus*. This essay is published with permission.

---

the importance of an external–internal dynamic in social movements: responses to political events and social structures pull people into a social change organization and then internal organizational strategies channel the emotions into activism.

Shoko Hara (2007) studied five elderly pet owners, seeing each one as a case and focusing on the way that these individuals use their pets to successfully navigate later life. He was able to ask and answer a series of questions, among them: "How does each person balance independence and dependence by having a pet? How does each depend emotionally on the pets, manage pet loss and cope with the limitations of the aging self through pet responsibilities?" (Hara, 2007: 101).

Jon Kraszewski (2008: 141) examined how displaced fans look to sports teams from their former places of residence as a way to understand "home." Kraszewski did a case study by becoming an observer and a participant at a Pittsburgh Steelers bar in Fort Worth Texas for two years. Using theories of diaspora, he examined how "sports fandom fits into the nexus of late capitalism, displacement and identity within the United States" (Kraszewski, 2008: 140). Describing many conversations in this bar in Texas that were about neighborhoods, roads, and towns in Western Pennsylvania, he argues that geographic discourse allows people to negotiate social tensions from their former home.

A major advantage of the case study is that it allows a "close reading" of social life and gives attention to the broader social context (Feagin, Orum, and Sjoberg, 1991: 274). Using this approach can be appropriate when working inductively, that is, moving from observations to theory, and can be particularly useful for exploratory and descriptive purposes. A disadvantage of the case study is its limited generalizability. Focusing on a sample of one or a few, with a specific phenomenon in its specific context, we cannot know if what is observed is unique or typical. In addition, sometimes doing a study at one point in time makes it hard to disentangle causal connections.

SUMMARY

 **STOP & THINK**   *Two researchers, Baker and Miller, are interested in studying changes in college students' attitudes about attending graduate school. Baker decides to do a cross-sectional study with a sample of freshmen and seniors at one university. Based on his interviews, Baker notes that the seniors have more positive opinions toward graduate school than the freshmen. In contrast, Miller selects a panel design. She interviews a sample of freshmen at one university and re-interviews them three years later when most are seniors. Miller finds that the students' attitudes as seniors are more positive toward graduate school than they were as freshmen. Baker and Miller both conclude that students' attitudes toward graduate education change, becoming more positive as students progress through the undergraduate years. In whose conclusions do you have more confidence, Baker's or Miller's? Why?*

Researchers begin with specific questions and purposes. They might be interested in exploring new areas, generating theories, describing samples and populations, seeing patterns between variables, testing a causal hypothesis, evaluating a program, or, as in the examples of Baker and Miller, documenting changes over time.

The purposes or reasons for the research influence the choice of study design. If, for example, we are interested in describing students' attitudes toward graduate school, selecting a cross-sectional design would be effective and sufficient. A cross-sectional design, like the study by Baker, allows us to describe the samples' attitudes toward graduate school and to *compare* the attitudes of freshmen and seniors. On the other hand, a panel design, like the research by Miller, although more expensive and time consuming, is better if we want to study *changes* in attitudes. A panel study allows us to ask about current attitudes at two times rather than depend on retrospective data, documents the time order of the variables, and lets us analyze changes in the attitudes of individuals not just aggregated changes.

Judging the appropriateness of a design for a given project involves a series of decisions. The answers to the following questions provide a basis for selecting one of the five study designs covered in this chapter and the experimental designs that will be discussed in Chapter 8:

- Is the focus on variables and the connections between them, or is the goal to analyze only one case or a small number of cases holistically?
- Is it useful and possible to collect data more than one time from the same sample?
- Is there a cohort that can be studied to answer the research question?
- Is it useful and possible to select a new sample each time if data will be collected more than once?
- If there is a causal hypothesis, is it possible, useful, and ethical to try to control or manipulate the independent variable in some way?

The focus of this chapter has been on the strengths and weaknesses of non-experimental designs as related to internal validity and generalizability. When selecting a study design, methodological concerns need to be balanced along with ethical considerations and practical needs. Although the summary shown in Figure 7.4 focuses on the methodological issues, ethical and practical criteria must also be considered when selecting a research design.

Summary of Cross-Sectional, Panel, and Trend Study Designs

Study Design	Design Features	Design Uses
CROSS-SECTIONAL STUDY	One sample of cases studied one time with data about one or more variables	Useful for describing samples and the relationship between variables
One Sample (divided up into categories of the independent variable during analysis)		Useful for explanatory purposes especially if the time order of variables is known and sophisticated statistical analyses possible

	time 1
category 1 of an independent variable	measure of a dependent variable
category 2 of an independent variable	measure of a dependent variable

Study Design	Design Features	Design Uses
PANEL STUDY	One sample of cases is studied at least two times	Useful for describing changes in individuals and groups over time (including developmental issues)
		Useful for explanatory purposes

	time 1	time 2
one sample	variables measured	variables measured

Study Design	Design Features	Design Uses
TREND STUDIES	Different samples from the same population are selected	Useful for describing changes in populations and for analyzing the effect of time period

	time 1	time 2
first sample	measure several variables	
second sample		measure the same variables

FIGURE **7.4** Summary of Cross-Sectional, Panel, and Trend Study Designs

EXERCISE 7.1

Identifying Study Designs

Find an article in a social science journal that reports the results of actual research. (Be sure it is not an article that only reviews and summarizes other research.) Answer the following questions about the article and the research.

1. List the author(s), title of article, journal, month and year of publication, and pages.
2. Did the researcher(s) begin with at least one hypothesis? If so, write one hypothesis and identify the independent and dependent variables. If there was no hypothesis, but there was a research question, write it down.
3. Identify the main purpose(s) of this study as exploratory, descriptive, explanatory, or applied.
4. Describe the population and the sample in this study. Is the population a cohort?
5. If this study used only one or a few cases or analyzed the cases as a whole rather than by focusing on variables, then it might be a *case study*. If you believe it is a case study, give support for this conclusion and skip Questions 6–9. If it's not a case study, skip this question and answer the rest.
6. How many times were data collected in this study?
7. If data were collected more than once, how many samples were selected?
8. Did the researchers try to control or manipulate the independent variable in some way?
9. Did the researchers use a case, cross-sectional, panel, trend, or cohort study design? If so, which one? Give support for your answer.
10. Was the design useful for the researcher's purposes and goals or would another design have been better? Give support for your answer.

EXERCISE 7.2

Planning a Case Study of Your Town or City

In his book *Code of the Street*, Elijah Anderson introduces his study by describing the 8.5-mile major artery in Philadelphia that links the northwest suburbs with the heart of the inner city. Anderson (1999: 15) says,

> Germantown Avenue provides an excellent cross section of the social ecology of a major American city. Along this artery live the well-to-do, the middle classes, the working poor, and the very poor—the diverse segments of urban society. The story of Germantown Avenue can therefore serve in many respects as a metaphor for the whole city.

In a case study of a geographic entity such as a city or a town, one or more streets, areas, organizations, groups, or individuals can be selected purposively. Think about the town or city in which you currently live or one in which you have lived in the recent past. Describe who or what your focus would be if you wanted to use a case study approach to study this town or city. Think of a research question you could answer with your case study.

1. Name the geographic area you want to focus on and what research question you would like to answer about it.
2. Describe who or what you would study. (You can identify one unit of analysis or you can propose to study more than one.)
3. Give support for your choices by discussing why it would be useful to do a case study of your city or town to answer your research question.

EXERCISE 7.3

Evaluating Samples and Study Designs

Janice Jones, a graduate student, designs a research project to find out if women who are very career-oriented are more likely to be mothers than women who are less career-oriented. Jones selects a random sample of women who work for large corporations and hold jobs as president or vice-president. Interviewing the women she finds that their average age is 45 and that 35 percent are mothers, 55 percent are childless by choice, and 10 percent are childless involuntarily. She concludes that strong career orientations cause women to choose childlessness.

1. Discuss some of the challenges to the internal validity of Jones's conclusions.
2. Think of a different way this study could have been done. Specifically describe a sample and a study design that you think would be useful in doing research to answer Jones's question.

© Andersen Ross/Jupiter Images

CHAPTER **8**

Experimental Research

185

INTRODUCTION

We're sure that you've been in many classrooms over the years, with dozens of teachers using a wide variety of techniques. Based on your experiences, you may have ideas about which classroom strategies are most effective. If you were planning a study, you could try to answer a research question about whether some techniques are more effective than others. That's the research question that Chris Caldeira, the author of this chapter's focal research, began with in order to create a thesis project as part of the requirements of her MA in Sociology. In this chapter we'll read about her study on whether using film clips in an introductory sociology class enhances student learning.

explanatory research, research designed to explain why subjects vary in one way or another.

causal hypothesis, a testable expectation about an independent variable's effect on a dependent variable.

We'll first continue our discussion of study designs by considering the designs that are especially useful for research with an explanatory purpose—research that seeks to explain *why* cases vary on one or more variables or *why* something is the way it is. **Explanatory research** begins with a **causal hypothesis** or a statement that one variable, the independent variable, *causes* another, the dependent variable. Frequently the researcher has deduced the hypothesis from an existing theory.

 **STOP & THINK** *Explanatory research can focus on any area of social life. If you were doing a study on some aspect of education, there are many research questions and hypotheses you could consider. Create one causal hypothesis that uses level of education or kind of education as a variable.*

Causal Hypotheses and Experimental Designs

There are many causal hypotheses that can be constructed using as an aspect of education as an independent, dependent, antecedent, or intervening variable. Doing a quick literature review of research in sociology, we found many studies with causal hypotheses about one or more aspects of education. Here are three of them:

1. Level of education contributes to a person's sense of control (Schieman, 2001).
2. Among Chinese students, their individual ideas of modernity affect their interest in attending universities that are outside of China (Du and Zhong, 2007).
3. For less-educated constituents, television increases basic knowledge about their U.S. Representative in Congress (but not about his or her challenger) because the less educated rely most heavily and exclusively on television for their political information (Prior, 2006).

As each of these is a causal hypothesis, a research strategy that considers the requirements for showing causality is necessary. As we noted in Chapters 2 and 7, causality cannot be directly observed but, instead, must be inferred from other criteria. You may remember that the factors that support the existence of a causal relationship include the following:

1. An independent and dependent variable that are associated, or vary in some way, with each other. This is the condition of *empirical association*.
2. An independent variable that occurs *before* the dependent variable. This is the condition of *temporal precedence* or *time order*.

3. The relationship between the independent and dependent variable is not explained by the action of a third variable (sometimes called an antecedent variable) that makes the first two vary together. That is, the relationship between the independent and dependent variables is not *spurious* or non-causal.

As in any study, it is necessary to make the series of choices we've discussed in the previous chapters. We'd need to find a way to measure the variables, select a population and sampling strategy, determine the number of times to collect data, and establish the number of samples to use. We might decide that it is appropriate to use one of the study designs that we presented in Chapter 7 and design a panel, trend, cohort, case, or cross-sectional study. Before making a final decision, however, we'd want to consider the possibility of using an **experimental design**. Although the study designs we considered in Chapter 7 typically allow the researcher to determine if there is a pattern or relationship between independent and dependent variables (and, in the case of longitudinal designs, temporal sequencing), experimental designs are especially helpful for determining the time order of variables and for minimizing the third-variable effects.

experimental design,
a study design in which the independent variable is controlled, manipulated, or introduced in some way by the researcher.

The Experiment: Data Collection Technique or Study Design?

As we discussed in Chapter 6, a critical set of decisions focuses on how to measure variables. That is, for each variable, the researcher has to figure out how to classify each unit of analysis into one of two or more categories. The uniqueness of the experimental design becomes apparent when we think about the issue of measurement because, in a traditional experiment, the independent variable is not measured at all! Instead, in an experiment, the independent variable is "introduced," "manipulated," or "controlled" by the researcher. That is, the independent variable does not occur naturally but is the result of an action taken by the researcher.

In the next few chapters, we will be describing methods of data collection, including asking questions, observing, and using available data. Because of the unique way that independent variables are handled in experiments, some textbooks define the experiment as a method of data collection or as a mode of scientific observation. Although we recognize the usefulness of this approach, we want to highlight the experiment's special treatment of the independent variable as a unique *design* element. For this reason, we include experiments in our discussion of designs and see this chapter as a natural transition between the consideration of study designs and the presentations on data collection techniques that follow.

In the first hypothesis listed earlier, level of education is the independent variable and sense of control is the dependent variable. Scott Schieman (2001) studied this hypothesis with data from a cross-sectional study using data from one large sample of respondents. The one-time survey asked many questions, among them one that measured education by asking how many years of education the respondent had completed. Four statements on the survey were used as indicators of sense of control as they asked respondents the extent to which they agreed or disagreed with the statements such as "Most of my problems are due to bad breaks." The cross-sectional design allowed Schieman to see if the two variables were associated in some way with each other but did not allow the determination of temporal sequencing—a critical issue for causality.

One way to determine the time order of the variables is to actually manipulate, introduce, or control the independent variable in some way rather than measuring it. If *practical* and *ethical*, a study can be designed so that the dependent variable is measured first, then the independent variable is introduced or manipulated, and, finally, the dependent variable is measured again. In this way, one can see whether the introduction of the independent variable comes *before* change in the dependent variable.

While it's not possible to do an experiment for Schieman's hypothesis since we can't manipulate the number of years of education that a person completes, there are many variables that can be manipulated. One of them—kind of teaching technique—can be manipulated and you'll see that in the experiment we highlight here, Chris Caldeira controlled the placement of the members of the sample into two categories of the independent variable. By controlling the independent variable and then measuring the dependent variable, the experimental strategy allows us to see the effect of the independent variable.

EXPERIMENTAL DESIGNS

stimulus, the experimental condition of the independent variable that is controlled or "introduced" by the researcher in an experiment.

placebo, a simulated treatment of the control group that is designed to appear authentic.

internal validity, agreement between a study's conclusions about causal connections and what is actually true.

pretest–posttest control group experiment, an experimental design with two or more randomly selected groups (an experimental and a control group) in which the researcher controls or "introduces" the independent variable and measures the dependent variable at least two times (pretest and posttest measurement).

Determining who receives which condition of the independent variable and controlling its timing are central elements in the experimental design. In experiments, the control group is exposed to *all* the influences that the experimental group is exposed to *except* for the **stimulus** (the experimental condition of the independent variable). The researcher tries to have the two groups treated exactly alike, except that instead of a stimulus, the control group receives no treatment, an alternative treatment, or a **placebo** (a simulated treatment that is supposed to appear authentic). For maximum internal validity, beyond introducing the independent variable, the experimental design tries to eliminate systematic differences between the experimental and control group. Remember that **internal validity** is concerned with factors that affect the internal links between the independent and dependent variables, specifically factors that support alternative explanations for variations in the dependent variable. In experiments, internal validity focuses on whether the independent variable or other factors, such as prior differences between the experimental and control groups, are explanations for the observed differences in the dependent variable.

Pretest–Posttest Control Group Experiment

The gold standard for experiments is the **pretest–posttest control group experiment,** which is also called classic controlled experiment. In this kind of experiment, there are two or more samples or groups with two measurements of the dependent variable—one before and one after the introduction of the

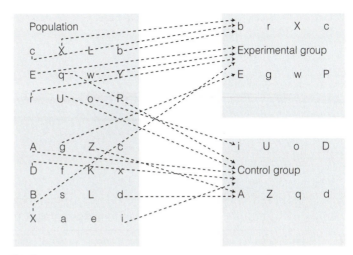

FIGURE **8.1** Use of Probability Sampling to Select Groups

---

pretest, the measurement of the dependent variable that occurs before the introduction of the stimulus or the independent variable.

posttest, the measurement of the dependent variable that occurs after the introduction of the stimulus or the independent variable.

probability sampling, a sample drawn in a way to give every member of the population a known (nonzero) chance of inclusion.

random assignment, a technique for assigning members of the sample to experimental and control groups by chance to maximize the likelihood that the groups are similar at the beginning of the experiment.

independent variable. These measurements are called the **pretest** and the **posttest,** respectively. The samples or groups are the experimental and the control group, groups that can be selected in one of several ways. The preferred method of picking the groups is to identify an appropriate population and, using **probability sampling,** to select two random samples from the population. Figure 8.1 illustrates the use of probability sampling to pick experimental and control groups. This method gives the greatest generalizability of results because selecting the samples randomly from a larger population increases the chances that the samples represent the population.

However, participants in experiments are rarely selected from a larger population usually because it's typically not practical and might not be ethical. Instead, participants are more typically a group of people who have volunteered or have been selected by someone in the setting. In such cases, the sample is really a convenience sample and might not be representative of a larger population.

There are two ways to sort a sample into experimental and control groups to minimize preexisting, systematic differences between the groups. By minimizing such differences, experiments can offer support for the conclusion that no third variable is making the independent and dependent variables appear to "go together"—the third requirement of causality.

One way is to assign members of the sample to the experimental and control groups using **random assignment,** the process of randomly selecting experimental and control groups from the study participants. This can be done by flipping a coin to determine which subject is assigned to which group or by assigning each subject a number and using either a random number table or a computer-driven random number generator to select members of each group (experimental or control). Figure 8.2 illustrates the use of random assignment to create two groups.

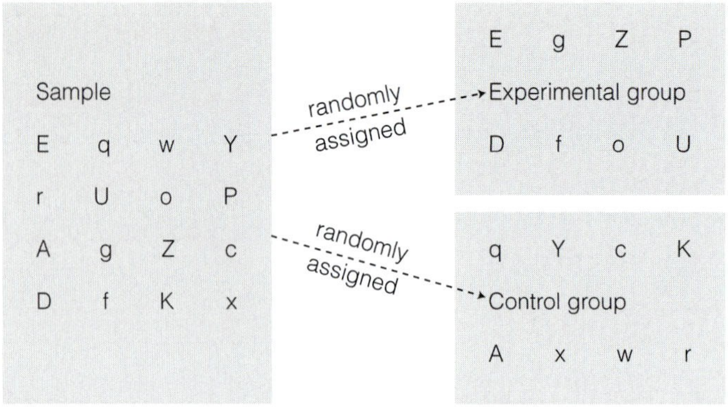

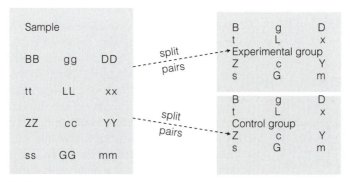

FIGURE **8.2** Use of Random Assignment to Select Groups

matching, assigning sample members to groups by matching the sample members on one or more characteristics and then separating the pairs into two groups with one group randomly selected to become the experimental group.

Another way to assign members of the sample to experimental and control groups is through **matching.** Using this method, the sample's members are matched on as many characteristics as possible so that the whole group is "paired up." The pairs are then split, forming two matched groups (see Figure 8.3). A concern is that typically several characteristics in the sample are matched while others are not. After the sample members are matched into groups, one group is randomly selected to become the experimental group and the other becomes the control group.

The pretest and posttest experiment has several elements that distinguish it from other study designs:

1. The study uses at least one experimental and one control group, selected using a strategy to make the groups as similar as possible (probability sampling, random assignment, or matching). Experimental groups are sometimes called treatment groups because an experiment will have at least one and possibly more experimental treatments.

FIGURE **8.3** Use of Matching to Select Groups

Groups	Time	
	1 (pretest)	2 (posttest)
Experimental	Measure dependent variable	Measure dependent variable again
Control	Measure dependent variable	* Measure dependent variable again

Stimulus

• The control group receives no stimulus, receives an alternative treatment, or receives a placebo.

FIGURE **8.4** Pretest–Posttest Control Group Experiment

2. One or more dependent variables are measured at least two times for the experimental and control groups. The first measurement is before and the second is after the independent variable is introduced. (These measurements are called the pretest and the posttest, respectively.)

independent variable,
a variable that is seen as affecting or influencing another variable.

3. The **independent variable** is introduced, manipulated, or controlled by the researcher between the two measurements of the dependent variable. While the experimental group receives the stimulus, the control group gets nothing, an alternative treatment, or a placebo.

dependent variable,
a variable that is seen as being affected or influenced by another variable.

4. The differences in the **dependent variable**(s) between the pretest and posttest are calculated for the experimental group(s) and for the control group. The differences in the dependent variable(s) for the experimental and control groups are compared.

A diagram of the pretest–posttest control group experiment is presented in Figure 8.4.

 STOP & THINK *Here's a causal hypothesis: Exposure to media images of "ideal bodies" negatively affects peoples' views of their own bodies. How could you design an experiment to test this hypothesis? Think about how you'd get an experimental and a control group, how you'd measure the dependent variable, and how you'd introduce the independent variable.*

Research by Hobza and Rochlen (2009) studied the effects of exposure to ideal masculine images on men's views of themselves. Other researchers had done cross-sectional studies measuring all the variables at one time, but without knowing the time order, it was hard to support causal conclusions. Hobza and Rochlen (2009: 122) chose a different design and conducted a pretest–posttest control group experiment in order to test the hypothesis that men exposed to ideal physical images of other men will report significantly lower body esteem when compared to men not exposed to these images.

Hobza and Rochlen (2009) selected a sample of 82 undergraduate men and randomly assigned them to a time to participate in the study. Each group arrived at the research lab where informed consent was obtained and participants were told that the purpose of the study was to examine advertisements to determine which ones were more likable and memorable. The participants then completed a questionnaire that included the pretest data for the dependent variable of body esteem. This variable was measured by 35 items on the questionnaire that asked respondents to rate their feelings about parts of their bodies, such as buttocks or chest, on a 5-point scale that went from "have strong positive feelings" to "have strong negative feelings."

The independent variable was then introduced by the presentation of a series of 25 slides of advertisements from print magazines. The experimental group saw the stimulus—slides depicting men with toned bodies (large muscles, flat stomachs, etc.) and broad chins. The control group saw slides of advertisements for household items such as detergents and toothpaste without any people in them. All of the participants were then asked to re-take the questionnaire, ostensibly to fill time and to provide data for another researcher's study. The researchers compared the pretest and posttest scores on the dependent variable for the two groups. There was support for the hypothesis that viewing men in the media with ideal physical features decreases men's body images as the men in the experimental group (who saw advertisements of ideal male bodies) had significantly lower body esteem scores at posttest than participants in the control group (who saw advertisements without people).

The advantages of the experimental design are evident in this pretest–posttest experiment. The researchers can see the association between the variables and also demonstrate evidence of time order as the dependent variable was measured both *before* and *after* the independent variable was introduced. However, demonstrating change in the dependent variable and a time order in which the independent variable precedes changes in the dependent variable offer *necessary* but not *sufficient* support for a cause and effect relationship. Another, unobserved factor could have caused the change in the dependent variable. Concern about the effects of other factors is the rationale for another essential design element in the controlled experiment—the use of two groups, an experimental and a control group, selected in a way (in this case random assignment) to minimize the chances of preexisting systematic differences between the two groups and the exposure of *only one group* to the stimulus (ideal male images) in the interval between the measurements. With two groups and the ability to control who is exposed to a stimulus, researchers using an experimental design can be fairly confident of separating the effects of the independent variable on the dependent variable from other possible influences.

Internal Validity and Experiments

We've already noted that having a pretest and posttest and an experimental and a control group can handle many concerns about internal validity by regulating and identifying systematic differences between the groups. The following

maturation, the biological and psychological processes that cause people to change over time.

testing effect, the sensitizing effect on subjects of the pretest.

history, the effects of general historical events on study participants.

selection bias, a bias in the way the experimental and control or comparison groups are selected that is responsible for preexisting differences between the groups.

are some other advantages of this kind of experiment that Donald Campbell and Julian Stanley (1963) identified:

1. The experiment is useful for separating the effects of the independent variable from those of **maturation,** the biological and psychological processes that cause people to change over time, because, in experiments, both the experimental and control groups are maturing for the same amount of time.
2. The experiment can control for some aspects of the **testing effect,** the separate effects on subjects that result from taking the pretest, because both groups take the pretest.
3. The experiment can control for some aspects of **history,** the effects of general historical events not connected to the independent variable that influence the dependent variable, because both groups are influenced by the same historical events.
4. The methods of selecting the control and experimental groups are designed to minimize **selection bias,** or preexisting differences between the groups.

★ STOP & THINK *Do you play video games or know people who spend a great deal of time playing them? Have you ever thought about what effect video games can have on people's behavior? Construct a hypothesis to test something about the effect of playing video games on some aspect of behavior and think of a study design that could test it.*

Posttest-Only Control Group Experiment

posttest-only control group experiment, an experimental design with no pretest.

Another experimental design, the **posttest-only control group experiment,** has no Time 1 measurement (no pretest), either because it is not possible to do a pretest or because of a concern that using a pretest would sensitize the experimental group to the stimulus (the testing effect). Otherwise, it has the same design elements as other experiments: control or manipulation of the stimulus and two or more groups using random selection or assignment. This design is illustrated in Figure 8.5.

Experiments using posttest-only designs have been conducted to study the effects of exposure to violent media on behavior. In one posttest experiment, Bushman and Anderson (2009) studied the desensitization hypothesis that

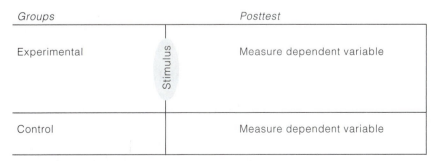

FIGURE **8.5** Posttest-Only Control Groups Experiment

links playing violent video games to decreased helping behavior. They used a sample of 320 college students, with equal numbers of men and women, who were told that the purpose of the study was to see what types of people liked what types of games. Students came to the research lab individually and were randomly assigned to the experimental and control groups. Within each group, the specific game was randomly assigned. Those in the experimental group played one of four violent video games while those in the control group played one of four nonviolent video games. Upon finishing the game, the student was handed a long questionnaire on video games to complete while the experimenter left the room. The actual purpose of the questionnaire was to keep the student busy while a recording of a staged fight was played outside the room. The six-minute recording had actors of the same gender as the participant sounding like they were fighting about one of them stealing the other's boyfriend or girlfriend—complete with lots of shouting and scuffling. At one point in the recording, the experimenter threw a chair on the floor outside the closed door of the lab and kicked the door twice. After one "person" in the fight left, the "other" groaned in pain. The measures of the dependent variable were if the participant tried to help the person in pain and, if so, the length of time he or she waited before trying to help. Bushman and Anderson (2009: 276) found that "Although in the predicted direction, there was no significant difference in helping rates between violent and nonviolent video game players, 21% and 25%…. When people who played a violent game did decide to help, they took significantly longer (M = 73.3 seconds) to help the victim than those who played a not violent game (M = 16.2 seconds)."

A major challenge to the internal validity of the posttest-only experimental design is the unverifiable assumption that a difference between the experimental and control groups on the dependent variable represents a *change* in the experimental group. Without a pretest, it is difficult to determine if the difference between the groups on the dependent variable is a change in the experimental group because of the stimulus or is a continuation of a preexisting difference between the groups. After all, though unlikely, it is possible in the example of the posttest-only experiment that the students in the control group were just nicer, more helpful people than those in the experimental group. We need to be cautious in reaching the conclusion that a causal hypothesis has been supported if there is no Time 1 measurement. However, when it is not practical to measure the dependent variable before the introduction of the dependent variable, as was the case in the video game experiment, the posttest-only experiment provides many of the advantages of the pretest and posttest design.

Extended Experimental Design

Solomon four-group design, a controlled experiment with an additional experimental and control group with each receiving a posttest only.

We will present one other experimental design that is useful in increasing internal validity. The **Solomon four-group design,** which is illustrated in Figure 8.6, has four groups—two experimental and two control groups, with only one set of groups having a pretest. It is most useful when there is a concern about a possible interaction effect between the first measurement of the dependent variable and the introduction of the stimulus. The compensation

Groups	Time		
	1 (pretest)		2 (posttest)
Experimental	Measure dependent variable	Stimulus	Measure dependent variable again
Control	Measure dependent variable		Measure dependent variable again
Experimental with no pretest		Stimulus	Measure dependent variable
Control with no pretest			Measure dependent variable

FIGURE **8.6** Solomon Four-Group Experiment

for the extra effort, time, and expense of this design is the ability to determine the main effects of testing, the interaction effects of testing, and the combined effect of maturation and history (Campbell and Stanley, 1963: 25). A Solomon four-group experiment was used by Divett, Critteden, and Henderson (2003) in a study of consumer loyalty and behavior among subscription patrons of a regional theater in Australia. They selected this design because of the possibility that the pretest would sensitize the respondents to the stimulus. In this study, four groups of subscribers were selected and the first experimental and control group received a pretest (a questionnaire focusing on the views of the theater's approachability and responsiveness and the respondent's loyalty and likelihood of expressing concerns directly to the theater staff). Then both experimental groups received the stimulus, which was a brochure outlining the way the theater could be contacted and examples of subscribers' experiences, while the two control groups received nothing. One year later, all four groups received the posttest (the same questionnaire as the pretest). In this experiment, the researchers found no pretest effect but found that the brochure did have an impact on subscriber attitudes.

 **STOP & THINK** *Let's say you want to evaluate a six-month "quit smoking" program. You know that you could evaluate its effectiveness by taking a sample of smokers who volunteer to join the program, divide them into two groups at random, and enroll only the experimental group in the program. At the*

end of six months, you could see how many in each group have quit smoking and what health changes, if any, have occurred in each group. But, what about the ethical concern of depriving potential "quitters" of the chance to stop smoking and the practical problem that members of the control group might get help elsewhere if asked to wait six months for help? What other study design could you use to study the impact of the "quit smoking" program?

Quasi-Experimental Design

quasi-experiment, an experimental design that is missing one or more aspects of a true experiment, most frequently random assignment into experimental and control groups.

In situations where it is not ethical or practical to do a true controlled experiment, the researcher can use a **quasi-experiment**. In this design, the researcher does not have full control over all aspects of the experiment. The missing element(s) could be control over timing, the selection of subjects, or the ability to randomize (Campbell and Stanley, 1963: 34). Most frequently it is random assignment into experimental and control groups that is missing. In the comparison group quasi-experiment, for example, there are two groups, but the groups are not selected by random selection, randomization, or matching to have pre-experimental equivalence. In this design, there is an experimental group and a comparison group in place of the control group (see Figure 8.7).

So, for example, if we were interested in the causal hypothesis that an anti-smoking program is effective in reducing smoking and improving health, we could do a comparison group quasi-experiment. We'd compare a group of volunteers who, on their own, decided to enroll in our program and find other similar people who weren't taking the program—say from a health center—and study them over the same time period. Because we didn't pick or assign the groups randomly, they'd serve as our experimental and comparison groups.

In a study of the impact of ethics instruction in the classroom, Heather Canary (2007) was not able to randomly assign students into an experiment and a control group. Instead she used a quasi-experimental design to test several hypotheses, one of which was that the degree to which ethics are

Groups	Time		
	1 (pretest)		2 (posttest)
Experimental	Measure dependent variable	Stimulus	Measure dependent variable again
Comparison	Measure dependent variable	*	Measure dependent variable again

• The comparison group receives no stimulus, receives an alternative treatment, or receives a placebo.

FIGURE **8.7** Comparison Group Quasi-Experiment

addressed in the classroom is positively associated with changes in students' moral-judgment competence (defined as the ability to formulate a moral course of action in a given situation). The study participants were 175 undergraduate students (102 males, 73 females) at a public university in the southwestern United States enrolled in two communication ethics courses and two conflict communication courses. The ethics courses covered a variety of topics including dominant ethics theories and ethical issues in interpersonal relationships, small groups, public speaking, organizational communication, and other topics. The conflict courses focused on conflict communication and specific strategies for communication in various relational contexts; toward the end of the course, there was a unit on ethics. For the experiment, the students in the ethics courses were the experimental group and those in the conflict courses were the comparison group. The independent variable varied as the experimental group had an entire course on ethics while the comparison group had one week of instruction on the topic.

The pretest was a questionnaire that included an assessment of moral-judgment competence using a series of questions about ethical dilemmas and behavior that was administered during the third week of classes. The posttest was administered at the end of the semester and covered the same moral-judgment items and some additional items on perceptions of the faculty member and ethics instruction. Although some of the other hypotheses that the researcher tested were supported, the hypothesis predicting a positive association between the degree to which ethics are addressed in the course classroom and moral-judgment competence was not supported (Canary, 2007).

There are several limitations to the internal validity of this quasi-experiment. Without a true experimental and control group, it is possible that preexisting differences between those in the different classes affected how the two groups responded to the independent variable. Canary (2007: 206) notes that "the use of similar recruitment strategies for all groups and assessments that the groups were equivalent on key indicators such as age, sex, grade point average, and pretest scores reduces threats to validity." She also considers a limitation of the sample as the study participants were juniors and seniors and suggests that future research on communication ethics instruction could productively investigate outcomes for lower division students (Canary, 2007: 206).

One final note about this study focuses on the choice of independent variable. Canary (2007: 205) comments on her findings as follows:

> Although it seems surprising that ethics students would actually experience fewer gains in moral reasoning competence than did students in conflict classes, these results also reveal the ethical value of curriculum and instruction in communication programs.... It is quite possible that many of the issues discussed [in the ethics course] were new to students and not applicable to their everyday lives.... The conflict course curriculum, on the other hand, emphasized conflict processes and outcomes within a variety of personal relationships, with a final unit on ethics of managing conflict. It could be that as students deal with situations that are familiar to them, such as interpersonal conflict, they increase their ability to reason through actions and consequences in ways that are transferable to other contexts, such as moral dilemmas.

In other words, Canary suggests a new independent variable to be considered—familiarity with specific ethical issues.

Experiments in the Field

field experiment, an experiment done in the "real world" of classrooms, offices, factories, homes, playgrounds, and the like.

Canary's (2007) quasi-experiment was also a **field experiment,** an experiment that takes place in "real-life" settings where people congregate naturally— schools, supermarkets, hospitals, prisons, workplaces, summer camps, parking garages, and the like. While the students in this study did volunteer to participate and signed informed consent forms before completing the pretest and posttest questionnaires, the critical issue is that the setting of the study was one that was familiar to them—a college classroom.

While all research is conducted within social settings, there is variability between settings and the degree to which the researcher has control over them. Some field experiments are actually built into the setting so that it is not obvious that an experiment is going on.

 **STOP & THINK** *In some experiments, the subjects do not know they are part of an experiment. What are some of the advantages and disadvantages of this strategy?*

We'll next turn to a field experiment that Chris Caldeira conducted in a classroom using a hypothesis about student learning that came from her work experience. She conducted the study in two university classes that were regular classes as far as the students were concerned.

FOCAL   RESEARCH

Student Learning: A Field Experiment in the Classroom

By Chris Caldeira*

I was searching for a topic to study in order to complete my M.A. in Sociology while I was working as a senior acquisitions editor for sociology at Wadsworth/Cengage Learning. In gathering feedback about textbooks, I was in contact with many faculty members and heard about the struggle that students often had in understanding and retaining sociology's key concepts. Some instructors talked about supplementing their classroom instruction with film and documentary clips in order to present the concepts in real world or simulated contexts and make them more easily remembered.

---

* Chris Caldeira conducted this research as part of the requirements for an MA in Sociology from San Francisco State University. This article is published with permission.

While few instructors expressed doubts that films are useful, I realized that there was little systematic research addressing the effectiveness of film in enhancing students' understanding and comprehension of sociological concepts. Because I was fortunate to find a faculty member who was willing to allow me to use her introductory sociology classes as the research setting, I decided to do a field experiment to test the effectiveness of using film and documentary clips in conjunction with text- and lecture-based explanations of concepts. The hypothesis I tested is that using film clips to teach sociological concepts in conjunction with the textbook and lecture is more effective in helping students learn core concepts than the lecture and textbook without the clips.

Working in consultation with a professor in a university setting made it possible for me to easily get access to data about student achievement and be able to complete my study over the course of one semester. Working in the field also meant limitations on my research. I could specify the independent variable (the intervention) and select the actual film clips to be used. On the other hand, I could not control many aspects of the research, including the composition of the classes or which students would be in the professor's afternoon or morning classes. I was, however, able to adapt my research strategy to the field requirements and use a quasi-experimental design and one of the course requirement results as the indicator of my dependent variable.

Methodology

The professor I collaborated with was teaching two introductory sociology classes at a mid-sized, public, open-enrollment university. As the professor used the same lectures, textbook, assignments, and examinations in both classes, I selected the students in them as my sample. Each class had approximately 50 undergraduate students who were enrolled to satisfy a general education or major requirement. The slight differences between the classes in their year at the university or average GPA were not statistically significant. The only statistically significant difference was the gender breakdown of the classes: 40% of students in the morning class were women as compared to 68% of the afternoon class. While studies have noted sex-related differences in learning and test performance, the experimental design described below compensates for this difference between the classes.

The independent variable was the set of film clips that was selected and showed. We selected a group of 12 concepts of that are typically covered in an introductory sociology course, including mechanical solidarity, organic solidarity, race as a social construction, culture shock, and stigma and labeling theory, and found clips for each of them. The clips were chosen not only to stimulate interest but because each presented a clear visual representations of a sociological concept. For example, for the concept "culture shock," a term used to describe the physical and emotional discomfort that occurs when people reorient themselves to a new culture or social environment, we selected 3 scenes from *Lost Boys of Sudan*, a film about a group of boys from Sudan who move to the U.S. as part of an immigrant resettlement program. The focus in the scenes is the difficult process the boys face in a culture that is foreign to them.

I selected a quasi-experimental design with two classes alternating as experimental and comparison groups. Over the course of a semester, the professor showed six clips on six concepts to the morning class and six different clips on six other concepts to the afternoon class. In other words, the independent variable (the film clips) was manipulated or controlled so that only one class saw them while the other did not. The morning class functioned as the experimental group for the six clips shown in that class while the afternoon class functioned as the comparison group (and vice versa).

Before showing each film clip to the experimental group, the instructor introduced the related concept, gave some general examples, and prepared students to view the clip with the concept in mind. After showing the clip, the instructor asked the students to write a paragraph describing how the film illustrated the concept, collected their writings, summarized the film's connection to the concept, and asked for questions or comments. For the comparison group, the instructor introduced the related concept with related examples and asked them to write a paragraph describing their understanding of that concept, collected their writings, and asked for questions or comments.

The dependent variable was student learning. The concept was operationally defined as student performance on multiple choice tests, the method by which student learning is actually assessed in most introductory sociology classes. The indicators were approximately five multiple choice questions for each sociological concept covered. The test questions were those that accompany the assigned textbook with additional multiple choice questions written by the professor. I received the grades on the tests with student names removed.

Findings

If film clips are an effective teaching tool, then the experimental group should perform better on the examination questions related to the concepts to which the film applied. The hypothesis was supported by the data. In 11 out of 12 cases, the experimental group did better on exams than the comparison group by between 2.6 and 16.8 points and by an average of 7.4 points (better by more than a half letter grade). For the one film clip where the experimental group did not score better, the experimental group scored lower than the comparison group by 1.1 points, a very small difference. For 6 of 12 concepts, the mean difference between the experimental and comparison group test scores was large enough to be statistically significant (using .05 level of significance). Of these six, the experimental group was the morning class for two and the afternoon class was the experimental group for other four.

Conclusion

The hypothesis for this study was that using film clips as an instructional supplement to traditional lectures and readings will help students learn core concepts. Carefully chosen scenes from films and documentaries were used to illustrate specific (rather than broad) sociological concepts. The quasi-experimental design was used with two introductory sociology classes, with the classes alternating as experimental and comparison groups. The students in each class saw six films that those in the other class did not. The experimental group tended to perform better on tests than those in the comparison group. Differences in sex composition of two classes had no significant impact on the results. This research offers support for the conclusion that film clips are effective in teaching core concepts in sociology.

In terms of this study's generalizability, it is important to note that this experiment involved a professor well-known for effective teaching as measured by student evaluations and teaching awards. Would film clips have the same impact on student achievement in other classrooms? Other research will need to be done to determine the impact of this kind of media when the instructor is less experienced and/or less effective in the classroom.

 **THINKING ABOUT ETHICS** *The project was not given a full ethics review because of the exemption for research conducted in educational settings involving normal education practices such as research on instructional strategies or classroom management methods. However, the research does meet ethical standards. The choice of study design was dictated in part by concern for not harming students. If, as the researcher hypothesized, watching film clips would in fact improve grades, those who saw the films would have received an advantage over the comparison group. By having students alternate between being in the experimental and the comparison groups, no group of students had an extra advantage and therefore there was no academic harm. In addition, all student data, including examination grades, were kept confidential.*

Another interesting field experiment was done in Rwanda. Elizabeth Levy Paluck (2009) obtained informed consent for the data collection in a field experiment she conducted in cooperation with a Rwandan nongovernmental organization that had produced a year-long "education entertainment" radio soap opera designed to promote reconciliation 10 years after that country's genocide. Paluck's (2009) hypothesis was that the soap opera, set in fictional Rwandan communities and featuring messages about reducing intergroup prejudice, violence, and trauma (reconciliation messages), would have an impact on listeners' beliefs, norms, and behavior. The experimental and control groups were in different communities with research assistants visiting each community and playing four soap opera episodes on a portable stereo for the participants.

The experimental group heard the stimulus of a reconciliation radio program while the control group heard a program focused on health. The dependent variables were measured in a variety of ways. One method used questions that included statements that asked about whether participants could "imagine the thoughts or feelings of" Rwandan prisoners, genocide survivors, poor people, and political leaders. Paluck (2009) found some support for her hypothesis as there were greater changes in the experimental group than the control group on the dependent variables of perceptions of intermarriage, open dissent, trust, empathy, cooperation, and trauma healing after the introduction of the independent variable.

generalizability, the ability to apply the results of a study to groups or situations beyond those actually studied.

The "realness" of field experiments is why they typically have good internal validity and **generalizability** (also called external validity). In real-world settings, the researchers can use samples that are more like (or actually may be) the actual students, patients, workers, customers, citizens, and other groups that they want to generalize their results to. The introduction of the independent variable (such as teaching technique, radio show, administrative decision, marketing strategy, medical treatment, or another variable) is quite natural. Paluck (2009: 576) points out that the experimental and control groups in her study heard the radio programs as part of village life with the distractions, social interactions, and emotional reactions of the real-world setting. On the other hand, experiments on media effects done in places other than their natural settings might pick up mostly short-term effects of simplified communications in artificial environments (Paluck, 2009: 583).

In Caldeira's study, the subjects didn't need to give informed consent as students don't usually get choices about teaching techniques and taking tests. Caldeira was able to introduce an independent variable, obtain the data on the dependent variable, and complete an experiment while, for the subjects, it was not a research project but just another class.

Although working in real settings has advantages, it can present difficulties. Researchers doing a field experiment are typically guests of the institutions or community settings and might not be able to control as much of the experiment as would be ideal. Flay and Collins (2005) looked at school-based evaluations of problem behavior (such as smoking) prevention interventions and found that although study designs have improved over the past 20 years, often there are difficulties in doing such experiments: school personnel resist randomization, appropriate control schools might not be used, the intervention (the stimulus) might not be fully implemented in some of the experimental group schools, permission for student participation is difficult to obtain, and biases can be introduced by data collection when the same people deliver the intervention and measure its outcome.

Experiments in the Laboratory

Another, perhaps more common setting for experiments is the "laboratory," which is a setting controlled by the researcher. The literature of psychology, social psychology, education, medical sociology, delinquency and corrections, political sociology, and organizational sociology is replete with reports of **laboratory research**, much of it conducted in universities using students as participants. For example, while both kinds of studies use experimental and control groups, a field experiment on job discrimination might have matched "applicants" of different races apply for real jobs in real company, while a laboratory experiment on the same topic might have undergraduate subjects rate hypothetical job applicants (Paver, 2007).

We've discussed two laboratory experiments in this chapter: the pretest–posttest control group experiment on the effects of exposure to ideal masculine images on men's views of themselves by Hobza and Rochlen (2009) and the Bushman and Anderson (2009) study on the desensitizing effects of violent media on helping others. The samples in these studies were groups of students who had volunteered to be part of a research project and participated in the project in a lab on a college campus. In these experiments the researchers had control over the setting and could easily randomly assign sample members to groups.

On the other hand, although realistic situations can be created in labs and subjects can be kept in the dark about the real purpose of the study, the participants in this kind of experiment know that they are involved in research. If a setting and the staging of the experiment, including the way the independent variable is introduced, do not feel real, the study's internal validity will be hampered. For example, Bushman and Anderson (2009: 275) used a pilot or trial study with 50 participants to see if the students found the "fight" realistic. They note that of the first 10 students, only 5 thought it was genuine. At that point, they added throwing the chair and kicking the door. The next 40 students reported believing that the fight was real.

It's important to evaluate a study's external validity or the ability to generalize the results from the lab to the "real world" by considering the following issues:

1. Was the situation very artificial, or did it approximate "real life?"
2. How different were study participants from other populations?
3. To what extent did the participants believe that they were up for inspection, serving as guinea pigs or play-acting, or have other feelings that would affect responses to the stimulus (Campbell and Stanley, 1963: 21)?
4. To what extent did the researcher communicate his or her expectations for results to the subjects with verbal or nonverbal cues?

One way to handle the issue of **experimenter expectations** is to use a design called the **double-blind experiment,** which involves keeping both the subjects and the research staff who directly interact with them from knowing which members of the sample are in the experimental or the control group. In this way, the researcher's expectations about the experimental group are less likely to affect the outcome.

laboratory research, research done in settings that allows the researcher control over the conditions, such as in a university or medical setting.

experimenter expectations, when expected behaviors or outcomes are communicated to subjects by the researcher.

double-blind experiment, an experiment in which neither the subjects nor the research staff who interact with them knows the memberships of the experimental or control groups.

Natural Experiments

What if you were interested in studying how employment affects marriage? For example, if you wanted to know if the amount of work-related travel that a spouse does has an impact on his or her chances of getting divorced, could you design an experiment to answer your research question?

There are many situations where it's just not possible to control the independent variable for practical and/or ethical reasons. So, even if we had an explanatory purpose, such as trying to find out if work schedules involving extended travel put stress on marriages and families, we couldn't do a traditional experiment because it's not possible for a researcher to control people's work schedules. Some researchers, however, do what they call **natural experiments** by finding real-world phenomena that can approximate the experimental design even though the independent variable is not controlled, manipulated, or introduced by the researcher. Wars, hurricanes, political events, and so on might affect some groups and not others, and these can be used as pseudo-experimental and pseudo-control groups.

natural experiment, a study using real-world phenomena that attempts to approximate an experimental design even though the independent variable is not controlled, manipulated, or introduced by the researcher.

For example, Joshua Angrist and John Johnson (2000) used data from the armed services to look at the effects of deployment away from home during the Gulf War on the marriages and families of armed services personnel. Although the *researchers* could not control deployment, in 1991, during the Gulf War, the military assigned some personnel to overseas duty, either in the Gulf area or elsewhere, while others remained at home. In addition, in 1992, the military conducted a survey of officers and enlisted personnel that included questions on marriage and family that can be used as a sort of "posttest." Using information obtained from the military, the researchers did a statistical analysis of soldiers by time away from home. They found that time away does have an impact—especially for women serving in the military, with each month away raising divorce probabilities by more than 1 percent. However, because this is not a true experiment, there are questions about internal validity. The researchers caution that, although there is a temptation to interpret the results as causal, the differences in divorce could be the result of other factors (Angrist and Johnson, 2000). Questions for natural experiments include the possibility of self-selection into experimental and control groups or other underlying differences between the groups and the issue of whether policy makers or others made interventions in anticipation of behavioral responses of the citizens in ways that are correlated with the behavior responses (Dunning, 2008). In other words, as in all explanatory studies, we need to be cautious about reaching conclusions about causality in natural experiments.

COMPARING EXPERIMENTS TO OTHER DESIGNS

Although the experimental study design is best suited for explanatory research and is a useful choice when you are considering causal hypotheses, many research questions cannot and should not be studied experimentally. For descriptive and exploratory research purposes, when large samples are needed, and in situations where it is not practical, possible, or ethical to control or manipulate the independent variable, researchers are wise to consider other study designs. As we discussed in Chapter 7, the panel, trend, cross-sectional, and case study designs are more appropriate for many research purposes.

 **STOP & THINK** *Can you think of hypotheses that do not lend themselves to being studied with an experiment? What is one such hypothesis? What design could be used to test your hypothesis?*

Social scientists often construct hypotheses using background variables like gender, race, age, or level of education as independent variables. Let's say that we hypothesize that level of education affects the extent to which young adults are involved in community service and political activity; more specifically, that an increase in education, the independent variable, causes increases in the dependent variables (amounts of community service and political activity). We can't control how much education an individual receives, so we'd need to use a non-experimental design to examine these relationships. For example, we could do a panel study and observe changes in the variables over time by starting with a sample of 18-year-olds. Following the sample for a number of years, we could measure the independent variable (years of schooling completed) and the dependent variables (involvement in community service and political activities) at least twice. We'd be able to see the pattern or relationship between the variables and their time order. We could use statistical analyses as a way to disentangle possible "third variables" and spuriousness.

Panel studies have been widely used to study the cause and effect relationship between behaviors such as drinking alcohol, smoking tobacco, and eating a diet rich in red meat and the onset and progression of illnesses such as heart disease and cancer. Health researchers can't and wouldn't randomly select groups of people and force them to drink alcohol, eat red meat, or smoke cigarettes for a number of years while prohibiting the members of other randomly selected groups from doing so, but they can follow people who make these choices themselves.

Panel studies like the Nurses' Health Study have followed people for decades. Started in 1976 when the youngest women were 25, and then expanded in 1989, the nurses' study has information provided by 238,000 nurse participants (NHS, 2008). Every two years the nurses enrolled in the study are asked for medical histories, asked questions about daily diet and major life events, and periodically provide blood, urine, toenail, and cell samples to the researchers. The response rate for the questionnaires is typically 90 percent for each two-year cycle. While the prevention of cancer is still a primary focus, the study has also produced landmark data on cardiovascular disease, diabetes, and many other conditions. Most importantly, these studies have shown that diet, physical activity, and other lifestyle factors can powerfully promote better health. The study is ethical because the researchers do not attempt to intervene in subjects' behavior and because they keep all data confidential. Although some are critical of the lack of an experimental design and of the study's expense, other public health officials argue that it is better than an experiment in identifying long-term or delayed effects (Brophy, 1999).

This panel study has contributed many important insights about the connections between behavior, lifestyle, and health, including support for explanatory hypotheses.

Let's consider another non-experimental design, the cross-sectional study, for its use in explanation. Assume we hypothesize that gender affects the degree of social isolation among senior citizens. While we can't "control" or randomly assign gender, we could measure this independent variable and any number of dependent variables, including social isolation in a sample of senior citizens. Because the time order of these independent and dependent variables is clear (the gender of a senior citizen obviously occurs before his or her social isolation), if we found an association between the variables, we could then look for other support for causality. We could, for example, use statistical analysis techniques to eliminate the effect of other variables that might make the original relationship spurious.

SUMMARY

This chapter adds to the researcher's toolbox by describing the study design options of controlled experiments and extended and quasi-experimental designs. Because experiments allow researchers to control the time order of the independent and dependent variables and to identify changes in the dependent variable, they are very useful for testing causal hypotheses.

Experiments can be done as field experiments in the "real world," or they can be conducted in laboratories. Although the field experiment might be less artificial, the laboratory experiment can provide a greater ability to regulate the situation. Another option is the natural experiment, which is really not an experiment in the traditional sense, but which attempts to approximate an experimental design.

Experimental designs can be extended by adding extra experimental and control groups or can be truncated into a quasi-experiment by eliminating one or more aspects of the classic design. Some design choices are more practical than others, whereas others have greater internal validity. Experiments typically use small, non-random samples and therefore usually have limited generalizability.

Ethical considerations and the practical concerns of time, cost, and the feasibility of controlling the independent variable are important issues in designing a study. Cross-sectional studies are typically the most practical, and when a large and representative sample is selected, the study will have good generalizability. However, the tradeoff for selecting a non-experimental design may be questions about internal validity if the time order is not clear or if there are preexisting differences between groups.

Each study design has strengths and weaknesses. As always, it's important to consider a project's goals against each design's costs and benefits and to balance methodological, ethical, and practical considerations when selecting a research strategy. Both experimental and non-experimental designs should be considered when evaluating options (see Table 8.1).

TABLE **8.1**

Summary of Design Studies

Study Design	Design Features	Design Uses
Pretest–Posttest Control Group Experiment	This has at least one experimental and one control group; the independent variable is manipulated, controlled, or introduced by the researcher; the dependent variable is measured at least two times in both groups (before and after the independent variable is introduced).	Useful for explanatory research testing causal hypotheses.
Posttest-Only Control Group Experiment	Same as above except that the dependent variable is measured only after the independent variable is introduced.	Same as above.
Extended Experimental Design	Same as the pretest–posttest control group experiment with an additional experimental and control group measured only at the posttest.	Same as above but can also sort out the effects of pretesting.
Quasi-Experiment	The researcher does not have complete control of experimental stimuli. The independent variable is introduced, controlled, or manipulated and a comparison group rather than a control group is used with an experimental group.	Same as experiment but with increased concerns about internal validity. May be more ethical or practical than a control group experiment.
Cross-Sectional Study	One sample of cases studied one time with data about one or more variables. Typically, the sample is divided into categories of the independent variable and compared on the dependent variable during analysis.	Useful for describing samples and the relationship between variables; useful for explanatory purposes especially if the time order of variables is known and sophisticated statistical analysis is possible.
Panel Study	One sample of cases is studied at least two times.	Useful for describing changes in individuals and groups over time (including developmental issues); useful for explanatory purposes.
Trend or Cohort Study	Two or more different samples are selected from the same population at least two times.	Useful for describing changes in populations.
Case Study	In-depth analysis of one or a few cases (including individuals, groups, and organizations) within their social context.	Useful for exploratory, descriptive, and, with caution, explanatory purposes; useful for generating theory and developing concepts; very limited generalizability.

EXERCISE 8.1

Reading about Experiments

Find a research article in a psychology or social science journal that reports the results of an experiment. Answer the following questions about the study.

1. List the author(s), title of article, journal, month and year of publication, and pages.
2. What is one hypothesis that the research was designed to test? Identify the independent and dependent variables.
3. Describe the sample.
4. How many groups were used and how were they selected? Is there a control group?
5. Was there a pretest? If so, describe it.
6. How was the independent variable (or the stimulus) manipulated, controlled, or introduced?
7. Was there a posttest? If so, describe it.
8. What kind of experimental design was used—pretest and posttest experiment, a posttest-only experiment, a Solomon four-group, a comparison group quasi-experiment, or some other kind? Give support for your answer.
9. Comment on the issues of internal validity and generalizability as they apply to this study.

EXERCISE 8.2

Conducting a Quasi-Experiment

What to do:

First, check with your professor about the need to seek approval from the Human Subjects Committee, the institutional review board, or the committee at your college or university concerned with evaluating ethical issues in research. If necessary, obtain approval before completing this exercise.

In this quasi-experiment, you will be testing the hypothesis that when there is a "conversational icebreaker," people are more likely to make conversation with strangers than when there is no icebreaker.

Working in "the field," select a situation with which you are familiar from your daily life where strangers typically come into contact with each other. Some examples of appropriate situations include riding in an elevator, being in a doctor's waiting room, waiting for a bus or a train, and standing in the check-out line in a supermarket.

Select one item to use as a conversational icebreaker. (Make sure it is something that could be commented on but is not something that could be perceived as threatening, hostile, or dangerous.)

Some examples of icebreakers that you can use are carrying a plant or a bouquet of flowers, holding a cake with birthday candles on it, pushing an infant in a stroller, carrying a fishbowl with water and tropical fish in it, or wearing a very unusual item of clothing or jewelry.

Over the course of several hours or days, put yourself in the same situation six times: three times with your icebreaker and three times without it. To the extent that you can, make sure that as much as possible, you keep everything else under your control the same each time (such as how you dress, the extent to which you make eye contact with strangers, and whether you are alone or with friends). But, recognize that the groups of people you interact with, with and without the icebreaker, are two convenience samples—with one an experimental and the other a comparison group.

Immediately after leaving each situation, record your field notes (detailed descriptions of what happened). In your field notes, record information about the number of strangers you interacted with, what they were like, and the way they interacted with you, specifically recording any comments that were made.

What to write:

After completing your experiment, answer the following questions:

1. What situation did you use for your experiment?
2. What icebreaker did you use?
3. Describe the dependent variable in the experimental group—that is, the kind of interactions that you had and any comments that strangers made in the three situations where you had an icebreaker.
4. Describe the dependent variable in the comparison group—that is, the kind of interactions that you had and any comments that strangers made in the three situations where you did not have an icebreaker.
5. Does your data support the hypothesis that an icebreaker is more likely to foster conversation and interaction with strangers?
6. Comment on the internal validity and generalizability of your study.
7. Comment on the experience of doing this exercise.

EXERCISE 8.3

Considering Non-Experimental Study Designs

In some situations, it won't be practical, ethical, or feasible to do a controlled experiment. Think about what you would do if, like Chris Caldeira, whose study you read in this chapter, you had a hypothesis that some kinds of teaching techniques are more effective than others, but you think it's not practical or ethical to do an experiment. Construct a specific hypothesis that has a teaching technique or curriculum approach as the independent variable and design a study using a non-experimental design to test it (see Figure 7.4 for these designs).

1. State your hypothesis, identifying the specific independent and dependent variables you have selected.
2. Identify, describe, and draw a diagram of a study design you could use to examine your hypothesis.
3. Describe how you would measure your independent and dependent variables in this study.
4. Comment on the issues of internal validity and generalizability for the study you propose.
5. Compare the internal validity, generalizability, and practicality of your proposed study design with the kind of study conducted by Caldeira.

CHAPTER **9**

Questionnaires and Structured Interviews

© Jeff Greenberg / Alamy

211

INTRODUCTION

 STOP & THINK *Which do you think is the most important issue for policy makers and citizens to focus on this year—strengthening the nation's economy and improving the job situation, working on improving access to health care, protecting the environment, or dealing with terrorism and crime? Do you feel that working toward one goal precludes working toward others? A poll by the Pew Research Center for People & the Press in January 2009 found that strengthening the nation's economy and improving the job situation were at the top of the American public's list of priorities for 2009 while issues such as the environment, crime, illegal immigration, and even reducing health care costs were seen as less important than they had been a year before. That same month, another poll indicated that the majority of American voters believe there is a conflict between economic growth and environmental protection (Rasmussen Reports, 2009). Do you think these polls do a good job of telling us what the American public thinks?*

questionnaire, a data collection instrument with questions and statements that are designed to solicit information from respondents.

structured interview, a data collection method in which an interviewer reads a standardized list of questions to the respondent and records the respondent's answers.

survey, a study in which the same data, usually in the form of answers to questions, are collected from one or more samples.

respondents, the participants in a survey who complete a questionnaire or interview.

self-report method, another name for questionnaires and interviews because respondents are most often asked to report their own characteristics, behaviors, and attitudes.

Have you ever received a phone call asking you to participate in a survey? Do you remember someone in your household completing a questionnaire from the U.S. Census Bureau? Have you heard or read about the results of recent polls that asked questions about government policies? Most of us are familiar with interviews and questionnaires; it's hard to avoid these methods of data collection because so many of them are conducted annually. **Questionnaires** and their first cousins, **structured interviews,** are the most widely used methods of data collection in the social sciences. Research projects that use structured interviews and questionnaires are sometimes referred to as surveys, because although these methods of data collection can be used in experiments or in case studies, they are most commonly used in surveys.

A **survey** is a study in which the same data are collected (most often a specified set of questions, but possibly observations or available data) from one or more samples (usually individuals, but possibly other units of analysis) and analyzed statistically. Surveys often use large probability samples and the cross-sectional study design, which is when data are collected about many variables at one time from one sample. However, surveys can also employ nonprobability samples and trend and panel study designs. See Box 9.1 for two examples of recent surveys. Although these examples are from the United States, survey research is conducted around the world.

The various modes of data collection that use communication between respondents and researchers can be seen as falling along a continuum from the least interactive to the most interactive—that is, from self-administered questionnaires to in-person interviews (Singleton and Straits, 2001). In all of the modes of data collection covered in this chapter, data are collected by asking study participants, typically called **respondents,** a set of predetermined questions. This technique is also called the **self-report method** because respondents are usually asked to provide information about themselves. The researcher can take the answers as facts or interpret them in some way, but the information provided by respondents is the basic data in questionnaires and interviews.

BOX **9.1**

Surveys are all around us. If we want to know what Americans think about an issue, we usually can find a survey that's asked questions on the topic. Two sources of interesting and methodologically careful surveys are Public Agenda (www.publicagenda.org) and the Pew Research Center for the People & the Press (http://people-press.org). Both of these nonprofit organizations conduct many studies annually. The work of Public Agenda has included studies on families, children, crime, corrections, education, health care, and the economy. The Pew Research Center focuses on attitudes toward the press, politics, and public policy.

 STOP & THINK *What's your opinion about America's health care system? Do you think substantial changes are needed? Should the government guarantee health insurance for all citizens? How about energy and the price of gas and fuel? Are you worried about the costs? Are you concerned about America's dependence on foreign oil or about global warming? Would you support increased taxes to fund developing renewable energy sources? Do you*

think your feelings in these areas are unusual or typical?

A national survey by the Pew Research Center for the People & the Press, conducted in March 2009, with a random sample of 1,308 adults found a large majority of Americans supporting sweeping changes in health care. Forty percent said the health care system needed to be completely rebuilt, while 36 percent thought fundamental changes were needed. About 60 percent favored government guaranteed health insurance. See more findings from this survey at http://people-press. org/report/500/support-for-health-care-overhaul

Public Agenda surveyed a national random sample of 1,001 adults over the age of 18 in January 2009 on energy. A great majority of the sample (89 percent) was concerned about the cost of gas and fuel and most (83 percent) worried a lot that the U.S. economy is too dependent on foreign oil. Seventy-one percent said they worried about global warming, but only 38 percent said they worried "a lot." Some of the proposals for alternative energy were supported by the public. See the details of this study at http://www.publicagenda.org/ pages/energy-learning-curve

Answering questions in a survey is a social act in the same way that attending a class or browsing in a shopping mall is. The social context of the data collection and the views of those we question are influenced by the surrounding culture. When we meet new acquaintances in our everyday lives, we often ask questions. In the same way, researchers use questions to elicit information about the backgrounds, behavior, and opinions of respondents. And, just as in our personal conversations, researchers try to figure out if the answers they've received are honest and accurate.

 STOP & THINK *The surveys by Public Agenda and Pew focused on attitudes toward national energy policies and health care, respectively. Suggest a list of other topics that you think you could ask questions about using a questionnaire. Would you be concerned about the accuracy of the answers on any of these topics?*

In the study described in the focal research that follows, Brandon Lang and Christopher Podeschi asked college students questions about environmental issues. They were interested in whether their attitudes were connected to some background characteristics and to behavior patterns. As you will see, these researchers were interested in a study with both descriptive and explanatory purposes.

Environmentalism among College Students: A Sociological Investigation

By K. Brandon Lang and Christopher W. Podeschi*

Introduction

In recent decades, Americans have become more aware of environmental issues such as air and water pollution, waste management, ozone depletion and global warming. Dozens of studies have shown, however, that environmental attitudes and behaviors vary according to a number of social and cultural variables. In this paper, we build upon previous work that considers these social and cultural predictors by examining the "social bases" of environmentalism using a sample of first-year students at a comprehensive public university in Pennsylvania. We also explore the relationship between the environmental worldview and behaviors of these students.

Previous Research

For decades, social scientists have examined environmental attitudes and people's willingness to engage in environmentally friendly behaviors. Researchers have also studied people's fundamental values and beliefs with regard to society's place in relationship to nature. Dunlap and Van Liere (1978) first examined adherence to either the "New Ecological Paradigm" or the "Dominant Social Paradigm." They identify the former as an ecologically-informed worldview, the latter a human-centered orientation that fails to acknowledge the need for ecological limits.

Early evidence and theory led scholars to predict a correlation between higher socio-economic positions and environmentalism (see Van Liere and Dunlap 1980). The "post-materialist" thesis says that people without economic concerns move on to considerations like beauty or ecology (Inglehart 1995). Recent studies, however, find mixed results concerning the relationship between environmentalism and social class (see Jones and Dunlap 1992; Klineberg et al. 1998; Marquart-Pyatt 2008).

Research on the relationship between environmentalism and residence also yields mixed results. An early hypothesis asserts that rural culture, because of dependence on occupations involving natural resources (e.g., farming, mining, logging), is more utilitarian than ecologically-inclined. In addition, urbanites have become concerned because they experience environmental problems like air pollution

---

* K. Brandon Lang and Christopher W. Podeschi are assistant professors of Sociology at Bloomsburg University of Pennsylvania. This article is published with permission.

(see Van Liere and Dunlap 1980). While there is some support for this position (see Jones and Dunlap 1992; Tremblay and Dunlap 1978), research by Jones et al. (1999) and others yields contradictory findings about residence (also see Jones et al. 2003). Perhaps environmental concern "diffused" or perhaps, as Bennett and McBeth (1998) suggest, economic diversification in rural areas has weakened the utilitarian aspect of rural culture.

More consistent predictors of environmentalism include political orientation and gender. Research on environmental attitudes and politics considers the effect of party identification and political ideology (e.g., attitudes about the government's role in the economy). Dunlap et al. (2001) provide an extensive review of this literature. They note that in the early stages of the modern environmental movement, an attitude distinction based on politics was not as apparent as it is today when Democrats and liberals are more pro-environment than Republicans and conservatives (see Dunlap et al. 2001). Also notable is the fact that while party identification and political ideology both predict environmental attitudes, the former has been a weaker predictor.

Much recent research finds women to be more environmentally oriented than men in terms of both attitudes and behaviors (see Zelezny et al. 2000; Davidson and Freudenburg 1996). Scholars feel that women tend to be more environmentally oriented because their socialization to be nurturing attunes them to risks to others and altruism (see Dietz et al. 2002). Furthermore, concern for others and consequently for collective goods like environmental health is thought to be fostered by women's disadvantaged position which is seen as leading to empathy (Kalof et al. 2002). Kalof et al. (2002) find that this "disadvantage" hypothesis is supported by data on both gender *and* race/ethnicity.

Finally, there is the relationship between environmental attitudes and behavior. While we might expect that these things would come together, research finds only modest or weak correlations (see Theodori and Luloff 2002; Tarrant and Cordell 1997). Why might the connection not be stronger? Bell (2004) identifies how the "social constitution of daily life" constrains our choices. For example, people drive despite their environmental concerns because we have developed car-centered communities. Similarly, Diekmann and Priesendorfer (2003) find strong support for the idea that attitudes and behaviors are more congruent when costs are low, but that relationships weakens as costs increase. For example, in a car-centered society, alternative transportation may have high time costs with the result that transport choices are less likely to be congruent with people's pro-environment views.

Research Questions and Hypotheses

Our study asks about the predictors of environmental worldview and about the link between attitudes and behavior. We test hypotheses in the settled and unsettled areas of research discussed above.

In the unsettled areas, we test two hypotheses:

1. Despite mixed results in the literature, higher class position will be associated with a pro-environment worldview because of "post-material" conditions and because parents of these students are more likely to have higher levels of education, a consistent predictor of environmentalism.
2. The difference in environmental worldview between rural and other will be small or non-existent because of the diffusion of environmentalism and changes in the rural occupation structure that has weakened the utilitarian aspect of rural cultures. Notably, coal mining has declined in rural Pennsylvania, and like the rest of the country, the population is shifting away from farming occupations.

In the more settled areas, we test three hypotheses:

3. Female students will have a more pro-environment worldview than male students.
4. Students that identify as Democrats or liberals will have a more pro-environment worldview than college students that identify as Republicans or conservatives, but party will be a weaker predictor than political ideology (i.e., liberal vs. conservative will be a stronger predictor).
5. Since recycling behavior and willingness to pay increased student fees for campus environmental programs are easy ways to enact one's environmental worldview, there will be a solid correlation between environmental worldview and these behaviors.

Methodology

The data were collected in August, 2008, at the end of orientation for new first-year students at a comprehensive public university in rural Pennsylvania. A short self-administered questionnaire was distributed to groups of approximately thirty students that had been grouped by major. Completed questionnaires were obtained from 1,226 out of the population of approximately 1,500 new students.

The sample is sixty-one percent female. Forty percent are politically liberal, 60.5 percent identify as Democrat, 52.9 percent are from a rural area or a small city, and seventy-one percent characterize themselves as wealthy or comfortable.

We measure environmental worldview with an index of ten items from the "New Ecological Paradigm" scale (NEP hereafter; see Dunlap et al. 2000). The appendix lists these and all other indicators used. In hypotheses one through four, environmental worldview is the dependent variable. For hypothesis five, it serves as an independent variable.

We measure environmental behavior in two ways: recycling and willingness to pay increased fees for campus environmental programs. Recycling behavior is measured with a three item index (alpha.755). Six items are indexed for the "willingness to pay" variable (alpha.706).

The independent variables are class, place of residence, gender, political party identification and political ideology. Rather than rely on multiple dummy variables, variables with multiple response categories have been recoded into dichotomies.

Results

The possible range of scores for NEP adherence extends from 10 to 40. The mean score on this measure for our sample is 27.8 (s.d. = 3.81). The possible range of scores for the recycle scale is 3 to 12 with an average score in the sample of 7.1 (s.d. = 2.98). The possible range of scores for the fees scale is 6 to 12 with an average score of 8.46 (s.d. = 1.79).

Eighty percent of those surveyed answered that they always or usually recycle bottles and cans. Roughly one third answered that they always or usually recycle paper while about half said they always or usually recycle cardboard. About half of the respondents answered that they would be willing to pay higher student fees in order for older buildings on campus to be renovated, for biodiesel to be used in campus buses and for recycled paper to be more widely used on campus. However, students were much less willing to pay higher fees for organic food, natural pesticides and low-flow shower heads.

We use means testing and OLS regression to test hypotheses one through four. Means tests reveal that students who consider themselves wealthy or comfortable do not score significantly higher on the NEP scale. The difference between rural/small city students and urban/suburban students is also not significant. By contrast, means tests show that females, liberals and those that identify as Democrats do have significantly higher NEP scores than males, conservatives and those that identify as Republicans, respectively.

In the regressions, we consider the effects of gender, politics (as ideology or party), social class and place of residence together upon NEP adherence. With regard to gender, female students score, on average, 1.50 units higher on the NEP than male students ($p < .001$). Our results also show that liberal students score 1.52 units higher on the NEP than conservative students ($p < .001$). In a second model, we substitute party identification for political ideology and find that the effect of considering oneself a Democrat is stronger than the effect of being liberal, producing a 2 unit increase in NEP ($p < .001$). Neither class nor residence has a significant impact upon NEP adherence.

We also use OLS regression to test hypothesis five. We consider the effect of NEP adherence upon two behaviors while controlling for all the above independent variables. Our results show a positive and significant correlation between NEP score and recycling behavior. Every unit increase in NEP adherence produces a .086 unit increase in recycling ($p < .001$). Similarly, each unit increase in NEP adherence produces a .132 unit increase in willingness to pay fees for campus environmental programs ($p < .001$). Simple bivariate correlations between NEP adherence and these behaviors are stronger than the standardized regression coefficients that show the relation between NEP adherence and these behaviors, controlling for demographics and politics.

Conclusions

We find support for all but one of our hypotheses. No significant differences in NEP adherence are found between rural/small city and urban/suburban students, lending support to the notion that environmentalism has diffused into rural areas and/or that the utilitarian worldview that comes with reliance on extractive industries has weakened because of economic changes. We find female students stronger adherents to the NEP than males, something that fits the existing literature. Our finding that students with a liberal political ideology and those that identify as Democrats have higher NEP scores as well also supports the findings of previous research. Contrary to findings in the literature, however, we find that party is a stronger predictor than ideology. These political findings should be interpreted with caution given the possibility that first-year college students may not be well-enough versed in politics to provide valid data in this area.

As expected, we also find the NEP is a solid predictor of both recycling behavior and willingness to pay fees for campus environmental programs. This is supported by both simple bivariate correlations and regressions that control for demographic and political factors. The fact that the bivariate correlations are stronger than the regression coefficients suggest that demographics and politics play a role in the link between environmental attitudes and behavior, something also found by Tarrant and Cordell (1997).

In addition, we find that reported recycling is significantly higher among urban and suburban students than among rural students. This suggests that ease of access plays a role since we expect that urban and suburban areas are more likely to have recycling programs.

The only unsupported hypothesis focuses on class differences. The literature is mixed, but we expected more comfortable classes to be more "post-materialist" and that those with more affluent parents would be more supportive of environmentalism. This is not the case, lending support to arguments about the diffusion of environmentalism in the class system or to contentions that there's been a dampening of environmentalism among those higher on the class ladder. As with the political variables, however, we are aware of validity concerns about our measure of class position.

References

Bell, Michael. 2004. *An Invitation to Environmental Sociology*. Thousand Oaks, CA: Sage.

Bennett, Keith and Mark McBeth. 1998. "Contemporary Western Rural USA Economic Composition: Potential Implications for Environmental Policy and Research." *Environmental Management*. 22,3: 371–381.

Davidson, Debra and William Freudenburg. 1996. "Gender and Environmental Risk Concerns: A Review and Analysis of Available Research." *Environment and Behavior*. 28,3: 302–339.

Diekmann, Andreas and Peter Priesendorfer. 2003. "The Behavioral Effects of Environmental Attitudes in Low-Cost and High-Cost Situations." *Rationality and Society*. 15,4:441–472.

Dietz, Thomas, Linda Kalof and Paul Stern. 2002. "Gender, Values and Environmentalism." *Social Science Quarterly*. 83,1:353–364.

Dunlap, Riley and Kent Van Liere. 1978. "The 'New Environmental Paradigm': A Proposed Measuring Instrument and Preliminary Results." *Journal of Environmental Education*. 9:10–19.

Dunlap, Riley, Kent Van Liere, Angela Mertig and Robert Jones. 2000. "Measuring Endorsement of the New Ecological Paradigm: A Revised NEP Scale." *Journal of Social Issues*. 56,3:425–442.

Dunlap, Riley, Chenyang Xiao and Aaron McCright. 2001. "Politics and Environment in America: Partisan and Ideological Cleavages in Public Support for Environmentalism." *Environmental Politics*. 10,4:23–48.

Inglehart, Ronald. 1995. "Public Support for Environmental Protection: Objective Problems and Subjective Values in 43 Societies." *PS: Political Science and Politics*. 28,1:57–72.

Jones, Robert and Riley Dunlap. 1992. "The Social Bases of Environmental Concern: Have They Changed Over Time?" *Rural Sociology*. 57:28–47.

Jones, Robert, Mark Fly and Ken Cordell. 1999. "How Green Is My Valley? Tracking Rural and Urban Environmentalism in the Southern Appalachian Ecoregion." *Rural Sociology*. 64,3:482–499.

Jones, Robert, Mark Fly, James Talley and Ken Cordell. 2003. "Green Migration into Rural America: The New Frontier of Environmentalism?" *Society and Natural Resources*. 16:221–238.

Kalof, Linda, Thomas Dietz, Gregory Guagnano and Paul Stern. 2002. "Race, Gender and Environmentalism: The Atypical Values and Beliefs of White Men." *Race, Gender and Class*. 9,2:112–130.

Klineberg, Stephen, Matthew McKeever and Bert Rothenbach. 1998. "Demographic Predictors of Environmental Concern: It Does Make a Difference How It's Measured." *Social Science Quarterly*. 79,4:734–753.

Marquart-Pyatt, Sandra. 2008. "Are There Similar Sources of Environmental Concern? Comparing Industrialized Countries." *Social Science Quarterly*. 89,5:1312–1335.

Tarrant, Michael and Ken Cordell. 1997. "The Effect of Respondent Characteristics on General Environmental Attitude-Behavior Correspondence." *Environment and Behavior*. 29,5:618–637.

Theodori, Gene and A.E. Luloff. 2002. "Position on Environmental Issues and Engagement in Proenvironmental Behaviors." *Society and Natural Resources*. 15:471–482.

Tremblay, Kenneth and Riley Dunlap. 1978. "Rural-Urban Residence and Concern with Environmental Quality: A Replication and Extension." *Rural Sociology*. 43,3:474–491.

Van Liere, Kent and Riley Dunlap. 1980. "The Social Bases of Environmental Concern: A Review of Hypotheses, Explanations and Empirical Evidence." *Public Opinion Quarterly*. 44:181–197.

Zelezny, Lynette, Poh-Pheng Chua and Christine Aldrich. 2000. "Elaborating on Gender Differences in Environmentalism." *Journal of Social Issues.* 56,3:443–457.

Appendix A: Group-Administered Questionnaire

Instructions read to respondents by the orientation workshop leader:

The University Campus Green Initiative has prepared a survey for the in-coming students to fill out during their orientation week. The survey should take about ten minutes to complete. Every student will be given a survey and an answer sheet. Please answer all of the questions on the answer sheet with a blue pen, black pen or pencil. Upon completing the survey, please return both sheets to the survey administrator. For the last questions, please write the answers directly on the answer sheet.

Please note that the surveys are anonymous. The information provided will not be linked with any individuals. Also, fully completed surveys are very helpful, but you may skip any questions that you would prefer not to answer. Thank you very much for completing these surveys. The members of the Campus Green Initiative value your opinions and appreciate that you are taking the time to answer these questions.

Thanks very much,

Dr. Brandon Lang, Department of Sociology, Social Work and Criminal Justice

Dr. Chris Podeschi, Department of Sociology, Social Work and Criminal Justice

2008 First-Year Student Environmental Survey (abridged)

Section 1:

With regard to what is said in the following statements, please answer whether you…

A: strongly agree B: agree C: disagree D: strongly disagree

1. Humankind is severely abusing the environment.
2. We are approaching the limit of the number of people the earth can support.
3. The earth has plenty of natural resources if we just learn how to develop them.
4. Plants and animals have just as much right as humans to exist.
5. The balance of nature is strong enough to cope with the impacts of modern industrial nations.
6. Despite our special abilities, humans are still subject to the laws of nature like other species.
7. The so-called 'ecological crisis' facing humankind has been greatly exaggerated.
8. Humans were meant to rule over the rest of nature.

9. Humans will eventually learn enough about how nature works to be able to control it.
10. If things continue on their present course, we will soon experience a major ecological catastrophe.

Section 2:

Would you be willing to pay increased student fees in order to...

1. Have organic food options on campus? A: yes B: no
2. Have 100% recycled paper for on-campus printing? A: yes B: no
3. Have only biodiesel and other low-emission fuels used in campus vehicles and buses? A: yes B: no
4. Have low-flow showerheads installed in dormitory bathrooms? A: yes B: no
5. Support the use of organic pesticides and fertilizers on campus? A: yes B: no
6. Renovate older buildings so they're more environmentally-friendly? A: yes B: no

Section 3:

How frequently do you recycle the following types of items...

1. Paper/Magazines/Newspapers?
   A: always B: usually C: sometimes D: never
2. Cardboard? A: always B: usually C: sometimes D: never
3. Bottles and cans? A: always B: usually C: sometimes D: never

Section 4:

1. I am from a... A: rural area B: small city C: large city D: suburb
2. I am... A: male B: female
3. Politically speaking, I am... A: very conservative B: conservative C: liberal D: very liberal
4. I support the... A: Republicans B: Democrats C: Green Party D: none (I'm an independent)
5. Growing up, my family was... A: very wealthy B: comfortable C: struggling D: poor
6. For work, my parents... PLEASE WRITE YOUR PARENTS' JOBS ON THE ANSWER SHEET
7. I was born in... PLEASE WRITE YOUR YEAR OF BIRTH ON THE ANSWER SHEET

 **THINKING ABOUT ETHICS** *The researchers submitted a proposal to their university's Institutional Review Board (IRB) where it was approved. The respondents were informed about the study's sponsor, were assured that the surveys were anonymous, and were told that they could skip any of the questions.*

THE USES OF QUESTIONNAIRES AND INTERVIEWS

Once researchers decide on the kind of information required, they'll typically consider if asking questions is an appropriate way to collect this information and then think about ways of administering those questions. Brandon Lang and Chris Podeschi selected a self-report method because it is very useful for collecting information about *attitudes*, *beliefs*, *values*, *goals*, and *expectations*, although data on these topics can sometimes be inferred from observed behavior. They asked their respondents to indicate their level of agreement with statements such as "Humankind is severely abusing the environment" and "The earth has plenty of natural resources if we just learn how to develop them."

Lang and Podeschi also asked questions about respondents' background characteristics and behavior. Although data about *social characteristics and past experiences of individuals* and information about *organizations or institutions* might be accessible through records and documents, they are suitable topics for questions. Questions are frequently employed to obtain information about *behavior*, even though observational methods will often yield more valid data. If you are interested in attitudes or behavior that typically occurs without observers (such as sexual victimization), or if you want to find out about behavior that would be unethical, impractical, or illegal to observe (such as voting behavior or getting tested for HIV), then a self-report method might be the only data collection option.

Finally, questions can be used to measure a person's level of *knowledge* about a topic. In Box 9.2 you'll see some examples of questions using the self-report method on a wide variety of topics.

 **STOP & THINK** *In addition to the one presented in this chapter, several other studies in the focal research sections of the book have collected data by asking questions. Can you remember any of the studies that have used this method? Do you have any concerns about the validity of the self-report data collected in any of the focal research studies?*

The self-report method has a wide variety of applications, including public opinion polls, election forecasts, market research, and publicly funded surveys. Questions can be asked in less structured ways, options that we will consider in the next chapter, or they can be asked using questionnaires and structured interviews, the formats we'll consider in this chapter. The more structured ways of asking questions can facilitate the collection of the same information from large samples, especially when the samples have been selected from regional, national, or multinational populations. The resulting analyses are often used by local, state, federal, and international organizations and are frequently the focus of articles in the mass media. For example, the Bureau of Justice Statistics (BJS) collects interview data each year about people's experiences with criminal victimization from a nationally representative sample of 76,000 households for the National Crime Victimization Survey (NCVS). The NCVS enables BJS to estimate the likelihood of victimization by rape, sexual assault, robbery, assault, theft, household burglary, and motor vehicle theft for the population as a whole as well as for segments of the population such as women, the elderly, members of various racial groups, city dwellers, or other groups.

BOX **9.2**

Topics and Examples of Questions

Questions can be asked about attitudes.

An example: Employers should be allowed to require drug testing of employees.

☐ agree strongly

☐ agree somewhat

☐ undecided

☐ disagree somewhat

☐ disagree strongly

Questions can be asked about organizations.

An example: Does your current place of employment have a policy concerning sexual harassment?

☐ no

☐ yes

Questions can be asked about an expectation.

An example: What is your best guess that you will graduate with honors from college?

☐ I have no chance

☐ I have very little chance

☐ I have some chance

☐ I have a very good chance

Questions can be asked about behavior.

An example: During a typical week, how much time do you spend exercising or playing sports?

☐ none

☐ less than an hour

☐ one to two hours

☐ three to five hours

☐ six or more hours

Questions can be asked about a social characteristic or attribute.

An example: How old will you be on December 31 of this year?

☐ under 18

☐ 18–20

☐ 21–23

☐ 24–26

☐ 27–30

☐ 31 and over

Researchers and policy makers benefit from the National Longitudinal Surveys of Youth, sponsored by the U.S. Department of Labor, which provide information about children's abilities, attitudes, and family relationships and the transition from school to work. The General Social Survey (GSS) and the International Social Survey Programs (ISSP) are two surveys that allow the documentation and comparison of attitudes and behavior over time. The GSS does so with samples of residents of the United States; the ISSP uses international ones. The numerous surveys make it possible to gather large amounts of information from mass publics that can be used descriptively and to inform public policy.

Although surveys are very widely used, there are important concerns about the *validity* of the information that they generate. The use of all self-report methods is based on the implicit *assumption* that people have the information being asked about and that they will answer based on their core values and beliefs. As Sudman and Bradburn (1982) have argued, the accuracy or validity of survey data is affected by memory, motivation, communication, and knowledge. It is problematic when the respondent wants to present a good image or tries to conform to what are perceived to be the researcher's expectations. It is critical to be aware of respondents' motivations to be perceived in positive ways.

Reliability concerns are that responses might represent casual thoughts rather than deep feelings and can reflect what the respondent heard on the news most recently rather than personally salient answers (Geer, 1991; Kane and Schuman, 1991: 82) and people in more positive moods have been found to give more favorable answers to questions than those who are feeling more negative (Martin, Abend, Sedikides, and Green, 1997).

It is fairly simple to design questions that ask respondents to describe their attitudes toward abortion, to share their views on laws that ban assault weapons, or to recall the number of sex partners they had in the past 12 months. However, as we noted in Chapter 6, we must consider **measurement error,** the giving of inaccurate answers to questions. In addition to the issue of question wording, which we will discuss later in this chapter, measurement error is affected by the way questions are administered, the respondents' level of knowledge, whether they have an opinion about the topic, their being able to recall events accurately from the past, and having the intent to answer honestly.

measurement error, the kind of error that occurs when the measurement we obtain is not an accurate portrayal of what we tried to measure.

 **STOP & THINK** *You agree to participate in a survey on driving behavior that includes the following questions: "Have you been pulled over by the police anywhere in this state in the past year?" "How many times in the last year were you pulled over?" "How many of these pull-overs were for speeding?" How accurate do you think your responses will be? What's your estimate about the accuracy of other respondents' answers?*

Recall of prior events by respondents has been found to be influenced by the time that has elapsed between the event and the data collection, the importance of the event, the amount of detail required to be reported, and the social psychological motivation of the respondents (Groves, 1989: 422). Researchers have known for decades that some survey respondents report having voted when they did not actually do so and that, in recent years, the gap between the percentage who actually vote and those reporting voting has increased to over 20 percent (Bernstein, Chadha, and Montjoy, 2001). The most common explanation for the gap is that respondents are motivated to give socially desirable responses so as to present themselves in a good light (Karp and Brockington, 2005). In one study, researchers counted attendees at worship, obtained Sunday school records, and then asked members of a large church if they had attended that past Sunday. They found more than a 20 percentage-point rate of over reporting church attendance (Hadaway and Marler, 1993: 127).

Other factors, such as gender and race, can also influence responses. When studying the play activities of fifth grade students, Janet Lever (1981) compared answers to the question "Where do you usually play after school, indoors or outdoors?" with activity logs that the children kept and found that boys overestimated and girls underestimated their outdoor play. In order to estimate the accuracy of respondents' answers, Tomaskovic-Devey, Wright, Czaja, and Miller (2006) surveyed a sample of drivers and asked them about having been stopped by the police while driving. They found that 23 percent of the white respondents and 29 percent of the African Americans, who according to official records had received one or more speeding citations in the previous six months, did not report any stops. Although studies that generate data by asking questions can be valuable, it's important to consider the specific questions carefully before accepting the results of analyses.

PARTICIPANT INVOLVEMENT

response rate, the percentage of the sample contacted that actually participates in a study.

One critical issue for all studies is the **response rate,** the percentage of the sample that is contacted and, once contacted, agrees to participate and completes the survey. The response rate is affected by the number of people who cannot be reached (non-contacts), the number who choose not to participate (refusals), and the number who are incapable of performing the tasks required of them (e.g., due to illness or language barriers). The response rates of studies using self-report methods vary greatly. An analysis of the response rates of 490 in organizational research published between 2000 and 2005 found that the average response rate was 53 percent in studies where data was collected from individuals and 36 percent where it was collected from organizations (Baruch and Holtom, 2008).

The in-person interview tends to have the highest response rate. For example, the in-person interview for the National Health Interview Survey, conducted by the Census Bureau since 1960, continues to get close to a 90 percent response rate (National Center for Health Statistics, 2009), and 80 percent of those approached for a national survey of adult sexual behavior agreed to be interviewed (Michael, Gagnon, Laumann, and Kolata, 1994: 33). On the other hand, some phone surveys have lower response rates. The California Health Interview Survey (2009), for example, the largest health survey conducted in any state, uses phone interviews, obtaining a response rate of 34 percent in 2003 and about 20 percent in 2007.

 **STOP & THINK** *What do you think has been happening to the response rates of surveys in recent years? Why do you think this is such a source of methodological concern?*

Refusal rates have grown in the past few decades, in the United States and internationally. Yehuda Baruch (1999) examined 175 published studies over three decades and found that the average participation rate of studies published in 1975 was 64 percent but only 48 percent in 1995. Curtin, Presser, and Singer (2005) note that the University of Michigan's Survey of Consumer Attitudes had experienced declines in response rates over the years when it was an in-person interview (1954–1976) but, when it became a phone interview, obtaining interviews became even harder. For example, they found that the number of calls to complete an interview doubled from approximately four to eight between 1976 and 1996 with the response rate dropping from 72 percent to 60 percent. Since 1996, the deterioration has become much greater, so that by 2003, it was 48 percent (Curtin et al., 2005: 90). In recent years, the rise in the non-contact rate, which is attributable to caller ID, growth in data transmission numbers, the increased numbers of cell phone–only households, and awareness of publicized identity scams and misuses involving personal data, are all likely contributors to lower response rates.

nonresponse error, an error that results from differences between non-responders and responders to a survey.

The major reason for concern about declines in response rates is **nonresponse error,** which, as we discussed in Chapter 5, is the error that results from the differences between those who participate and those who do not. In comparing response rates, some studies have found higher refusal rates among older persons, non-blacks, and those with less than a high school education (Groves, 1989; Krysan, Schuman, Scott, and Beatty, 1994). When participants and non-participants differ in social characteristics, opinions, values, attitudes, or behavior, then generalization from the sample to the larger population becomes problematic. Recently, however, some have questioned whether higher rates of refusal automatically mean more bias (Curtin et al., 2005; Groves, Presser, and Dipko, 2004). Keeter, Miller, Kohut, Groves, and Presser (2000) have found that, with particular topics, there aren't necessarily very large differences in the findings of probability samples with substantially different levels of response. Moreover, Groves (2006) and Groves and Peytcheva (2008) demonstrated that nonresponse errors are only indirectly related to nonresponse rates and are particularly likely to exist when the survey's variable of interest is strongly associated with the likelihood of responding. Thus, if people are simply more averse than they once were to *surveys of all kinds,* lower response rates, generally, don't necessarily mean greater levels of nonresponse bias or error.

While typically greater participation is found among those interested in the topic of a survey, some researchers note that there are many other influences on participation and that only factors linked to survey statistics should be causes of concern (Groves et al., 2004). Noncooperation is most likely to be associated with error if respondents' lack of cooperation is systematically related to their attitudes toward the subject of the survey, as, for example, when people with conservative social values refuse to participate in a survey on sexual behavior (Weisberg, 2005: 160).

 **STOP & THINK** *Have you ever been asked to complete a questionnaire or an interview? What factors did you consider when deciding whether to participate? Think about what you would do if you were on the other side. How would you try to convince someone to participate in a survey?*

Participation in studies can best be understood within a social exchange context. That is, once potential sample members are contacted, they must decide about cooperation after thinking about the costs and benefits. Respondents are asked to give up their time, engage in interactions controlled by the interviewer, think about issues or topics that might cause discomfort, and take the risk of being asked to reveal embarrassing information. Even though they are usually assured that the information they provide is anonymous or will be confidential, potential participants might worry about privacy and lack of control over the information. On the other hand, potential respondents might want to participate because of an interest in the topic, a desire to share their views, or knowledge that their information will be useful to science and society.

Advance mailings or other forms of contact may be used to combat the decline in response rates. However, because pre-contact tends to disproportionately raise participation rates among some segments of the population, potential bias needs to be considered (Link and Mokdad, 2005).

One common method for increasing a response rate is to use incentives. For example, in the focal research in Chapter 5, Keeter and his colleagues at the Pew Research Center gave $10 to each cell phone respondent to defray the respondent's cost of the cell phone time. Studies of incentives find that small cash incentives (such as $1 or $2) are better at increasing the response rate to a mailed survey than no incentive or a non-cash incentive (Mann, Lynn, and Peterson, 2008; Rose, Sidle, and Griffith, 2007; Teisl, Roe, and Vayda, 2006); larger cash incentives ($100) also work for increasing participation, especially among those with less education (Guyll, Spoth, and Redmond, 2003).

In addition, focusing on the interesting aspects of the study, re-contacting possible participants who have not responded, making it clear that the research is legitimate with a bona fide sponsor, and minimizing the costs of participation (such as time and possible embarrassment) can all be effective in increasing participation. See Box 9.3 for a discussion of one researcher's efforts to recruit a sample for her study.

Lang and Podeschi had a high response rate (82 percent of all first-year students at one university) because of the way they administered their questionnaire. Both researchers are members of the Campus Green Initiative, a college environmental group, and were interested in the amount of student support for initiatives such as recycling, low-flow shower heads, and keeping rooms at lower temperatures in the winter. They approached the college's orientation coordinator who agreed to have all student orientation leaders administer the questionnaires to each group of entering students as part of the university's summer orientation program (Lang and Podeschi, personal communication). The distribution of the questionnaire during an official university event by the orientation leaders, the fact that the respondents were a "captive audience," the relative brevity of the questionnaire, and the anonymity of the responses are all likely to have contributed to the high response rate.

BOX **9.3**

Recruiting a Sample

Australian sociologist Jo Lindsay found that it wasn't easy to find a sample of nonprofessional workers for a survey on sexual health, alcohol, and drug consumption. Once she gained access to workplaces, she put up fliers about the project, but didn't find it a successful strategy. Instead, she did better when she made in-person contact with potential participants at lunchtime. She says, "In the pitch I made to the young workers, I presented myself as their advocate. I told them it was their right to have good health information and that I would give them a voice in the health issues affecting them. I emphasized the low impact of the research again. I talked about the limited time the survey would take and the confidentiality of the research" (Lindsay, 2005: 124).

SELF-ADMINISTERED QUESTIONNAIRES

self-administered questionnaires, questionnaires that the respondent completes by himself or herself.

interview, a data collection method in which respondents answer questions asked by an interviewer.

There are different methods or modes of data collection that ask questions of respondents. A variety of new technologies have resulted in more ways of asking and answering surveys. **Self-administered questionnaires** can be offered in paper-and-pencil versions or in computer-assisted formats (CASI) such as web-based designs, versions that use touch screen responses, or ones that have respondents answer computer-controlled questions by pressing the keypad on a touch-tone telephone. An **interview** can be done by phone or in person, with the interviewer recording the responses for later data entry and transcription, or by using computer-assisted technology to enter data as the interview proceeds. Time, money, population, kind of sample, and response rate are issues to consider when selecting a self-report mode.

One analysis of 48 studies that compared respondents' answers on various methods of administering questions found that, overall, computer administration did not have an overall effect on how respondents answer questions (Richman, Kiesler, Weisband, and Drasgow, 1999), but other studies have found differences due to method. One study that focused on risk-taking behavior among adolescents found more reports of health-risk behaviors in self-administered computer questionnaires than in either face-to-face interviews or written questionnaires (Rew, Horner, Riesch, and Cauvin, 2004). A study of child abuse with a sample of female college students compared three modes—paper and pencil questionnaire, computer-based assessment, and face-to-face interview. The study asks us to consider ethical issues because although there was little variation in reported victimization rates among the different modes, the abuse victims experienced more emotional distress after taking the computerized version of the survey than after the other versions (DiLillo, DeGue, Kras, Di Loreto-Colgan, and Nash, 2006: 416).

The setting for taking a survey can also influence answers. One study of self-reported cigarette smoking found that 20 percent of youths who had reported having smoked in the past year when answering the school survey reported never having smoked when completing their household survey (Griesler, Kandel, Schaffran, Hu, and Davies, 2008).

Group-Administered Questionnaires

group-administered questionnaire, questionnaire administered to respondents in a group setting.

Lang and Podeschi used a paper questionnaire administered in group settings—the classrooms where the orientation workshops for new students were held. If you want to conduct a survey using a questionnaire and if, like Lang and Podeschi, you could locate a group setting for an appropriate sample (such as church-goers, club members, or students) and were able to obtain permission to recruit participants, then you could use a **group-administered questionnaire.** Administering a questionnaire to a group of people who have congregated gives the researcher the opportunity to explain the study and answer respondents' questions, provides the researcher with some control over the setting in which the questionnaire is completed, allows the respondents to participate anonymously, and usually results in a good response rate.

Group-administered questionnaires are typically inexpensive. Lang and Podeschi volunteered their time to develop the instrument and analyze the data and did not have to pay the orientation leaders for administering the questionnaire. Once they got approval from the university staff, they did not consider any other method. They were pleased with their response rate and kept their expenses to a minimum: $90 to print the double-sided, one-page questionnaire. The cost of 7 cents per completed questionnaire demonstrates the bargain that this method of data collection can be.

On the other hand, group-administered questionnaires have drawbacks, beyond the fact that there might be no group setting for the population the researcher wants to study. The extra pressure to participate that people might feel when in a group setting raises an ethical concern about voluntary participation, and the limit to the length of time that groups will allot to data collection mandates using a relatively short questionnaire. Lang and Podeschi (personal communication) feel that they obtained enough data for their purposes using many short questions. They estimate that their questionnaire took about 10 minutes to complete.

Mailed and Individually Administered Questionnaires

If mailing addresses are available, another way to administer questionnaires is to send them to respondents' homes or, less commonly, to their workplaces. The **mailed questionnaire** remains a commonly used method, with advantages to recommend it. Michele Hoffnung, in the panel study of college graduates described in the focal research in Chapter 4, used a brief mailed questionnaire for some data collection.

mailed questionnaire, questionnaire mailed to the respondent's residence or workplace.

Mailed questionnaires have a long history. The biggest U.S. survey is the U.S. Census. The census of the population is required by the U.S. Constitution and has been conducted every 10 years since 1790. For the most recent version, census employees went door-to-door in 2009 to update address lists nationwide, and then between February and March 2010, every American household received the Census questionnaire, most by mail (U.S. Census Bureau, 2009b). The estimated cost of the 2010 Census is $11.3 billion over its life cycle, making it the most expensive in the nation's history (GAO, 2007).

Mailed questionnaires are widely used for several reasons: people are familiar with them, they are fairly inexpensive and the researcher can cover a wide geographic area. There is a wide range in the response rates of mailed questionnaires, with one meta-analysis finding that individual studies have between a 10 and 89 percent response rate (Shih and Fan, 2008: 265). Some research suggests that mailed questionnaires can get better response rates than phone interviews (Link, Battaglia, Frankel, Osborn, and Mokdad, 2008).

A new option for researchers is to use database technology, such as the Delivery Sequence File (DSF) used by the U.S. Postal Service, a computerized file that contains all delivery-point addresses with the exception of General Delivery that offers the potential of covering 97 percent of all U.S. households for selecting national and local probability samples (Link et al., 2008).

If respondents can be encouraged to complete and return the questionnaire, the mailed questionnaire can be useful as filling it out without time deadlines in their homes or workplaces permits respondents to check their records and not feel rushed. One way to increase the response rate is to personalize a mailed survey, such as by including names and addresses onto letters and the use of salutations and real signatures or postscripts. An analysis of 9 studies that contained 17 experimental comparisons of personalization found that response rates increased from 3 percent to 12 percent with increased personalization (Dillman et al., 2007).

Those who receive mailed questionnaires can answer them in private, which means facing fewer social pressures to conform to the expectations of others and eliminating the worry about trying to please an interviewer. This might explain some of the research that compares the same question using different modes of data collection. Susan Gano-Phillips and Frank Fincham (1992), for example, found that married couples expressed greater satisfaction with marital quality over the phone than in their written responses; Krysan and her colleagues (1994) found that mail respondents were more likely than phone respondents to indicate that they were having problems in their neighborhoods and less likely to express support for affirmative action (both for women and for blacks) or for open housing laws. A study of attitudes in Los Angeles found that phone respondents were more likely than those who completed a mailed questionnaire to voice approval for the police and express fear of crime but were less likely to perceive social cohesion in their neighborhoods (Hennigan, Maxson, Sloane, and Ranney, 2002).

individually administered questionnaires, questionnaires that are hand delivered to a respondent and picked up after completion.

There are also **individually administered questionnaires.** In this method, a research associate hands out or delivers the instrument and makes sure that it is returned. Approaching respondents in person is similar to using a mailed questionnaire, although it is more expensive and often has a better response rate. For example, in a study of the attitudes that gay, lesbian, bisexual, and transgendered persons have about the police and community policing tactics, Wayne Gillespie (2008) used rented space at a three-day Atlanta Pride Festival Market to obtain completed individually administered questionnaires from a convenience sample of 179 passersby.

STOP & THINK *What do you think of the new ways of administering questionnaires—such as posting them on websites or e-mailing them to respondents? Are they likely to be equally effective with all populations? Better for some topics than others?*

Internet and Web Surveys

The rapid rise in the use of computers has created an explosion in the surveys using the Internet. The U.S. Census Bureau (2009c) finds that about two thirds of American households have Internet access in the home; households in Australia and Europe are not far behind (Internet World Stats, 2009). So much of the population can be reached electronically that many questionnaires can be e-mailed or put on websites for respondents to complete. Doing a **web or Internet survey** is becoming more widespread as researchers seek survey cooperation from respondents, a trend in line with the more general culture of the web where people are encouraged to be content producers and not just passive recipients of information (Couper and Miller, 2008). (See Box 9.4 for ideas about how to conduct your own Internet survey.)

web or Internet survey, survey that is sent by e-mail or posted on a website.

BOX **9.4**

It's easy to create and distribute your own web survey. Check out http://www.surveymonkey.com, http://www.zapsurvey.com, http://www.formsite.com, or http://www.zoomerang.com. Sites like these allow you to create a survey without any financial expense. They have survey templates and questionnaires that can be adapted to your needs, offer a choice of languages, a wide variety of question formats, and advice to increase response rate. If you do a web survey, consider the issues of the validity, reliability, and generalizability of your data.

Researchers have found using the Internet to be cost effective when collecting data on large samples, can reduce the time involved in collecting data, and is useful when samples are geographically dispersed (Fox, Murray, and Warm, 2003; Fricker and Schonlau, 2002). In addition, the web-based survey can be interactive and can be programmed to vary the questions and keep respondents from skipping questions.

When a percentage of a population does not have Internet access, an issue for Internet surveys is coverage error, the kind of sampling error that arises when the sampling frame is different from the intended population. However, probability samples can be obtained by using other methods to sample and recruit respondents and Internet access can be provided to those without it (Couper and Miller, 2008).

If a list of e-mail addresses of a population is available and accessible, such as from a workplace, a school, or an organization, then a random sample or the whole population of those with access can be invited to participate. Five e-mail surveys of members of the Evangelical Lutheran Church, for example, were conducted with response rates ranging from 10 percent to 58 percent (Sims, 2007). Convenience samples of self-selected Internet users can be recruited by posting invitations to participate on websites or sending them to Internet discussion groups, social networking sites, chat rooms, e-mail lists, and other venues.

One concern is that Internet surveys tend to have low response rates, especially in comparison to other methods. Diment and Garrett-Jones (2007) found that when given the choice of completing either a web- or a paper-based survey, the great majority of their respondents selected the paper-based mode. Comparing 39 recent studies, Shih and Fan (2008: 264) found considerable variation between studies, but that overall, web surveys had, on average, a 10 percent lower response rate than mailed questionnaires. Follow-ups, such as e-mail reminders to complete the Internet questionnaire, have been found to increase response rates (Klofstad, Boulianne, and Basson, 2008). However, Couper and Miller (2008: 833) note that response rate issues affect "all types of Web surveys, from list-based samples to pre-recruited probability panels and opt-in or volunteer panels ... [as] an ever increasing number of panel vendors and surveys appear to be chasing an ever-dwindling group of willing respondents, raising future concerns about the selectivity of respondents in such surveys."

Some researchers have found relatively high rates of abandonment (i.e., when the participants begin but do not complete the study) for web surveys (Crawford, Couper, and Lamias, 2001), but other studies have found their completion rates to be acceptable (Denscombe, 2006).

Cover Letters and Follow-Ups

cover letter, the letter accompanying a question-naire that explains the research and invites participation.

Whether distributed in person, by mail, or electronically, self-administered questionnaires require reading and language skills and are less likely to be completed and returned than interviews. Without an interviewer, a researcher using a questionnaire often depends on an introductory or **cover letter** to per-suade respondents to participate. The cover letter provides a context for the research and information about the legitimacy of the project, the researcher, and the sponsoring organization (see Box 9.5). The physical appearance of the questionnaire and the difficulty of completing it can affect the response rate. The format should be neat and easy to follow, with clear instructions, including statements about whether one or more than one answer is appropri-ate, and sufficient room to answer each question. The question order should be easy to follow and should begin with interesting and non-threatening ques-tions, leaving difficult or "sensitive" questions for the middle or end.

Multiple contacts, both before and after the questionnaire has been distrib-uted, and for mailed questionnaires, including a return envelope with first class postage, can improve response rates. Advanced notification can help with data collection (Kaplowitz, Hadlock, and Levine, 2004) as can follow-ups. Response rates tend to increase by 20 to 25 percent after the first follow-up with smaller returns after additional efforts (Dillman, Sinclair, and Clark, 1993). Converse, Wolfe, Huang, and Oswald (2008) compared first responses and those after follow-ups with various combinations of mail, e-mail, and web-based question-naires with a sample of teachers. When used initially, the mailed-based ques-tionnaire was more effective than an e-mail-directing participant to a questionnaire on the web (81 versus 70 percent response rate), but a follow-up using the other method brought in additional responses for both versions.

When Hoffnung uses annual mailed questionnaires in her longitudinal study described in Chapter 4, she finds that about 50 percent are returned

---

BOX **9.5**

Desirable Qualities of a Cover Letter

A good cover letter should

- Look professional and be brief
- State the purpose of the study
- Tell how the respondent was selected
- Describe the benefits of participation
- Appeal to the potential participant to respond

- Explain that participation is voluntary
- Describe the confidentiality or anonymity of responses
- Include information about the researcher or a con-tact person in case of questions or concerns
- Explain how to return the questionnaire

within a month, another 35 percent come in after follow-up letters, e-mails, or phone calls, and a few additional responses are returned months later. She gives her study participants the choice of filling out the questionnaire and mailing it back or using e-mail.

Unless everyone in the sample will be getting a follow-up letter, the researcher needs to know who has not returned their questionnaires. This can be accomplished by using some sort of identification on the instrument. This means, however, that replies will not be anonymous. Although respondents can be assured of confidentiality, some respondents might be less willing to participate when the survey is not anonymous.

 **STOP & THINK** *If given the choice, would you rather complete a questionnaire yourself or have the questions read to you by an interviewer? Would it matter to you what the topic was? How about the gender, race, and age of the interviewer—would that affect your responses?*

INTERVIEWS

An alternate way of getting answers to questions is by having an interviewer ask them. In a structured interview, an interviewer reads a standardized set of questions and the response options for closed-ended questions. The interview has some similarities to a conversation, except that the interviewer controls the topic, asks the questions, and does not share opinions or experiences. In addition, the respondent is often given relatively few options for responses. The set of instructions to the interviewer, the list of questions, and the answer categories make up the **interview schedule,** but it is basically a questionnaire that is read to the respondent.

interview schedule, the list of questions and answer categories read to a respondent in a structured or semi-structured interview.

Interviewers are supposed to be neutral and are trained to use their interpersonal skills to encourage the expression of attitudes and information but not to help *construct* them (Gubrium and Holstein, 2001: 14). The goal is to make the questioning of all respondents as similar as possible. The assumption is that differences in answers are the result of real differences among respondents rather than because of differences in the instrument (Denzin, 1989: 104).

The use of a structured interview allows for some flexibility in administration, clarification of questions, and the use of follow-up questions. Interviews can be useful for respondents with limited literacy and education. Although rates have dropped for all surveys in recent years, interviews, especially in-person interviews, tend to have good response rates.

interviewer effect, the change in a respondent's behavior or answers that is the result of being interviewed by a specific interviewer.

On the other hand, interviews are more expensive than questionnaires because interviewers need to be hired and trained. In addition, using an interviewer adds another factor to the data collection process—the **interviewer effect,** or the changes in respondents' behaviors or answers that result from some aspect of the interview situation. As we noted previously, some studies have found that answers to the same question will vary depending on how the question is administered. The interviewer effect can be the result of the interviewers' personal and social characteristics and of the specific way each interviewer presents questions. Moreover, the same interviewer might present questions differently on different days or in different settings.

The majority of paid interviewers in many large surveys are women, and interviewers and respondents frequently differ from each other on important background traits including age, ethnicity, gender, and/or race. Characteristics such as race and gender have been found in some studies to affect responses for at least some topics (Davis, 1997; Hill, 2002; Kane and Macaulay, 1993). For example, in a study of attitudes on race and race-related issues, using both African American and white respondents and interviewers, the results varied by question topic. The researchers found no race-of-interviewer effects for some questions, but for other questions, African Americans provided less liberal racial attitudes to white interviewers, while white interviewers seemed to reduce the negative attitudes of white respondents (Krysan and Couper, 2003).

In-Person Interviews

in-person interview, an interview conducted face-to-face.

The **in-person interview** was the dominant mode of survey data collection in this country from 1940 to 1970 (Groves, 1989: 501). Depending on the sampling procedure—as, for example, when the researcher has a list of home addresses—and when there is a need for privacy in conducting the interview, interviews can be conducted in respondents' homes. Interviews can also be conducted in public places, such as an employee cafeteria or an office. The in-person interview is a good choice for questions involving complex reports of behavior, for groups difficult to reach by phone, for respondents who need to see materials or to consult records as part of the data collection (Bradburn and Sudman, 1988: 102), when the interview is long, and when high response rates are essential. The GSS, for example, collects data by means of an in-person interview of approximately 3,000 people every other year and 4,500 in 2006. The GSS response rate in 1975 was 76 percent and has stayed fairly constant, although it did drop to around 71 percent during data collections from 2000 to 2006 (Davis, Smith, and Marsden, 2007).

rapport, a sense of interpersonal harmony, connection, or compatibility between an interviewer and a respondent.

Another benefit of the in-person interview is **rapport,** or sense of interpersonal harmony, connection, or compatibility between the interviewer and respondent. Increased rapport might be why people are more willing to provide information on topics such as income in face-to-face interviews than on the phone (Weisberg, Krosnick, and Bowen, 1989: 100). Furthermore, the in-person interviewer can obtain visual information about respondents and their surroundings and a sense of the honesty of the answers especially if the interviewer pays attention to verbal and nonverbal cues.

On the other hand, beyond the interviewer effect, there are other disadvantages of in-person interviewing, including the inability of interviewers to make unscheduled contact with respondents in some buildings, such as in apartment buildings with security personnel at the entrances, and the reluctance of interviewers to go into rough neighborhoods (Bradburn and Sudman, 1988: 102). In addition, although it is standard practice to instruct interviewers to do interviews in private, some researchers report the presence of a third party during interviews, most typically a spouse, which has been found to affect some responses (Zipp and Toth, 2002).

In-person interviews are the most expensive of the self-report methods because they involve time to locate and contact respondents, the cost of interviewer's salaries, and travel time and expenses to the interview. Some interview studies now use computer-assisted personal interviewing (CAPI) methods where the interviewer uses a handheld computer to collect, store, and transmit interview data. As this method can be cost effective, it is likely to increase in the future. However, some research indicates that the addition of a computer into the interview affects the interaction in the way the interviewer asks questions, the way the interviewer enters responses, and the interviewer–respondent interaction when the computer demands take the interviewer's attention from the respondent (Couper and Hansen, 2001). Childs and Landreth (2006) found that interviewers using handheld computers obtained responses of acceptable quality for the most part, especially when short questions tailored to this mode of delivery were used.

Phone Interviews

telephone interview, an interview conducted over the telephone.

Before the 1960s, the proportion of households with telephones in the United States was too small to do interviews by phone for national probability samples. By the 1970s, 90 percent of U.S. households had telephones, and the phone interview was commonplace (Lavrakas, 1987: 10). Today, the **telephone interview** is a dominant mode of survey data collection in the United Sates (Holbrook, Green, and Krosnick, 2003). It became the preferred approach because it can yield substantially the same information as an interview conducted face-to-face at about half the cost (Groves, 1989; Herzog and Rogers, 1988), can closely supervise interviewers to aim for interview standardization, and, using random-digit dialing, can be conducted without names and addresses. Perhaps the greater feeling of anonymity is why phone interviews generated higher estimates of self-reported harm caused by alcohol use than did in-person interviews (Midanik, Greenfield, and Rogers, 2001). Phone interviews are especially useful for groups that feel they are too busy for in-person interviews, such as physicians, managers, or sales personnel (Sudman, 1967: 65). The interviewers do not have to travel, can do more interviews in the same amount of time, and require fewer supervisors (Bradburn and Sudman, 1988: 99). In addition, newer technology makes it cost effective to do computer-assisted telephone interviews (CATI), in which data are collected, stored, and transmitted during the interview.

On the other hand, disadvantages of phone interviews include the previously mentioned lower response rates when compared to in-person interviews and even mailed questionnaires and the elimination of those without landline phones from the sample. One estimate is that 7 percent of American households have neither landlines nor cell phones (Belinfante, 2005: 2). An additional concern is that about one fifth of Americans now have cell phones only (Blumberg and Luke, 2009), which typically means greater costs and higher refusals rates (Brick, Edwards, and Lee, 2007).

Qualitative interviews that asked people about their willingness to participate in telephone interviews indicate that potential respondents base their decisions to participate on the time they have available, their interest in the interview topic, and their evaluation of the purpose and sponsor of the study (Kolar and Kolar, 2008). This mode of interviewing means limiting the length of the interview and the kinds of questions that can be asked. Phone interviews tend to have a quicker pace and elicit shorter answers than the in-person version (Groves, 1989: 551). It is unlikely for respondents to stay on the phone for more than 20 or 30 minutes, whereas in-person interviews often last 40 or more minutes. Complicated questions are more difficult to explain over the phone than in person. The phone interviewer can't show something to the respondent and may not be able to judge the respondent's comprehension. Phone interviews are more likely to generate higher proportions of "no opinion" answers (Holbrook et al., 2003). "Sensitive" or threatening questions (questions that are perceived as intrusive or raise fears about the potential ramifications of disclosure) are answered less frequently and less accurately on the phone (Aquilino, 1994; Johnson, Hougland, and Clayton, 1989; Kreuter, Presser, and Tourangeau, 2008). Finally, the increasing use of technology designed to screen calls means a greater inability to contact potential respondents.

Because the request for a phone interview is frequently the first contact with a respondent, and requests for interviews are often mistaken for solicitations or telemarketing, the interviewer's introductory statement is crucial. It's up to interviewers to convince potential respondents that they aren't soliciting contributions or selling anything. The interviewer should make it clear that the research is legitimate and be open about the time the interview will take, the topic, and the sponsor of the study.

In interviews where an appointment is scheduled, the rest of life is "put on hold" to some extent while the interview takes place. But, in most surveys, interviewers call or ring doorbells while respondents are in the midst of daily life. Interruptions, like a baby crying or a neighbor stopping over, are often part of the interview process (Converse and Schuman, 1974).

For phone interviews, the interviewer must often call back to contact the household or find the person specified by sampling decisions. From 1979 to 1996, for example, the interviewers for the Survey of Consumer attitudes had to double the number of call backs to obtain the same response rate (Curtin, Presser, and Singer, 2000). Trying to reach potential respondents over a period of weeks rather than over a few days can result in a much higher response rate (Keeter et al., 2000).

Whether they are volunteers or paid staff, interviewers are trained so that they understand the general purpose of the study, are familiar with the introductory statement, the questions, and the possible answers. For the 2010 Census, classes for interviewers typically used a verbatim approach with a focus on reading interview scripts during a week's training session (GAO, 2007). In interviewing, the interviewer should be polite, open, and neutral, neither critical nor overly friendly. He or she should try to establish rapport or a sense of connection with the respondent, although this is more critical for qualitative interviewing, the topic of Chapter 10. In keeping with ethical standards of research, all interviewers should enter into an agreement not to violate the confidentiality of the respondents.

Interesting new developments are blurring the line between the interview and the questionnaire as researchers experiment with a variety of options, including virtual interviewers, human-like elements in questionnaires such as computer-generated faces asking questions, and doing interviews by instant messaging (IM). Fontes and O'Mahony (2008) did interviews about online social networks using IM and obtained a 60 percent response rate. They note that on a practical level, IM interviewing is a low-cost option with respondents from diverse locations that results in transcribed data.

Two studies focused on new ways of delivering questions. Krysan and Couper (2003) compared responses on racial issues using live and "virtual" interviewers (i.e., interviewers who delivered questions via pre-recorded videos) and found some race-of-interviewer and mode-of-interview effects. For some questions, the answers of African American respondents were less liberal when there was a white interviewer (both live and virtual) while white respondents gave more conservative answers to virtual black interviewers. Gerich (2008) compared questions delivered in video clips to some respondents to those that were self-administered and those asked by a real interviewer face-to-face. He found few differences by mode of administration, with some evidence that the "virtual" interviewers produced responses least affected by social desirability issues. As technology changes, it's likely that we'll see new modes of interviewing as well.

CONSTRUCTING QUESTIONS

Researchers make a series of decisions based on assumptions about how respondents read or hear the questions that are asked. Theories of answering questions assume that the stages are understanding and interpreting the question, retrieving or constructing an answer, and then reporting the answer using the appropriate format (Schaeffer and Presser, 2003: 67).

Types of Questions

open-ended questions, questions that allow respondents to answer in their own words.

Questions can be asked with or without answer categories, as in the examples in Box 9.6. **Open-ended questions** allow the respondents to answer in their own words. On a questionnaire, one or more blank spaces or lines can be used to indicate an approximate length for the answer as Lang and Podeschi did when they asked "I was born in …. PLEASE WRITE YOUR YEAR OF BIRTH ON THE ANSWER SHEET." In an interview, the interviewer waits for the response and can ask for an expanded answer or specific details.

BOX **9.6**

One Question—Two Formats

Open-Ended Version:
Thinking about your job and your spouse's, whose would you say is considered more important in your family?

Probe Questions for Follow-Up:
Could you tell me a little more about what you mean?
Could you give me an example?

Closed-Ended Version:
Whose job is considered more important in your family? Would you say your spouse's job is considered more important, both of your jobs are considered equally important, or your job is considered more important?

closed-ended questions, questions that include a list of predetermined answers.

Closed-ended questions use a multiple-choice format and ask the respondent to pick from a list of answer categories. One common practice is to use open-ended questions in a preliminary draft and then develop closed-ended choices from the responses. Answer categories for closed-ended questions must be exhaustive and, if only one answer is wanted, mutually exclusive. Exhaustive answers should have sufficient answer choices so that every respondent can find one that represents or is similar to what he or she would have said had the question been open-ended. (Sometimes "other" is included as one of the answers to make the list exhaustive.) Answer categories should be mutually exclusive so that each respondent can find *only one* appropriate answer. For example, some surveys offer the less useful choice of "single" rather than "never married" when asking about marital status on their questionnaire. Some divorced and widowed respondents might define themselves as single in addition to divorced or widowed and be tempted to check two answers. Another example is the common error made by beginning researchers to list overlapping categories, such as when age groups are used as closed-ended answers to a question about age. If ages are listed as 15 to 20, 20 to 25, and 25 to 30, the category set will not be mutually exclusive.

STOP & THINK *Lang and Podeschi used almost all closed-ended questions in their survey. They included four answer categories, from "strongly agree" to "strongly disagree," as responses to each of the statements on the environment and from "always" to "never" on questions about recycling behavior. Why do you think they used closed-ended rather than open-ended questions? Do you think the answer choices were sufficient?*

Selecting whether to use open- or closed-ended questions involves several issues. Answer choices can provide a context for the question, and they can make the completion and coding of questionnaires and interviews easier. Surveys that use closed-ended questions are typically less expensive to administer. On the other hand, respondents might not find the response that best fits what they want to say, and possible answer categories (such as "often" or "few") can be interpreted differently by different respondents. In addition, respondents' "presentation of self" might be influenced by using the response alternatives to infer which behavior is "usual," if they assume that the "average" person is represented by the middle category of a response scale (Schwarz and Hippler, 1987: 167).

Although a small percentage of respondents will not answer open-ended questions (Geer, 1988), and other respondents give incomplete or vague answers, leaving the answers open-ended does permit respondents to use their own words and frames of reference. As a result, open-ended questions typically encourage fuller and more complex answers that closed-ended categories miss. One time-saving and cost-effective alternative is to use mostly closed-ended questions and include a few open-ended ones where necessary. In questionnaires and interviews, all open-ended responses must be categorized before the researcher does statistical analyses. A limited number of answer categories must first be created for each question so that the data can be coded. **Coding,** the process of assigning data to categories, is an expensive and time-consuming task that can result in the loss of some of the data's richness.

coding, assigning observations to categories.

In the focal research, Lange and Podeschi restricted their instrument to closed-ended questions with a few open-ended questions. They felt that their closed-ended questions and answer categories adequately measured the variables of interest while minimizing the time necessary to complete the questionnaire and code the data.

screening question, a question that asks for information before asking the question of interest.

A useful type of question is the **screening question,** a question that asks for information before asking the question of interest. It's important to ask if someone voted in November, for example, before asking for whom they voted. Screening questions can help reduce the problem of "nonattitudes" (when people with no genuine attitudes respond to questions anyway). By asking "Do you have an opinion on this or not?" or "Have you thought much about this issue?" the interviewer can make it socially acceptable for respondents to say they are unfamiliar with a topic; when this type of question is used, some studies have found a sizable number saying they have no opinion (Asher, 1992). A screening question is typically followed by one or more **contingency questions,** which are the questions that are based on (or are contingent upon) the answer to the previous question (see Box 9.7). For example, in an interview, those who answer "yes" to the question, "Do you work for pay outside the home?" can then be asked the contingency question, "How many hours per week are you employed?" On a questionnaire, the respondent can be directed to applicable contingency questions with instructions or arrows.

contingency questions, questions that depend on the answers to previous questions.

Most self-report methods use direct questions to ask for the information desired, such as "What is your opinion about whether people convicted of murder should be subject to the death penalty?" Other self-report techniques are less direct.

 **STOP & THINK**

How would you feel about keeping a 24-hour time diary or a weekly calendar to record your activities? Do you think the data would be valid and reliable?

time diary, a self-report method that asks about amount of time spent on particular activities in time blocks.

One way to find out how people spend their time is to ask respondents to keep a **time diary** of daily events or create an event history calendar by listing information reconstructing their lives in specific time increments. Research

---

BOX **9.7**

Using a Contingency Question

Have you applied to a graduate or professional program?

☐ no

☐ yes →   If yes, please list the program(s)

_____

_____

_____

indicates that this approach produces higher quality data than a conventional, standardized questionnaire for some variables such as cohabitation, employment, unemployment, and smoking history (Belli, Smith, Andreski, and Agrawal, 2007). Hiromi Ono and Hsin-Jen Tsai (2008) studied computer use time among young school-age children using a time diary. A sample of 6- to 11-year-olds or a caregiver completed a 24-hour chronology for one randomly selected weekday and one weekend day. The researchers coded the data into several hundred activities and then aggregated them into 20 categories including computer activities, excluding school use. Ono and Tsai (2008) found that of all activities, home computer use time increased the most in percentage terms among 6- to 11-year-olds between 1997 and 2003 and that while the black–white difference in home computer use time was similar across years, the racial gap declined in percentage terms.

Occasionally researchers use indirect questions, especially when there is a socially desirable answer. One way to do this is to ask what others think as an indicator of the person's own opinion. For example, in 2005, polls found that 72 percent of Americans said that they would be willing to vote for a woman for president, but only 42 percent said that most of *their family*, *friends*, and *co-workers* would be willing to vote for a woman for president (Rasmussen Reports, 2005). More recently, Streb, Burrell, Frederick, and Genovese (2008) suggested another indirect way to measure the same attitude. Asking a respondent which of a list of things makes them "angry or upset," they found that 26 percent expressed anger over the statement "A woman serving as president."

vignettes, scenarios about people or situations that the researcher creates to use as part of the data collection method.

Using **vignettes**, scenarios or stories about people or situations that the researcher creates, is another approach. The use of vignettes may be less threatening for sensitive issues than direct questions. In study of friendship norms, Felmlee and Muraco (2009) used vignettes with short descriptions about hypothetical friends in specified circumstances. For example, one vignette was "Suppose that you have plans to go to a movie with a friend, Jim, next Saturday. On Friday, Jim calls you and says that he has plans to do something with his girlfriend and that he has to cancel plans with you." The respondent was first asked to rate the appropriateness of the friend's behavior, using a Likert-type scale, with answer categories ranging from 1 (*extremely inappropriate*) to 7 (*extremely appropriate*), and then asked to answer the open-ended question, "Why?" The researchers believe that vignettes allow respondents to examine norms in situational contexts and identify situations with contradictions regarding normative expectations (Felmlee and Muraco, 2009).

How to Ask Questions

Keep questionnaires and interviews as short as possible, asking only the questions that are necessary for the planned analysis. Each concept or variable is measured by one or more questions so consider the issues of validity and reliability to make sure that questions asked of respondents are accurately and consistently measuring what they are supposed to be measuring. Lang and Podeschi, for example, measured students' social class backgrounds with two questions, one open-ended, "For work, my parents … PLEASE WRITE YOUR PARENTS' JOBS ON THE ANSWER SHEET"), and one closed-ended, "Growing up, my family was… A: very wealthy B: comfortable C: struggling D: poor." But, after analyzing their data, they concluded that these were not adequate. They intend to do similar surveys in the future and will ask additional questions about social class (Lang and Podeschi, personal communication).

If you will be constructing scales or indices from answers to a series of questions, consider the approach Lang and Podeschi used in constructing their New Ecological Paradigm (NEP) index. They varied the ends of the four-point answer categories so that "strongly agree" was not always the more environmentally concerned answer. This meant that respondents couldn't just find a pattern—like choosing "strongly agree" and follow it—but had to think about each question. In calculating the NEP scores, the researchers had to reverse the coding for some of the statements. This meant that, for example, for the statement "Humans were meant to rule over the rest of nature," the answer "strongly disagree" was the most environmentally concerned answer and received a code of "4," but for the statement "Plants and animals have just as much right as humans to exist," the answer "strongly agree" was coded as "4."

If you plan on doing a survey, it's useful to start by making a list of the variables that are of interest and then do a literature search to see if another study's measurement strategy can be used or adapted. You can adapt questions that other studies have used by checking out sources like *Handbook of Research Design and Social Measurement* (Miller and Salkind, 2002) for ways to operationalize variables.

pilot test, a preliminary draft of a set of questions that is tested before the actual data collection.

A preliminary draft or **pilot test** should be used with a small sample of respondents similar to those who will be selected for the actual study. Giving the interview or questionnaire in this kind of pre-test will help determine if the questions and instructions are clear, if the open-ended questions generate the kinds of information wanted, and if the closed-ended questions have sufficient choices and elicit useful responses. Lang and Podeschi tried out drafts of their questionnaire and made numerous changes based on the feedback from the pilot tests.

 **STOP & THINK** *An interviewer was thinking of asking respondents, "Do you think that men's and women's attitudes toward marriage have changed in recent years?" and "Do you feel that women police officers are unfeminine?" Try to answer these questions. Do you see anything wrong with the wording?*

Guidelines for Question Wording

Interviews and questionnaires are based on language, so it's important to remember that words can have multiple meanings and can be interpreted differently by different respondents. Keeping questions concise and using carefully selected words is important, as the specific choice of words can affect the results. There will always be advantages and disadvantages in asking questions in specific ways. A study of affirmative action, for example, found 35 percent agreeing that where there has been job discrimination against blacks in the past, "preference in hiring and promotion" should be given, while 55 percent said they favored "special efforts" to help minorities get ahead (Verhovek, 1997: 32). Studies with questions about public spending priorities found much more support for "providing assistance to the poor" than "spending for welfare" (Smith, 1989). One study found 45 percent saying they favored physician-assisted suicide when the question was "Do you favor or oppose physician-assisted suicide," while another found 61 percent saying "yes" to the question "When a person has a disease that cannot be cured and is living in severe pain, do you think doctors should be allowed by law to assist the patient to commit suicide if the patient requests it, or not?" (Public Agenda, 2002). In cross-national studies and within countries with large immigrant populations, translation becomes important. The choice of words is particularly significant because there might be no functional equivalents of some words in other languages. In one study, for example, Japanese researchers concluded that no appropriate word or phrase in Japanese approximated the Judeo-Christian-Islamic concept of God (Jowell, 1998).

Our suggestions for question construction include

1. Avoid *loaded words*, words that trigger an emotional response or strong association by their use.
2. Avoid *ambiguous* words, words that can be interpreted in more than one way.
3. Don't use *double negative* questions, questions that ask people to disagree with a negative statement.
4. Don't use *leading questions*, questions that encourage the respondent to answer in a certain way, typically by indicating which is the "right" or "correct" answer.
5. Avoid *threatening* questions, questions that make respondents afraid or embarrassed to give an honest answer.
6. Don't use *double-barreled* or *compound* questions, questions that ask two or more questions in one.
7. Ask questions in the language of your respondents, using the idioms and vernacular appropriate to the sample's level of education, vocabulary of the region, and so on.

Now review the poorly worded questions and the suggested revisions in Box 9.8 to see how these general guidelines can be applied to specific question wordings.

 **STOP & THINK** *Did you identify what was wrong with the questions in the previous "Stop and Think"? Did you see that the first question is double-barreled in that it asks two questions about marriage, one about men's attitudes and one about women's? Did you notice that the word "unfeminine" in the question about women police officers is really a loaded word?*

Response Categories Guidelines

When response categories are used, a large enough range of answers is needed. Some common closed-ended responses are listings of answers from strongly agree to strongly disagree, excellent to poor, very satisfied to very dissatisfied, or a numerical rating scale with only the end numbers (such as 1 and 10) given as descriptors. For some issues, it's better to use a continuum rather than two answer choices. Lee, Mathiowetz, and Tourangeau (2007) compared the yes–no format to continuums and found the rates of disability

BOX **9.8**

Examples of Poorly Worded Questions and Suggested Revisions

Ambiguous question:

In the past year, about how many times have you seen or talked with a doctor or other health professionals about your health?

Revised:

We are going to ask about visits to doctors and getting medical advice from them. How many office visits have you had this year with all professional personnel who have M.D. degrees or those who work directly for an M.D. in the office, such as a nurse or medical assistant?

Double negative question:

Do you disagree with the view that the President has done a poor job in dealing with foreign policy issues this past year?

Revised:

What do you think of the job that the President has done in dealing with foreign policy issues this past year?

Double-barreled question:

What is your opinion of the state's current economic situation and the measures the governor has taken recently?

Revised:

What is your opinion of the economic situation in the state at present?

What do you think of the measures that the governor has taken in the past month?

Leading question:

Like most former presidents, the President faces opposition as well as support for his economic policies. How good a job do you think that the President is doing in leading the country out of the economic crisis?

Revised:

Are you aware of the President's economic policies?
     If so, what is your opinion of them?

Threatening question:

Have you been arrested for driving under the influence of alcohol in the past year?

Revised:

In the past year, have you driven a vehicle after having had too much to drink?

---

to be higher when respondents had answer choices with a range (such as "no difficulty at all" to "completely unable") for questions like "How much difficulty do you have performing basic physical activities such as walking or climbing stairs?" For questions with more than one possible answer, a list can be used and the respondent can be asked to identify all that apply, by asking, for example, "Which of the following television shows have you watched in the past week? Check all that apply." Some, however, argue against a "mark all that apply" approach. Research finds that, although it takes more time, deeper processing and more options are selected when questions are broken into separate questions that are followed by a forced choice format such as "yes" or "no" (Rasiniski, Mingay, and Bradburn, 1994; Smyth, Dillman, Christian, and Stern, 2006).

Another debate is whether a "no opinion," "don't know," or "middle category" should be offered. (For example, "Should we spend less money for defense, more money for defense, or continue at the present level?") Some argue that it is better to offer just two polar alternatives, whereas others claim it is more valid to offer a middle choice. Studies have found people to be much more likely to select a middle response if it is offered to them (Asher, 1992; Schuman and Presser, 1981) and that listing the "middle" option last rather than in the middle makes it more likely to be selected (Bishop, 1987). Lang and Podeschi decided to use four answer categories ranging from strongly agree to strongly disagree for one set of questions after "really grappling with the inclusion of a fifth response to our Likert scales. We considered putting in a 'neutral' or 'neither agree nor disagree' response. In the end, we did not include that choice because we were worried that most respondents would gravitate towards it as an answer" (Lang and Podeschi, personal communication).

Question Order and Context

Responses to questions can be affected by the question order as earlier questions provide a context for later ones and people may try to be consistent in their answers to questions on the same topic. For example, one study found that reports of alcohol consumption were higher when questions were included on the Semi-Quantitative Food Frequency Questionnaire than when they were used on a questionnaire exclusively targeting alcohol use (King, 1994). Another study found that support for gender-target affirmative action decreased when the questions followed those about race-targeted affirmative action; however race-targeted affirmative action support increased when the questions followed those about affirmative action based on gender (Wilson, Moore, McKay, and Avery, 2008). Some research on question order in surveys of attitudes on privacy issues indicates that when order effects are found, they are not especially large and may be relatively local, that is, restricted to a few nearby items that are related conceptually (Tourangeau, Singer, and Presser, 2003).

There are no easy solutions to the issue of question order because each choice involves tradeoffs. Consider a logical order that makes participation easier for respondents. To encourage respondent participation, start with interesting, non-threatening questions, and save questions about sensitive topics for the middle or near the end. If questions about emotionally difficult topics are included, it's preferable to have some "cool down" questions follow them so as to minimize psychological discomfort.

SUMMARY

Self-report methods are widely used in social research. They can be used profitably to collect information about many different kinds of variables, including attitudes, opinions, knowledge, social characteristics, and some kinds of behavior. Clear introductory statements and instructions and the judicious use of carefully worded open- and closed-ended questions are critical if questionnaires and interviews are to provide useful results. As we saw in the focal research article on environmentalism, the researchers were able to collect enough data inexpensively with clearly worded questions and answer choices and a sample of respondents who were willing to participate.

Questionnaires and structured interviews are less useful when the topic under study is difficult to talk about or when asking questions about a topic is unlikely to generate honest or accurate responses.

In choosing between the various kinds of questionnaires and interviews, factors like cost, response rate, anonymity, and interviewer effect must be considered. Questionnaires are typically less expensive than interviews. Group-administered questionnaires, like the one used by Lang and Podeschi, are among the least expensive and typically have good response rates. In addition, questionnaires can usually be completed anonymously and privately, which makes them a very good choice for sensitive topics. If the questionnaire is delivered on the web, by mail, or by e-mail, the researcher will need to think of ways to encourage participation, as these methods tend to have lower response rates than group-administered ones.

Interviews, especially in-person ones, usually have good response rates and little missing data. They can be especially helpful with populations that have difficulty reading or writing and when topics are complex. Phone interviews have lower response rates but allow researchers to collect self-reports from national samples in a less costly way than in-person interviews, but the latter allow for better rapport. For both kinds of interviews, interviewer effect should be considered.

EXERCISE 9.1

Open- and Closed-Ended Questions

Check with your instructor about the need to seek approval from the Human Subjects Committee, the IRB, or the committee at your college or university concerned with evaluating ethical issues in research. If necessary, obtain approval before doing the exercise.

1. Make five copies of the list of questions that follows. Find five people who are working for pay and are willing to be interviewed. Introduce yourself and your project, assure potential participants of the confidentiality of their answers, and ask them if they are willing to participate in a brief study on occupations. Write down their answers to these open-ended questions as completely as possible. At the end of each interview, thank the respondent for his or her time.

   a. Are you currently employed? (If yes, continue. If no, end the interview politely.)
   b. What is your occupation?
   c. On average, how many hours per week do you work?
   d. Overall, how satisfied are you with your current job?
   e. In general, how does your current job compare to the job you expected when you took this job?
   f. [*If the job is different from what was expected*]
      In what ways is the job different from what you expected?
   g. If it were up to you, what parts of your job would you change?
   h. What parts of your job do you find most satisfying?

2. Before data can be analyzed statistically, it must be coded. Coding involves creating a limited number of categories for each question so that each of the responses can be put into one of the categories. From your interview data, select the answers to three questions from d to h and for each respondent, *code* the data using *no more than five* response

categories. We've started a table below that you can use as a model for your coding.

Respondent	Answers to Question d coded	Answers to Question __ coded	Answers to Question __ coded
1	She is very satisfied with her job		
2	He thinks that his job is better than most so is somewhat satisfied		
3	She thinks her job is awful and is very dissatisfied		
4	He thinks there is an equal number of good and bad things about this job		
5	_____		

3. Based on your respondents' answers, your coding schemes, and other possible answers to the questions, rewrite and include the three interview questions you selected for Question 2 and make them closed-ended by including the answer categories.
4. Describe your response rate by noting the number of people you asked to participate to get five respondents. Based on your experience, comment on the reasons you think some people agree to participate in a short in-person interview and others decline.
5. Attach the five completed interviews to your exercise.

EXERCISE 9.2

Asking Questions (or Have I Got a Survey for You!)

Check with your instructor about the need to seek approval from the Human Subjects Committee, the IRB, or the committee at your college or university concerned with evaluating ethical issues in research. If necessary, obtain approval before doing the exercise.

1. Select between two and four concepts or variables to study. Do not choose "sensitive" or threatening topics. (Appropriate topics include opinions about candidates running for office, career aspirations, involvement in community service, marriage and family expectations, social characteristics, and so forth.) List the variables you have selected.

2. Construct between one and five questions to measure each variable. (Feel free to use or adapt questions from articles you've read or those listed in sources like *Handbook of Research Design and Social Measurement* (Miller and Salkind, 2002).) Develop a list of no more than 20 questions.
3. Make six copies of the interview questions, leaving enough space to write down respondents' answers. Contact potential respondents to find six people who are willing to be interviewed. Introduce yourself and your project, assure potential participants of the confidentiality of their answers, and ask them to participate. Do three of the interviews over the phone and three in person.
4. Turn the list of interview questions into a questionnaire, complete with instructions

and a cover sheet explaining the research and its purpose. (Introducing yourself as a student and the survey as a part of the requirements of your course would be appropriate.) Give the questionnaire to three new people to complete.

5. Compare the information you obtained using the three methods of data collection. Comment on the validity and reliability of the information you received using each method. Which did you feel most comfortable using? Which was the most time-consuming

method? If you were planning an actual study using these questions, which of the three methods would you select? Why?

6. Based on your experience, select one of the methods, and make any changes to the questionnaire or list of questions that you think would make the answers more valid indicators of your variables in an actual study.

7. Attach the six completed interviews and the three completed questionnaires to this exercise as an appendix.

EXERCISE 9.3

Direct and Indirect Questions

Check with your instructor about the need to seek approval from the Human Subjects Committee, the IRB, or the committee at your college or university concerned with evaluating ethical issues in research. If necessary, obtain approval before doing the exercise. In this exercise, you'll be comparing asking direct questions versus asking indirect questions using a vignette.

1. Make five copies of the questionnaire that follows, leaving enough space after each of the first four questions for the respondent to write a few sentences. Get five envelopes to give with the questionnaires.

2. Using a convenience sample, find five people who are willing to answer this questionnaire anonymously. Have them complete the questionnaire while you wait, and ask them to seal it in the envelope before returning it to you.

3. Analyze the information on the completed questionnaires. Specifically, compare the information from the first four direct questions to the information from Question 5, an indirect question. Comment on the validity and reliability of the information you received using each way of asking questions. Which way of asking questions do you think is better? Why?

4. Attach the five completed questionnaires to this exercise as an appendix.

Questionnaire on Family Roles

Thank you for agreeing to participate in a study on family roles for a course on research methods. Please don't put your name on this questionnaire so that your responses will be anonymous. Answer the first four questions with a few sentences. For the last set of questions, place an X on the spot on the scale that best reflects your point of view.

When you have completed your responses, please put this questionnaire in the envelope provided, seal it, and return it to me.

1. All couples in long-term relationships need to make decisions about family roles. How do you think married couples should divide household and economic responsibilities?

2. If there are children in a two-parent household, do you think that the mother's and father's roles should be different in some ways when it comes to caring for the children?

3. Do you think husbands should be expected to make career sacrifices to accommodate their families? If so, what kinds of sacrifices?

4. Do you think wives should be expected to make career sacrifices to accommodate their families? If so, what kinds of sacrifices?

5. Susan and Steve have been married for five years and both have worked full-time during this time. Susan makes a slightly higher salary than Steve and is expecting a promotion soon. Steve's company has been bought by

another company, and he has been told that his job will disappear after the two companies merge. Susan is pregnant with their first child, and both she and Steve agree that it will be better for their child if one parent stays home for the first few years of the child's life. After discussing their situations, Steve and Susan agree that she will continue to work full-time while he will stay home with the baby.

What's your impression of Susan and Steve? Put an X on each of the following scales.

Susan is

1	2	3	4	5

dedicated to family   not dedicated to family

Susan is

1	2	3	4	5

a caring person   not a caring person

Susan is

1	2	3	4	5

ambitious   not ambitious

Steve is

1	2	3	4	5

dedicated to family   not dedicated to family

Steve is

1	2	3	4	5

a caring person   not a caring person

Steve is

1	2	3	4	5

ambitious   not ambitious

EXERCISE 9.4

Using Structured Interview Questions from the GSS

The GSS, conducted by the National Opinion Research Center (NORC), uses a structured interview schedule. The GSS was administered annually from 1972 and then biennially from 1994 to a national probability sample. Some questions are asked on every survey, other questions have rotated, and some have been used only in a single survey.

The GSS measures a great many variables using answers to questions as indicators. The variables included on past surveys cover a wide range of behaviors, among them drinking behavior, membership in voluntary associations and the practice of religion, and numerous attitudes, such as opinions about abortion, affirmative action, capital punishment, family roles, and confidence in the military.

Browse through the archives of the GSS to find questions asked between 1972 and 2008, which is available at http://sda.berkeley.edu/archive.htm and select a few variables of interest.

1. List the names of the variable and the questions on the GSS that are the indicator(s) of those variables.
2. Comment on the specific questions that were used. Do you think they are valid indicators for the variables? Are the questions clear and easy to understand? What do you think of the answer categories?

EXERCISE 9.5

Doing a Web Survey

There are many websites that allow users to create and distribute their own web-based surveys. Some of them are free of charge. Check out this process by using a website to construct and implement a questionnaire. Afterward, evaluate the data you obtained. Some sites that you can check out are

http://www.zapsurvey.com/
www.zoomerang.com/index.zgi
www.formsite.com
http://www.surveymonkey.com/

When you have completed the survey, write a brief report about this experience including the following:

1. What was the topic you were interested in?
2. Print a copy of your questionnaire.
3. Evaluate the quantity and quality of the data you obtained using the Internet survey method.
4. Based on your experience, what do you think of web-based surveys?

CHAPTER 10

Qualitative Interviewing

© Somos/Veer/JupiterImages

INTRODUCTION

Suppose a researcher called you and asked you to spend an hour or two talking about your college experiences and your plans for the future. Would you agree to be interviewed? Would you feel flattered that someone wanted to hear some of your life story or annoyed that you were being asked to spend the hour or two? Would you feel intimidated by having your views recorded? On the other hand, consider being the interviewer. How would you feel about conducting a series of interviews with your peers?

In this chapter, we'll focus on qualitative interviewing, a social science tool for more than a century—and a method whose uses have been debated for almost as long (Lazarsfeld, 1944; Thurlow, 1935). The less structured interview, which has been called "the art of sociological sociability" (Benney and Hughes, 1956: 137), is an important method of data collection. These more intensive interviews are opportunities to learn a wide variety of things, among them people's backgrounds and experiences, their attitudes, expectations, and perceptions of themselves, their life histories, their sense of the meaning of events, and their views about groups of which they are a part and organizations with which they interact.

 **STOP & THINK** *We discussed the structured interview in the last chapter. How do you think it differs from the qualitative interview that we'll be discussing in this chapter?*

qualitative interview, a data collection method in which an interviewer adapts and modifies the interview for each interviewee.

structured interview, a data collection method in which an interviewer reads a standardized list of questions to the respondent and records the respondent's answers.

The **qualitative interview**, also called the in-depth or intensive interview, has much in common with the **structured interview** discussed in the previous chapter. Both kinds of interviews rely on self-reports and anticipate that the interviewer will do more of the asking and the respondent will do more of the answering. Both kinds of interviews, "far from being a kind of snapshot or tape-recording … in which the interviewer is a neutral agent who simply trips the shutter or triggers the response" (Kuhn, 1962: 194), are opportunities for data to emerge *from the social interaction* between the interviewer and interviewee.

Despite these commonalties, the two kinds of interviews are distinct. In the structured interviews that we discussed in Chapter 9, the interviewer uses a standardized set of questions, heavily weighted toward closed-ended questions. The questions and closed-ended answer choices are delivered to each respondent with as little variation as possible to achieve uniformity of questioning. In addition, some kinds of structured interviews, such as phone interviews using random-digit dialing, can be done anonymously without identifying the respondent.

In contrast, the qualitative interview is designed to allow the study's participants to structure and control much more of the interaction. Some researchers use the interview to obtain information about the interviewees' lives and social worlds; others describe it as an opportunity for participants to tell their stories or construct narratives. In the qualitative interview, as Box 10.1 illustrates, the interviewer asks open-ended questions (either preformulated or constructed during the interview), frequently modifies the order and the wording of the questions, and typically asks respondents to elaborate

BOX **10.1**

What Would You Say?

You're seated across from the interviewer and she continues the interview by asking you, "So, how has your family been of help as you've pursued your undergraduate education?" Perhaps while you think about what to say, you wonder exactly who you should talk about in your answer. You might think she means for you to talk about your mother or father, who have saved over the years to help pay your tuition. But you wonder if she also means for you to talk about your great-aunt, whom you rarely see but who sometimes sends you little notes of encouragement. Then you consider if she can possibly mean for you to tell about your "cousin," who is not a relative at all, but the family friend who has helped you find a summer job each year. You might even consider mentioning your roommate, the person who gives you emotional support when you most need it. If you're like most participants in an interview, you have your own unique understanding of what is meant by each word and question. Fortunately, the in-depth interview can provide an opportunity to foster the kind of give-and-take where the participants can ask for clarifications about the questions and discuss their meanings.

rapport, a sense of interpersonal harmony, connection, or compatibility between an interviewer and an interviewee.

or clarify their answers. One of the major advantages of the more conversational tone of the qualitative interview is that it is useful for allowing **rapport** or a sense of connection to develop between the interviewer and interviewee.

Before the interview, the interviewer typically knows the topics the interview is intended to cover and the kind of information that is desired, but has the opportunity to tailor each interview to fit the participant and situation. In the in-depth interview, the answers are usually longer; the interviewees can impose their own structure, ask for clarification about questions, and answer in their own terms; and the interviewers can follow up by asking additional questions. Because these interviews are almost always scheduled in advance and usually occur face-to-face, the interviewer typically knows the identity of the interviewee and, ethically, must keep confidential all information obtained.

Qualitative interviewing has been used as the sole data collection technique in a wide variety of studies, including a classic study of white women's social constructions of race (Frankenberg, 1993), research on friendship, trust, and love in Latino marriages (Harris, Skogrand, and Hatch, 2008), and a study of how pregnant women perceive ultrasounds in dealing with fetal health anxiety (Harpel, 2008). Qualitative interviewing has also been used in conjunction with **observational techniques** (collecting data by observing people in their natural settings, which we will discuss in detail in Chapter 11) and other methods of data collection. For example, Gayle Sulik (2007) completed 60 interviews and did participant observation while working for six months as an administrative assistant and volunteer with a community-based breast cancer organization to learn how women with breast cancer balance illness with caring for themselves and for others. To study how adoptive parents validate their children's ethnocultural origins, Amy Traver (2007) conducted semi-structured interviews with 91 Americans adopting children from China, read first-person accounts of adoption from China, attended cultural celebrations with Families with Children from China, and spent weekends at Chinese culture camps.

observational techniques, methods of collecting data by observing people, most typically in their natural settings.

QUALITATIVE VERSUS STRUCTURED INTERVIEWS

The techniques of the structured interview are quite appropriate when studying homogeneous populations not too different from the investigator (Benney and Hughes, 1956: 137), when the focus is on topics that lend themselves to standardized questions, and when large samples and cost containment are necessary. As we noted in the last chapter, structured interviews can be useful when the researcher wants to collect data on many topics, especially when surveying representative samples.

On the other hand, structure in an interview can limit the researcher's ability to obtain in-depth information on any given issue. Furthermore, using a standardized format implicitly assumes that all respondents understand and interpret questions in the same way. Another concern is that the interaction in such interviews is very asymmetric and hierarchical with interviewers having control of the interaction and not sharing their views. Some critics have gone so far as to argue that the approach used by the structured interview "breaks the living connection with subjects so that it is forever engaged in the dissection of corpses" (Mies, 1991: 66).

For researchers who are less interested in measuring variables and more interested in understanding how individuals subjectively see the world and make sense of their lives, the less structured approaches are more appropriate. In qualitative interviewing, the interviewer can "break the frame of the interview script" and shift her or his perspective to that of the interviewee and adapt the questions for the person being interviewed rather than using standardized wordings.

Encouragement to elaborate can increase the interviewee's personal investment in the interview and decrease the sense of being a machine producing acceptable answers to questions (Suchman and Jordan, 1992: 251). Qualitative interviews allow the researcher an insight into the meanings of their participants' everyday lives by exploring with them their behavior, roles, feelings, and attitudes. Interviewees can tell their stories in ways that are meaningful to them and to use their own words rather than selecting from a set of responses. Although the interviewer and interviewee are usually strangers, the qualitative interview can provide vibrant data. Such interviews can facilitate research with greater depth and breadth and allow us to understand expectations, experiences, and worldviews of interviewees without imposing the external structure of standardized questions.

Perhaps because of its engaging nature, the qualitative interview typically has a high response rate. For example, in a study of heroin and methadone users, 90 percent of those approached in two methadone clinics agreed to complete a one- to two-hour interview (Friedman and Alicea, 1995: 435). In the focal research that follows, most of those asked by Sandra Enos agreed to participate and complete a qualitative interview.

Sandra Enos, a sociologist whose research interests included both families and corrections, combined these interests with a study on how the social processes of mothering are worked out when mothers are in prison. She first collected data while working as a volunteer in the prison parenting program, observing and talking informally to the participants. In the excerpt that follows, she draws on her qualitative interviews for her analyses of "mothering" while in prison.

Managing Motherhood in Prison

By Sandra Enos*

Introduction

The United States has experienced an incarceration boom. By mid-year 2000, the population behind bars had grown from 732,000 in 1985 to just under 2 million prisoners (Beck and Karberg, 2001). The United States has one of the highest incarceration rates in the world, which means that proportionately we send more individuals to prison than does almost any other country in the world (Mauer, 1997). Although males are more likely to be imprisoned, a growing number of women are finding themselves serving terms behind bars. While the "imprisonment boom" increased the male population behind bars by 165 percent, the number of sentenced women and those awaiting trial increased by over 300 percent during the same time (Beck and Karberg, 2001).

The children of these inmates are also affected by imprisonment. It is estimated that there were 1.5 million children with parents in prison in 1999 (Mumola, 2000). Research has shown that children of inmates are likely to follow their parents' criminal careers; their grades are likely to suffer in school; they display behavioral problems; and they become early users of drugs and alcohol (Johnston, 1995). The impact of having either parent in prison is likely to be detrimental, but because mothers are more often the primary caretakers of children, the incarceration of a female inmate is especially burdensome for children and for families. Most women in prison are mothers, with approximately 65 to 80 percent of them having children under the age of 18 (Baunach, 1985). Because arrest, pretrial holding, and court processing are usually unplanned, the placement of children may be haphazard; imprisonment means that existing family and extended family units must reorganize to adjust to life without the offender. Nationally, 90 percent of children who have fathers in prison live with their mothers, while only 25 percent of those with mothers in prison reside with their fathers (Snell, 1994). The children of inmate mothers are more likely to experience changes in their living arrangements when their mothers come to prison and may, in fact, be significantly more affected by the incarceration of their mothers than by that of their fathers.

Inmates as Mothers

Mothering under normal conditions is challenging and often stressful. A variety of myths and ideologies about the "family" and about

*This article has been revised by Sandra Enos from her paper of the same name that appeared in the first edition of this book and is published with permission.

mothering carry enormous weight in influencing our values, norms, and behavior patterns. These ideas about what it means to be a family and what it means to be a mother have also affected the research that social scientists have done (Baca Zinn and Eitzen, 1993). Traditional sociological work has proposed one normative family form and defines other forms as deviant or underdeveloped (O'Barr, Pope, and Wyer, 1990).

Some of the most important myths that underlie contemporary American views of mothering include the following: mothers are totally absorbed in mothering, the mother is the only figure who can give the child what s/he needs, the mother has resources to give the child what s/he needs, and the mother should be able to manipulate the child's behavior toward successful adulthood (Caplan and Hall-McCorquodale, 1985; Chodorow and Contratto, 1982). These assumptions do not make room for mothering styles outside the middle class (Dill, 1988; Maher, 1992) or consider that motherhood is experienced differently by women in different race/ethnic categories or life cycle positions. There is only a small body of work (Collins, 1990; Rubin, 1994; Stack, 1974; Young and Wilmott, 1957) that has investigated how mothers under strain and economic stress "do mothering." There is little room for models of mothering where caretakers share mothering responsibilities, such as "other mothers" identified by Collins (1990). Few have studied mothering in a way that makes room for interpretations of the mothers themselves. It is important to analyze mothering, including race and class differences, in a way that goes beyond overly idealistic images or negative categorizations of mothers as evil or neglectful caretakers. In other words, it is important to "unpack" motherhood if we wish to learn what is really involved in this work.

Because images of mothers are idealized, we are likely to look harshly upon women who violate our expectations. Women's involvement in crime and drug use is viewed through a gendered lens. Women are strongly condemned for these behaviors because they suggest ignoring the welfare of their children. These expectations seldom come into play for male offenders. When women who are mothers are sent to prison, this brings the case not only to the attention of criminal justice agencies but also to the attention of child welfare authorities. In some instances, a sentence of longer than one year can be grounds to move to terminate a mother's right to a child. So, the importance of motherhood carries more weight for a female inmate than does fatherhood for a male inmate.

Several research questions guided this project: How mothers remain connected to their children, how they maintain a place and position in their families, how they arrange and evaluate child-care options and the impact on their long-term careers as mothers are elements of the social process of "managing motherhood" in prison.

The Sample and Method

A women's prison in the Northeast served as the study site. The facility is small in size, holding a maximum of 250 women awaiting trial or serving sentences. Approximately half of the mothers in the prison were involved with the state's child welfare agency as a result of complaints of child abuse, neglect, or the presence of drugs in infants at birth. In other cases, the women sought help from child welfare to place children when they felt they had no other viable option.

The parenting program in this prison is operated off-site and allows considerable freedom from usual security constraints. The program facility is filled with toys and more closely resembles a school than a correctional site. Child welfare authorities view participation in the program as an indication of inmate mothers' interest in rehabilitation. Participation in the program is a way for inmate mothers to demonstrate their "fitness" as parents. This may help them to regain custody and resume care of their children upon release. For these reasons, participation in the program is a welcome opportunity.

To participate in the parenting program, the women had to be mothers or guardians of minor children, have either minimum or work release security status, and be free from work obligations at the time of the program. The participants in the parenting program were serving sentences similar to those of the general population—ranging from six months to a few years for nonviolent offenses such as embezzlement, fraud, soliciting for prostitution, drug offenses, breaking, and entering.

I was granted access to the parenting program by the Warden and the Director of the parenting program in 1993. I assumed two roles— observer and volunteer/program aide. The mothers gave me permission to sit in on the parenting group (an hour or more discussion group designed to help mothers work out issues related to reunification with children) and to observe the children's visits. Attendance at the program varied from week to week, attracting from 7 to 12 women and between 12 and 25 children. There was turnover in the population; approximately 20 women participated during the four months I was an observer.

I attended 16 sessions and held informal conversations with both the women and their children during the parent–child visits. The conversations were typically about the child and the women's path to prison. Although I was able to interview some women fairly completely during this time, the informal conversations with the women and children were typically limited to a few minutes at each session so that my research interests did not detract from the real purposes of the parenting program— to allow contact and enhance communication between parents and their children. During this time, I took extensive field notes, and generated methodological and analytical memos, a method that supports the grounded theory approach (Charmaz, 1983).

I observed two differences during this preliminary work: Both the placement of children and their mothers' paths to prison were different for African Americans and whites. An interest in the patterns and possible connections among them led me to select qualitative interviewing as my main method of data collection. I conducted 25 additional interviews,

most with women who had participated in the parenting program. I
gathered a purposive sample consisting of women serving short sentences and those with longer terms; those with involvement with child
welfare and those without; and with women who were recidivists and
those serving their first terms in prison. Because I was interested in
the effect race and ethnicity played on the management of motherhood
in prison, the sample reflected the racial/ethnic composition of the
prison population. Finally, because I was investigating where children
were placed, I located mothers whose children were living in a variety
of placements: with grandmothers, with fathers, in foster care, and with
other relatives. The 25 women had given birth to 77 children, 18 of whom
had been adopted by relatives or other individuals.

I identified myself as a researcher and told the mothers that the interview was about the challenges female inmates faced in maintaining
relationships with their children while in prison. I made it clear that unlike
other conversations about their children with representatives from child
welfare, counseling agencies, drug treatment facilities, school departments, and so on, their conversations with me would not directly affect
their status at the prison. All signed informed consent forms (see
Appendix B) and were assured that our conversations were confidential.
Most were curious about my study and wanted to share their stories,
hoping it might help others or lead to changes in the parenting program.

In the lower security levels, the interviews were held in a hallway or program area. In the higher security level, a correctional official stood outside the
room where I was allowed to conduct interviews. Some of the participants
allowed me to tape-record the interview and all but one of the rest agreed to
my taking notes. One woman insisted on neither notes nor tape, which required reconstructing the interview from memory immediately after leaving
the site. The interviews ranged from one to two and one-half hours.

The Emotional Content of Interviewing

I was concerned about how forthcoming the women would be. Given the
dynamics related to the presentation of self and to the pressures for women
to conform to idealized conceptions of mothering, I was concerned about
the fruitfulness of certain lines of questioning. For the most part, I directed
conversations along paths initially suggested by women themselves. In
later interviews, I occasionally paraphrased the responses from previous interviews as a way of giving the women something to respond to. In very
few instances did I ask questions that would result in a yes or no answer.

My aims were to strike a balance between protecting each woman's
privacy, getting an accurate picture of how she viewed her role and
place as a mother, and hearing her story in her own voice. To establish
this balance, I asked very broad questions in a semi-structured interview
format, such as "What brought you to prison?" (See Appendix A for the
interview schedule.) The answers often required additional clarification.
Some women had been to prison several times and were confused

about whether I was referring to their first incarceration or their most recent term. The question about where their children were living was too vague, as some women had been caring for children who were not their biological children and others had already lost custody of their children to the state or other caretakers before imprisonment. Many issues were brought into the conversation by the women themselves. For instance, several women, more typically the middle-class women or those in prison for the first time, brought up problems with the conditions of confinement, such as the lack of meaningful programming and the arbitrary nature of the rules, although I did not ask about them.

In the interviews, the women were direct and forthcoming. Some interviewees stated very simply that they were not emotionally bonded to certain of their children; others expressed overly optimistic views of the possibility of reestablishing good and easy relations with their children; others were not in touch with their children because current caretakers prevented contact; some discussed in detail strategies for maintaining a drug habit while care taking their children. In other words, while the emotional context of the interviews was rich and the conversation occasionally difficult, the mothers rarely presented an idealized view of themselves as mothers or a view of mothering as nonproblematic.

The women expressed an appreciation for the time spent speaking about their children and their hopes for the future. Another rewarding aspect for the mothers was that I offered to photograph their children, giving them both the photographs and the negatives. Since many of the women didn't have recent pictures of themselves or their children, this opportunity was appreciated.

I transcribed the interviews and coded them using a software program called HyperRESEARCH. The social situations that affect inmate mothers' careers as mothers that emerged from the analysis include arranging care and managing caretakers, demonstrating fitness, negotiating ownership of children, constructing and managing motherhood, and balancing crime, drugs, and motherhood. Because of space limitations, my discussion of results will be brief.**

Findings

Arranging Care and Managing Caretakers

A recent survey has reported that 35 percent of state and 15 percent of federal women inmates have not lived with their children prior to their incarceration (Mumola, 2000). For those who co-reside with their children, child placement becomes a challenge upon imprisonment. As in previous research (Snell, 1994), there were differences in white, African American, and Hispanic child placements (see Table 10.1). While children most frequently reside with grandparents, white and Hispanic mothers are

---

**For the complete findings, see S. Enos, *Mothering from the Inside: Parenting in a Women's Prison*, Albany: State University of New York Press, 2001.

much more likely to leave their children with husbands; African American women are more likely to place their children with their mothers or other relatives. Making a decision about placing a child is not easy. Stacey,*** a white woman in her twenties, wanted to protect her child but also wished to minimize her involvement with child welfare. Her four-year-old son was living with his father, and she had heard from friends that the father had resumed a drug habit and the selling of narcotics. Stacey knew that contacting the child welfare authorities would prompt an investigation of the father. However, this could also mean having an agency remove the child from the home and place him in foster care. As she noted, "Most of the time, trying to get help means getting into trouble."

TABLE **10.1**
Race/Ethnicity of the Sample of Female Inmates (N = 25) and Current Residences (N = 29) of Their Children*

	Grandparent	Foster	Husband	Other Relative	Total
White	4	1	5	–	10
African American	7	2	–	5	14
Hispanic	2	–	2	–	4
Native American	–	–	1	–	1
Total	13	3	8	5	29

*Several siblings do not reside in the same residence.

Constructing and Managing Motherhood

Maintaining presence as a mother is difficult when a woman is absent from the home and is even more problematic when she has few resources. Bernice, a pregnant African American mother of a five-year-old son who had been in many foster homes, echoed Stacey's concern. She said, "I feel like I'm having babies for the state. Once they [the child welfare agency] get in your life, they're always there. They stick with you forever." This mother remarked that foster care was the worst place for an African American child to be. Her child was placed with a foster family only because her family and friends wouldn't (or couldn't) respond to her need for a place for him to live during her incarceration.

Other women were more positive about putting their children in the care of foster parents. Some white women characterized it as a "life saver." Louise talked about her satisfaction with foster care, and she was careful to make a distinction between being a child's mother and the care taking of the child. Her comments indicate that she was able to separate the biological identity of mother and the social and psychological work of "mothering."

***To protect confidentiality, Stacey and all the other names of interviewees are pseudonyms.

Enos:	So, you had a baby just before you came to jail. Where is he now?
Louise:	He's with foster parents who really love him a lot. They are just great with him.
Enos:	Does he visit you here?
Louise:	Sure he does, a little, the foster parents bring him to me.
Enos:	And does he know you?
Louise:	Well, he knows who I am. But, he doesn't know me know me.
Enos:	Can you explain that?
Louise:	Well, he knows I'm his mother, but he doesn't know me that way.

The distinction Louise makes may imply special challenges when the mother resumes care taking of the child or when she reflects upon how she may handle the tasks associated with care taking. Pam, a white middle-class woman, noted that her husband had taken up much of the responsibility for their developmentally delayed son. She felt that some activities can be delegated, while others are better for mothers to do. As an example, she described what happened when the father took the son shopping for school shoes. She was dismayed at his choice, saying "I looked at what they picked out and thought to myself, 'My God, I would have never bought such a thing for him. No mother would have.' There is a difference in what a mother would have done."

Unlike other women, Pam had no difficulty maintaining her position in her family. She was still defined as the family's major caretaker. She reported that her husband visited often and that they maintained the close working relationship that parenting a special child required.

Negotiating Ownership

Family members who assume caretaking responsibilities immediately after a child's birth were viewed differently than those who became involved later in the child's life. Vanessa, an African American woman, said, "Just after I had the baby, my mom took him. Then, I got a short bid. But ever since he was born, he was her baby. I never had a chance." Vanessa is suggesting she has a marginal role in the child's life as his mother—either because others have asserted ownership in her absence or because of her lack of effort.

Other women noted the same development, but suggested that this may have turned out for the best. Kate, a white woman, said, "I don't know what was wrong with me then. I just never had any feeling for my first. I just had it and then my mom took over. We used to do drugs together and party all the time and then she went straight and was preaching to me all the time. Then, the baby came and she took over." Kate conveyed a simpler delineation of responsibility by putting her child in her mother's hands with little regret.

Significant differences were also revealed as women described their paths to prison. White women were much more likely to attribute blame to their families of origin for their problems, suggesting that they were pushed out of their homes or subjected to early sexual and physical abuse and neglect. Few white women relied on their families to take care of their children, and most were wary of incurring "unpayable debts" if they utilized family resources to help them with child care.

On the other hand, African American women tended to trace their paths to prison to their attraction to the "fast life," arguing that their families had tried unsuccessfully to divert them from a life of crime. Placing children with family and friends was expected and acceptable. In many cases, the mothers had previously prevailed upon relatives to care for their children when they were engaged in criminal activities. These children were taken care of by a variety of adults before the women's imprisonment, and most African American women expected to continue using these shared child caring arrangements after release.

There were exceptions to these patterns. Not all children of white women were in the care of husbands, the state, or other non-kin, and not all children of African American mothers lived with relatives or other kin. White women with access to deviant life styles had placements similar to the African American inmates. That is, white women with parents or relatives involved in law-breaking behavior, or who lived in areas where entry into this lifestyle was easy, did place children with their own mothers. On the other hand, African American women who traced their imprisonment to deficient families of origin and were estranged from them placed their children in foster care.

Family "expectations of trouble" occur in white and African American families. Histories of conflict with the law and with other institutions create capacities within families to extend resources when needed. Some women suggested that racial and ethnic differences characterized families of white women and those of color. They stated that expectations for girls in white families were more rigid and punitive, with young women expected to "stay out of trouble," and to follow gendered prescriptions for behavior. Most of the white women did report that their families had expected them to follow the rules, and handled deviations by constricting the family's resources to protect the other members of the family and to keep the offender isolated.

Balancing Crime, Drugs, and Motherhood

Relationships with family before and during incarceration, the quality of those relationships, whether they were supportive or undermined the relationship between mothers and their children, the extent of the involvement of women in crime and drugs, all affect the long-term career of inmates as mothers. Research indicates that about one-third of these mothers will not maintain custody of their children (Martin, 1997). Some mothers maintained that their involvement in criminal behavior (such as larceny, selling drugs, or assault) supported the work they did as mothers by supplementing their meager incomes or by protecting children from physical danger. Others stated that they managed to keep their long-term involvement in drugs and crime from their partners and from their small children. There was disagreement among the women, however, on the issue of balancing drug use and motherhood. Some inmate mothers claimed that one could separate what one did as a drug user or an offender and what one did as a mother by confining each activity to a separate part of the day and making certain that the drug use did not grow out of control. Other women challenged this perception, as demonstrated by the comments of Tee and Margaret.

> You think you are doing right, but you are really not. You are doing so many things, you really don't know what you're doing really. You think you're being good, but you're not. You're neglecting them because you're taking a chance by going stealing and risking yourself. You might not come back to them knowing that they love you.

> Can't nobody say they didn't do it [use drugs] in front of their kids, because they're lying. That addiction is always there.

There were a variety of mother-careers among incarcerated mothers. Some were primary caretakers of children before incarceration; some shared care with others; some did not live with children or assume major responsibilities for their care. Upon release, some inmate mothers will resume the major responsibility of caretaker. Others will assume this role for the first time as other caretakers terminate their roles as the child's primary care giver. Still others will lose their rights to children in formal judicial hearings. Finally, some will continue to share child care with others.

Summary and Conclusion

Imprisonment provides an important vantage point for sociologists to examine family responses to women in trouble and how these responses are affected by larger structural forces. The availability of families to assist women in prison is very much influenced by normative expectations. The financial resources and capabilities of families to lend a hand—here evidenced by their taking care of children—appear to be less important than cultural beliefs about what members do for each other in crisis. Race and ethnicity have a powerful impact on the resources women inmates have in arranging care for their children. Few child welfare organizations recognize these differences and few provide support that matches the needs of families from different racial and ethnic groups.

This examination has also focused on the complex nature and meanings of motherhood. Motherhood might mean a biological connection to child, taking care of a child, maintaining overall responsibility for a child's welfare (even if not directly engaged in the care), or sustaining a unique and irreplaceable relationship with a child. We gain insight into the management of motherhood by seeing how women in prison understand motherhood and meet its challenges. While parenting programs may provide opportunities to visit with children, the variety of mother careers supports the need for a range of programs to help inmate mothers with the challenges they'll face upon release from prison.

References

Baca Zinn, M. and D. S. Eitzen. 1993. *Diversity in families*, 3rd ed. New York: HarperCollins College.

Baunach, P. J. 1985. *Mothers in prison*. New Brunswick, NJ: Transaction.

Beck, A. J. and J. Karberg. 2001. *Prison and jail inmates at midyear 2000*. Bureau of Justice Statistics: Washington, D.C.: U.S. Department of Justice.

Caplan, P. J. and I. Hall-McCorquodale. 1985. Mother-blaming in major clinical journals. *American Journal of Orthopsychiatry* 55: 345–353.

Charmaz, J. 1983. The grounded theory method: An explication and interpretation. In *Contemporary field research: A collection of readings*, edited by R. M. Emerson, pp. 109–126. Boston: Little, Brown.

Chodorow, N. and S. Contratto. 1982. The fantasy of the perfect mother. In *Rethinking the family*, edited by B. Thorne and M. Yalom. New York: Longman.

Collins, P. Hill. 1990. *Black feminist thought: Knowledge, consciousness, and the politics of empowerment*. New York: Routledge.

Dill, B. T. 1988. Our mothers' grief: Racial ethnic women and the maintenance of families. In *Race, class and gender: An anthology*, edited by M. L. Anderson and P. H. Collins, pp. 215–237. Newbury Park, CA: Sage.

Johnston, D. 1995. The effects of parental incarceration. In *Children of incarcerated parents*, edited by K. Gabel and D. Johnston, pp. 59–88. New York: Lexington.

Maher, L. 1992. Punishment and welfare: Crack cocaine and the regulation of mothering. In *The criminalization of a woman's body*, edited by C. Feinman, pp. 157–192. New York: Harrington Park.

Martin, M. 1997. Connected mothers: A follow-up study of incarcerated mothers and their children. *Women and Criminal Justice* 8(4): 1–23.

Mauer, M. 1997. *Americans behind bars: U.S. and the international use of incarceration*. The Sentencing Project. Washington, D.C.

Mumola, C. J. 2000. *Incarcerated parents and their children*. Bureau of Justice Statistics: Washington, D.C.: U.S. Department of Justice.

O'Barr, J., D. Pope and M. Wyer. 1990. Introduction. In *Ties that bind: Essays on mothering and patriarchy*, edited by J. O'Barr, D. Pope, and M. Wyer, pp. 1–14. Chicago: University of Illinois Press.

Rubin, L. B. 1994. *Families on the fault line*. New York: HarperCollins.

Snell, T. L. 1994. *Women in prison: Survey of state inmates, 1991*. Bureau of Justice Statistics. Washington, D.C.: U.S. Department of Justice.

Stack, C. 1974. *All our kin: Strategies for survival in the Black community*. New York: Harper & Row.

Young, M. and P. Wilmott. 1957. *Family and kinship in East London*. London: Routledge & Kegan Paul.

Appendix A: Interview Questions

Information was first collected from the correctional staff and official records and checked in the course of the interview, as appropriate. This information included the woman's race and ethnicity, living arrangement of child(ren), length of sentence, recidivism status, and involvement of child welfare.

Paths to Prison

Can you tell me what brought you to prison?

How did you get involved in drugs/crime?

Did your family and friends provide an entry? Did a boyfriend?

How would you describe your family when you were growing up?

What has been the hardest thing about being in prison?

Children

Where are your children living right now?

Were you living with your children before you came to prison? If not, who was taking care of them?

How old are your children?

Is DCF (child welfare) involved? If so, in what way?

What are your plans after you are released? Will you live with your children? Immediately? Eventually?

Have any of your children been adopted?

People say that crime and getting into trouble with the law might be passed down through generations. Are you concerned about your children getting into trouble? What can be done, if anything, to prevent this?

Caretaker Characteristics

How did you decide where to place your children?

Did you feel you had some good options?

What are the pros and cons of placing your children with your parents?/foster care?/your husband?/relatives?

Do you think your child's caretaker is doing a good job? A better job than you can just because of your situation?

Are you involved in making decisions about your children, like where they will go to school? If they need to go to the doctor?

What kind of burden do you think it is for other people to take care of your children? Do you think it is hard for them?

What kinds of obligations do you think families have for each other?

What are family members supposed to do if somebody in the family needs help?

Are you comfortable asking your family for help?

Management of Motherhood

Women who are serving a long sentence must have to make lots of arrangements for their children. What are some of these and how is that different from women who are serving a short sentence?

Are there things that only mothers can give and do for their children? If so, what are these?

Are there things that you think your children are missing because someone else is taking care of them?

Do your children understand that you are in prison? What do you think they think about this?

How often do you see your children? How often are you in touch with them?

Do your children understand that you are their mother even though someone else is taking care of them?

Have you heard of instances where children call women other than their mother "Mom?"

What are some things mothers are supposed to do for their children?

How do you think women in prison try to make up for the fact that they are in prison?

What are some of the things that make it hard to be a good mother to your children?

In terms of your family and friends, do you need to prove to them that you are a good mother? How do you do that?

In terms of child welfare, do you need to prove to them that you are a good mother? How do you do that?

Women's Understanding of Other Mothers

There are some racial differences in where children live when their mothers come to prison. African American and Hispanic children seem to be more likely to live with relatives while white kids seem more likely to go into foster care or live with husbands. Can you explain why you think that happens?

Can you tell when women in prison are ready to make a change in their lives and go straight? What are some of the signs?

There has been a lot of talk about making it easier for the state child welfare agency to terminate parental rights if there has been a child death in the family or if the children have been exposed to drugs. What do you think about these new laws?

Do you think it is possible to tell if someone is a good mother by seeing how she acts with her children in prison?

Do you think the courts are easier on women who have children or harder? Why does this happen?

Appendix B: Excerpts from Informed Consent Form Used by Sandra Enos

What will be done: If I decide to take part in the project, I will be asked about my children, their living arrangements, what brought me to prison, and how imprisonment is affecting my relationship with children and other family members. I may also be asked about involvement with the child welfare agency. My part in the study will involve an interview which will last about 1–2 hours and which will be tape recorded. My name will not be on the tape and after the interview is typed, the tape recording will be destroyed.

Risks: The possible risks in the study are small. I may feel some discomfort as I talk about the past. There is a chance that some of the interview questions may result in my feeling uncomfortable or anxious. If that is the case, the interview may be suspended at that point if I wish. The researcher will ask several times during the interview if I want to stop. The decision to participate in the study is up to me. I may terminate the interview at any time. Whatever I decide will not be held against me. I understand that the researcher is not affiliated with the Department of Corrections and that my participation in the interview will not have an impact on my treatment, criminal processing, or any other matter.

Reportable child abuse: If, while talking with Sandra Enos, I tell her about some abuse or neglect of a child that I say has not been reported to DCYF, the researcher will inform me that (1) she is required by law to report the abuse/neglect to DCYF and (2) that we must terminate the interview. The tape recording of our interview will be immediately destroyed. I understand that the purpose of the study is not to track reportable incidents of abuse or neglect.

Benefits: There are no guarantees that my being in this research will provide any direct benefit to me. I understand that taking part in this research will have no effect on my parole, classification status, and/or inmate record. My taking part will provide important information for people who are trying to understand the impact of prison on women and their families.

Confidentiality: My participation in the study is confidential to the extent permitted by law. None of the information collected will identify me by name. All information provided by me will be confidential. Within two weeks after the interview, the tape will be transcribed, and the recording will be destroyed. No information that is traceable to me will be on the transcript. Transcripts will be maintained in a locked cabinet in a secure location available only to the researcher. No information collected by the research that identifies me will be given to the Department of Corrections.

 **THINKING ABOUT ETHICS**

Enos submitted her proposal to her university's Institutional Review Board (IRB) where it was approved. All of the women interviewed in this research volunteered and gave informed consent before being interviewed. All information was kept confidential, although each participant was cautioned about the legal limits of confidentiality, and no participant's actual name is used by Enos.

VARIATIONS IN QUALITATIVE INTERVIEWS

Number and Length of Interviews

Qualitative interviews can vary in several ways: the *length of each interview*, the *number of times* each member of the sample is interviewed, the *degree of structure* in the interview, the *number of interviewees and interviewers* participating in the session, and whether interviews are used alone or with other techniques. In any given study, there is typically variation, sometimes quite a lot, in how long each interview takes. The interviews in Emily and Roger's study on retirement typically took between 1 hour and 90 minutes; the Enos interviews lasted between 1 and 2½ hours.

Interviews can be used in longitudinal research, with multiple interviews as in Hoffnung's long-term study of college graduates (in Chapter 4) and in Emily and Roger's panel study of the transition to retirement (in Chapter 7). However, in studies like the Enos's, where participants are not followed over time, each member of the sample is typically interviewed only once. Despite the additional cost, however, some researchers suggest dividing the material to be covered into sequential interviews. Irving Seidman (1998: 11–12), for example, suggests a three-session series, with the first focused on the past and background to provide a context for the participant's experience, the second to cover the concrete details of the participant's present experience in the topic area, and the third directed to the participant's reflections on the meaning of the experience. Other researchers advocate multiple interviews because they feel that the additional contact will give interviewees more confidence in the procedure and increase their willingness to report fully. In one study, Robert Weiss (1994) found that only in a fourth interview did one man talk about his wife's alcoholism.

 **STOP & THINK** *Sometimes an interviewer doing a structured interview feels foolish reading the questions almost rigidly, not being able to modify, add, or delete questions. How do you think you'd feel in the opposite situation—without a "script," and needing to spontaneously construct questions as the interview proceeded?*

Degree of Structure

semi-structured interview, an interview with an interview guide containing primarily open-ended questions that can be modified for each interview.

interview guide, the list of topics to cover and the order in which to cover them that can be used to guide less structured interviews.

Qualitative interviews can vary from unstructured to semi-structured interactions. **Semi-structured interviews** are designed ahead of time but are modified as appropriate for each participant. They begin either with a list of interview questions that can be modified or with an **interview guide,** which is a list of topics to cover in a suggested order. Although some semi-structured interviews use at least a few closed-ended questions, more typically there is a core set of open-ended questions, supplemented liberally with probes and questions tailored to the specific person. Such an approach will generate some quantifiable data and some material that allows for in-depth analysis. Constructing questions ahead of time makes the interviewer's job easier because there is a "script" to ensure coverage of all the topics in each interview. The interviewer must judge whether the questions are appropriate or not, re-order and re-word them if necessary, and use follow-up questions and encouragement to help the interviewee answer fully.

In her study of college graduates, described in Chapter 4, Hoffnung chose the semi-structured interview for two of her data collections to make the interview more like a conversation, to be able to identify new issues as they came up, and, given her heterogeneous sample, to have the ability to re-word questions for each individual respondent. Similarly, Emily and Roger in their study on the retirement transition decided that this method was best suited to understanding the changes in people's lives as they moved from full-time employment to other activities. Sandra Enos decided to use the semi-structured interview because pre-formulated questions helped her organize her thoughts and paths of investigation, but left things open enough to pursue "surprises"—unexpected but interesting avenues of investigation (Enos, personal communication).

The semi-structured approach is most useful if you know in advance the kinds of questions to ask, feel fairly sure that you and the interviewees "speak the same language," and plan an analysis that requires the same information from each participant. It allows the study participants to be actively engaged and have an impact on the direction of the interview. In addition, the questions for a semi-structured interview can be made available to other researchers, allowing them the opportunity to evaluate the questions and to replicate the interview in other settings.

unstructured interview, a data collection method in which the interviewer starts with only a general sense of the topics to be discussed and creates questions as the interaction proceeds.

At the other end of the qualitative interview continuum is the **unstructured interview.** Researchers doing totally unstructured interviews start only with a sense of what information is needed and formulate questions as the interview unfolds. Although some interviewers develop an interview guide with a list of topics, they are not bound by the list. In all unstructured interviews, interviewee "digression" tends to be valued as much as core information. Flexibility in questioning can provide insight into the participant's viewpoint and the meaning behind statements. In their study of disaster survivors, Ibañez et al. (2003: 7) began each interview with the broad and non-threatening question, "It [the disaster] must have been awful, how did the disaster affect you and others who lived here?" and followed up with "example" and "experience" questions, such as "Can you give me an example of how others helped you?" and "Tell me about your experiences in the shelter."

life story interview, a short account of an interviewee's life created in collaboration with an interviewer.

The unstructured or narrative interview lends itself to situations where the researcher wants people to tell about their lives as a whole from their own perspectives, as in life history or **life story interview** when the researcher wants to understand the social context of the participant and when the researcher is developing hypotheses or theories during data collection. Roma Hanks and Nicole Carr (2008) used this approach in their study of turning points in the lives of women currently in jail. They began each interview by asking each woman to talk about her life and, as they shared their life stories, used a visual aid—a sheet of paper with a single line printed across the page with the word "Birth" on the left and "Now" on the right—to help the participant identify important life events (Hanks and Carr, 2008). A similar interview approach, also with inmates, was used by Jennifer Schlosser (2008) in her focus on "identity moments," which are situation-specific, contextual, life-changing phenomena. She elicited these with open-ended questions about things like a first experience with drugs and notes that the narrative evolves and becomes more complete as each moment is discussed (Schlosser, 2008: 1516).

As understandings or hypotheses are formulated, additional questions can be added to the interview. Sulik (2007: 862) was interested in the worldviews of women with breast cancer and also in constructing theories of coping. As she developed models of the women's coping strategies, she modified the interview questions to incorporate and evaluate her understandings of the way the women accept and ask for help and their views of entitlement to care.

Shulamit Reinharz (1992) labels the unstructured approach an "interviewee-guided" interview because the focus is more on understanding the interviewee than on getting specific questions answered. This method may be most useful when the researcher does not know in advance which questions are important to ask, when interviewees are expected to find different meanings in questions or possess different vocabularies (Berg, 1989: 16), or when the researcher wants to work inductively. Westervelt and Cook (2007) used this approach in their study of people who had been on death row before being exonerated from the crime to collaborate with the study's participants. They felt their primary research role was facilitation to allow the participants' voices to be heard as authentically as possible to "explore and understand the flesh-and-blood realism of their experiences" rather than reducing their stories to statistical patterns (Westervelt and Cook, 2007: 28). The less structured interview can also be used *first* to develop themes and topics that are then covered in more structured ways, as Garza and Landeck (2004) did in their study exploring risk factors for dropping out of college, or can be used *after* a survey to obtain more complex narratives as Fine et al. (2003) did in their study of urban youth and their experiences with and attitudes toward adult surveillance.

 **STOP & THINK** *The interview methods we've discussed so far have relied solely on written or verbal responses. How about more visual data? Can you see a way that drawings or photographs can be used to collect data?*

Interviews Using Visual Methods

Some researchers argue that traditional methods of data collection, including interviewing, are improved by the addition of visual methodologies. One such approach asks interviewees to make drawings as another source of data. For example, in a study of women with heart disease, Marilys Guillemin (2004) first completed a series of individual interviews and then handed each participant blank cards and colored pens, asking them to make a drawing of how they visualized their condition. After the drawings were completed, the researcher asked questions about the images. While the request for a drawing was often met with nervous laughter and a disclaimer about ability, Guillemin (2004) found most participants drew images that were vivid embodiments of their illnesses. A study of children's use of and attitudes toward tobacco also used drawings. Worried about the possible lack of validity and reliability in verbal reports, Mair and Kierans (2007) added a "draw and write" technique in their study by asking students to draw or write about situations where smoking occurs. They feel this technique allowed participants to give insight into how they framed tobacco use within everyday social life (Mair and Kierans, 2007: 131).

photo elicitation, a technique where photographs help reveal information and encourage discussion of events and meanings in the participant's life.

Another approach is to add photographs to an interview. The most common of these methods is **photo elicitation,** a technique where photographs help reveal information and encourage discussion of events and meanings in the participant's life. A related technique is photo voice or photo novella (picture stories) where photographs are used to encourage people to talk about events and daily routines, to focus on the meaning or significance of the events or the representation of the participants' communities (Hurworth, 2003).

In these methods, participants take or bring photographs of themselves or their surroundings and are asked to talk about them. For example, in Coralee McLaren's (2009) study of the experiences of hospitalized children, research assistants escorted the study participants on a walking tour of the hospital's lobby atrium from the floor above. The children were asked to take photographs of architectural and design features that interested them and these photographs were used to generate thoughts about how the patients perceived and navigated the hospital space. Anamika Majumdar's (2007) primary method was the life history approach in her study of marriage in South Asian communities, but she also asked the women to bring photographs of the houses and areas where they lived before marriage as memory triggers for discussions of emotionally close relationships they had before marriage. The photographs grounded the women's narratives in specific places and helped the participants discuss personal experiences (Majumdar, 2007). In a study of rural women's lives in China, photo novella activities were used along with group interviews (Heinonen and Cheung, 2007). Women in five counties in China were trained and given cameras for five days to photograph their daily lives. They discussed these photographs in groups, focusing on issues that included why they chose to take specific pictures and what made the person, place, or event important to them.

These visual methods can be used in combination with interviews or with the observational methods described in the next chapter. Dave Snell and Darrin Hodgetts (2007) combined observational, visual, and verbal methods to document the efforts of a group of heavy metal fans to negotiate a sense of community with shared history, identity, and sense of belonging. The researchers used direct observations at a heavy metal bar, a series of individual interviews with patrons and bar staff, and a photo-voice project where the fans were asked to take and discuss photographs that represented their experiences of the music and community.

photo-interviewing, a data collection technique using photographs to elicit information and encourage discussion usually in conjunction with qualitative interviewing.

Photo-interviewing has some limitations, such as the ethical concerns of privacy and questions of probability sampling, but its advocates argue that it is very effective in a number of ways. It can be used at any stage of the research process, assists in building rapport with study participants, promotes longer, more detailed interviews, and may elicit unexpected information (Hurworth, 2003). In addition, photo-interviewing can be effective in eroding some of the power differential inherent in traditional interviews where the interviewer has primary control over what questions, topics, and information are relevant; photographs can allow the participant more control over shaping the discussion, defining what is significant, and interpreting meaning (Frohmann, 2005).

 **STOP & THINK** *So far, all the interviewing we've discussed has been one-to-one. Are there any situations you can think of when it might be better to have more than one interviewer or more than one interviewee?*

Joint Interviewers

The use of more than one interviewer is fairly uncommon. When used, it works best when the person being interviewed is not easily intimidated, and the perspective of two or more interviewers is important. For example, in the study of the women who served in the Rhode Island legislature (Adler and Lemons, 1990), one of us, Emily, interviewed several of the more than 50 women jointly with her co-researcher, historian J. Stanley Lemons. This enabled the researchers to develop similar interviewing styles and to elicit information that was of interest both sociologically and historically. After the joint interviews, we moved to one-to-one interviewing to be able to complete the interviews (each lasting two to three hours) in a timely fashion.

Group and Focus Group Interviews

group interview, a data collection method with one interviewer and two or more interviewees.

It is more usual to have more than one interviewee as in the joint or **group interview**, where one interviewer or moderator directs the inquiry or interaction in an interview with at least two respondents. The individuals in the group are selected because they have something in common. They might know each other (such as a married couple or members of the same church) or be strangers (such as teachers from schools in different towns or patients in a given hospital). Such interviews can be based on a predetermined set of questions or can use an unstructured format. Group sessions are less time consuming than interviewing the same number of participants individually and have "the advantages of being inexpensive, data rich, flexible, stimulating to respondents, recall aiding, and cumulative and elaborative over and above individual responses" (Fontana and Frey, 1994: 365).

Group interviews are useful when the group is the unit of analysis. It can provide either a shared or a disparate view of people's experiences and allow the researcher to observe interaction between participants while the interview is in progress. Racher, Kaufert, and Havens (2000: 367) interviewed frail, rural, elderly couples jointly and, to maximize understanding the couple as a unit, used the interview itself as an opportunity to observe the couple's verbal and nonverbal interaction while focusing on the content of their answers. When the unit of interest is not the group itself, the group interview can have the advantage of releasing the inhibitions of individuals who are otherwise reluctant to disclose what for them are private matters. Because some "are more willing than the others to speak of personal experiences and responses, they tend to be among the first to take active part in the discussion. As one ventilates his experiences, this can encourage others to ventilate theirs ... and establishes a standard for the rest who progressively report more personalized responses" (Merton, Fiske, and Kendall, 1956: 142–143). In addition, the interaction in the group might remind each individual of details that would otherwise not be recalled (Merton et al., 1956: 146).

Group interview data can suppress negative attitudes and may be biased toward the views of one participant. William Aquilino (1993: 372) found that even though the absolute magnitude of effects was small, the presence of a spouse influenced subjective assessment of marriage in the direction of more positive assessment. Some argue that it is better to interview husbands and wives separately to enable them to talk more freely about their feelings and views of each other (Adler, 1981; Rubin, 1983) and when the researcher is interested in the individual's rather than the couple's experience. In one study of married couples, women's perspectives were found to be so prominent in the joint interviews that the researchers suggested interviewing men on their own to find out about the husbands' experiences of illness and fatherhood (Seale et al., 2008). Even among those who are strangers, some participants might take over the group, monopolize the interaction, or inhibit other members.

focus group interview,
a type of group interview where participants converse with each other and have minimal interaction with a moderator.

A special kind of group interview is the **focus group interview**, a research tool that uses group interaction on a topic to obtain data. Instead of asking questions of each participant in turn, the focus group has a moderator or facilitator who encourages the group's participants to talk to and ask questions of each other. The goal of the focus group is to have participants interact within the group to hear many points of view. Focus groups were originally used in social sciences in the 1940s and 1950s (Merton et al., 1956), but then became more of a market research for decades. Rediscovered by social scientists, more than 200 articles a year using focus groups appeared in academic journals by the end of the 1990s (Morgan, 2001).

Focus groups can be used alone or in combination with other methods; they can provide useful qualitative data or can precede or supplement a questionnaire or structured interview. Some focus groups are more structured, with the moderator being more involved in controlling the group dynamics, keeping it on topic and focused on the researcher's interests while others are less structured, much more conversational, and more self-managed (Morgan, 2001: 147).

Some argue that focus group interaction highlights issues and concerns that would have been neglected by questionnaires (Powell, Single, and Lloyd, 1996) and encourages participation by those who would be more likely to decline one-to-one interviews (Madriz, 2000). But, the use of focus groups is not without problems. Like other studies that use small and nonrandom samples, focus group studies will have limited generalizability. In addition, in all group interviews, including the focus group, it is very important ethically for participants to keep confidential the information provided by others. An agreement to maintain confidentiality can be included on an informed consent form, but it is hard for the researcher to monitor compliance.

Examples of focus group research include an exploration of Koreans' perceptions of what makes them happy (Kim, Kim, Cha, and Lim, 2007) and a study on the way that alcohol serves as a central marker for social recognition among teenagers in Denmark resulting in considerable social pressure on those who drink very little or not at all (Demant and Jarvinen, 2006).

Focus groups are designed to have between 3 and 12 participants, selected because they are homogeneous on the characteristic for which the researcher recruited them, such as people who have been recently widowed or have specific health concerns. The participants usually don't know each other before the group interaction. Projects vary in the number of focus groups that are used, ranging from just a few to dozens of groups. The groups are usually conducted with participants seated around a table with a moderator who starts the interaction with one or more questions. In the study on Korean happiness, only three questions were asked: "What makes you happy? What could make you happier than now? In general, who is a happy person?"

BOX **10.2**

A Focus Group on Interracial Dating

Erica Chito Childs used focus groups as one source of data in a study of black college students' attitudes toward interracial dating. The participants were women between the ages of 18 and 23 who were active in black/African American student organizations on three college campuses. Chito Childs selected the questions to generate discussion between the women. She decided to use focus groups because she was most interested in the collective responses of the women and the attitudes toward interracial dating that they would discuss as a group. In a focus group interview, the discussion allows the respondents to build upon other's comments, either adding to what another said in agreement or offering a different perspective that allows for interesting debate. Chito Childs (personal communication) notes that while it is possible that respondents may alter what they say in a group context, one of the reasons to use the focus group is when you *want* to know the collective attitudes, experiences, or beliefs respondents will express in a group.

Chito Childs has very generously provided us with a description of her interview questions. Notice that she began with general questions before asking more specific ones. She used follow-up questions to encourage participants to expand their answers.

I began by asking about their general views on race relations with the question "How would you characterize race relations between blacks and whites?" I started with very general questions about societal views to encourage the discussion to flow, and then would follow up with more specific questions about their own views/experiences, encouraging them to talk. I next asked "Tell me about interracial dating. Are there certain ideas about these relationships?" I followed this up with questions about their own views. I also asked them about the idea that black women as a group are opposed to interracial dating and if they agreed with that view. After the women talked about their own feelings, some discussed their views of how other black women felt. I followed up with questions, such as "Where do these views come from?" "Explain that a little more"; "Why do you think that is?" "Could you give me some examples of a situation like that?" Since much of the discussion was on black men dating white women, I also asked "What about black women dating white men?" and followed up with questions to have them explain in more depth. Finally, I asked "How do you think whites feel about interracial dating?" I found using informal, conversational questions that did not steer the discussion too much worked best for me. (Chito Childs, personal communication)

How would you respond to questions like these? Do you think your answers would be different depending on whether the questions were asked in a focus group or in a one-to-one interview?

(Kim et al., 2007). See Box 10.2 for the questions Erica Chito Childs used in her study of black college women's views of interracial dating and her thoughts on focus groups.

While the face-to-face interaction is an important feature of the focus group, some researchers have begun to use "virtual focus groups." Examples include research projects that used a software conferencing technique to conduct a series of one- to two-hour focus group interviews—one as part of a study on parenting support systems for new mothers and the other on the role of the Internet. One of the new mothers said she enjoyed participating in the virtual focus group because "It's better than a Q [questionnaire] as sometimes someone else's answer can make you think of something you wouldn't have thought of by yourself" (Madge and O'Connor, 2006: 203).

ISSUES OF VALIDITY AND PRACTICALITY

As with the other self-report methods, there are concerns about the validity of data generated by qualitative interviews. Inaccurate memories, misunderstandings, and miscommunications and misrepresentations must be considered in evaluating the information obtained. As early as 1935, researchers debated the "questionable value" of the interview for securing reliable data when people become defensive concerning their private and personal lives (Thurlow, 1935). In the qualitative interview, the interviewer is not a passive listener. The way an interviewer questions, responds, and acts can affect the way a participant responds. It's possible that a different interviewer or even the same interviewer using different wording would be told a different account, raising concerns about reliability as well. Furthermore, as all qualitative interviews need to be analyzed, an issue that we'll discuss later in the chapter, it's important to consider if different researchers would interpret responses in the same ways.

Specific validity concerns for focus groups are that some voices might not be heard and that there is pressure to conform to group attitudes. Michell (1998: 36) found, when comparing focus groups to interviews, that the lowest-status teen girls in the study were silent and withdrawn in the focus groups, but willing to reveal feelings and personal information in individual interviews. In a focus group, "the emerging group culture may interfere with individual expression, the group makes it difficult to research sensitive topics, 'group think' is a possible outcome, and the requirements for interviewer skill are greater because of group dynamics" (Fontana and Frey, 1994: 365). In selecting a specific kind of qualitative interview, the researcher might select a one-to-one interview to elicit individual answers about specific behaviors and experiences but a focus group to find out about a community. Sometimes, researchers find it interesting to compare the data from the group and individual interviews. See Box 10.3 for an example.

The counterpoint to concerns about qualitative interviews is that standardizing it might be no better at producing comparable data because different respondents can hear the same question in different ways. The ability to reword and restate questions in the more qualitative method can give the researcher a chance to present them in a way that's more understandable by each participant. In addition, a relatively natural interactional style allows the interviewer to better judge the information obtained. Qualitative researchers can try out what they think they have come to know as a result of interviewing others, so that interviewees can say "That's what I meant" or "No. What I meant was …" to validate and self-correct understandings (Gold, 1997). Enos and many others who have used the qualitative interview believe that, overall, interviewees tell the truth as they understand it and rarely offer false information knowingly.

Conducting and transcribing interviews is a time- and labor-intensive task. As a result, qualitative interviewing is an expensive method that can sometimes mean a small sample with limited generalizability. As we saw in the focal research for this chapter, Enos only conducted 25 semi-structured interviews in her study; Emily and Roger did 44 in theirs. However, other

Constructing Boyhood in Focus Groups and Individual Interviews

In a study of boyhood in London, researchers conducted individual and focus group interviews with 14-year-old boys and obtained different data. In the focus groups, the comments demonstrated hierarchies between the boys, included a great deal of ridicule of girls, and demonstrated the boys' humor and naughtiness. In the individual interviews, the boys were quieter, more serious, and talked about their close relationships with parents, grandparents, pets, children, and girls. Rather than privileging one account over the other, the researchers, Pattman, Frosh, and Phoenix (2005: 560), conclude that "What the very different and contradictory accounts of boys in the different modes of research suggested was that boys were *both powerful and vulnerable.*"

researchers have been able to use larger samples. Lillian Rubin (1994), for example, conducted 388 separate in-depth interviews in her study of 162 working-class and lower-middle-class families. Because of the benefits and the costs, some argue that the qualitative interview is most appropriate for exploratory research, for research that seeks an understanding of interviewees' worlds and meanings, and when the researcher is interested in generating "grounded theory"—that is, theory developed in intimate relationship with data, without preconceived notions, and with researchers aware of themselves as instruments for developing theory (Strauss, 1987: 6).

LOCATING RESPONDENTS AND PRESENTING THE PROJECT

In all interviews, the researcher must decide on the population and the kind of sample before locating potential interviewees and contacting them. If names and addresses or phone numbers are available, the approach can be a letter or a phone call. Hoffnung, whose panel study of women college graduates was discussed in Chapter 4, began with a list of seniors who had been selected at random from each of five colleges. She called each woman, introduced herself, and asked her to participate in a study of women's lives, starting with a one-hour in-person interview. In her study of working parents and family life, Hochschild (1989) first contacted a random sample of employees at a large corporation; she then asked the interviewees for the names of friends and neighbors to contact others using snowball sampling.

 **STOP & THINK** *What about groups of people for whom there are no lists? For example, what would you do if you wanted to talk to homeless people, noncustodial fathers, or people planning to retire in the next few years? How could you locate samples like these? How could you encourage them to participate?*

Qualitative researchers are frequently interested in studying groups of people for whom there are no lists—such as mothers in prison, fathers who do not live with their children, and grandfathers raising grandchildren—so selecting a random sample might not be possible. One strategy is to contact participants

gatekeeper, someone who can get a researcher into a setting or facilitate access to participants.

through **gatekeepers,** that is, people who control access to others. Gatekeepers include the parents and guardians of children under 18 and, as in Enos's study in this chapter, the heads of institutions, community organizations, agencies, or groups whose members are the desired sample. Some researchers use their personal networks, notices on bulletin boards, announcements at meeting, e-mails, and the like to recruit participants. Madge and O'Connor (2006) found a convenience sample of new mothers through a hotlink to their cyberparents website from the home page of the largest parenting website in the United Kingdom. Sulik (2007) began with one community-based breast cancer organization and then added cancer support groups, two treatment centers, cancer-related community events, bulletin boards, and snowball sampling to recruit her sample of 60 breast cancer survivors.

The more "political" or "deviant" one's topic is, the more difficult it is to get access to and participation of potential respondents. For example, when Ruth Frankenberg, a researcher interested in the social construction of race, told people that she was "doing research on white women and race," she was greeted with interest in some circles, but suspicion and hostility in others. She realized that her approach was "closing more doors—and mouths—than it was opening" (Frankenberg, 1993: 32). She was more successful when she told white women that she was interested in whether they interacted with people of different racial or cultural groups and whether they saw themselves as belonging to an ethnic or cultural group (Frankenberg, 1993: 35).

Other difficulties include the concern of potential interviewees that the information they provide will be used against them or that the researcher is not who she says she is. Enos presented herself as someone who was "outside the system." The women who Enos approached knew that she was not part of the prison administration or the child welfare system, and they believed her pledge of confidentiality with limitations, including her promise not to use their real names. Some researchers find that potential participants view them with suspicion—perhaps as vice officers or narcotics agents in disguise (Miller, 1986) or as investigative reporters, union organizers, or industrial spies (Harkess and Warren, 1993: 324).

Some studies provide interviewees with incentives to encourage participation. Ciambrone (Chapter 3) gave participants a small gift certificate to a local store. In their study of how Latino parents transmit their culture to their adolescent children, Umaña-Taylor and Bámaca (2004: 268) did the same thing, noting that they felt that monetary compensation was useful for participants, many of whom were making a great effort to participate while working double shifts or more than eight hours a day. Campbell and Adams (2009) recruited rape victims from many Chicago neighborhoods by advertising their study through posters, fliers, and in-person presentations, noting that there would be financial compensation of $30 for participating. In response to the question about how they decided to participate, only 14 percent of the 102 women they interviewed specifically mentioned the $30 compensation as a primary motivating factor. Much more common was the desire to help other women (38 percent), the feeling that being part of the study would be helpful to themselves and their recovery (34 percent), and a general interest in supporting research (25 percent) (Campbell and Adams, 2009: 399–402).

Most researchers don't offer financial incentives, feeling that payment doesn't have a big influence on the decision to participate. Emily and Roger found that many interviewees commented on their interest in talking about their lives and retirement plans and noted, afterward, that doing so helped them to clarify their thinking. The fathers Jennifer Hamer (2001: 10) interviewed for her study "What It Means to Be Daddy" lived apart from their children and expressed gratitude for the opportunity to talk about parenting and their children. In Enos's study, the women appreciated her offer of photographs of their children, but participated because they enjoyed being listened to with interest and respect and because they thought the study could help others in their position.

PLANNING THE INTERVIEW

Using Consent Forms

informed consent form, a statement that describes the study and the researcher and formally requests participation.

Anonymous surveys like those discussed in the previous chapter are often exempted from the requirements of using **informed consent forms**. In some studies using face-to-face interviews, researchers rely on verbal rather than written permission because it is possible in an interview to explain the purpose of a study, offer confidentiality of answers, and assume that the decision to participate implicitly means giving consent. However, as discussed in Chapter 3, IRBs, the groups explicitly designed to consider ethical issues, usually require interviewers to use a written informed consent form. The researcher provides a document that informs potential participants about the purpose of the study, its sponsor, the benefits to participants and others, the identity and affiliation of the researcher, the nature and likelihood of risks for participants (such as the possibility of raising sensitive issues), and an account of who will have access to the study's records and for what purposes. With such information, potential participants can weigh the benefits and risks in deciding whether to participate. Chapter 3 has the informed consent form that Ciambrone used in her study of women's experiences with HIV/AIDS and excerpts from Enos's form are presented in this chapter. Both forms were approved by a university's IRB and each researcher gave a signed copy of the form to each interviewee. Each form offers confidentiality to the extent provided by law, although some question the impact of this limitation on the validity and reliability of the data (Palys and Lowman, 2001).

Constructing an Interview Guide or Schedule

Once the study's objectives are set, the researcher determines the information that is essential to collect from participants. If the interview is to be unstructured, a list of topics, an introductory statement, and a general sense of the order of discussion is sufficient. The interviewer typically starts with general questions and follows up participants' comments. More information can be sought to provide the context, chronology, or additional meaning of the answer. Usually, it's better to wait until rapport is developed before asking questions about sensitive issues.

If you're doing semi-structured interviewing, you'll need to construct a list of questions, both basic and follow-up questions to gather information. Sometimes "filler questions" are needed to provide a transition from one topic to another. In less structured interviews, starting with broad, interesting questions is important. The guidelines for question construction provided in Chapter 9—such as staying away from double-barreled, double-negative, or threatening questions and not using wording that is ambiguous or leading—also apply to qualitative interviews. Doing practice interviews helps to prepare for the actual experience and allows the interviewer to gain experience covering issues and developing conversation generators. Questions that might elicit more information include "Can you tell me more about that?" "When did that happen?" "Who did you go with?" and "What happened next?" Noticing not only what is mentioned, but what is *not* mentioned, can lead to additional questions.

The intended analysis affects the choice of specific questions. Reviewing Sandra Enos's questions at the end of the focal research article, you'll see that they invite the participants to tell their stories. Susan Chase takes a similar approach in her case study of one university (discussed in Chapter 7). Although she is interested in the issue of diversity on campus, she doesn't start with direct questions on that topic. Instead she first asks students about the past (such as "Tell me a bit about your high school years. What were your interests then?"), before moving to the college career ("What kinds of transitions did you go through during your first semester and first year at here?" "Who has been most influential in your life during your college years?"). She notes that most students bring up diversity issues "naturally" as they tell their stories about transitions to college and through their college years. At that point, she asks a series of questions about diversity (such as "Do you have friends who are 'different' from you in important ways?") (Chase, personal communication).

Once an interview guide or schedule is constructed, it should be pilot tested with people similar to those who will be interviewed during the actual data collection. Trying out questions can be helpful in deciding which will be most useful for focusing the conversation on desired topics. The interviewees in these preliminary interviews can be asked about the interview experience—what was effective and what was difficult to understand. In a qualitative study, the list of questions can continue to change during the course of data collection, especially as the focus of the study evolves.

 **STOP & THINK** *As a college student, you use words and language in certain ways. Are there any groups of Americans that you think might use language in ways that are so different from your usage that you might initially have difficulty communicating with them?*

Speaking the Same Language

Obviously it is essential for the interviewer and interviewee to literally speak the same language, but for participants who speak more than one language, language of preference should be considered. For example, Jeong-eun Rhee found some of her interviewees wanted to speak Korean while others used English or a combination. In her interviews, she came to see that some participants thought that English was the appropriate language for public and academic settings and others saw their language choice as a statement about identity (Subedi and Rhee, 2008).

It is necessary to be familiar with the cultural milieu of study participants and to understand what they mean by terms. Umaña-Taylor and Bámaca (2004: 265), in their research with Latino parents, for example, note their participants sometimes switched between English and Spanish, perhaps because some words had no direct translation and others, like "abuelita," had more sentimental value than the English equivalent, "grandmother." They also point out that while Latinos share Spanish as a common language, there are expressions and words whose meanings are unique to specific nationalities. Enos found that she'd picked up enough "prison lingo" during the observation part of her study that she knew what the interviewees meant when they reported "getting jammed up" (getting in trouble), "catching a bid" (getting sentenced), and the like. If she didn't understand something, she'd usually ask that it be repeated or explained. For example, one woman reminisced about her life, saying that "back in the day," things were great. Enos wondered about those words and what they meant. Finally, she asked another inmate mother about the phrase and was told that it was when you were young and free or out on the streets—getting high, having fun, or whatever. In the context of her study, it isn't surprising that Enos sometimes had to ask the women to explain family relationships because some of the children they talked about "mothering" were not their biological children, but those for whom they were informal guardians.

CONDUCTING THE INTERVIEW

Where and How to Interview

Interviews can be held in offices, in the interviewee's home, or elsewhere. Ching Yoon Louie preferred to conduct her interviews of Chinese, Korean, and Mexican sweatshop workers who had become leaders in the movement to improve work conditions wherever they were working: "camped out at movement offices, sandwiched between pickets and workshops, while driving across the state to demonstrations, leafleting at factory gates, and fighting with government officials" (2001: 8).

If possible, individual interviews are conducted in private. But sometimes others are present. In a study of married couples, when Emily was interviewing the wives in their homes (while her associate was interviewing the husbands), she found that, on occasion, children would be present for part of the interview—sometimes commenting on the woman's comments! If privacy is important for an interview, it's important to consider that when scheduling the time and place.

Although the interview might not follow all the conventions of conversation, it *is* a social interaction. The interviewer should be aware of all that the interviewee is communicating: verbal and nonverbal messages and any inconsistencies in them. The interviewer should strive to be nonjudgmental in voice tone, inflection, phrasing of questions, and body language. A goal is to communicate openness to whatever the interviewee wants to share, rather than conveying expectations about his or her answers. Following up answers, the interviewer can ask for details, clarification, and additional information. One useful skill is in knowing when to keep quiet because interested and active listening can help in obtaining complete responses.

Recording the Interview

If the interview covers a large number of topics, a recording—either audio or video—is invaluable; the more unstructured the interview, the more necessary recording becomes. If there is no strict question and probing is an important part of the process, the interviewer may find it difficult to attend to things what the interviewee is saying while trying to write it down. Recording will also allow later review of specific words, voice tone, pauses, rephrasing of answers, and other subtle information. But the impact of recording an interview is hard to assess. On the one hand, it can help rapport by allowing the interviewer to make eye contact, nod, and show interest and help the interviewer concentrate on follow-up questions. On the other hand, some interviewees refuse permission to record the interview and others are intimidated or inhibited by the recording process.

 **STOP & THINK** *After describing his relationship with his father, the person you are interviewing asks you how you get along with your father. What would you do? Would you tell him that what you think really isn't important because he's the one being interviewed, answer that you'll tell about your father after the interview is over, describe your relationship with your father honestly but briefly, or launch into a detailed description of how you get along with your father?*

Being "Real" in the Interview

In more traditional qualitative interviewing, the interviewer maintains social distance from the interviewee. This means using a style that gives evidence of interest and understanding in what is being said (nodding, smiling, murmuring "uh-huh"), but that prohibits judgment, reciprocal sharing, or "real conversation." The interviewer is advised not to share opinions or any personal information with respondents because it can increase the chance of leading subjects to say what they think the interviewer wants to hear and shift attention from the interviewee.

Critics of the traditional method dispute the view that the traditional interviewer response is really neutral. Frankenberg (1993: 31) feels that "evasive or vague responses mark one as something specific by interviewees, be it 'close-mouthed,' 'scientific,' 'rude,' 'mainstream,' 'moderate,' or perhaps 'strange'—and many of those are negative characterizations in some or all of the communities in which I was interviewing." Some researchers argue that the interviewer and interviewee should treat each other as full human beings. Fontana and Frey (2000: 658) believe that there is more support now for researchers' attempt to minimize status differences and show their human side by answering questions and expressing feelings, especially as this can allow for the development of a closer relationship or greater rapport between the two interview participants.

Although Enos was only asked an occasional personal question about herself or her opinions, quite a few interviewers report interviewees asking numerous questions—about the study, about their personal lives, or for advice on the topic under study (Acker, Barry, and Esseveld, 1991; Oakley, 1981; Westervelt and Cook 2007). Westervelt and Cook (2007: 31), for example, were willing to reveal relevant details of their personal lives and histories believing that they could not gain openness and honesty from others if they were closed and guarded about themselves. For the research on women living with HIV/AIDS (Chapter 3), Desiree Ciambrone fielded questions about her personal life, typically if she was dating, married, or had children and if she thought it was difficult dating or meeting people and, less frequently, about her sexual or drug-use history. She was open and honest, in part because she is comfortable talking about these topics and in part because she believes in disclosure and in minimizing power imbalances in the interview. She thinks her receptiveness and candor fostered rapport and ultimately helped the women to feel comfortable and be more forthcoming (Ciambrone, personal communication). Joan Acker and her associates not only disclosed information about their lives but also formed friendships with many of those in the study. "However, we recognized a usually unarticulated tension between friendships and the goal of research. The researcher's goal is always to gather information; thus the danger always exists of manipulating friendships to that end. Given that the power difference between researcher and researched can not be completely eliminated, attempting to create a more equal relationship can paradoxically become exploitation and use" (Acker et al., 1991: 141). Anticipating that interviewees may ask personal questions allows the interviewer to consider how they'll respond.

Listening to someone describe an intimate, personal story can be a moving experience. Being "real" as an interviewer can mean acknowledging that at least some of what you hear is painful, difficult, or upsetting. The demanding nature of interviewing is not often discussed. Interviewers, even those with training, might not be prepared for or trained to handle the emotional impact of the interview process. In her work on juvenile prostitution, Melrose describes how she often felt distressed by the accounts she had heard: "[A]lone in a hotel in an unfamiliar place, struggling with my own feelings of anger and despair … I was forced to contain these feelings in order to carry on with the fieldwork the next day … I also imposed the expectation on myself that I should be able to 'manage' these feelings" (2002: 325). The two full-time interviewers in a study on low income families found the research process to be emotionally stressful as they heard about experiences of sexual abuse, rape, domestic violence, suicide attempts, and other problems; the interviewers provided each other with emotional support through daily phone calls (Gorin, Hooper, Dyson, and Cabral, 2008).

Interviewing across the Great Divides

STOP & THINK

How effective do you think you would be as an interviewer if you were interviewing someone considerably older than you are? How about if the person were of the opposite gender, of a different race, or from a different ethnic group? Do you think the topic of the interview would make a difference in interviewing someone whose background was very different from yours?

interviewer effect, the change in a respondent's behavior or answers that is the result of being interviewed by a specific interviewer.

As we discussed in Chapter 9, researchers using interviews need to think about **interviewer effect**, the change in a participant's behavior or the answers given to questions as the result of being interviewed by a specific interviewer. Now we'll add the issue that the researcher's identities, such as gender, age, social class, race, and sexual orientation, can affect all aspects of the research process, including data collection. Because each interviewer brings unique qualities and characteristics to the interview situation, an ongoing debate concerns the desirability of matching interviewers and interviewees on specific social characteristics, mostly significantly gender, race, ethnicity, age, and class.

The most common perspective is that it's advisable to match the participants in an interview because people of similar backgrounds are thought to develop better rapport. One concern is that interviewers who differ from interviewees on class, ethnicity, gender, or race tend to get different responses and less honest answers than when the participants have the same backgrounds. Sandra Enos believes that her gender was helpful in the interview process. "A man in that setting could have been problematic, especially since the women would have had some trouble understanding why a man would be interested in mothering. Also a number of the women expressed exasperation with relying on men and others were still looking for Prince Charming" (Enos, personal communication). In Box 10.4, Nicole Banton argues that interviewing within your own group makes the most sense as the interviewer knows the language and traditions of the group and is more likely to be trusted by interviewees.

On the other hand, it is possible for interviewers and interviewees of different backgrounds to have a good research relationship and for those who are alike to have problems. The researchers in charge of the *Sex in America* study (Michael, Gagnon, Laumann, and Kolata, 1994: 32) asked people of different races in focus groups who they would feel most comfortable talking to. They were surprised to hear that almost everyone—men, African Americans, and Hispanics—preferred middle-aged white women as interviewers. Similarly, in her study of male clients of female prostitutes, Grenz (2005) asked prospective interviewees whether they preferred to be interviewed by a man or a woman and found that none wanted to be interviewed by a man and many indicated a preference to be interviewed by a woman. Perhaps, as Robb (2004: 402) found in his study of fathers, masculinity should be seen as something that is "worked on" in the interview process and that men being interviewed by men will be partly motivated by a desire

BOX **10.4**

On My Team? Why Intragroup Interviewing Is Better by Nicole Banton, PhD*

What groups do you belong to? What makes your group unique—language? customs? In some groups, like those based on race, ethnicity, and/or gender, the socialization on things like language, traditions, and ideas about right and wrong begin at birth. Now, imagine an outsider asking you questions about yourself as a representative of your group. What if that person is a member of a group that has historically (and continues to) oppress(ed) yours? I consider these issues when I plan to collect data through interviews because they can affect how the interviewee acts during an interview.

When people talk to members of their own group, they are more likely to be comfortable because they are more likely to share a common language, including slang and jargon, as well as customs and traditions. In intragroup interviews, the interviewees are not as likely to think that they have to represent their entire group

because they are talking to one of their own. The interviewee is more likely to feel a kinship, and the interviewer is more likely to understand the interviewee. For example, when African Americans identify each other to other blacks, they don't just say that someone is black. They describe skin tone, for example, Nicole is light-skinned. They do this because African Americans are aware that "blackness" is a rainbow of shades, while non-blacks are less sensitive to this issue and the politics surrounding it. I want people to be relaxed during an interview, to understand them, and to get truest information possible. The best situation is when the researcher and the researched are members of the same team(s).

*Nicole Banton has just completed a research project on the politics of infant feeding among African American mothers. This essay is published with permission.

---

to prove their masculinity. For an interesting commentary on this topic, see Don Naylor's discussion in Box 10.5.

Hall (2004), a white British woman, described both the positive and the negative outcomes of being an outsider while interviewing women of South Asian heritage about their immigration experiences. As an outsider, she was sometimes viewed with suspicion, but also was judged to be a neutral researcher who was safe to talk to because she had few community links. In thinking about the issue, Enos feels that it might have been better to have worked with both African American and Hispanic co-researchers, but she didn't have the option of hiring interviewers. She believes that she was able to communicate effectively across race, class, and ethnic lines.

However, even when matching on key characteristics, there can still be differences. Interviewing members of parliament (MPs) in Britain, Puwar (1997: 9.4) found that because of occupational differences, rapport did not necessarily follow when she interviewed MPs with whom she shared gender, ethnicity, parental background, and even regional accent. Practically, it's typically not possible to match interviewer and interviewees on more than one or two key characteristics. In any event, *all* interviewers have personal characteristics and social identities that need to be considered in evaluating who does the interviewing and the resulting data.

BOX **10.5**

Reflections on Matching the Gender of the Interviewer and Interviewee by Donald C. Naylor*

In planning a study in which I'd be interviewing men about gender, I knew that the men in my sample might not be completely honest with me. But, I didn't think I needed to be concerned about the interviewer effect or be worried about getting the polite answers that sometimes result when people are interviewed by someone who is perceived as different and possibly unreceptive to their views. In my study, the interviewees and I would be the same gender and would be similar in other ways—age, class, and sexual orientation. Of the many problems associated with interviewing, interviewer effect was not one I thought I would have to consider.

Upon reading transcripts of my interviews though, I noticed that what was said seemed to have been affected by something happening between myself and the men I was interviewing. In one instance, I noticed that the man's answer seemed like something he felt he had to say to appear masculine. I wondered if he believed what he'd said or if it was just what one guy thought he was "supposed" to say to another. While the literature on interviewing says how much better it is when the interviewer and interviewee are similar, I wondered if there were also some problems with similarity.

Reflecting on the interview process, I've had some thoughts about what was occurring. First is that there is an effect even when the interviewer and interviewee are similar. It is a different dynamic than when the two are different, but it is still there. Similarity can affect the content of the interview, the types of questions asked, the answers given, and the way the interviewer and interviewee interact. In addition, it seems harder to observe critically a familiar interaction.

Second, I began to see that there was more than interviewer effect going on. The interviewee also had an active part in constructing the interview. This was especially true since I was doing an unstructured interview. When the men varied in what they said and how they said it (amount of openness, attempts to control, emotions displayed), I would change my style and demeanor. They were affecting me as much as I was affecting them. I began to think not so much in terms of interviewer effect, but about interviewer–interviewee effect as we were jointly constructing the interview.

Third, it began to be apparent to me that the interview process is a gendered one. We were "doing" an interview not just as two people but as two men. We were "doing gender." The dialog being created was typical of how men often converse. Of course, variations resulted since there are a lot of masculinities (and femininities) and these men (and myself) were not identical. But the more I read the transcripts, the more apparent it became that we had been "doing gender." I noted that some men tried to control the situation (as did I) and most avoided topics with much emotional content (as did I some of the time). They all tried to present themselves as competent, and each wanted to avoid a discussion of problems, perhaps seeing these as failures rather than as normal difficulties (did I collude in this?). The ways I responded to these men and the ways they reacted to me seemed to be typically "male" ways of interacting.

As I thought about the basic structure of the kind of interview I used, it became obvious that this was a method that did not fit easily with the style of many men. Having a stranger walk into their living rooms and ask personal questions is not what men typically do. Many men are hesitant about disclosing personal information, especially to strangers in unfamiliar situations.

I found it most useful to try not to press the men for very personal information. I found that when I was accepting, uncritical, and non-threatening, the men did not clam up or shut down. Oddly, at times, the less I asked, the more they told me. Perhaps unconsciously I was trying to conduct an interview in a way that was "guy like." This is "doing gender." But it also, I believe, allowed me to obtain more information than if I had tried a different style.

Interviewer–interviewee effect and "doing gender" cannot be avoided. The question for me was, given the inevitability of this, what do I consider the effects on the information obtained to be? We can acknowledge and understand these effects, and perhaps even profit from them.

*Permission granted by Donald C. Naylor, who wrote this when he was a graduate student in the Department of Sociology, University of Southern California.

AFTER THE INTERVIEW'S OVER

 **STOP & THINK** *Imagine that you're almost at the end of an interview on college students' relationships with significant others. After describing how the most recent love relationship ended, the student you're interviewing looks up and says, "I'm so depressed, I feel like killing myself." What would you do?*

At its end, the interviewer typically has the desired outcome—a completed interview—yet there may be many different responses from the interviewee. The respondent might rush off as soon as possible, "turn the tables" and interrogate the interviewer, bring up his or her own agenda, or continue talking about the topic after the official end of the interview (Warren et al., 2003: 98). When covering emotionally difficult topics, researchers might need to prepare for an emotional response or a request for help. At the very least, researchers usually include a series of "cool down" questions at the end of an interview so the interview doesn't end immediately after talking about sensitive subjects. Some researchers prepare something to leave with participants, most typically a list of local organizations that provide services in the area under discussion.

In a study of low income families, some parents became distressed during the interview (Gorin et al., 2008). Offered the option of ending the interview, most expressed a desire to continue to tell their stories. The researchers were prepared with materials about national and local support services that they offered to all participants; when distress was evident, the interviewer stayed with the participant for a while afterward and found additional information about services and support when a request was made (Gorin et al., 2008).

Although some interviewers will continue to have contact with interviewees, most researchers limit their interaction beyond the interview and try to make clear the difference between an interview and therapy. In one innovative approach, researchers who interviewed rape victims offered to locate or provide counseling or therapy sessions for the interviewees after the interviewing process (Bart and O'Brien, 1985).

Analyzing Interview Data

If interviews have been recorded, they are usually transcribed. This is a time-consuming task. Enos found that, on average, it took her more than four hours to transcribe an hour interview (personal communication). While speech recognition software should make the process of transcription less expensive and time consuming in the future, analyses of current versions of the software suggest that using such software takes about the same amount of time as the traditional methods of transcription for research purposes (Dresing, Pehl, and Lombardo, 2008; Park and Zeanah, 2005).

Interviewing qualitatively generates a great deal of text and the process of analyzing such data is typically more inductive than deductive. That is, although the researcher might come to the data with some tentative hypotheses or ideas from another context, the most common approach is to read with an open mind, while looking for motifs.

Here, we concentrate on some of the possible approaches to interview data and save the more technical aspects of data reduction and analysis for Chapter 15. One general approach to data analysis is to take the topics that the participants have talked about and to use them as a sorting scheme and looking for patterns. Enos, for example, found common pathways to prison (the "run-away path," involvement with drugs, family involvement with crime, etc.) and used it to categorize the women in her study. In a study of women's perceptions of abuse, Levendosky, Lynch, and Graham-Bermann (2000) asked "How do you think that the violence you have experienced from your partner has affected your parenting of your child?" After reading the responses, they developed 14 answer categories that four coders used with 90 percent interrater reliability. At that point, they combined similar responses into seven categories, including "no impact on parenting" and "reducing the amount of emotional energy and time available for children," and coded the interviewee's answers.

A different approach is to construct life histories, profiles, or "types" that typify the patterns or "totalities" represented in the sample. Narratives such as life stories can be read as a whole (Atkinson, 2001) or general phenomena can be described by focusing on their embodiment in specific life stories (Chase, 1995: 2). Some caution that a "battle of representation" can arise because the researchers ultimately write in their own voices, but also present long and edited narratives drawn from informants (Fine, Weiss, Wesson, and Wong, 2000). There is also a concern about reliability. One study used open-ended questions in a qualitative interview and had six analysts read verbatim transcripts of three interviews using the technique of reconstructing interpretative frames; there was only moderate agreement among the analysts on the issues identified in the interviews and the problems the respondents felt should be prioritized (Moret, Reuzel, van der Wilt, and Grin, 2007).

Cases can also be examined one at a time to see if most fit a hypothesis, and then the "deviant cases"—those that don't fit the hypothesis—can be examined more closely (Runcie, 1980: 186). Enos originally interviewed to hear people's stories, but as she heard them, she saw variables emerge. She was then able to see patterns in relationships between variables and then look for exceptions or the cases that didn't fit (personal communication). By suspending judgment and focusing on the conditions under which the patterns did not hold, she was able to look for other explanations and consider additional independent variables.

If a description fits a series of respondents, the investigator can propose a more general statement as a theory. This is similar to what Glaser and Strauss (1967) call theoretical saturation. In these approaches, the researcher determines that he or she is getting an accurate picture when successive interviewees repeat similar things.

SUMMARY

The qualitative interview is an important and useful tool for researchers interested in understanding the world as others see it. These less structured interviews can be especially useful for exploratory and descriptive work. They allow the researcher to develop insights into other people's worlds and lend themselves to working inductively toward theoretical understandings.

The qualitative interviewer typically uses either a list of topics or questions. Using mostly open-ended questions, the interview is modified and adapted for each interview, and participants are encouraged to "tell their stories." In such interview settings, rapport may develop between the participants, and the flexibility of the qualitative approach lends itself to good response rates, complex topics, and interviewing for more lengthy sessions. However, because it is an expensive and time-consuming method of data collection, the qualitative interview is most frequently used with relatively small samples.

There are several choices to be made when using qualitative interviews. Among the most important are the amount of structure in the interview, the possible use of visual materials, doing the interviews one-to-one or in groups, whether to interview or moderate, whether to "match" interviewer and interviewee, and the extent to which the interviewer shares opinions and information with interviewees during the interview.

When the interviewing process is completed, the creative process of looking for patterns and themes in the transcribed accounts begins. Data analysis can be a rewarding task, but is often very time consuming. Qualitative interviews can't provide comparable information about each member of a large sample as easily as more structured methods can, but they can be important sources of insight into people's realities and lives.

EXERCISE 10.1

Doing an Unstructured Interview

1. Pick an occupation about which you know something, but not a great deal (police officer, waiter, nurse, veterinarian, letter carrier, high school principal, or any other occupation). Find someone who is currently employed in that occupation and is willing to be interviewed about her or his work.

2. Following the ethical guidelines discussed in Chapter 3, conduct an unstructured interview of about 30 minutes in length, finding out how the person trained or prepared for her or his work, the kinds of activities the person does on the job, approximately how much time is spent on each activity, which activities are enjoyed and which are not, what the person's satisfactions and dissatisfactions with the job are, and whether he or she would like to continue doing this work for the next 10 years.

3. Transcribe at least 15 minutes of the interview, including your questions and comments, and the interviewee's replies. Include the transcription with your exercise.

4. Consider this a case study and write a brief account of the person and his or her occupation based on your interview. If you can, draw a tentative conclusion or construct a hypothesis that could be tested if you were to do additional interviewing. Include a paragraph describing your reactions to using this method of data collection.

EXERCISE 10.2

Constructing Questions

1. Construct a semi-structured interview guide for a research project on work in an occupation of your choice.

2. Following the ethical guidelines discussed in Chapter 3, write an introductory statement that describes the research project and write at least 12 open-ended interview questions.

Include both basic and follow-up questions. Focus on the same kinds of information called for in Exercise 10.1, including how the person trained or prepared for her or his work, the kinds of activities the person does on the job, the amount of time spent on each activity, which activities are enjoyed and which are not, what the person's satisfactions and dissatisfactions with the job are, and so on. Do a pilot test of your questions and then make modifications as appropriate. Turn in the introductory statement and the final version of the interview questions.

EXERCISE 10.3

Evaluating Methods

In her research with incarcerated mothers, Sandra Enos used an in-person semi-structured interview. Two other self-report methods are the in-person structured interview and the mailed questionnaire (both discussed in Chapter 9). Compare these other two methods to the method Enos used for her study.

1. Discuss the advantages and the disadvantages for Enos's study had she chosen to use a *mailed questionnaire* given her topic and the population she was interested in.
2. Discuss the advantages and the disadvantages for Enos's study had she chosen to use an *in-person structured interview* given her topic and the population she was interested in.

EXERCISE 10.4

Writing Part of Your Life Story

One kind of unstructured interview that we've discussed in this chapter is the life history interview. The Life Story at the University of Southern Maine archives life stories of people of all ages and backgrounds. To get a sense of life stories, see the archives at the Center's website at http://usm.maine.edu/olli/national/lifestorycenter. Then be both interviewer and interviewee to write a part of your life story by answering these questions which we've adapted from *The Life Story Interview* by Robert Atkins.

1. What was going on in your family, your community, and the world at the time of your birth?
2. Are there any family stories told about you as a baby?
3. Are there any stories of family members or ancestors who immigrated to this country?
4. Was there a noticeable cultural flavor to the home you grew up in?
5. What was growing up in your house or neighborhood like?
6. What family or cultural celebrations, traditions, or rituals were important in your life?
7. Was your family different from other families in your neighborhood?
8. What cultural values were passed on to you, and by whom?

EXERCISE 10.5

Photo-Interviewing

Find someone who has at least one pet and who has one or more photographs of the pet(s). Following the ethical guidelines discussed in Chapter 3, get permission to interview this person for about 15 minutes about his or her life with the pet(s). Ask the interviewee to bring one or more photographs to the interview and use the photograph(s) as part of an unstructured interview as you ask the interviewee to tell you about his or her pet(s). Focus on the family and social life with the pet(s), how the person handles the responsibilities of having a pet, how children, spouses, and/or neighbors relate to the pet, the quality of the emotional interaction the interviewee has with the pet, and any other issues you would like to cover. Record the interview in some way and transcribe it to attach to this exercise. Consider this a case study and write a brief account of the person and his or her relationship with his or her pet. Discuss whether using photograph(s) was helpful in conducting the interview.

CHAPTER **11**

Observational
Techniques

© Howard Grey/Getty Images

INTRODUCTION

observational techniques, methods of collecting data by observing people, most typically in their natural settings.

participant observation, observation performed by observers who take part in the activities they observe.

nonparticipant observation, observation made by an observer who remains as aloof as possible from those observed.

Do you do anything special for Halloween, perhaps no longer as a Trick-or-Treater, but maybe as a participant at a costume party? Have you ever seen anyone dress up as a person of a different race? What did you think he or she meant by doing so? One of the authors of this chapter's focal research, Leslie Houts Picca, asked college students to make journal entries about their observations of events in which race was a central feature. One of her colleagues, Danielle Dirks, noticed that a number of these journal entries focused on observations about Halloween costumes and experiences. Subsequently, Houts Picca and Dirks and their co-author, Jennifer Mueller, asked other students, as an extra-credit assignment, to observe and make journal entries about "Halloween as an event of sociological interest with regard to race/ethnicity, gender, sexuality, social class, and age." None of the authors had a very good idea what they'd find through this research: it was a good example of exploratory research. This was also true for Chiu (2009) when he examined the reasons why some skateboarders in New York City persisted in skating on the streets even after the city had provided skate parks and attempted to discourage, through regulation, street skating; for Kimberly Huisman and Pierrette Hondagneu-Sotelo (2005) when they became interested in the meaning of dress practice, such as wearing headscarves, among Muslim refugee women in the United States; and for Katherine Frank (2005) when she wanted to find out about the motivations of men who go to strip clubs. All these authors practiced a brand of research that we broadly call **observational techniques,** and under which rubric we include methods that are sometimes called participant and nonparticipant observation. **Participant observation** is performed by observers who take part in the activities of the people they are studying; **nonparticipant observation** is conducted by those who remain as aloof as possible. Mueller, Dirks, and Houts Picca's students were engaged in a kind of participant observation, as they observed what they themselves and other students did when they dressed as people of different races for Halloween.

FOCAL RESEARCH

Unmasking Racism: Halloween Costuming and Engagement of the Racial Other

By Jennifer C. Mueller, Danielle Dirks, and Leslie Houts Picca*

In 2003, Louisiana State District Judge Timothy Ellender arrived at a Halloween party costumed in blackface makeup, afro wig, and prison jumpsuit, complete with wrist and ankle shackles. When challenged, he said the costume was "a harmless joke" (Simpson, 2003). In 2002,

---

*From *Unmasking Racism: Halloween Costuming and Engagement of the Racial Other* by Jennifer C. Mueller, in *Qualitative Sociology.* Reprinted by permission of the Copyright Clearance Center.

Massachusetts-based Fright Catalog marketed and sold the Halloween mask "Vato Loco," a stereotyped caricature of a bandana-clad, tattooed Latino gang thug, while retail giants Wal-Mart, Party City, and Spencer Gifts began sales for "Kung Fool," a Halloween ensemble complete with Japanese kimono and a buck-toothed, slant-eyed mask with head-band bearing the Chinese character for "loser" (e.g., Unmasking Hate at Halloween, 2002). Additionally, there have been several Halloween party-related blackface incidents documented at universities across the United States over the past several years. White college students have donned blackface and reenacted images of police brutality, cotton pick-ing, and lynching at such parties, invoking degrading stereotypes and some of the darkest themes in our nation's racial past and present.

Collectively, these incidents indicate that Halloween may provide a unique opportunity to understand contemporary racial relations and racial thinking in the United States. Given the relevance of a sociolog-ical study of holidays (Etzioni, 2000) and that very little work has criti-cally addressed Halloween as a social phenomenon reflective of the broader society, the current research studies how racial concepts are employed during Halloween.

Halloween as a Uniquely Constructive Space for Engaging Racial Concepts

Holidays have been theoretically described as socializing agents, acting on members of the society to reinforce shared beliefs and re-affirm commitments to values (Durkheim, [1912] 1995; Etzioni, 2000). While many holidays, such as Christmas and Easter, are thought to directly enforce shared commitments, holidays such as Halloween arguably serve as tension-management holidays. Such holidays fulfill the socialization process *indirectly*, by managing the tensions that result from close conformity to societal mores and their concomitant behavioral prescriptions (Etzioni, 2000).

Rogers (2002) argues that Halloween was eventually promoted to national status in the United States in part because it fostered a context for social inversion during the mid- to late-nineteenth-century era, when other holidays became more institutionalized and focused on the values of family, home, and respectability. As Skal (2002, p. 17) contends, the tradition grew that for one night each year, individuals could enjoy "a degree of license and liberty unimaginable—or simply unattainable—the rest of the year." This context of free license often creates the impres-sion among revelers that all potential for insult is suspended.

Given the long legacy of racism in the United States and the ways in which it continues to structure "the rhythms of everyday life" today, it is important to discover how American holidays—particularly Halloween—remain sites where racial concepts and images are passed down and racist actions occur (e.g., Feagin, 2000, p. 2; Rogers, 2002; Skal, 2002). In light of the notable examples described above, an examination of the current relationship between the Halloween costuming ritual and the social reproduction of racism is a critical undertaking.

Participant Observation Journals

The current study began with data collected in 2002 and 2003 by Houts (2004) from college students. She collected a total of 663 individually written, participant observation journals on racial events. Following the pioneering qualitative research of McKinney (2000), Miller and Tewksbury (2001), and Myers and Williamson (2001), using journals to gather data offers a unique alternative to survey and interview methods for studying racial views, attitudes, and actions. In the journals, Halloween costumes and experiences emerged as a recurring theme among the racial events recorded.

Stimulated by Halloween as a potential racial event, a supplementary sample was gathered during the Fall 2004 semester among a demographically similar group of undergraduate students. This theoretical sample—a non-representative sample collected with the primary goal of capturing themes and developing an analytic framework (Glaser & Strauss, 1967)—was collected from students at a single, large Southeastern state university. Students in this sample were asked to specifically address Halloween as an event of sociological interest with regard to race/ethnicity, gender, sexuality, social class, and age. As with the first sample, these students received both oral and written instructions on how to do participant observation (while maintaining anonymity of those they observed) while writing their reactions and perceptions in their journals the two weeks before and after Halloween. Specifically, students were instructed to record the "what," "where," "when," and "who" of their observations—while indicating the age, gender, and race/ethnicity of the people around them. Importantly, all students were assured that "even 'no data' is data" in sociology to encourage writing about even the most mundane events around this time. Students in both samples were offered the "extra-credit" opportunity to engage in unobtrusive participant observation, recording their notes in journals. To ensure voluntary participation, students were told they could participate in the journal writing, but without having their journals used as data. For our analysis, we included as data both these observations and students' own reflections.

The original sample included the journals of 626 white students (68% female and 32% male). While Houts actually collected journals from a larger sample of 934 that included students of color, her primary analytical focus was limited to white respondents' journals. It is from this subset of 626 white students' journals that the Halloween theme was initially coded, and our subsequent analysis of Houts's original data for the current project also utilized these journals exclusively. The majority of students in this sample were between the ages of 18 and 25, although there were many students in their late 20s and

30s, and a small minority of students in their 40s and 50s. Despite aggressive efforts to collect journals from a geographically diverse sample of U.S. college students, 63% of these 626 journals came from students at five colleges and universities in the Southeast; 19% were from students in the Midwest; 14% in the West; and 4% in the Northeast. Students came from both small and large schools; private and public schools; and rural and urban settings in 12 states.

The supplementary "Halloween-only" theoretical sampling yielded 37 additional journals collected and approved for use by participating students. Seventy percent of this sample was women, 30% men, the majority between the ages of 18 and 25, with only one student in her late 20s. By race, nearly half of the participants self-described as white American/Caucasian, nearly 19% Hispanic/Latino/a, close to 11% African American/black American, nearly 13% Asian American/Pacific Islander, and just over 8% multiracial. Numerically, 19 students of our 663 total participants (3%) in our supplementary sample were people of color. Unlike the original geographically inclusive sample, this smaller, theoretical sample included students from just one of the Southeast universities included in the original sample. Given the collection methods described, most of the students in our sample attended schools in the Southeast.

Journals were analyzed using a grounded theory methodology for qualitative data analysis (Glaser & Strauss, 1967). Although journals were evaluated for relevant racial themes in regard to the costumes, rituals, and celebrating of Halloween, there was no preexisting coding scheme applied, remaining true to the emergent nature of qualitative research requiring that categories of analysis be shaped by the data (Lofland & Lofland, 1995).

Halloween License: Setting the Stage for Engagement with Racial Concepts

One of our primary arguments is that Halloween provides a uniquely constructive space for engaging race, in part because of the holiday's intuitive license, such that revelers assume the right to do, say, and be whatever they want. Indeed, college students in our sample consistently described Halloween as a holiday affording them freedom and a license to "take a break from" or even "defy" social norms. As one student observed of her friend's enjoyment of holidays like Halloween, "He calls them 'breaks from reality where he can just go wild'" (white female). In addition, for many college students, the freeing experience of Halloween costuming is intimately tied to breaking from their everyday roles: "Halloween is a way for people to see themselves as something different and uninhibited, if only for a day. Instead of being tied by how they expect others to interact with them" (Hispanic/Latino male). Such comments suggest that being "tied" to a certain identity in everyday life creates limits and inhibitions that one feels compelled to abide by, and from which Halloween provides an appropriate release.

Stereotyping as the Predominant Guide in Cross-Racial Costuming

While students discussed and employed cross-racial costuming in a variety of ways, our analysis of journals reveals the near universal guide of racial stereotype in directing their efforts. Student commentary suggests that capturing race, both "physically" and "behaviorally," is the core criterion for determining cross-racial costuming success and, as a result, most portrayals play to stereotypical ideas about the racial other. Our analysis of this phenomenon within the journals led to an emergent typology, such that the cross-racial costuming discussed, described, and engaged in by our participants tended to fall within three categories: celebrity portrayals, "role" portrayals, and generic/essentialist portrayals. In some respects, these "discrete" types capture overlapping concepts. Most critically, all three types rely on stereotype to guide their portrayals. As such, it is useful to conceptualize these categories as something of a continuum in this regard.

Celebrity Portrayals

In some cases, cross-racial/ethnic dressing occurred as a function of students masquerading as celebrities, television/movie personalities, and otherwise notorious individuals. For instance, one Asian American woman recorded seeing a black man dressed as the white rapper, Eminem. A white woman found two white male friends "covered in black paint from head to toe" in preparation for their costuming as Venus and Serena Williams, describing the scene as "the funniest thing [she] had seen in a long time." Yet another student wrote of dressing with two friends as "Charlie's Angels." She, a white woman, dressed as Asian American actress Lucy Liu's character; her African American roommate dressed as white actress Cameron Diaz's character; and another friend, who is white, dressed as white actress Drew Barrymore's character.

While cross-racial celebrity costumes tend toward the seemingly more "innocuous" end of the range—focused primarily on embodying the physical attributes of real "characters" while attempting to capture race as the most important or salient feature—it is important to remark on celebrity portrayals that involve distinctly more behavioral and stereotypical prescriptions. Consider the following student's recollections regarding a dinner conversation he had with friends over their ideas for Halloween costumes:

> My friend, who is white, well educated, and comes from a prominent upper-class family, immediately told us his plans. He planned on being, "The Black Girl from (the movie) Coyote Ugly." He then elaborated, "All I'll have to do is paint my skin and smell bad, oh and it'll help if I act like I don't know how to swim." Everyone got a good laugh out of it. (White male)

It is not clear whether this young man's comments are solely meant to communicate a joke rather than his actual plans for costuming; yet what is distinctive about this disclosure is that the young man draws on a celebrity identity ("The Black Girl from Coyote Ugly" being African American actress/model Tyra Banks), but reaffirms strictly raced conceptions of that identity, essentially negating her person-hood. In other words, it would appear that it is fundamentally critical to capture race in cross-racial costuming not only in the most obvious "physical" way, through skin color, but also through behaviorally pre-scribed ways such as smelling "bad" and acting like one cannot swim, two degrading stereotypes of African Americans.

"Role" Portrayals

One student responded to a friend's use of blackface paint, saying, "[His] outfit would be perfect if he went out and stole something before we left" (white male). Costumes such as this are indicative of racial "role" portrayals, and highlight attempts to embody race through the use of demeaning stereotypical notions about people of color. Unlike celebrity portrayals, however, "role" portrayals have no person-specific or "real" reference, leaving much room for white imagineering of racial others.

Mass marketing of items such as "Vato Loco," "Kung Fool," and numerous pimp, thug, and American Indian-themed costumes suggest the prevalence of racial caricatures in the larger culture. Rather than purchasing ready-made costumes, however, most journals documen-ted students employing their own creativity in fashioning stereotypical cross-racial/ethnic identities, a finding that echoes McDowell (1985). Particularly plentiful were descriptions of "gangstas," "thugs," pimps, and Mafiosos. While some might contend that such representations are not fixed to one particular race or ethnicity, in reality they are typi-cally connected to stereotypical racial caricatures, a finding supported by the students' journals. When whites costume in "ghetto" dress (with low-slung baggy pants and thick gold chains) or as pimps (com-plete with gold teeth, afro-like wigs, and velvet suits), they are argu-ably attempting to parody stereotypical images of blacks, even if they do not make use of blackface. Many students were clear about this in their responses: "one of my white friends, Eric, wanted to be a ghetto pimp. He defined ghetto as acting or being black" (black female).

Generic/Essentialist Portrayals

While stereotypical cross-racial costuming most often drew on carica-tured images of the racial other, a number of students described costumes that represented completely generic representations, such that simply portraying "race," usually blackness, was considered cos-tume. Such portrayals represent the most extreme employment of stereotypes guiding cross-racial costuming. Consider a non-celebrity example: one young woman recalled a discussion over costumes prior to Halloween, "We were all getting dressed up and one person said that they wanted to paint themselves black and wear a diaper and be a black baby" (white female). It would appear that such generic ideas represent whites' most fundamental attempts to strip all unique identity from people of color, to reveal race as the only relevant marker of those they claim to represent in costume. It is also significant that all generic representations in our sample referred to blackness. Arguably, generic and essentialist portrayals such as these tap into the most debased of representations, invoking the historical and archetypal consideration of the racial other in the white mind—that of the inferior black (Feagin, 2000, 2006).

Collectively considered, each cross-racial costume type helps us understand how such costumes serve as vehicles for transmitting racial judgments about people of color, particularly in light of the fact that stereotype guides each to a greater or lesser degree. From the relatively "innocuous" celebrity portrayal, to the "role" portrayal, to the fundamentally degrading generic/essentialist portrayal, cross-racial costuming represents the effort to create inhabitable representations (McDowell, 1985) of the racial other and to, indeed, engage costume as a metaphor for those depictions.

Responses in the Halloween Context

The journal-writing opportunity provided a regular way for students to reflect on the cross-racial costuming and other racialized Halloween rituals and serve as some of the most interesting points for analysis. For example, one student, observing the tendency of "ethnic costumes" to reflect stereotypes, posed the rhetorical question, "are these stereotypical costumes offensive, or merely observing that there are differences between people that can be parodied?" (white female). This was a concern addressed both explicitly and implicitly by other students, and indeed, how they responded formed a basis for thematically organizing students' reflections.

Active Participation/Unquestioning Support

For most students, notions of sensitivity or social or political correctness should be put aside for the holiday, as cross-racial costuming is afforded by the Halloween license for fun. Students adhering to this rationale unsurprisingly invoked the racial other in their own costuming, or provided minimal critique of those who did. Dressing up for a day, "as anything … or anyone you want" is entitled by the holiday: "Consciously, I can't think of a time when I have placed a limitation on Halloween costumes or decorations based on race/ethnicity, gender, sexuality, social class or even age" (white female). Similarly, another white woman wrote, "Halloween is a holiday in which people like to believe that the lines of race and gender are blurred, with no one truly caring exactly where they lie." She appears confident that her belief is universal, and as such, concerns about choosing a gendered or racially offensive costume need not be entertained.

Among students who uncritically embrace cross-racial costuming, a belief in the "fun factor" of dressing across racial/ethnic lines emerges. Frequently, students who emphasize the fun of costumes translate that impression into an equation where humor "trumps" offensiveness—in other words, as long as a costume is perceived as funny, onlookers should take no offense.

Dubious Curiosity/Questioning Support

A number of students expressed dubious curiosity and questioning support for cross-racial costuming. Responses in this group reflect a sense of ambivalence, with students appearing less confident about the social permissibility of cross-racially dressing. However, critical thinking about doing so is often coupled with statements that invalidate the offensiveness of the practice, as the following quotation demonstrates:

> We were all talking about what we should dress up as for Halloween. My boyfriend's friend Mike (22-year-old white male) is having a huge costume party and my boyfriend thought it would be fun to go as a black couple. I think it would be really fun and funny to do that but I'm afraid that black people would be offended. I don't get offended when people dress up as "Indians" for Halloween and I don't see why black people would care if we dress as black people. I asked my mom what she thought and she said we choose another costume because even though it's silly, black people probably would be offended and we shouldn't do things that could hurt someone's feelings intentionally. (Native American/white female)

While this woman recognizes such costumes might offend black people, she discredits their potential "silly" objections, particularly given her estimation that masquerading in this way would be both "really fun and funny," a clear return to the theme of the "fun factor" litmus test. Interestingly, we also note that relying on her identity as a racial minority, in one sense, led her to give pause in contemplating the issue, but in another sense served as validation for cross-racially dressing as a generic "black couple."

Firm Objection—Antiracism versus White Supremacy

Finally, we reach the other end of the pole—those responses grouped as firm objections to cross-racial costuming. One student was particularly clear: blacks and whites dressing as the "other race" represented a way to "mock each other" (multiracial male). While his analysis represents a form of "equal opportunity racism" and neglects the stereotypical and highly degrading ways in which people of color are often portrayed by whites, it does represent a firm belief that the practice is offensive.

Given the disturbing abundance of negatively racialized incidents in our sample, it is important to acknowledge those students who expressed their own or detailed others' antiracist thinking regarding cross-racial costuming. Typically these students chose not to cross-racially costume, or to critically evaluate this practice based on beliefs that it is offensive and degrading to people of color. One of the most hard-lined unequivocal excerpts was the following:

> I saw the most disturbing thing tonight. I went with a group of friends to a Halloween party. As we were leaving, I saw two white people, a male and a female, standing outside the other party who were dressed in blackface, as what I can only assume was their rendition of Jamaicans. I could do nothing but stand there with my mouth, literally, open. I was so shocked. I have never seen anything in person as horrifically blatantly racist and offensive as I found that to be. Who comes up with an idea like that? I don't know if they were doing it as a joke or what their costume purpose was, but I don't find that funny at all. (White female)

While expressed antiracism was the exception and not the rule among white students, students of color were more universal in their critique of cross-racial costuming, as well as in their willingness to challenge others, particularly when they observed highly stereotypical portrayals. It is important to give voice to the frustration and hurt they expressed, as well. One Latina woman skeptically attended a "ghetto party" with a black friend. As she detailed:

> When we arrived at the party we were shocked at what we saw. First, we did not see one black or Hispanic person. Blonde hair, blue eyed kids were walking around with aluminum foil on their teeth, bandanas on their head, fur coats, big huge earrings, and shirts that said "Project Chick" or "Ghetto Fabulous." … when I was younger my family was pretty poor and our living situation was very bad. We lived in what people refer to as the "ghetto," and it wasn't fun and it sure wasn't what those kids were portraying it to be. They were glamorizing it and at the same time almost making fun of it. My friend was insulted because this is how a lot of white people view black people and it is sad that this is true. (Hispanic/Latina female)

Conclusion

While Skal (2002) suggests that "tasteless" Halloween costumes might simply represent an extension of Halloween's historical pranking tradition, it is perhaps more fitting to draw on a different relic from Halloween's historical tradition—that of departed spirits returning to wreak mischief and even harm. As Bonilla-Silva (2003) documents, the societal norms of the post-Civil Rights era have disallowed the open expression of racial views. In this way, for many, Halloween has become a culturally tolerated, contemporary space for the racist "ghost" to be let out of the box. Indeed, our findings support the thesis that Halloween's combination of social license, ritual costuming, and social setting make the holiday a uniquely constructive context for negative engagement of racial concepts and identities.

With respect to theorizing what activates the cross-racial costuming behavior of our respondents, it is useful to draw upon the concept of "rituals of rebellion"—culturally permitted and ritually framed spaces (like New Year's Eve and Mardi Gras) where the free expression of countercultural feelings are tolerated, and protected to some degree by the agents of the official culture (Gluckman, 1963). Interestingly, although the Gluckman framework might predict the use of cross-racial costuming among people of color, it is not immediately apparent that students of color use Halloween as an opportunity to create costume performances that subvert the racial and/or social hierarchy. Even in the very few cases where cross-racial costuming among respondents of color did occur, costumes were most frequently celebrity portrayals, and none appeared to pose an indictment of whiteness *per se* (as particularly opposed to the clear degradation of blackness revealed). To be sure, the relatively small proportion of students of color in our sample limits our ability to fully explore this theme, and future research is needed to examine this phenomenon in greater depth.

In contrast, there does appear to be a unique, ritually rebellious form of performance that occurs among many white students. In the "colorblind" post-Civil Rights era, it has become commonplace for whites to express frustration and resentment toward color-conscious racial remediation programs, such as affirmative action (Feagin, 2000, 2006; Wellman, 1997).

Although in truth white students occupy the dominant racial social identity group, we posit that many may entertain if not a sense of "oppression," at minimum a sense of normative restriction by a social code which prescribes "nonracist" presentations, and for which racialized Halloween "rituals of rebellion" afford some release. One white student praised Halloween as "great" because it eliminates the need to worry about racial offense. For those whites who actively endorse the idea that whites are now victimized by the preferencing of people of color (e.g., in employment, admissions), Halloween may ironically signify a suspension of this imagined "hierarchy."

We must also examine the needs left unfulfilled by contemporary approaches to multiculturalism and political correctness, some of which have become dogmatic. The confused critiques of many students reflect the ways in which we have become a society reproducing what Bonilla-Silva (2003) refers to as "racism without racists." He suggests that mere education will not lead to racial tolerance.

Indeed, as Johnson (1997) points out, the social reproduction of racism does not require people explicitly acting in racially hostile ways, but simply those who will uncritically acquiesce in the large culture order. While our data indeed reveal the explicit intentions of some students to degrade blackness through costume, the majority of white respondents actively suspended their criticisms or behaved in wholly uncritical ways. It is highly significant that regardless of intention, each of these response "types" share the outcome of reproducing stereotypical racist images, thereby supporting the racial social structure.

Rogers (2002) notes that while "Halloween is unquestionably a night of inversion," the holiday's context probably provides little substantial opportunity to actually challenge how society operates in a determined or sustained way—"At its best, Halloween functions as a transient form of social commentary or 'deep play'" (p. 137). With respect to race, we would argue that the holiday provides a context ripe for reinforcing existing racialist concepts. In particular, it provides an implicitly approved space for maintaining the privilege that whites have historically enjoyed, and to define and caricature African Americans and other people of color in degraded and essentialist ways. Ultimately, the white privilege to racially differentiate supports both material and ideological benefits and disadvantages built into the systemic racial structure. In the United States this system has deep historical roots and is well-formulated and ingrained into the everyday rhythms of life. As such, Halloween social commentary which engages race can hardly be described as transient, and actually reflects the dominant racist ideology, coupling contemporary imaging with racist conceptualizations as old as the country itself.

References

Bonilla-Silva, E. (2003). *Racism without racists: Color-blind racism and the persistence of racial inequality in the United States.* Lanham, MD: Rowman and Littlefield.

Durkheim, E. ([1912] 1995). *The elementary forms of religious life.* Translated by K. E. Fields. Reprint, New York: Free Press.

Etzioni, A. (2000). Toward a theory of public ritual. *Sociological Theory*, 18, 44–59.

Feagin, J. (2000). *Racist America: Roots, current realities and future reparations.* New York: Routledge.

Feagin, J. R. (2006). *Systemic racism: A theory of oppression.* New York: Routledge.

Glaser, B. G., & Strauss, A. L. (1967). *The discovery of grounded theory: Strategies for qualitative research.* Hawthorne, NY: Aldine de Gruyter.

Gluckman, M. (1963). Order and rebellion in tribal Africa: Collected essays with an autobiographical introduction. London: Cohen and West.

Houts, L. A. (2004). Backstage, frontstage interactions: Everyday racial events and white college students. Ph.D. dissertation, Department of Sociology, University of Florida, Gainesville, FL.

Johnson, A. G. (1997). *Power, privilege and difference.* New York: McGraw Hill.

Lofland, J., & Lofland, L. H. (1995). *Analyzing social settings: A guide to qualitative observation and analysis.* Belmont, CA: Wadsworth Publishing Company.

McDowell, J. (1985). Halloween costuming among young adults in Bloomington, Indiana: A local exotic. *Indiana Folklore and Oral History*, 14(1), 1–18.

McKinney, K. D. (2000). Everyday whiteness: Discourse, story and identity. Ph.D. Dissertation, Department of Sociology, University of Florida, Gainesville, FL.

Miller, J. M., & Tewksbury, R. (2001). *Extreme methods: Innovative approaches to social science research*. Boston, MA: Allyn and Bacon.

Myers, K. A., & Williamson, P. (2001). Race talk: The perpetuation of racism through private discourse. *Race and Society*, 4, 3–26.

Rogers, N. (2002). *Halloween: From pagan ritual to party night*. New York: Oxford University Press.

Simpson, D. (2003). White La. judge draws fire for costume. The Associated Press, November 10, Dateline: New Orleans, Domestic News. Retrieved June 5, 2005. Available: LEXIS-NEXIS Academic Universe, News Wires.

Skal, D. J. (2002). *Death makes a holiday: A cultural history of Halloween*. New York: Bloomsbury.

Unmasking Hate at Halloween (2002). Tolerance.org, October 24. Retrieved June 10, 2005. (http://www.tolerance.org/news/article_print.jsp?id=629).

Wellman, D. (1997). Minstrel shows, affirmative action talk and angry white men: Marking racial otherness in the 1990s. In R. Frankenberg (Ed.), *Displacing whiteness: Essays in social and cultural criticism* (pp. 311–331). Durham, NC: Duke University Press.

THINKING ABOUT ETHICS

Houts Picca (then Houts) consulted with the Institutional Review Board (IRB) at her university before the study that sparked interest in the Halloween costume study. Since she had professors collect the journals in courses at some 25 universities, she had to complete an IRB protocol at some of the others as well, though many of them accepted her own university's protocol. Then she, Dirks, and Mueller consulted their IRB again before the follow-up study, focusing exclusively on Halloween. Both times all participating students were given a five-page detailed instruction sheet on how and what to write in their journals (see Box 11.1), as well as informed consent sheets, explaining, among other things, that they had the option of completing the assignment for their class, but not having their data used in either Houts's first study or the current one (Houts Picca and Dirks, personal communication). Participation in the current study was therefore both voluntary and informed. Before they engaged in their observations, students were informed about the need to maintain the confidentiality of the people they observed.

OBSERVATIONAL TECHNIQUES DEFINED

"Observational techniques" are sometimes referred to as "qualitative methods" and "field research." But both "qualitative methods" and "field research" connote additional methods (e.g., qualitative interviewing and using available data) that we cover elsewhere. Here we focus on "observations" alone. Mueller, Dirks, and Houts Picca's students gave their own accounts allowing the researchers to classify them as unquestioning supporters, questioning supporters, and firm

BOX **11.1**

Excerpt from the Five-Page Information Sheet Given Out by Houts Picca for Her First Study of Race and Re-used by Mueller, Dirks, and Houts Picca for Halloween Costume Study

How Will I Be Graded? Your instructor will determine what credit you will receive for participating in this project. She or he will also determine for how long you will be asked to keep a journal and the length required of each journal entry to merit credit. Again, you will not be graded on "what" you see (so please don't make up any data!), rather you will be evaluated on the quality and detail of your systematic observations and field notes.

Is This Voluntary? Yes. All students who participate in the journal-writing activity and who agree to share their journals with me for the dissertation project will be asked to sign an IRB informed consent form. You have the option of participating in the journal-writing assignment for your class, but not signing an IRB informed consent form, in

which case your journals will not be used or analyzed for the project, ensuring voluntary participation. In other words, even if the journal-writing assignment is an assignment for your course grade, you are still given the choice whether or not to participate in the study, with no penalty for your decision.

What Will the Data Be Used For? This study is part of a larger dissertation project involving students across the nation. The data collected in this project may lead to publications such as articles and books on racial issues.

What If I Have Questions? You will have the opportunity to meet regularly with your instructor to ensure all your questions and concerns are answered. If you have additional questions at any time, please contact Leslie Houts.

---

objectors, but they also reported things like the two white male friends "covered in black paint from head to toe" in preparation for their costuming as Venus and Serena Williams. We'd like to stress that a variety of methods *are* and *should be* used in the field, but here we want to focus on those field observational methods that have not been covered in other chapters.

Of course, the data from questionnaires, interviews, content analysis, experiments, and so on *are* all based on observations of one kind or another. For questionnaires and interviews, the observations are about responses to questions. For content analysis, which we'll cover in Chapter 13, the observations are about communication of one sort or another. In experiments, however, observations are frequently of the sort that closely resemble those used by participant and nonparticipant observers. In most cases, experimental observations are of a more controlled sort than those used in the field. **Controlled (or systematic) observations** involve clear decisions about just what kinds of things are to be observed. In their experiment on the effects of exposure to violent media, Bushman and Anderson (2009) observed whether or not students who thought they'd heard someone injured in a fight would offer help and the amount of time it took to offer. But such controlled observations are not the primary focus of this chapter.

Mueller, Dirks, and Houts Picca's students' immersion in the natural setting of their subjects distinguishes their work from these other approaches. Rather than merely read about or ask about "Halloween cross-racial costuming," they observed such costuming in person, perhaps even participated in

controlled (or systematic) observations, observations that involve clear decisions about what is to be observed.

it themselves. Moreover, because they attempted to keep their observations unobtrusive, their intervention was minimal. We here confine our attention to observational techniques, then, that are used to observe actors who are, at least in the perception of the observer, acting naturally in natural settings.

 **STOP & THINK** *Could Mueller, Dirks, and Houts Picca have used only questionnaires to study cross-racial costuming for Halloween? What kinds of insights might have been "lost" if they had?*

The origins of modern observational techniques are the subject of some current debate. Conventional histories attribute their founding to anthropologists like Bronislaw Malinowski, with his *Argonauts of the Western Pacific* (1922), and sociologists like Robert Park, who, around the same time, began to inspire a generation of students to study city life as it occurred naturally (see, for example, Wax, 1971, and Hughes, 1960). However, Shulamit Reinharz (1992) has argued that Harriet Alice Fletcher, studying the Omaha Indians in the 1880s, and Harriet Martineau, publishing *Society in America* in the 1830s, promoted essentially the same techniques that Malinowski and Park received fame for in the 1920s, thus suggesting the earliest "social science" observers might have been American and British women.

REASONS FOR DOING OBSERVATIONS

Perhaps the most important reason for using observational techniques is that they offer a relatively unfiltered view of human behavior. You can, of course, ask people about what they've done or would do under given conditions, but the responses you'd get might mislead you for any number of reasons. People do, for instance, occasionally forget, withhold, or embellish. Houts Picca knew, from waiting tables at a Florida steakhouse to help pay her way through graduate school, that racist and sexist events took place in the "back-stage" kitchen of the restaurant she worked at, and she also knew that the agents of these events frequently hid their racism and sexism in more public places. She'd also read enough sociology to know that whites frequently are more forthcoming about racist attitudes in interviews than in questionnaire surveys. But she believed that only through the observation of more private settings could she get a more unfiltered view of behaviors. Moreover, she thought she'd get more valid data if she let people who had unquestioned access to those private settings (i.e., other students) do the observing (personal communication). Thus one of her students reports that a friend claimed that, to portray "The Black Girl from (the movie) Coyote Ugly," all he'd have to do is "paint my skin and smell bad, oh and it'll help if I act like I don't know how to swim." And that "Everyone got a good laugh out of it." Upon reading this report, even if the friend was joking rather than describing an actual costume, we feel we are getting more intimate insight into relevant behavior than if we simply asked questions about Halloween behaviors.

Observation can be useful when you don't know much about the subject under investigation. A classic technique in anthropology, the ethnography, or study of culture, is primarily an observational method that presumes that the culture under study is unknown to or poorly known by the observer. Margaret Mead's classic investigation, *Sex and Temperament* (1933), of gendered behavior among three tribes in New Guinea, for instance, found evidence that sex-role definitions were more malleable than convention held at the time. Similarly, many well-known sociological investigations have been of things less than well known beforehand. More recently, Mears and Finlay (2005) examined how fashion models manage bodily capital and Muir and Seitz (2004) studied rituals within the male collegiate rugby subculture. Little was known by social scientists about any of these topics beforehand. Mueller, Dirks, and Houts Picca justify their investigation of cross-racial dressing for Halloween, in part, because nobody else had written about it. Dirks claims, in fact, that even as she was helping Houts Picca organize her data into an electronic format for analysis, she might have missed the prevalence and significance of accounts dealing with Halloween "if Halloween were not my own favorite holiday" (personal communication). Certainly nothing she'd read had prepared her to focus on them.

thin description, bare-bone description of acts.

thick description, reports about behavior that provide a sense of things like the intentions, motives, and meanings behind the behavior.

Observational techniques also make sense when one wants to understand experience from the point of view of those who are living it or from the context in which it is lived. Norman Denzin makes a distinction between **thin** and **thick description**, the former merely reporting on an act, the latter providing a sense of "intentions, motives, meanings, context, situations, and circumstances of action" (1989: 39). Mueller, Dirks, and Houts Picca are pretty much limited to thin description, because their student observers, as a result of their promise not to interfere with the situations they observe, can't ask other students why they act in certain ways. In fact, they were explicitly told NOT to ask other students about the actions observed (personal communication). But this isn't true when observers permit themselves to ask questions of others in the situation. In an example of how visual sociology can generate observations and thick description of those observations, Radley, Hodgetts, and Cullen (2005) studied how homeless people in London understand their lives in shelters and on the streets. They gave homeless participants cameras to photograph typical scenes from their days. Afterward, Radley et al. interviewed the homeless about the photographs they'd made. Different participants used the opportunity to show how they'd adapted to their homelessness in different ways. For instance, one woman photographed the public places she slept in and stressed how she could make a "normal" life despite being a "rough sleeper," while another woman photographed friends and other homeless people, using them to show how she used her relationships with others to adapt to her life without permanent shelter (Radley et al., 2005: 280).

Observational techniques are also useful when you want to study quickly changing social situations. Kai Erikson largely based his account in *Everything in Its Path* (1976) of the trauma visited on an Appalachian community by a disastrous flood in 1972 on descriptions gleaned from legal transcripts and personal interviews. He, as a sociologist, had been asked by a law firm to assess the nature and extent of the damage to the Buffalo Creek community by a flood that had been caused, in part, by the neglect of a coal company. His book presents the legal case he made then, using content analyses and unstructured interviews. When he presented the effects of the favorable judgment on the community he studied, he described a gathering "in a local school auditorium to hear a few announcements and to mark the end of the long ordeal." He interprets the gathering as "a kind of graduation ... that ended a period of uncertainty, vindicated a decision to enter litigation, and furnished people with sufficient funds to realize whatever plans they were ready to make. But it was also a graduation in the sense that it propelled people into the future at the very moment it was placing a final seal on a portion of the past" (1976: 248–249). Erikson then used direct observation to mark a momentous occasion: the moment (two years after the flood) when Buffalo Creek began, as a community, to heal itself after its calamitous ordeal. If he hadn't been there when he was, this moment of significant change would necessarily have gone unnoticed by himself and his readers.

There are other reasons why one might prefer observational techniques. Questionnaire surveys can be very expensive, but participant or nonparticipant observation, although time consuming, can cost as little as the price of paper and a pencil. Ethically, studying homeless people in their natural settings, as Liebow (1993) did, raises fewer questions than would studying, say, the effects of homelessness by creating an experimental condition of homelessness for a group of subjects. As you can imagine, the reasons for doing observation can have complex interactions with one another, with the salience of some undermining the feasibility of others. Rebekah Nathan (2005), a college professor, became a college student for a year to gain a deep, unfiltered understanding of what it means to be a college student today, partly because she didn't think she'd get the same kind of high-quality data from simply asking her students. (Rebekah Nathan is a pseudonym that the anthropologist adopted for her study.) Nathan didn't want to accept grants from funding agencies to pay for her tuition, books, housing, and so on because she feared that would give the agency too much influence over what she could report. So she paid for these things out-of-pocket, making the research for *My Freshman Year: What a Professor Learned by Becoming a Student* the most expensive one she'd ever done.

complete participant role, being, or pretending to be, a genuine participant in a situation one observes.

OBSERVER ROLES

participant-as-observer role, being primarily a participant, while admitting an observer status.

observer-as-participant role, being primarily a self-professed observer, while occasionally participating in the situation.

complete observer role, being an observer of a situation without becoming part of it.

We suggested earlier that one of the defining characteristics of observational techniques is their relative unobtrusiveness. But this unobtrusiveness varies with the role played by the observers and with the degree to which they are open about their research purposes. Raymond Gold's (1958) typology of researcher roles suggests a continuum of possible roles: the **complete participant**, the **participant-as-observer**, the **observer-as-participant**, and the **complete observer**. It also suggests a continuum in the degree to which the researcher is open about his or her research purpose. As a complete participant, the researcher might become a genuine participant in the setting under study or might simply pretend to be a genuine participant. The student observers for Mueller, Dirks, and Houts Picca were obliged to remain complete participants as they made their observations about other students' behavior in relation to Halloween. They were explicitly instructed to "use unobtrusive research techniques so that the person(s) [they wrote] about in [their] journal[s] will not be aware that they are being studied" (from the instructions given to students, personal communication). As already mentioned, they were also instructed NOT to interview the people they observed and to "interact with people as [they] usually would." One ethical dilemma associated with complete participant observation is informed consent. Obviously, telling the people whom they observed in the act of cross-racial costuming that they were being observed for a possible research report might have changed their behavior, and perhaps especially what they would have said about the ways they dressed or proposed to dress. (It's possible that the student who made demeaning comments about the African American character in *Coyote Ugly*, for instance, might have done more self-censoring if he'd known he was being observed.) The American Sociological Association (ASA) Code of Ethics provides guidance for dealing with the issue of informed consent while conducting observational research (see Box 11.2). In general, researchers who would observe people in "private" places owe observed individuals informed consent unless "no more than minimal harm" is likely to befall such individuals and the research becomes impracticable under the conditions of informed consent and an IRB confirms both of these conditions obtain. Mueller, Dirks, and Houts Picca (and Houts Picca before them) did receive IRB agreement that no more than minimal harm could befall participants and that the research would have been impracticable if the observed students knew they were being observed.

The complete participant, in Gold's typology, is defined by withholding the fact that the observer is also a researcher. Thus, all Mueller, Dirks, and Houts Picca's student observers collected data, while participating in dormitory

---

BOX **11.2**

Excerpts from the ASA Code of Ethics on Informed Consent in Observational Research

12.01 Scope of Informed Consent

(a) Sociologists conducting research obtain informed consent from research participants or their legally authorized representatives ... when behavior of research participants occurs in a private context where an individual can reasonably expect that no observation or reporting is taking place.

(b) Despite the paramount importance of consent, sociologists may seek waivers of this standard when (1) the research involves no more than minimal risk for research participants, and (2) the research could not be practicably be carried out were informed

consent to be required. Sociologists recognize that waivers of consent require approval from institutional review boards or, in the absence of such boards, from another authoritative body with expertise on the ethics of research...

(c) Sociologists may conduct research in public places ... without obtaining consent. If, under such circumstances, sociologists have any doubt whatsoever about the need for informed consent, they consult institutional review boards...

Source: ASA Code of Ethics, p. 14, www.asanet.org/galleries/default-file/CodeofEthics.pdf

life, school work, and Halloween-costuming preparations, telling no one that they were doing research.

 **STOP & THINK** *Do you agree that not informing observed students in Mueller, Dirks, and Houts Picca's study was ethical?*

The opposite of the complete participant role is that of the complete observer: one who observes a situation without becoming a part of it. In principle, complete observers might ask for permission to observe but they usually don't. Harrell and his colleagues (cited by Bakalar, 2005) made more than 400 anonymous observations of children and their parents in 14 supermarkets to see if more attractive children were treated better by their parents than others, coding not only the attractiveness of the children but also whether the parents belted their youngsters into grocery cart seats, the frequency with which parents lost track of their children, and the number of times children were permitted to do dangerous things like standing up in the cart. Harrell and others remained socially invisible to those whom they observed. They didn't tell those they watched that they were observing them. A rather more expensive version of Harrell's basic strategy was employed by observers trained at the National Opinion Research Center (NORC) in preparation for, among others, the work of Robert Sampson and Stephen Raudenbush (1999) on disorder in urban neighborhoods. These observers were driven in a sport utility vehicle at a rate of five miles per hour down every street in 196 Chicago census tracts, videotaping from both sides of the vehicle while speaking into the videotape audio. Martyn Hammersley and Paul Atkinson (1995: 96) note the paradoxical fact that complete observation shares many of the disadvantages and advantages of complete participation. Neither is terribly obtrusive: Harrell, the NORC researchers, and Mueller, Dirks, and Houts Picca's students were not likely to affect the behavior of those they watched. On the other hand, neither complete participation nor complete observation permits the kind of probing questions that one can ask if one admits to doing research. In addition to the issue of (not being able to ask scientifically interesting) questions, the lack of openness (about one's research purposes) implicit in pure participation or pure observation has raised, for some, the ethical issue of whether deceit is ever an appropriate means to a scientific end—an issue, you will recall, that also has come up in our discussion of other methods. Like complete participants, complete observers can find guidance in the ASA Code of Ethics (see Box 11.2 again), which suggests that, most of the time, observations in public places do not require the informed consent of those observed. The same applies in more private places, as long as no more than minimal harm is likely to befall those observed, the research becomes undoable when informed consent is requested and when an IRB agrees that these conditions exist.

Both the participant-as-observer and the observer-as-participant make it clear that they are doing research; the difference is the degree to which they emphasize participation or observation. The participant-as-observer is a participant in the setting who admits he or she is also an observer. The observer-as-participant is a self-professed observer who occasionally participates in the situation. Chiu's (2009) study of skateboarders in New York City, mentioned earlier, is an example of research conducted by a participant-as-observer. Chiu himself entered each of the four major sites of his study—two public spaces, one of which even has an old sign, attached to a lamppost, that declares, "No rollerblading and skate boarding allowed," and two parks built for skateboarding—as a skateboarder. He became engaged in informal conversations and interactions with other skateboarders and gained a feel for their everyday lives. He then asked 15 other skateboarders and one police officer for permission to conduct semi-structured interviews. He was forthcoming about his research purpose when others asked in the informal interactions and explicitly announced his purpose as he requested the interviews. He is able, as a result of his interviews, to provide a thick description of, among other things, why some skateboarders continue to employ public spaces to skate, despite regulations against skating and the presence of designated areas for skating. Street skaters turn the built environments created by architects and urban planners into something other than what the planners had in mind. As one skater put it:

> You see everything different with different eyes. Someone sees something and says oh, yes, it is just up there, you don't even notice that, but as a skater, you say yeah, I can skate that, if I hit it like this, I can get all the buzz out. (Chiu, 2009: 33)

Skateboarding in public spaces affords skateboarders, Chiu asserts, the opportunity for an "urban journey, which is more important than the destination" (34).

Elliot Liebow, in his study of homeless women, is more explicit about his alternation between being primarily participant and primarily observer in the women's shelter called the Refuge and defines the time when he was primarily observing as his research time:

> I remained a volunteer at the Refuge [throughout the study], and was careful to make a distinction between the one night [a week] I was an official volunteer and the other nights when I was Doing Research (that is, hanging around). (1993: x)

Thus, Liebow, as researcher, would fall into Gold's observer-as-participant role.

Liebow clearly announced his interest and even records the announcement of his purpose, and his request for permission:

> "Listen," I said at the dinner table one evening, after getting permission to do a study from the shelter director. "I want your permission to take notes. I want to go home at night and write down what I can remember about the things you say and do. Maybe I'll write a book about homeless women." (1993: ix)

And the book he did write, *Tell Them Who I Am*, has become a classic study of women's homelessness, and especially of the desperation and hope with which the women led their lives (Liebow, 1993: 1). He watched, for instance, as Elsie, one of the women who lived at the Refuge, fasted for three days, because, as she reported, her pastor had told his parishioners to do so if they "wanted a special favor from God." And Elsie, she told Liebow, wanted a job "sooo bad" (Liebow, 1993: 73).

BOX 11.3

The Continuum of Observer Roles: The Chances Being Obtrusive and Asking Probing Questions Vary by Role

Complete Participant	Participant-As-Observer and Observer-As-Participant	Complete Observer

⟶ +Chances of Being Obtrusive+ ⟵

⟶ +Chances of Asking Probing+ ⟵
Questions

One thing Liebow's, and even Chiu's, study makes clear is that within any given research project, researchers can play several different observational roles. There were times when they were primarily participants and times when they were primarily observers.

One danger of both the participant-as-observer and the observer-as-participant roles is that they are more obtrusive than those of the pure participant or the pure observer and that those being studied might shift their focus to the research project from the activity being studied, much as they might in responding to a questionnaire or an experimental situation. Those being studied might have been aware of being such projects, but the verisimilitude and believability of what Chiu and Liebow report tends to dispel one's doubts about the overall validity of their reports.

The danger of obtrusiveness and the value of thick description aside, there are times when announcing one's research intention offers the only real chance of access to particular sites. The only other way participant observers Becker, Geer, Hughes, and Strauss could have gained access to the student culture in medical school for their classic study *Boys in White* (1961) would have been to apply and gain admission to medical school themselves. Think of all the science courses they would have had to take!

Different research purposes require different roles. It's hard to imagine, for instance, that Mueller, Dirks, and Houts Picca's study would have worked if their students had announced their research intentions. There are no hard and fast rules about what observer role to adopt in a given situation. As in all research, you should be guided by methodological, practical, and ethical considerations.

GETTING READY FOR OBSERVATION

We've seen how researchers using other techniques (e.g., experiments or surveys) can spend much time and energy preparing for data gathering, in formulating research hypotheses, and developing measurement and sampling strategies, for example. Such lengthy preparations are usually not as vital for studies involving observational techniques, where design elements can frequently be worked out on an ad hoc basis. The exception to this rule, however, is controlled, or systematic, observations, which are defined by their use of explicit plans for selecting, recording, and coding data (McCall, 1984). In their controlled observations, Harrell and his colleagues (cited in Bakalar, 2005), as you know, studied parents children in supermarkets. Because this team began with a more or less well-defined hypothesis (parents will take better care of pretty children than they will of ugly ones), one can imagine them filling out a code sheet like the one we've constructed for you in Figure 11.1.

Far more often, however, observers begin their studies with less clearly defined research questions and considerably more flexible research plans. Adler and Adler (1994b) suggest that an early order of business for all observational studies is selecting a setting, a selection that can be guided by opportunistic or theoretical criteria. The selection is made opportunistically if, as the term suggests, the opportunity for observation arises from the researcher's life's circumstances (Riemer, 1977). Liebow studied shelters for homeless women because, after he'd been diagnosed with cancer and quit his regular job in 1984, he decided to work in soup kitchens and shelters, only to find their clients extraordinarily interesting. Jeffrey Riemer (1977: 474) advocates opportunistic settings for ease of entry and of developing rapport with one's subjects and for the relatively accurate levels of interpretation they can afford.

On the other hand, Mueller, Dirks, and Houts Picca asked students to observe others as they cross-racially dressed for Halloween primarily for theoretical reasons rather than because it was so easy for them to do so. In fact, Dirks reports that once she noticed the frequency with which students had

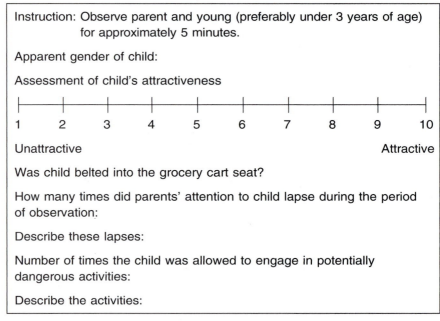

FIGURE **11.1** A Potential Codesheet for Recording Observations in the "Care of Pretty/Ugly Children Study" by Harrell and others (cited in Bakalar, 2005)

mentioned Halloween events in Houts Picca's previous study, "Fall was quickly approaching," and they had to hurry to line up professors who might ask students to do journals on their Halloween experiences and to get IRB permission to proceed (personal communication). This was anything but opportunistic or easy. But it was theoretically motivated. As Mueller reports, all three authors are "heavily influenced by 'macro,' structural theories of privilege/oppression," but they wanted to see how racism was enacted in the everyday lives of students. After reading through Houts Picca's initial data, they felt it was appropriate to ask students to report on racialized performances surrounding Halloween (personal communication).

 **STOP & THINK** *Do you remember Laud Humphreys's (1975) study of brief, impersonal, homosexual acts in public restrooms (mentioned in Chapter 3)? Do you imagine that Humphreys's decision to study public restrooms was informed primarily opportunistically or purposively? What makes you think so? (Humphreys's answer appears in the next "Stop and Think" paragraph.)*

The suggestion that setting selection can be made on either opportunistic or theoretical grounds might be slightly simplistic, however, insofar as it implies that one chooses only one site, once and for all, and that it is primarily opportunity or theory that informs the choice. Actually, Adler and Adler (1994a), in the study of adult-organized afterschool activities for children, were not so very unusual in their decision, initially, to immerse themselves in settings that were available (their own children's afterschool activities), but then to seek out interviews that spoke to ideas they developed on their own:

> [W]e derived our ideas about what to write *about* from our observations of and casual interaction with kids, … but to get quotes from kids themselves and to see how they felt or if their experiences jibed or differed from our perspectives, we conducted interviews.… These interviews were selected by theoretical [or purposive] sampling to get a range of different types of kids in different activities, age ranges, and social groups. (personal communication)

Especially in seeking their interviews, then, Adler and Adler were guided by the effort to test their own theoretical notions against those of other participants, an approach they recommend elsewhere (Adler and Adler, 1987). By emphasizing their efforts to "get a range of different types of kids," in fact, they suggest they attempted to approximate the ideal of **theoretical saturation** (Adler and Adler, 1994b; Glaser and Strauss, 1967)—that is, the point where new interviewees or settings looked a lot like interviewees or settings they'd observed before. Opportunity, then, might have inspired their early choices; some mix of opportunity and theory influenced their later choices.

theoretical saturation, the point where new interviewees or settings look a lot like interviewees or settings one has observed before.

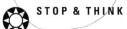 **STOP & THINK** *Humphreys says he embarked on his study of sex in public restrooms after he'd completed a research paper on homosexuality. He states that his instructor asked about where the average guy went for homosexual action and that he (Humphreys) articulated a "hunch" that it might be public restrooms. His further comment, "We decided that this area of covert deviance, tangential to the subculture, was one that needed study" (1975: 16), suggests that his decision to study public restrooms was purposively, rather than opportunistically, motivated.*

purposive sampling, a nonprobability sampling procedure that involves selecting elements based on the researcher's judgment about which elements will facilitate his or her investigation.

Moreover, Mueller, Dirks, and Houts Picca's study suggests what seems to be generally true of sampling in studies using observational techniques: Observations are most often done in a nonprobabilistic fashion. Their students practiced, as most participant observers do, **purposive** (or, as they call it, theoretical) **sampling**—they observed situations that would give them insight into the major question they were assigned: Halloween events "of sociological interest with regard to race/ethnicity, gender, sexuality, social class, and age." Devising a sampling frame of the kinds of things studied by participant and nonparticipant observers (e.g., afterschool activities, occupations involving emotion work, homeless women, bathrooms used for homosexual encounters) is often impossible, though again, this might be less true of the (relatively infrequent) studies involving systematic observations. The researchers from NORC did, in fact, select a stratified probability sample of Chicago census tracts "to maximize variation by race/ethnicity and SES" (Sampson and Raudenbush, 1999) for its sports utility van–driven study of public spaces.

One other decision one should make before entering the field is how much one will tell about yourself and your research. *Disclosure* of your interests (personal and research) can help develop trust in others, but it can also be a distraction from, even a hindrance to the unfolding of, events in the field. What you're willing to disclose about your interests is obviously related to the research role you will play. Researchers who feel that disclosure will seriously endanger the validity of the data they collect will almost surely choose complete participant or complete observer roles. Mueller, Dirks, and Houts Picca's students, complete participants, weren't permitted to disclose the purposes of their observations for fear that other students' behaviors would be affected in a misleading way. Harrell and his colleagues, the ones who observed children and their care in supermarkets, were complete observers and also wanted a genuinely unfiltered view of human behavior. But researchers like Chiu (2009), a participant-as-observer in New York City's skateboarding scene, and Liebow (1993), an observer-as-participant in the Refuge's shelter for homeless women, clearly felt that disclosure of their research purposes would help build the trust necessary for participants to give valid accounts of their behavior and not affect significantly the behavior they observed.

We would like to make two more general recommendations about preparing yourself for the field. First, we think potential observers (and, perhaps, especially if they are novices) should review as much literature in advance of their observations as possible. Such reviews can sensitize the researchers to the kinds of things they might want to look for in the field and suggest new settings for study. You should know, however, that our advice on this matter is not universally accepted. In fact, the advice of Jack Douglas (1976) is that researchers immerse themselves in their settings first and avoid looking at whatever literature is available until one's own impressions are, well, less impressionable. Moreover, Adler and Adler suggest that researchers should begin to "compare their typologizations to those in the literature" only once they begin to "grasp the field as members do" (personal communication).

Second, whatever decision you make about when to review the literature on your own topic, we strongly recommend spending some time leafing through earlier examples of participant or nonparticipant observation, just to see what others have done. Particularly rich sources of such examples are the *Journal of Contemporary Ethnography* (which was, in earlier avatars, *Urban Life* and *Urban Life and Culture*) and *Qualitative Sociology*.

Gaining Access and Ethical Concerns

Although choosing topics and selecting sites are important intellectual and personal tasks, gaining access to selected sites is, as Lofland, Snow, Anderson, and Lofland (2005) point out, an important *social* task. At this stage, a researcher must frequently use all the social skills or resources and ethical sensibilities he or she has at his or her command.

Lofland et al. (2005) suggest that the balance between the need for social skills or resources and ethical sensibilities, however, can depend, as much as anything, on whether or not the observer plans to announce his or her presence as observer. If the observer plans not to reveal the intention to observe, the major issues entailed in gaining (or using) access are often ethical ones (see Chapter 3), especially because of the deceit involved in all covert research. This can be a particularly knotty problem when one seeks "deep cover" in a site that is neither public nor well known, as was the case for Patricia Adler (1985) in her study of drug dealers and smugglers, but it exists nonetheless even when one plans to study public places (as Humphreys did while watching impersonal sex in public restrooms) or to study familiar places, as Mueller, Dirks, and Houts Picca's students did when they observed other students engaged in Halloween activities. Seeking the advice and consent of an IRB in such situations is particularly important.

Gaining access to, and then getting along in, a site becomes much more a matter of social skill or resources when the observer intends to announce his or her observer role. The degree to which such skills or resources are taxed, however, is generally a function of one's current participation in the setting. Furthermore, for Adler and Adler (1994a), who were already participants in many afterschool settings, at least early in their study, their pre-study participation not only made access easy, but also made it relatively easy to decide whom they needed to tell about their research, whom they needed to ask permission from, and whom they needed to consult for clarifications about meanings, intentions, contexts, and so on. Chiu (2009) found access to sites where skateboarders rode particularly easy because he himself was a skateboarder. Ease of access, then, is certainly one major reason why Adler and Adler advocate "starting where you are."

 **STOP & THINK** *Suppose you wanted to study a musical rock group ("The Easy Aces") on your campus, to which you did not belong. How might you gain access to the group for the explicit purpose of studying the things that keep them together?*

Access particularly taxes social skills or resources when you want to do announced research in a setting outside your usual range of accustomed settings. In such cases, Lofland et al. (2005: 41–47) emphasize the importance of four social skills or resources: connections, knowledge, accounts, and courtesy. If, for instance, you want to study a rock group ("The Easy Aces") comprising students on your campus, you might first discuss your project with several of your friends and acquaintances, in the hopes that one of them knew one of the "Aces" well enough to introduce you. In doing so, you'd be employing a tried and true convention for making contacts—using "who you know" (*connections*) to make contact. In general, it's a good idea to play the role of "learner" when doing observations, but such role-playing can be overdone on first contact, when it's often useful to indicate some *knowledge* of your subject. Consequently, before meeting "The Easy Aces," it would be a good idea to learn what you can about them—either from common acquaintances or from available public-access information. (If, for instance, "Aces" have a website, you might want to listen to their music and refer to their songs, perhaps even admiringly, at your first meeting.) You'd also want to have a ready, plausible, and appealing explanation (an **account**) of your interest in the group for your first encounter. The account needn't be terribly clever, long-winded, or highbrow (something reasonably direct like "I'd like to get an idea of what keeps a successful rock group together" might do), but it might require some thought in advance (blurting out something like "I'm doing a research project and you folks seem like good subjects" might not do). Finally, and perhaps most commonsensically, a bit of common *courtesy* (such as the aforementioned expressions of appreciation for the "Aces" music) can go a long way toward easing access to a setting.

account, a plausible and appealing explanation of the research that the researcher gives to prospective participants.

All the aforementioned skills and resources—connections, knowledge, accounts, and courtesy—can be more difficult to muster effectively when there are gender, class, language, or ethnic differences between the researcher and other actors. Convincing **gatekeepers**, or persons who can let you into a setting, can be hard work, as Michelle Wolkomir discovered when she tried to gain access, as a heterosexual Jewish woman, to an ex-gay ministry in a southern, conservative Christian church. She spent five months meeting sporadically with the group leader and the church board to convince them that she would not undermine the security of the men using the ministry to "cure" their homosexuality and that she would not inhibit group talk (2001: 309–310).

gatekeepers, someone who can get a researcher into a setting.

Still, the eventual success of Wolkomir's project, and that of other female observers in mixed-gender settings, as well as that of white male observers, like Liebow, who, for *Tally's Corner* (1967), studied poor, urban, African American males, suggests that even gender, race, class, and sexual-orientation divides are not necessarily insurmountable.

Participant and nonparticipant observations, perhaps more than any other techniques we've dealt with in this book, lend themselves poorly to "cookbook" preparations. We can't tell you precisely what to do when you come up against bureaucratic, legal, or political obstacles to your observations, but we can tell you that others have encountered them and that authors like Lofland et al. (2005) provide some general strategies for surmounting them.

Gathering the Data

Perhaps the hardest thing about any kind of observation is recording the data. How does one do it?

STOP & THINK *Quick. We're about to list three ways of recording observations. Can you guess what they are?*

Three conventional techniques for recording observations are writing them down, recording them (perhaps with a digital camera, a phone, or an IPod), and recording them in one's memory, to be written down later (only shortly later, one hopes). Given the notorious untrustworthiness of memory, you might be surprised to learn that the third technique, trusting to memory for short periods, has probably been the most common. Despite gaining the explicit permission of his subjects, Liebow (1993) waited until he got home at night to write down what he'd observed about homeless women. Mueller, Dirks, and Houts Picca's students also waited until they'd left the scene to write in their journals. Liebow, in particular, was cognizant that even his relatively well-trained powers of observation and memory were subject to later distortion (whose aren't?) and so was at pains to record his observations as soon after he'd made them as possible. His goal, the goal of all observers, was to record observations, including quotations, as precisely as possible, as soon as possible, before the inevitably distorting effects of time and distance were amplified unnecessarily.

Why not record them on the spot? Joy Browne, in her reflections on her study of used car dealers, gives one view when she advises prospective observers,

> [D]o not take notes; listen instead. A panicky thought, perhaps, but unless you are in a classroom where everyone is taking notes, note taking can be as intrusive as a tape recorder and can have the same result: paranoia or influencing the action. (1976: 77)

The intrusiveness of note-taking, recording, and picture-taking, then, make them all potentially troublesome for recording data.

On the other hand, shorthand notes and recordings are certainly appropriate when they can be taken without substantially interrupting the natural flow. Frances Maher and Mary Kay Tetreault (1994), for instance, taped classes of professors to discern varieties of feminist teaching styles. Chiu (2009) recorded notes on his fieldwork experiences and informal conversations with other skaters after his site visits, but he also recorded 16 semi-structured interviews. A problem of recording, however, is that one can gather daunting quantities of data: It can take a skilled transcriber three hours to type one hour of recorded conversation (Singleton, Straits, Straits, and McAllister, 1988: 315). And even if you're able to take notes or record, it is still wise to transcribe and elaborate those recordings as soon as possible, while various unrecorded impressions remain fresh enough that you also can add them.

Visual Sociology

visual sociology, an approach to studying society and culture that employs images as a data source.

Even more data become available, however, when one uses cameras or video equipment. For much of the history of sociology, pictures and images were used, when used at all, to illustrate points in an argument (e.g., Grady, 2007b). The field of **visual sociology**, which uses images to study society and culture, really didn't begin, however, until the publication of Howard

Becker's "Photography and Sociology" (1974). A visual sociologist doesn't just use images for illustrative purposes, but as a source of data. Perhaps you recall that in Chapter 6 we asked you to analyze the front cover of this book for an idea of what we, Emily and Roger, were trying to tell you about research methods through the pictures. You were then practicing visual analysis, a technique employed by visual sociologists, by using the pictures to decipher what we were trying to convey. You were also clearly practicing an observational technique in doing so. But there are actually several distinct ways in which the images from cameras and video equipment can be used in observational studies.

Three techniques that visual sociologists use are:

1. analyzing visual documents
2. subject-image making
3. photo and video ethnography (see, for instance, Grady 2007c)

In analyzing visual documents, one takes pictures or video of a place one is studying and supplements field observations in this way. Chiu (2009), for instance, provides pictures of the four sites in which he does his fieldwork for the skateboarding study in New York City. An image of one park, for instance, shows an elevated plaza and wide stairs that, Chiu credibly asserts, make the park attractive to "[b]eginners [who] can practice elementary tricks on these low steps and smooth ground" (2009: 28).

In the practice of subject image-making, researchers typically provide participants with cameras to record images that are of interest to them and/or to the researcher. When combined with the process of photo-elicitation (see Chapter 10), in which participants are then interviewed about the images they (or others) have made (or found), subject image-making can make for a particularly rich source of observational data and thick description. Radley et al.'s (2005) study of homeless people, mentioned earlier, is one example. So is Glover, Stewart, and Gladdys's (2008) attempt to discover democratic ways of soliciting the views of various stakeholders in the ways land use might change in a community. Glover et al. gave all participants disposable cameras and asked them to photograph places that were "important within their daily lives" (389). They then asked each person to talk about the photos they'd taken and to expand on what was important about the place they'd photographed. One woman, for instance, took a photo of a farmer's market because, she discovered through the process of photography, it represented to her "a great social time" (391). Then all participants were asked to bring their photos and their "discoveries" to a civic forum where many of them shared their newly articulated land use values. Glover et al. believe that they've uncovered a democratic way, via subject image-making and photo-elicitation, to help stakeholders articulate and share the ways in which they want land in their communities used in the future.

Visual sociologists also practice photo and video ethnography. Ethnography, of course, is the study of human culture. It's been the defining approach of cultural anthropology since its inception. In a relatively new development for sociologists, however, some researchers have begun to use images not only to augment their note-taking and memory, but to capture emotion and

video ethnography, the video recording of participants and the reviewing of the resulting footage for insights into social life.

feel for situations. **Video ethnography**, a cutting-edge approach to studying social life, is the video recording of participants and the reviewing of the resulting footage for insights into social life. Tutt (2008) recorded three living-room situations in which people interacted with relatively new (to them) media to get a feel for the emotional content of such interactions.

Participant and nonparticipant observers commonly supplement their observations with interviews and available data. Chiu (2009) refers to the 16 semi-structured interviews he conducted; Mueller, Dirks, and Houts Picca, to a newspaper story of a Louisiana judge who "costumed in blackface makeup, afro wig, and prison jumpsuit, complete with wrist and ankle shackles." Interviewing other participants, known as **informants**, frequently provides the kind of in-depth understanding of what's going on that motivates researchers to use observation techniques in the first place. One concern raised by feminist approaches to observation, in fact, is how to retain as much of the personal perspective of informants as possible (e.g., Reinharz, 1992: 52). How, in effect, to give "voice" to those informants? Some observers (e.g., Maher and Tetreault, 1994) have even asked permission of their respondents to report their names and have thus abandoned the usual practice of offering confidentiality or anonymity. We recommend, however, that confidentiality or anonymity be offered to all interviewees whether in traditional interviews or to those in the field. Informants who later see their words used to substantiate points they themselves would not have made, for instance, might regret having given permission to divulge their identities in print, no matter how generous the intention of the researcher.

informants, participants in a study situation who are interviewed for an in-depth understanding of the situation.

ANALYZING THE DATA

The distinction between data collection and data analysis is generally not as great for participant and nonparticipant observation as it is for other methods (e.g., questionnaires, experiments) that we've discussed. (See Chapter 15, where we discuss qualitative data analysis in greater depth.) For, although these other techniques tend to focus on theory verification, observational techniques are more often concerned with theory discovery or generation. Thus, Mueller, Dirks, and Houts Picca generate a theory of Halloween activities that suggests, among other things, that, for "those whites who actively endorse the idea that whites are now victimized by the preferencing of people of color (e.g., in employment, admissions, etc.), Halloween may ironically signify a suspension of this imagined 'hierarchy.'" This is not to say that observational studies can't be used for theory verification or falsification. Perhaps the primary theoretical upshot of Harrell and his colleagues's (cited in Bakalar, 2005) study of parent–child interactions in supermarkets is to falsify the notion that the attractiveness of the children has no effect.

Nevertheless, more studies based on observation techniques are like Mueller, Dirks, and Houts Picca's students' than like Harrell and his colleagues—more concerned, that is, with theory generation or discovery than with theory verification or falsification. Important questions are these: When does the theory building begin and how does it proceed? The answer to the first question—when does theory building begin?—seems to be soon after you record your first observation. We know, from personal testimony, that Dirks focused on the data associated with Halloween costuming in Houts Picca's original data involving much more than Halloween costuming and that this focusing was the taking off point for this chapter's focal research. The very act of focusing on those data, rather than whatever else she might have homed in on (e.g., conversations in dorm rooms or cafeterias), implies that the observer attaches some importance or relevance to the observation. This choice is likely to be based on some preconceived notion of what's (theoretically) salient, a notion that the sensitive researcher will, some time during or after the initial observation, try to articulate.

Once articulated, these notions become concepts or hypotheses, the building blocks of theory. At this point, the researcher can begin to look for behaviors that differ from or are similar to the ones observed before. The pursuit of similarities can lead to the kind of generalizations on which **grounded theory** (Glaser, 2005; Glaser and Strauss, 1967), or theory derived from the data, is based. Hence, the observation, repeated many times, that some of their student observers actively endorsed, some passively endorsed, and some actively criticized the cross-racial costuming of other students led Mueller, Dirks, and Houts Picca to the three-part typology of the observers as unquestioning supporters, questioning supporters, or firm objectors.

The process of making comparisons, of finding similarities and differences among one's observations, is the essential process in analyzing data collected during observations. It is also the essential ingredient for making theoretical sense of those observations. Through it, Humphreys (1975) came up with his typology of those who used public restrooms for impersonal sex and Erikson (1976) came up with the concept of "collective trauma."

grounded theory, theory derived from data in the course of a particular study.

ADVANTAGES AND DISADVANTAGES OF OBSERVATIONAL TECHNIQUES

The advantages of observational techniques mirror the reasons for using them: studying behavior, getting a handle on the relatively unknown, obtaining an understanding of how others experience life, studying quickly changing situations, saving money (see the section labeled "Reasons for Doing Observations" earlier in this chapter).

Adler and Adler (1994b) suggest that one of the chief criticisms of observational research has to do with its validity. They claim that observations of natural settings can be susceptible "to bias from [the researcher's] subjective interpretation of situations" (Adler and Adler, 1994b: 381). They are surely right. But compared with measurements taken with other methods (e.g., surveys and interviews), it strikes us that measures used by observers are especially likely to measure what they are supposed to be measuring and, therefore, be particularly valid. We can imagine few combinations of questions on a questionnaire that would elicit as clear a picture of the ways in which some students use Halloween as an opportunity to reveal the racial stereotypes they have than the report of the white man who mused about what it would take to dress up as the black women in *Coyote Ugly*.

A final advantage of observation is its relative flexibility, at least compared with other methods of gathering social data. Observation is unusually flexible for at least two reasons: one, because a researcher can, and often does, shift the focus of research as events unfold and, two, because the primary use of observation does not preclude, and is often assisted by, the use of other methods of data gathering (e.g., surveys, available data, interviews). Thus, once Chiu's focus shifted from how skateboarders used various urban spaces to why certain skateboarders preferred one type of space over another, his primary method of data acquisition changed from observation to interview.

Generalizability is a perpetual question for participant and nonparticipant observation alike, however. This question arises for two main reasons. The first has to do with the reliability, or dependability, of observations and measurements. One example will stand for the rest. In Erikson's study of the effects on a flood on one community, he concludes that the community experienced "collective trauma" because all the individual presentations given by survivors were "so bleakly alike" in their accounts of "anxiety, depression, apathy, insomnia, phobic reactions, and a pervasive feeling of depletion and loneliness" (1976: 198). But it remains a question whether another observer of the same accounts would have detected the bleak sameness that Erikson did, much less come to the conclusion that sameness was an indicator of "collective trauma," rather than of many individual traumas. One way to check whether one's observations are reliable, of course, is to ask other participants whether they seem right to them. This is what Chiu did as much as possible in his study of urban skateboarding. But sometimes such a strategy can't work, as, for instance, when one doesn't disclose his or her research purpose to others (as, of course, Mueller, Dirks, and Houts Picca's students couldn't do) or when one thinks one might elicit inaccurate interpretations of the situation by asking other participants about it (as, for instance, Liebow [1993: 321] did in his study of homeless women).

The second reason that questions of generalizability arise is a direct result of the nonrandom sampling procedures used by most observers. In the focal research, we see examples of negative racial stereotyping that are persuasive. But whether it frequently occurs among some larger population of college students in the United States depends on how generalizable their sample was. And because that sample was not randomly selected (how could it have been?), the question of its generalizability is one that, of necessity, must wait another day (and replication [by one of you?] or, perhaps, the use of another method).

STOP & THINK *Can you imagine how the generalizability of the racial stereotyping associated with cross-racial costuming at Halloween might be studied with another method? Would you be inclined to use a survey? A cross-sectional or longitudinal design? A probability or nonprobability sample?*

demand characteristics, characteristics that the observed take on simply as a result of being observed.

A third problem associated with observational methods, especially when they are announced, is the bias caused by **demand characteristics**, or the distortion that can occur when people know (or think) they are being observed. One wonderful example of demand characteristics in action comes from D. L. Rosenhan's classic study entitled "On Being Sane in Insane Places"

(1971). Rosenhan was concerned about the accuracy of psychiatric diagnoses and so, with seven accomplices, he gained admission as a pseudo patient to eight psychiatric hospitals. In this first instance, all pseudo patients took the role of complete participants and, although all were as sane as you and us, none was diagnosed as sane once he or she was admitted as a patient. In a second instance, however, Rosenhan told the staff in one hospital that, during a three-month period, one or more pseudo patients would be admitted. Of the 193 patients who were admitted in the next three months, 41 were alleged by at least one member of the staff to be pseudo patients; 23 of them by attending psychiatrists. Because no pseudo patients, to Rosenhan's knowledge, had attempted to be admitted, the results of this second "experiment" suggest that even the threatened presence of observers can distort what goes on in a setting. Thus, despite the effort of most observers to create a situation in which action goes on as if they weren't there, there's little reason to believe that these efforts can be totally successful—especially when their presence is announced.

Three other problems of observational methods are worth emphasizing. First, in comparison with virtually every other technique we present in this book, observation can be extremely time consuming. Adler and Adler (personal communication) typically spend from 18 months to 2 years on most studies and 8 years on the observations for the book, *Peer Power*; Liebow saw about nine years pass between the conception and publication of his book on homeless women. Second, and clearly related to the issue of time consumed, is the problem of waiting around for events "of interest" to occur. In almost every other form of research (administering a questionnaire, an interview, an experiment, or collecting available data), the researcher exercises a lot of control over the flow of data. In the case of field observations, control

---

BOX **11.4**

Advantages and Disadvantages of Observation

Advantages

1. Provides relatively unfiltered view of human behavior
2. Can provide an in-depth understanding of a natural situation or social context
3. Permits the study of quickly changing situations
4. Can provide insight into relatively poorly understood aspects of human life
5. Can be a relatively flexible mode of data collection

Disadvantages

1. Generalizability is a perpetual question, thanks partly to the possible unreliability of observations and partly to the typical use of nonprobability samples
2. Those observed may distort or change their behavior if they know they are being observed
3. Ethical issues may arise about informed consent, especially if the researcher feels that providing adequate information will distort the behavior under observation
4. Observation studies are typically time consuming
5. One may have to wait for "things to happen"
6. Each stage of an observational study—choosing appropriate settings, gaining access to those settings, analyzing the data—can be fraught with unexpected difficulties

is handed over to those being observed—and they might be in no hurry to help events unfold in an intellectually interesting way for the observer. Third, although observation can be fun and enlightening, it can also be terribly demanding and frustrating. Every stage—from choosing appropriate settings, to gaining access to settings, to staying in settings, to leaving settings, to analyzing the data—can be fraught with unexpected (and expected) difficulties. Observation is not for the unenergetic or the easily frustrated (see Box 11.4).

SUMMARY

Although observation is often used in experiments, the observational techniques we've focused on here—those used by participant and nonparticipant observers—are those used to observe actors who are, in the perception of the observer, acting naturally in settings that are natural for them. Mueller, Dirks, and Houts Picca's students, for instance, observed other students as they prepared for, participated in, and talked about cross-racial dressing in relation to Halloween. Observational techniques are frequently used for several purposes: to gain relatively unfiltered views of behavior, to get a handle on relatively unknown social, to obtain a relatively deep understanding of others' experience, to study quickly changing situations, to study behavior, and to save money.

A sense of the variety of roles a participant or nonparticipant observer can take is suggested by Gold's (1958) classic typology. A researcher can become a complete participant or a complete observer, both of which imply that the researcher withholds his or her research purposes from those being observed. Although both of these roles entail the advantage of relative unobtrusiveness, they also entail the disadvantages of incomplete frankness: for example, of not being able to ask probing questions and of deceiving one's subjects. The ethical issues associated with observation almost always merit the attention of an IRB, except, in most cases, when the observation occurs in public places. Both the participant-as-observer and the observer-as-participant do reveal their research purposes, purposes that can prove a distraction for other participants. On the other hand, by divulging those purposes, they earn the advantage, among others, of being able to ask other participants detailed questions about their participation.

Observers can choose their research settings for their theoretical salience or their natural accessibility or for some combination of theory and opportunity. Generally, participant and nonparticipant observation begin without well-defined research questions, measurement strategies, or sampling procedures, though systematic observations can involve some combination of these elements. In all cases, however, researchers will want to decide how much of themselves and their research they will want to disclose to others in the research setting. At some point they may also need to employ social skills to gain access to the setting from gatekeepers and to learn what they need to learn from other actors.

Once in a setting, researchers can record observations by writing them down, by using some mechanical device (e.g., tape recorder or camera), or by trusting to memory for short periods. Most often observers have preferred trusting to memory because alternative techniques involve some element of intrusiveness. In all cases, the effort to record or transcribe "notes" should be made soon after observations are made, improving the chances for precise description and for recording less precise impressions. Increasingly, visual sociology has pressed the case for collecting or using images for making observations and studying society. The analysis of observational data involves comparing observations and searching for important differences and similarities.

The advantages of participant and nonparticipant observation mirror the purposes for using them and include the likelihood of valid measurement, and of considerable flexibility. Both types of observation permit relatively unfiltered views of behavior and enable in-depth understandings of particular settings or situations. On the other hand, disadvantages of observational techniques include almost unavoidable suspicions that they might yield unreliable and nongeneralizable results. Observation can also affect the behavior of those observed, especially when the observation is announced, and contains an element of deceit, especially when the observation is not announced. Observation can also be a particularly time consuming and sometimes frustrating path to social scientific insight.

EXERCISE 11.1

Observing Gestures in a Film or Television Program

For this exercise, which we've adapted from ones suggested by Tan and Ko (2004: 117–118), you'll need to find a film or television program that you can watch to observe mannerisms (gestures) and facial expressions. We want you to prepare a two-page essay in which you discuss the expression of emotions through gestures and facial expressions in one scene of the film or text. Be sure your essay mentions the title of the film or television program and gives a one- or two-sentence synopsis of the characters and action in the scene. Then outline the emotions you observed in the various characters involved. What are the actions/expressions that lead you to your conclusions about these emotions? You need to clearly distinguish the mannerisms and expressions from what they mean. (For instance, you might point out that a character raised his/her voice or motioned with an index finger [expressions or mannerisms you observe] and you might say the person looked angry [your conclusion about the emotional meaning of the expressions].) This means you need to distinguish the action you observe from how you interpret what the particular action means. **Please do not write about the storyline beyond your brief synopsis of the action.**

EXERCISE 11.2

Practicing Subject Image-Making and Photo-Elicitation

For this exercise, you'll need a digital camera and the capacity to print two pictures. You'll be replicating the subject image-making and photo-elicitation that Glover, Stewart, and Gladdys (2008) asked of their participants (see section on visual sociology).

1. Take some pictures of places that are "important within your daily life." You choose the places; you take the pictures. Then, print two of the images. (You will now have completed the subject image-making phase of the exercise.)
2. Now practice photo-elicitation on yourself.
   a. Explain, in your own words, what the places are that you've taken pictures of.

b. Ask yourself why the places you've depicted are so important to you and write down your responses.

c. Now ask yourself what details from the pictures you've taken might help your reader understand why the places are so important and write down your responses. (You will now have completed the photo-elicitation phase of the exercise.)

3. Turn in both the images and your responses.

EXERCISE 11.3

Participant Observation

(Can be done during class time, if class is about 90 minutes or more. If class time is less than 90 minutes, students should work on "Planning the Data Collection," mentioned later, before class.)

In this exercise, you will pick **either** the variable *helpfulness* or the variable *friendliness* to study. Create ONE research question that you could answer about the variable you picked that can be answered on campus using an observation method. If you don't want to create your own research question, you can use one of these:

- Are women or men on campus *friendlier*?
- Are students *more helpful* than faculty or staff?
- Are people in groups *less helpful* than people who are alone?
- Do people act *more friendly* to people with whom they share one or more background characteristics (like gender or race)?

You will have 30 minutes to collect data to answer your research question. Working in teams of two or three, you will be *participant observers* and will interact with strangers on campus. At least one person in the group should take the role of **complete participant** and play the role of a new visitor to the campus rather than being a current student who already knows his or her way around campus.

Planning the Data Collection

1. Decide on *the concept* you want to study (either helpfulness or friendliness) and write down a specific *research question* you intend to answer on the page you will turn in.

2. Create a conceptual definition for the concept that is in your research question (either "helpfulness" or "friendliness").

3. Decide about the population(s) you will interact with on campus during your data collection. The population can be students, faculty, staff, or some of each.

4. Create an **operational definition** of helpfulness or friendliness using **indicators that are observations**. As your indicators, you can focus on the words, the body language, the voice tone, and/or the behavior of the sample members that you select to interact with.

5. Decide what you will do to see how people will interact with you. See some examples of things you can say to begin an interaction with the members of the sample below.

6. Based on your research question, decide what variables you will need data on and how you will collect data for EACH of the variables. [Note: If your research question includes two variables, you need to collect data on BOTH variables in each interaction.]

For each variable you can collect data with **field notes and/or by using code sheets**. For example, you could take field notes describing each interaction that includes the information about each variable OR you could come up with a code sheet that you'll use that also includes both variables. If you are using code sheets, create one and make copies for the group now.

YOU CAN WRITE DOWN THE ANSWERS TO SECTIONS 1 AND 2 ON THE REPORT NOW IF YOU WANT TO. THEN COLLECT DATA AND WRITE DOWN THE REST OF YOUR REPORT.

Collecting Data during 30 Minutes in the Field

Each group will initiate and collect data on between 8 and 10 interactions **on campus** during the 30-minute data collection period. To start the interaction, one person in the group should *approach someone they don't know and initiate interaction with that person.* The other person or people in the group should observe the interaction (and be close enough to hear what is said). *[Be aware that there are likely to be other class members nearby also doing this exercise so don't approach the same people whom they do.]*

1. In each interaction, approach a person and initiate interaction to see how the person responds. Some possible things you can say are to ask the following:
   - "I'm parked illegally, I think. Do they really ticket here? Where should I park that's not too far away?"
   - "Where is [name of some building on campus]?"
   - "I'm thinking of transferring to [name of college] next year. Do you like it here?"

   (The first two questions work for students, faculty, or staff, but the last one seems most appropriate to ask of students.)

   Feel free to create other appropriate questions to use when playing the role of complete participant.

2. While one person in the group approaches someone, the other one or two people in the group should be observers (without being too obvious about it). (Take turns being the person who asks the questions). IN ADDITION to observing the interaction and focusing on the indicators of friendliness or helpfulness, if you need the information to answer your research question, you should record at least one or two pieces of information about the *backgrounds* of the people you observe and of the *setting* to answer

your research question. Some background variables you could collect data on are the person's gender, approximate age, race, or status (student, faculty, or staff). Some things you could note about the setting could be whether the person was alone or not and the location on campus.

3. After the interaction is over, record your field notes or fill in the code sheet you created with information about the degree of "friendliness" or "helpfulness" AND the other data that is necessary to answer your research question (such as information about the person's characteristics and information about the setting).

Writing the Research Report

Return to the classroom and finish writing down your report using the exercise format outlined below. Attach your data to the page (field notes, code sheets, etc.).

Writing down your research report
Your names:

1. **Introduction:** State your research question. Make sure that one of the variables in your question is helpfulness or friendliness.
2. **Methodology:**
   A. State your conceptual definition by writing a sentence that says: We defined friendliness as _____ or We defined helpfulness as _____.
   B. Describe your operational definition by listing the indicators you decided to use to measure this concept.
   C. How many people did you approach? How many observations did you make?
3. **Findings:** Give your answer to your research question. Give support for your answer by summarizing the data you collected AND giving some examples from your field notes or code sheets to illustrate your findings.

CHAPTER 12

Using Available Data

© Ariel Skelley/Getty Images

INTRODUCTION

You've probably heard immigration blamed for a lot of what ails us as a nation: the competition for skilled and unskilled jobs, demands on educational and other social services, and violent criminality, to name but three. Suppose you wanted to test one of these assertions. How might you do it? Let's say you wanted to focus on the connection between immigration and criminal violence. You might try participant observation (see Chapter 11), but the chances of seeing anything of interest in any particular immigrant (or non-immigrant) community probably aren't that great. You might use a questionnaire (see Chapter 9) or even qualitative interviews (see Chapter 10) to determine whether immigrants are more likely than non-immigrants to admit to having committed violent crimes, but the chances of receiving reliable responses (see Chapter 6) are pretty small (for all groups). Who's going to volunteer, after all, that they've committed a violent crime?

A concern with the crime/migrant connection moved Ramiro Martinez and Matthew Lee, the authors of this chapter's focal research. The two concerns, about crime and migration, didn't develop at the same time, though. Martinez asserts that he first became interested in crime, then in migration: "When in graduate school, about ten years ago, I asked somebody a question about Latino (Hispanic) homicides.... At the time we knew that homicide was the leading cause of death among young African American males but little attention was paid to Latinos. Only once I began to collect Latino data did the question of immigration status occur to me" (personal communication).

available data, data that are easily accessible to the researcher.

Martinez and Lee's paper illustrates the use of **available data** in the social sciences. Available data are, as the term suggests, data that are available to the researcher. Available data are sometimes actually statistics, or summaries of data, that have been computed from data collected by a large organization— for example, governments (as when they publish census or crime statistics) or businesses (as when they prepare for annual reports). Such statistics, when they have been made available in statistical documents, are called **existing statistics**. If you're sitting at your computer, for instance, you can try the website www.fedstats.gov/, "the gateway to statistics from over 100 U.S. Federal agencies," and see how quickly you can find the population of the country or of your state.

existing statistics, summaries of data collected by large organizations.

secondary data, research data that have been collected by someone else.

primary data, data that the same researcher collects and uses.

Available data are sometimes actual *data*, as opposed to statistics, collected by others, data that the researcher analyzes in a new way. Data that have been collected by someone other than the researcher are called **secondary data**. Secondary data are frequently contrasted with **primary data,** or data that are collected and used by the same researcher. Most of the focal research that we've featured uses primary data, such as the data collected by Mueller, Dirks, and Houts Picca for the focal piece on Halloween costuming (in Chapter 11) and the data collected by Adler and Clark on retirement (in Chapter 7). Survey data that have been made available for others to reanalyze are perhaps the best known kind of secondary data. The General Social Survey (GSS), conducted now every other year by the National Opinion Research Center (NORC) to collect data on a large number of social science variables, is probably the most popularly used set of survey data by social researchers, but it's just one of hundreds of sets of survey data that have been subjected to secondary analysis. Yet secondary data don't need to be survey data; they just need to have been collected by someone other than the current researcher. Thus, one of Martinez and Lee's main sources are the "logs" Miami police created to help them solve homicide cases. These became secondary data as soon as Martinez and Lee adapted them for their research purposes. We've chosen to focus on Martinez and Lee's type of secondary data analysis, not because it's especially typical of research based upon available data, but because it's a particularly interesting version of it. The use of available data generally, however, and of secondary survey data in particular became the dominant mode of social research toward the end of the twentieth century. The increasing popularity is suggested because although Roger (Clark, 1999) found that only about 12 percent of a probability sample of sociology articles published in 1956 involved the analysis of data that had been collected by governments, organizations, or private researchers (i.e., available data), 36 percent did so in 1976 and almost 50 percent did so in 1996. Fully 39 percent of articles published in 1996 were based on survey data collected by one group and reanalyzed by the authors of the articles (i.e., secondary data).

A particularly accessible form of available data (for research purposes) is that which is presented in published reports of government agencies, businesses, and organizations. Many studies of crime and violence have relied on national statistics gathered and forwarded to the Federal Bureau of Investigation (FBI)'s *Uniform Crime Reports* (UCR). One such study, reported by Martinez and Lee, was the U.S. Commission on Immigration Reform (1994) investigation that, among other things, compared the city of El Paso's overall crime and homicide rates to those of similar-sized cities in the United States. El Paso is a town along Mexico's border and therefore has a high percentage of migrants within its population. The Commission's finding that El Paso's overall crime and homicide rates were much lower than those in cities of comparable size supports the belief that migrants don't commit more crimes, and particularly more homicides, than nonmigrants. The Commission's evidence, however suggestive, is nonetheless subject to a famous criticism: that it commits the **ecological fallacy**. A fallacy is an unsound bit of reasoning. The ecological fallacy is the fallacy of making inferences about certain types of individuals from information about groups that might not be exclusively composed of those individuals. The Commission wanted to see whether cities with higher proportions of migrants had higher crime rates than cities with lower proportions of migrants, and so using cities as a unit of analysis made a great deal of sense. So far so good. But does its finding that a city with higher proportions of migrants had a lower crime rate than comparable cities with lower proportions of migrants mean that you should conclude that migrants are less likely to commit crimes than nonmigrants? Not legitimately. Such a conclusion risks the ecological fallacy because it might be that the migrants are committing most of the crimes in the cities with lower proportions of migrants. The problem is that we have looked at cities as our units of analysis and that, whatever the proportions of migrants, all cities are composed of both migrants and nonmigrants. Another example of the ecological fallacy

ecological fallacy, the fallacy of making inferences about certain types of individuals from information about groups that might not be exclusively composed of those individuals.

would occur if we inferred that individuals who are unemployed are more likely than those who are employed to suffer from mental illness from a finding that a city with a particularly high unemployment rate also has a high rate of mental illness. (It could be, for instance, that it is the employed that are mentally ill.)

STOP & THINK *You might remember that Émile Durkheim (in his classic sociological study, Suicide) found that Protestant countries and regions had higher suicide rates than Catholic countries and regions. He inferred from this the conclusion that Protestants are more likely to commit suicides than Catholics. Is this a case of the ecological fallacy? Explain.*

Martinez and Lee, you will see, have devised an especially clever way of avoiding the ecological fallacy. To do so, they had to dig a little deeper than many researchers do using available data: in this case, into the "homicide logs" of the Miami Police Department. In using these logs, however, they show us many of the virtues of using available data.

FOCAL   RESEARCH

Comparing the Context of Immigrant Homicides in Miami: Haitians, Jamaicans, and Mariels

By Ramiro Martinez, Jr., and Matthew T. Lee*

Are new immigrants more crime prone than native-born Americans? This issue has been debated since the early 1900s and continues to fuel controversy on the "criminal alien" problem in contemporary society (Hagan and Palloni, 1998; Short, 1997). Some scholars speculate that immigrants engage in crime more than other population groups by using the growing presence of the foreign born in the criminal justice system as evidence of heightened criminal involvement (Scalia, 1996). Other writers encourage legislation designed to rapidly deport immigrant criminals, such as California's Proposition 187 and the U.S. Congress 1996 Immigration Reform bill, despite any systematic evidence of an "immigrant crime problem" (Beck, 1996; Brimelow, 1995). Still others maintain that large labor pools of young immigrant males willing to work for lower wages than native-born Americans, in particular inner city residents, exacerbate ethnic conflict and contribute to urban disorders (Brimelow, 1995). The consequence of immigration, according to this view, is a host of social problems, especially violent crime (Beck, 1996).

*This article is adapted with permission from Ramiro Martinez and Matthew T. Lee, "Comparing the Context of Immigrant Homicides in Miami: Haitians, Jamaicans, and Mariels," *International Migration Review*, 2000, Vol. 34, 794–812.

This study probes the immigration and crime link by focusing on where immigrants live and how immigration might increase or decrease violent crime, at least as measured by homicide in a city that serves as a primary destination point for newcomers. Immigrants are undoubtedly influencing urban crime, but they are also potential buffers in stabilizing extremely impoverished areas (see Hagan and Palloni, 1998). Our primary argument is that while immigration shapes local crime, it might have a lower than expected effect on violent crime rates. The result is areas with high proportions of immigrants but low levels of violence, relative to overall city rates. The key question is whether immigrants have high rates of violence or if immigration lessens the structural impact of economic deprivation and thus violent crime.

It is, of course, preferable to study the impact of immigration and social institutions on violence across time and in diverse settings. While we do not directly test these variables with a multivariate analysis, we provide a starting point for this research agenda by conducting one of the first contemporary studies of immigrant Afro-Caribbean (Haitians, Jamaicans, and Mariel Cubans) homicide in a major U.S. city.

Previous Immigration and Crime Studies

Evidence of high immigrant crime rates is mixed (Shaw and McKay, 1931; see Short, 1997, for review of the literature). Lane (1997) notes that white newcomers (e.g., Irish, Italian) were often disproportionately involved in violent crime in the turn of the century Philadelphia and New York City. Crime rates declined, however, as settlers from abroad assimilated into mainstream society and integrated into the economy (Short, 1997). This finding is particularly instructive, since most immigrant victims and offenders were young males without regular jobs and presumably fewer ties to conventional lifestyles, suggesting that, even a hundred years ago, gender and economic opportunities played a major role in violent crime (Lane, 1997).

More contemporary studies on immigrant violence, however, have been scarce (Short, 1997). Many crime and violence studies rely on national statistics gathered from police agencies and forwarded to the FBI's Uniform Crime Reports (UCR). The UCR typically ignores the presence of ethnic groups such as Latinos or Afro-Caribbeans and directs attention to "blacks" or "whites" in compiling crime data. As a result, comparisons beyond these two groups are difficult to examine in contemporary violence research (Hawkins, 1999).

Furthermore, the handful of recent studies that have explored the connection between immigrants and crime have focused on property crimes or criminal justice decision-making processes (*cf.* Pennell, Curtis, and Tayman, 1989). And, despite studies (e.g., Hagan and Palloni, 1998) that suggest metropolitan areas with large immigrant populations have lower levels, or at least no higher levels, of violent and property crime than other metropolitan areas, the impression remains that immigrants are a major cause of crime (Hagan and Palloni, 1998: 378).

Research Setting

The city of Miami, Florida, provides a unique opportunity to examine the issue of whether or not immigration influences violent crime. Miami is one of the nation's most ethnically diverse cities and prominently known for its hot weather, natural disasters, waves of refugees, and tourist killings. It is also well known for its record high rates of violence. At various points throughout the 1980s, the violent crime rate was over three times that of similarly sized U.S. cities, and the city was widely characterized as a "dangerous" city in the popular media (Nijman, 1997).

This perception is linked, in part, to the presence of its large immigrant population (Nijman, 1997). After several distinct waves of immigration from Latin America since 1960, Latinos, mostly Cubans, comprised almost two thirds of the entire population in 1990 and had grown to almost one million in surrounding Dade County (Dunn, 1997). In 1980 alone, the number of Cubans increased dramatically as the Mariel boatlift brought 100,000 persons to the United States, and most of these settled in the greater Miami area (Portes and Stepick, 1993). The arrival of the Mariel refugees generated a moral panic in which the newcomers were labeled as deviants allegedly released from Cuban prisons and mental hospitals (Aguirre, Saenz, and James, 1997: 490–494).

Unlike the Mariels, who were welcomed by government officials (at least initially), Haitian immigrants were alternately detained or released by the Immigration and Naturalization Service (INS) in the early 1980s while fleeing political terror and economic despair in their home country (Portes & Stepick, 1993). Haitians, like the Mariels, however, were also stigmatized in the popular press, not only as uneducated and unskilled boat people, but also as tuberculosis and AIDS carriers (Martinez and Lee, 1998).

Miami was also a leading destination point for another Afro-Caribbean group—Jamaicans. Thousands of Jamaicans emigrated throughout the 1970s and 1980s, after open access to England was closed. The Jamaican population in Miami-Dade County tripled between 1980 and 1990, becoming the second largest Caribbean black group in the area (Dunn, 1997: 336).

As the newcomers converged in Miami, speculation arose that crime, especially drug-related types surrounding the informal economy, was growing within immigrant and ethnic minority communities (Nijman, 1997; see also Inciardi, 1992). *The Miami Herald*, for instance, reported that throughout the early 1980s (Buchanan, 1985; Schenker, 1984) Jamaican gangs operated freely in south Florida, controlling a large share of the local crack-cocaine trade and engaging in widespread violence (see also Maingot, 1992).

In sum, it stands to reason that immigration should increase absolute crime rates in Miami if it brings in a large number of people,
without a corresponding emigration to surrounding areas (see Hagan
and Palloni, 1998). Yet, despite the important empirical and theoretical
reasons to anticipate that immigration should influence urban violence,
this link has been largely ignored in the social science literature. In this
article, we extend previous studies on the social mechanisms of immigration and crime by considering the violence patterns of three distinct immigrant groups in the city of Miami.

Data Collection

Data on homicides from 1980 through 1990 have been gathered directly
from files in the Homicide Investigations Unit of the City of Miami
Police (MPD). These "homicide logs" contain information on each
homicide event, including a narrative of the circumstances surrounding
each killing not readily available from other data sources. Most importantly, extensive data on victim and, when known, offender or suspect
characteristics such as name, age, gender, race/ethnicity, and victim/
offender relationship are also noted by the investigators on the logs.

Starting in 1980, homicide detectives began to note whether the
victim or offender was white, black, or Latino. As the year progressed,
and in response to media inquiries, MPD personnel would also note
whether the victim or offender was a Mariel Cuban refugee, Haitian, or
Jamaican. These designations were based on a variety of sources. Some
Mariels were identified through INS documentation such as a "Parole
ID" discovered at the scene of the crime. A temporary tourist visa or INS
green card was also frequently found on the victim or at his/her residence. Still other ethnic clues were gathered from witnesses, family
member, friends, or neighbors including language (e.g., speaking Creole
or English with West Indian accents), providing country of birth on death
certificate, or names less than common to African Americans ("Michel
Pierre" or "Augustine Seaga"). All of these clues taken together helped
provide a more precise definition of victim and offender ethnicity.

Each homicide type is examined, including the level of gun use and
circumstances surrounding each killing. The nature of the prior contact
and the circumstances of offense type are often distinct and vary by
ethnicity (Hawkins, 1999). Types of homicide are therefore placed into
five categories. First is "other felony" or a type of homicide committed
during the course of a felony other than robbery. The second category
consists of "robbery" homicides. Third is "primary non-intimate" or killings among acquaintances, neighbors, friends, or co-workers. Fourth is
"family intimate homicides" or killings between spouses, lovers, estranged partners, and immediate family members. Finally, an "unknown
category" is included to account for homicides not cleared with an arrest.

Since our focus is on comparing and contrasting immigrant homicides, we concentrate our account on immigrant Caribbean killings and

their characteristics across two time points, 1980 and 1990. Specifically, the average number of homicide suspects (offenders) per 100,000 population from 1980 through 1984 is used for the first, and the average over 1985 to 1990 for the second. Total city homicides are also included as a baseline comparison. We begin by directing attention to suspect rates, but, given the current limitation of missing data, victim and offender relationships will be more fully explored.

We also use the 1980 and 1990 decennial censuses to provide estimates of the population in each city. Reliable population figures for our three immigrant groups are, of course, notoriously difficult to provide. We did, however, employ several strategies. First, Portes and colleagues (1985) estimated about 75 percent of the Mariel population (roughly 125,000) settled in the greater Miami area after the 1980 census was taken across the United States. Of that percentage, most (61.1 percent) moved into the city of Miami, at least as we can best determine (Portes, Clarke, and Manning, 1985: 40). Thus, we estimated roughly 57,281 Mariels were in the city limits between 1980 and 1990.

The number of Haitians and Jamaicans was equally difficult to estimate. Unfortunately, neither the 1980 nor the 1990 census provides specific information on Jamaicans and Haitians at the city level, forcing us to use county estimates by Marvin Dunn (1997: 336) on Caribbean-born blacks in 1980 and 1990. The following group-specific rates are therefore based on the number of Caribbean blacks in Miami-Dade County born in Jamaica and Haiti. While these data are useful due to their detail regarding geography and ethnicity, we acknowledge that an undercount problem exists, especially among immigrants. It is therefore likely that the level of immigrant homicides is even lower because the denominator in the calculation of the rates is likely to be larger. With these cautions in mind, we believe that the following results are reliable guides on immigrant homicides.

Results

Between 1980 and 1990, over 1,900 homicides were recorded by the Miami Police Department; 28 percent of these victims were Caribbean immigrants and 15 percent of all suspected killers were identified as Haitian, Jamaican, or Mariel. While the percentage of Mariel victims was higher than Haitians and Jamaicans (78 percent : 22 percent), the number of Mariels settling in the city of Miami (57,281) in the early 1980s was almost twice the number of Caribbean blacks in the entire greater metropolitan Miami area.

To further illustrate the distinct experiences of each group, we examine the homicide offending rate in Table 12.1. These results are especially important because much of the current anti-immigrant sentiment focuses on newcomers as killers rather than as victims. But, with a lone exception, the immigrants were not high-rate killers. In Miami during the early 1980s, the total city homicide offender rate was 69.9. This rate was higher, and in one case substantially higher, than the group specific rates for Mariels (55.5) and Haitians (26.4). One prominent contrast is the Jamaican rate of offending (102.2)—a level one and a half times the total city rate.

The offender data cannot be understood outside of their historical context, as the early 1980s proved to be a time of both social strife and social change in Miami. Social and political frustration among the city's African American population, for instance, erupted into violent riots touched off by questionable police killings in 1980, 1982, and 1989 (Dunn, 1997). In general, however, as the local turmoil subsided throughout the 1980s, the Caribbean violator rate decreased dramatically and quickly, relative to the city as a whole. The Jamaican killer rate displayed the largest decrease of all three groups (84 percent), but the Mariel and Haitian rates also declined sharply by 1990. These decreases are even more noteworthy after considering the substantial Haitian and Jamaican population growth throughout metropolitan Miami. In 1980, less than 13,000 Haitians lived in the country, but ten years later the number quadrupled to over 45,000. Similarly, the Jamaican population tripled over the same time span to more than 27,000 (Dunn, 1997: 336–337).

TABLE **12.1**
Immigrant Homicide Suspect Rates: 1980 and 1990

	1980	1990
Haitian	26.4	12.8
Jamaican	102.2	15.9
Mariel	55.5	10.5
City Total	69.9	49.9

Sources: City population estimates from 1980 and 1990 Bureau of the Census. Haitian and Jamaican population estimates cited in Dunn, 1997: 336. Mariel estimates described in Martinez, 1997: 120 and 121.

If the stereotype of Miami in the popular media as a high crime, drug-infested area, overrun by dangerous immigrants, was correct, then immigrant Caribbeans should have rates consistently higher than the city average. Apparently, immigrant murderers, at least in most cases, were engaged in killing at a much lower level than the total city rate, and the lone exception was confined to a single time point.

Table 12.2 describes two variables linked to type of homicide characteristic—gun use and victim–offender relationship. The gun distribution of victims varied in some cases. Gun inflicted homicides for Caribbeans were always higher than for the rest of Miami at both time points, except for the Mariels in 1990, although percentages for Haitians and Jamaicans dropped that same year.

The circumstances surrounding the homicide victim levels also varied. Table 12.2 reports that most Haitian victims were killed during the course of a robbery or by a non-intimate, regardless of the time frame. Jamaicans were usually murdered during the course of a felony, robbery, or by a non-intimate, but some variation occurred. For example, felony and robbery-related homicide increased in 1990, similar to the rest of Miami, while non-intimate killings dropped by half. Mariels, typically, were fatally assaulted by non-intimate acquaintances, friends, or neighbors.

One finding worthy of special attention is that of domestic-related homicides between family members and intimates. A popular myth exists that immigrant Latinos and Afro-Caribbeans are more likely to engage in domestic violence, in part because of "traditional values" or "machismo" that encourage greater control at home, including physical responses to interpersonal problems (see Buchanan, 1985). Contrary to this image, we discover lower family/intimate proportions for immigrants than for the city as a whole.

TABLE **12.2**

Characteristics of Immigrant Homicide Offense Type: 1980–1990

	Victim			
Offense	**Haitian**	**Jamaican**	**Mariel**	**Miami Total**
1980				
Gun	95.2	85.7	86.0	78.3
Felony	8.7	17.9	10.9	11.2
Robbery	39.1	25.0	9.2	17.0
Non-intimate	34.8	50.0	60.9	48.7
Family/Intimate	17.4	3.6	15.5	18.7
Stranger/Unknown	0.0	3.6	3.4	4.4
1990				
Gun	81.6	85.7	74.0	74.1
Felony	22.2	33.3	18.2	22.3
Robbery	31.5	33.3	15.6	18.8
Non-intimate	37.0	26.7	55.8	46.7
Family/Intimate	7.4	6.7	6.5	10.3
Stranger/Unknown	1.9	0.0	3.9	2.0

Note: All numbers in percentages

Discussion and Summary

A popular perception exists that immigrants are responsible for crime and violence in U.S. cities with large foreign-born populations. It is important to bring empirical evidence to bear on the policy debates surrounding this concern. To address this issue, we examined Afro-Caribbean homicides in the city of Miami. The picture that emerged undercuts claims that immigrants significantly influence violent crime.

We found that, almost without exception, Afro-Caribbeans had a lower homicide rate than the city of Miami total population. Comparing the early 1980s to the late 1980s, we also found a strong pattern of declining violence, especially for Jamaicans and Mariel Cubans, while Haitians continuously maintain an overall low rate of violent crime. As immigrant groups became more established, grew in size, and were less dominated by young males, the most violence-prone group, the homicide rate rapidly dissipates.

To illustrate this process, recall the high 1980 Jamaican offending level—the only immigrant rate to exceed the city average. It is possible that a high proportion of unattached and young Jamaican males migrated to south Florida and were employed in the informal economy. As the decade progressed, they left or were removed from the drug industry, and Jamaicans with more conventional ties to society followed the earlier entrants in search of legitimate work. Although our explanation is speculative and data are not available to directly examine this proposition at the city level, we believe this is a plausible account for our findings.

The results of this study have relevance for policies in vogue, such as targeted sweeps by the INS or legislation designed to deport criminal aliens. Many of these policies are motivated by the notion that immigrants threaten the order and safety of communities in which they are concentrated (Beck, 1996; Brimelow, 1995). We discovered, however, that, overall, Afro-Caribbeans are involved in violence less than the city of Miami population. Thus, the current INS strategy to remove "criminal aliens" in an effort to combat violent crime is questionable and is a misguided basis for limiting and restricting immigration.

The question of comparability to other areas naturally arises. We stress the caveat that our findings are, perhaps, unique to Miami for a host of demographic and historical reasons. It is possible that crime differences might arise relative to other cities with large immigrant populations (e.g., New York City, Chicago, Los Angeles, San Diego). Thus, local conditions, social class in the home country, and manner of reception might be important determinants of immigrant well-being. We invite further inquiries along these lines.

Our findings suggest that some of the key propositions of social disorganization theory must be revised to account for contemporary relationships among the immigration and homicide patterns found in cities such as Miami. Rapid immigration might not create disorganized communities but instead stabilize neighborhoods through the creation of new social and economic institutions. For example, Portes and Stepick (1993) describe how immigrant-owned small business revitalized Miami by providing job opportunities for co-ethnics. The low rate of involvement of the three immigrant groups we studied demonstrates the value of exploring intra-ethnic distinctions in future research on neighborhood stability, economic conditions, and violent crime. Increased attention to the issues we have raised will likely lead scholars to challenge misleading and deleterious stereotypes and result in more informed theory, research, and policy on urban immigrant violence.

References

Aguirre, B. E., R. Saenz, and B. S. James. 1997. Marielitos ten years later: The Scarface legacy, *Social Science Quarterly*, 78: 487–507.

Beck, R. 1996. *The case against immigration*. New York: Norton.

Brimelow, P. J. 1995. *Alien nation*. New York: Random House.

Buchanan, E. 1985. Dade's murder rate was on the rise again in 1984, *Miami Herald*, January 2, B1.

Dunn, M. 1997. *Black Miami in the twentieth century*. Gainesville: University Press of Florida.

Hagan, J., and A. Palloni. 1998. Immigration and crime in the United States. In *The immigration debate*. Ed. J. P. Smith and B. Edmonston. Washington, DC: National Academy Press. pp. 367–387.

Hawkins, D. F. 1999. African Americans and homicide. In *Issues in the study and prevention of homicide*. Ed. M. D. Smith and M. Zahn. Thousand Oaks, CA: Sage.

Inciardi, J. A. 1992. *The war on drugs II: The continuing epic of heroin, cocaine, crack, crime, AIDS, and public policy*. Mountain View, CA: Mayfield.

Lane, R. 1997. *Murder in America: A history*. Columbus: Ohio State University Press.

Maingot, A. P. 1992. Immigration from the Caribbean Basin. In *Miami Now!* Ed. G. J. Grenier and A. Stepick. Gainesville: University of Florida.

Martinez, R., and M. T. Lee. 1998. Immigration and the ethnic distribution of homicide in Miami, 1985–1995, *Homicide Studies*, 2: 291–304.

Nijman, J. 1997. Globalization to a Latin beat: The Miami growth machine, *Annals of AAPSS*, 551: 164–177. May.

Pennell, S., C. Curtis, and J. Tayman. 1989. *The impact of illegal immigration on the criminal justice system*. San Diego: San Diego Association of Governments.

Portes, A., J. M. Clarke, and R. Manning. 1985. "After Mariel: A survey of the resettlement experiences of 1980 Cuban refugees in Miami," *Cuban Studies*, 15: 37–59.

Portes, A., and A. Stepick. 1993. *City on the edge: The transformation of Miami*. Berkeley: University of California Press.

Scalia, J. 1996. *Noncitizens in the federal justice system, 1984–1994.* Washington, DC: U.S. Department of Justice, Bureau of Justice Studies.

Schenker, J. L. 1984. Police: Rastafarian gangs setting up shop in Florida, *Miami Herald*, March 20, B8.

Shaw, C. R., and H. D. McKay. 1931. *Social factors in juvenile delinquency. Volume II of report on the causes of crime.* National Commission on Law Observance and Enforcement, Report No. 13. Washington, DC: U.S. Government Printing Office.

Short, J. F. 1997. *Poverty, ethnicity, and violent crime.* Boulder, CO: Westview Press.

THINKING ABOUT ETHICS *The data about individuals that Martinez and Lee obtained were collected by others and were used without any personal identification. Although the American Sociological Association (ASA) Code of Ethics states that confidentiality is not required in the case of information available from public records, the researchers acted appropriately by collecting the data about victims and offenders from the records without names.*

STOP & THINK *A good question to ask after reading a piece of research is "What was the unit of analysis, or type of subject, studied by the researchers?" What was Martinez and Lee's unit of analysis?*

STOP & THINK AGAIN *Other good questions are "What were their major independent and dependent variables?" "How did they measure those variables?" "How did they expect those variables to be related?" "How were they related?" Try to answer as many of these questions about Martinez and Lee's article as you can.*

KINDS OF AVAILABLE DATA

Martinez and Lee take advantage of at least four sources of available data in their study: "homicide logs" from the Miami Police Department (MPD) for information about offenders and victims and their relationships; decennial census reports for information about the population of Miami; Portes, Clarke, and Manning's (1985) estimates of the number of Mariels in the city of Miami; and Marvin Dunn's (1997) estimates of the number of Caribbean-born blacks born in Jamaica and Haiti. The imaginative combination of a variety of data sources is quite common in analyses involving available data. It can make these analyses a great deal of fun, as well as illuminating. Useful in the art of such analyses is understanding the kinds of available data. Three kinds we'll introduce you to are secondary data, existing statistics, and the ever-popular "others."

Secondary Data

Secondary data are data that have been collected by others and have been made available to the current researcher. The "homicide logs," created by the Miami Police while they tried to solve murder cases, but then used by Martinez and Lee to identify the ethnicity of offenders and victims, are an unusual instance of secondary data. Martinez and Lee's access to actual police crime logs is an interesting story, indicating that access to available data isn't always just a matter of finding the right report or downloading the right computer tape. Their accounts of this story are slightly different, but both point to possible stumbling blocks in accessing "available" data. They both agree that, in the beginning, Martinez was a new assistant professor at the University of Delaware with few research contacts. He wanted to study Latino homicides but had no real access to information about Latinos. "There are few Latinos in Delaware," he reports (personal communication), "but I had a colleague with research projects in Miami. He introduced me to somebody on his research staff who knew the MPD Chief. Introductions were made and I eventually received permission to enter the police department" and access its homicide logs. His co-author, Lee, suggests that Martinez's account is overly "modest." Lee points out that "police are understandably reluctant to give 'outsiders' unrestricted access to study their files" and that "only by logging countless hours accompanying the detectives on calls, and demonstrating that he was trustworthy, could Ramiro [Martinez] gain the kind of unrestricted access that I [Lee] witnessed when I visited the police department with him: the detectives greeted him enthusiastically and treated him as an insider" (personal communication).

 **STOP & THINK** *While the information that Martinez and Lee collected from homicide logs certainly counts as secondary data, it is not typical of the kind of secondary data most used by social scientists to produce research articles. What kind of secondary data, do you recall, provides the basis for around 40 percent of all articles published in the social sciences?*

By far the greatest amount of social science research based on the analysis of secondary data, however, employs survey data collected by others. In fact, Clark (1999) estimates that about 40 percent of all research articles published in 1996 were based on such data and that the percentage is increasing. Thus, Berg (2009) used data from the Bureau of Labor Statistics, the U.S. Census, and the GSS to determine how white public opinion toward undocumented immigrants had changed between 1996 and 2004. He found that whites living in urban areas were less likely in 2004 to favor government action against such immigrants than they had been in 1996. Ren (2009) analyzes 2000 Public Use Microdata Samples (PUMS) of the U.S. Decennial Censuses to show that immigrant Chinese to the United States with a Cantonese background have higher fertility levels than Mandarin counterparts, thus apparently retaining norms for higher fertility characteristic of the Cantonese in China, even after the immigrant experience. And Hofferth (2005) offers an overview of research in family studies that has relied on secondary analysis of survey data: data from, among others, the National Longitudinal Survey of Youth, the National Survey of Families and Households, the Survey of Program Dynamics, the Early Childhood Longitudinal Survey, the Panel Study of Income Dynamics (PSID) Child Development Supplement, the National Survey of Adolescent Health, the National Survey of Child and Adolescent Well-Being, the Health and Retirement Study, the Fragile Families and Child Wellbeing Study, the Early Head Start (EHS) Research and Evaluation Project, and the Welfare, Children, and Families: a Three-City Study. You can find hundreds of such data sets at the Inter-University Consortium for Political and Social Research (ICPSR) at the University of Michigan (http://www.icpsr.umich.edu/) and at the Roper Center for Public Opinion Research (http://www.ropercenter.uconn.edu/), both clearinghouses for survey data, mainly, but not exclusively, from the United States. Again, the point is that a large and increasing percentage of the research articles and books published in the social sciences are based on secondary data, data collected by others, and that secondary data sets are increasingly accessible to those of us interested in using them.

Both the ICPSR and the Roper Center place some restrictions on access to secondary data, through either fees or the requirement of institutional membership. If you are planning a major research project, you might want to consult your professor about what data sets your college or university already has or whether there might be another convenient way for you to gain access to secondary data you've seen referred to in the literature. One of the most delightful sources of easy-access secondary data, however, is the Social Documentation and Analysis (SDA) Archive at the University of California, Berkeley (http://sda.berkeley.edu/archive.htm). Here you can browse a variety of data sets (see Box 12.1) and analyze the data online.

Suppose you have a research project in social psychology in which you're studying the connection between gender and love. You might want to find out whether, for instance, males or females are more likely to say they would rather suffer themselves than let their loved one suffer.

 **STOP & THINK** *Quick. Take a guess. Who do you think is more likely to say they'd rather suffer themselves than let loved ones suffer: males or females?*

BOX **12.1**

Some Data Sets Available at SDA Archive
(http://sda.berkeley.edu/archive.htm)

GSS Cumulative Datafile (1972–2008)
National Election Study (1948–2004)
Multi-Investigator Survey 1994, 1998–1999
Census Microdata: 2000–2003 American Community
    Surveys (3.8 million cases)

1990 and 2000 U.S. 1 percent PUMS
National Race and Politics Survey 1991
National Health Interview Survey 1991
Health Studies from Brazil

Well you can use the most recent GSS, which just happens to ask respondents whether they'd rather suffer themselves than let the one they love suffer, as well as what their gender is. One way to analyze this association would be to access the SDA Archive by going to web address http://sda.berkeley.edu/archive.htm.

 **STOP & DO**   *If you have easy access to a computer why don't you try this? Then click on the GSS Cumulative Datafile 1972–2008 or whatever GSS file that contains the 2008 survey data. Then place "sex" in the column box and "agape1" in the row box, before clicking on "run the box." What do you find? Are males or females more likely to say they strongly agree with the statement "I would rather suffer myself than let the one I love suffer?"*

It might not surprise you as much as it did us that males are substantially more likely than females (69.1 percent to 57.3 percent) to strongly agree with the statement "I would rather suffer myself than let the one I love suffer," but perhaps you will be interested to know how easily, using secondary survey data, you found out. Think how impressive it would be to add such a finding to your social psychology paper!

The *unit of analysis* for studies based on secondary analysis of survey data is almost always the individual. We've learned something about people from our foray into GSS data on love: that men are more likely than women to say they strongly agree with the statement about suffering for the one they love. Robinson and Martin (2008) used GSS data to discover something else about individual people: that happier people differ from unhappier people in their daily activities, specifically that they report being more engaged in social activities, like religion and newspaper reading, and being less engaged in television watching. People respond to the surveys and people, therefore, tend to be the units of analysis of the research based on those surveys. The units of analysis in Martinez and Lee's study are also individuals. They are focusing on individual people, for instance, when they use the homicide logs to determine how many murders were committed by Mariel Cubans, Haitians, and Jamaicans. And when they use the various estimates of the Mariel, Haitian, and Jamaican populations in Miami, they are simply trying to supply the denominators necessary to calculate homicide rates, much as we needed to use the total number of males and females in the GSS survey to calculate the percentage of each that strongly agreed with the statement about love. In general, a big difference between studies based on secondary analysis of, say, survey data and studies based on existing statistics is the unit of analysis, as we'll see later.

 **STOP & THINK**   *Emile Durkheim's classic study, Suicide ([1897] 1964), is based, in part, on comparisons of suicide rates in more or less Catholic provinces in Europe. He finds the provinces with high percentages of Catholic residents, for instance, have lower suicide rates than the provinces with low percentages of Catholic residents. Can you tell what the unit of analysis of this part of Durkheim's study was?*

Existing Statistics

 **STOP & THINK** *Did you ever wonder what is of interest in other countries? One source of available data is the webpage "Zeitgeist around the World" where the most common queries to, and trends in such queries to, Google are posted. In May 2009, for example, interest in "Obama" worldwide, or at least in the 40 countries covered by "Zeitgeist," over the previous 12 months had peaked around November 4, 2008, and again around January 20, 2009. Can you guess why? If you check out the site at www.google.com/intl/en/ press/intl-zeitgeist.html, look at the data presented and think about what the unit of analysis is.*

Existing statistics are statistics provided by large organizations. Martinez and Lee employ one of the most frequently employed sources of existing statistics: the U.S. Bureau of the Census to gather information about the overall population of Miami. Census products come in printed reports, on computer tapes, on microfiche and CD-ROM laser disks, and now via websites. But they found that they needed to combine statistics from one source with data or statistics from another (in their case, data from the homicide logs). Most users of existing statistics find the need to make similar combinations. Thus, Frank, Hardinge, and Wosick-Correa (2009) used data presented by Frank, Camp, and Boucher (2008) on rape laws and their amendments in 51 countries and data on reported rapes per capita presented in a variety of other sources to show, among other things, that the reform of rape laws was strongly and positively associated with elevated police reporting of rapes between 1945 and 2005. Similarly, Hirsch (2009) matched data from one report by the U.S. Equal Employment Opportunity Commission (EEOC) on whether employers had been formally charged with data from another report from the EEOC on employment discrimination and the conditions of work for women and racial minorities. She found that the enforcement of anti-discrimination laws did, in fact, change the industrial and legal environments of the workplaces in question.

In general, the *unit of analysis* in studies based on existing statistics is not the individual. For Durkheim's study of suicide rates by province, the unit of analysis was the province. In Frank et al.'s, as well as in "Zeitgeist around the World," it was nations. In Hirsch's study, it was workplace organizations. Statistics, by their nature, summarize information about aggregates of people (countries, regions, states, cities, counties, organizations, etc.), and when statistics are the data or information we have to work with, we're almost surely focused on comparing aggregates of people.

The real trick to doing research with existing statistics (after, of course, you've determined what your research question is) is often to find sources of data about your unit of analysis that permit you to address your research question. Roger, along with a student, Christine Arthur (Arthur and Clark, forthcoming), wanted to know what contributes to domestic violence in a society. We'd come across, through our readings, population-based indicators of the percent of women physically assaulted by an intimate partner as reported by Heise, Ellsberg, and Gottemoeller (1999). We found other indicators by following up leads in literature we read. But some we found by Googling things like "date women's vote." In this way we discovered the Inter-Parliamentary Union's (2004) *Women's Suffrage* webpage (http://www. ipu.org/wmn-e/suffrage.htm) and found the years in which women got the vote in different countries and the United Nations Statistics Division (2000) *The World's Women 2000: Trends and Statistics* (http://unstats.un. org/unsd/demographic/ww2000/) and found the level of women's labor force participation. We did, in short, what is very common in research based on existing statistics: We tried to find sources of statistics that provided information about our units of analysis (in our case, nations) that measured our concepts of interest.

Roger and Christine actually located other sources that yielded statistics enabling them to measure variables of interest about most countries of the world. And what is true of their work is true of much work involving existing statistics: one searches for information about one's units of analysis (whether they be countries, regions, states, organizations, etc.), often from different sources and combines that information into a single data set about those units (see Box 12.2). Creating the combination often involves listing units of analysis (as Arthur and Clark do their countries in the left-hand column of Box 12.2) and then listing values of the variables of interest (as they do in subsequent columns).

BOX **12.2**

A Portion of Arthur and Clark's Data Set on Countries: Finding Data Sources That Complement One Another

Countries	From Heise et al. (1999) ↓ Percent of Women Physically Assaulted by and Intimate Partner	From Inter-Parliamentary Union (2004) ↓ Year Women Got Vote	From United Nations Statistics Division (2000) ↓ Women's Labor Force Participation Rate (1999)
Australia	8	1902	55
Bangladesh	47	1972	56
Barbados	30	1950	62
Canada	27	1917	61
.	.	.	.
.	.	.	.
.	.	.	.

 **STOP & THINK** *Can you use Box 12.2 to determine the year when women in Australia gained the vote? What was the women's labor force participation rate in Bangladesh in 1999?*

Once you've collected statistics from various sources, you can begin analyzing them using statistical techniques like the ones outlined in Chapter 15's section on quantitative data analysis.

The sources of existing statistics are innumerable. If you have an idea of the kind of topic you'd like to work on, and the unit of analysis, you can find usable data in either of two ways: do a literature review to see what others have used or do a web search. You could, of course, do both. If you choose to search the web, finding key words is crucial. Are you interested in homicide rates in American cities? Try Googling "homicide rates in American cities." You'll get there. If you remember that the FBI warehouses information about crime in the United States in its UCR, you might Google "Uniform Crime Reports." You get right to the UCR website (http://www.fbi.gov/ucr/ucr.htm), and you'll soon be looking at interesting information.

 **STOP & DO** *If you're sitting at your computer, why don't you see if you can find the website for the U.S. Census Bureau.*

If you wanted the U.S. Census Bureau, you could Google "U.S. Census Bureau" and find its web address http://www.census.gov/.

We won't try to anticipate your needs, but we will provide you with some web addresses most frequently used for sources of existing statistics (Box 12.3) and the assurance that, even as recently as the last decade, finding existing statistics (which used to mean a trip to the right library and a hunt for the right volume) was never so easy.

Of course, the use of existing statistics constitutes one of the richest research traditions in the social sciences. It provided grist, as we've already pointed out, for Émile Durkheim's classic study, *Suicide* ([1897] 1964), and for Karl Marx's argument for economic determinism in his *Das Kapital* ([1867] 1967). It is hard to imagine what these early social scientists would make of the ever increasing and increasingly accessible amounts of statistics available today—and, frankly, there's no way we could do justice to them here—but it's almost a certainty that Marx and Durkheim would have done a lot with them.

Other Kinds of Available Data

physical traces, physical evidence left by humans in the course of their everyday lives.

erosion measures, indicators of a population's activities created by its selective wear on its physical environment.

accretion measures, indicators of a population's activities created by its deposits of materials.

You don't need to depend on outside agencies to make your "available data" available. In their now classic list of alternative data sources, Webb, Campbell, Schwartz, and Sechrest (2000) mention, for instance, **physical traces** and personal records. Physical traces include all kinds of physical evidence left by humans in the course of their everyday lives and can generally be divided into what Webb and his associates call **erosion measures** and **accretion measures.** An erosion measure is one created by the selective wear of a population on its physical environment. Webb and associates cite Frederick Mosteller's (1955) study of the wear and tear on separate sections of the *International Encyclopedia of the Social Sciences* in various libraries as a quintessential

BOX **12.3**

Some Good Sources of Existing Statistics

Units of Analysis: American states, cities, counties, and so on

- *The Bureau of the Census Home Page* (http://www. census.gov/) allows users to search census data on population. In particular, the bureau's *Statistical Abstract* (http://www.census.gov/statab/www/) provides useful information on crime, sex and gender, social stratification, and numerous other topics.
- *FBI UCR Page* (www.fbi.gov/ucr/ucr.htm) offers access to recent crime data.
- *Bureau of Labor Statistics Data Home Page* (www.bls.gov/data/) has, among other things, data on men's and women's labor force participation.
- *Department of Education Home Page* (http:// www.ed.gov/index.jhtml) leads to much information about educational achievement, and so on.

- *The National Center for Health Statistics Home Page* (http://www.cdc.gov/nchs/) is a wonderful source of information about health and illness.

Units of Analysis: Countries

- *The United Nations Yearbook Statistical Division* (http://unstats.un.org/unsd/default.htm) provides basic statistical information, including data on population (in its *Demographic Yearbook*), trade, health, education, and many other topics.
- *The International Labor Office* (http://www.ilo. org/) is a great source of information about work worldwide.
- *CIA World Factbook* (http://www.cia.gov/cia/ publications/factbook/) provides a wide variety of information on most countries in the world.
- *The World Bank* (http://www.worldbank.org) offers basic information and much financial information about countries of the world.

---

unobtrusive measures, indicators of interactions, events, or behaviors whose creation does not affect the actual data collected.

example of the use of erosion measures, in this case to study which sections of those tomes received the most frequent examination. An accretion measure is one created by a population's remnants of its past behavior. The "Garbage Project" at the University of Arizona has studied garbage since 1973. By sorting and coding the contents of randomly selected samples of household garbage, William Rathje and his colleagues have been able to learn a great deal, including what percentage of newspapers, cans, bottles, and other items aren't recycled and how much edible food is thrown away (Rathje and Murphy, 1992: 24). Because there are "certain patterns in which people wrongly self-report their dietary habits" (Rathje and Murphy, 1992: 24), the proponents of "garbology" argue that this type of available data can sometimes be more useful than asking people questions about their behavior. One great advantage of using physical traces, like garbage, is that their "collection" is **unobtrusive** (or unlikely to affect the interactions, events, or behavior under consideration).

 **STOP & THINK** *Suppose that, like Kinsey and associates (1953), you wanted to study the difference between men's and women's restrooms in the incidence of erotic inscriptions, either writings or drawings. Would you be taking erosion or accretion measures?*

personal records,
records of private lives,
such as biographies, letters,
diaries, and essays.

Personal records include autobiographies, letters, e-mails, diaries, and essays. With the exception of autobiographies, which are often published and available in libraries, these records tend to be more difficult to gain access to than the public records we've already discussed. Frequently, however, personal records can be used to augment other forms of data collection. Roger has used autobiographies, for instance, to augment information from biographies and questionnaire surveys to examine the family backgrounds of famous scientists, writers, athletes, and other exceptional achievers (Clark, 1982; Clark and Ramsbey, 1986; Clark and Rice, 1982). Private diaries, when they are accessible, hold unusually rich possibilities for providing insight into personal experience. R. K. Jones (2000) suggests ways in which these can be used in contemporary medical research, and historians like Laurel Thatcher Ulrich have used them to create classics, such as her *A Midwife's Tale* (1990), to piece together accounts of times gone by. Letters can be an especially useful source of information about the communication between individuals. Lori Kenschaft (2005), for instance, has studied the ways in which two important nineteenth-century educators, Alice Freeman Palmer and George Herbert Palmer, who happened to be married to each other, affected each other's thinking and American education by carefully studying their written correspondence (available at the Harvard University library). Similarly Karen Lystra's *Searching the Heart: Women, Men and Romantic Love in Nineteenth-Century America* (1989) uses letters almost exclusively to build an argument that nineteenth-century conventions of romantic love required people to cultivate an individual, unique, interesting self that they could then present to the beloved other—that is, that romance is an historical and psychological root of individualism.

Kenschaft's and Lystra's works point to one great advantage of available data: the potential for permitting insight into historical moments that are inaccessible through more conventional social sciences techniques (such as interviews, questionnaires, and observation). Other sources of such historical data are newspapers and magazines, and organizational and governmental documents. Beverly Clark (2003) has scrutinized nineteenth- and early twentieth-century newspapers and magazines for insight into the popular reception of (children's) books written by, among others, Louisa May Alcott. Clippings from the *Cleveland Leader* and the *Albany Daily Press and Knickerbocker* indicate that the Cleveland public library required 325 copies of *Little Women* in 1912 to satisfy constant demand; the New York City branch libraries required more than a thousand copies.

Texts, like the diary studied by Ulrich (1990), the letters studied by Kenschaft (2005), and the newspapers and magazines studied by Clark (2003), are amenable to both quantitative and qualitative data analysis. Have a look at the next chapter, on content analysis, for an idea of how others have done these kinds of analyses.

Sociologists have frequently used archives of unofficial and official records as sources of data, but they've tended to neglect visual materials (Grady, 2007b: 66). Increasingly, though, visual sociologists have taken advantage of such archives and repositories, just as the number of easily accessible archives and repositories has grown. These repositories can constitute an excellent source of primary data for social research. Grady (2007a), as mentioned in Chapter 6, studied his own archive of all advertisements containing a black figure in *Life* magazine from 1936 to 2000 and found in them evidence of increasing "white commitment to racial integration," despite also finding areas of social life where whites are still wary of blacks. Anden-Papadopoulos (2009) investigated home-made videos uploaded to YouTube by coalition soldiers in Afghanistan and Iraq and found considerable evidence that their firsthand experiences contradicted the official narrative of the war offered by the U.S. government. Excellent sources of still and video images are the "image" and "video" functions of Google.

 **STOP & DO**   *Check out Google video. You might type in something like "Iraq war YouTubes by soldiers" and watch two or three. Do you agree with Anden-Papadopoulos (2009) that these YouTube videos created by soldiers tend to be critical of the government's conduct of the war?*

Grady (2007b: 66) suggests you might get fascinating social insights from studying family albums, yearbooks, postcards, greeting cards, magazine advertisements, and newspaper advertisements. You could also look at television advertisements, television programs, and movies. Students in Roger's classes have done interesting research on the changing portrayal of female characters in animated films by Disney, in motion pictures winning the Academy award for best actress, and in television situation comedies. The possibilities are endless.

The larger point here is, however, this: You really don't want to be guided in your search for available data by any of the "usual suspects" we've suggested. A much sounder approach is to immerse yourself in a topic by reading the available literature. In the course of your immersion, look at the sources other researchers have used to study the topic. If you discover that they've typically used the FBI *UCR*, you might want to use that too. If they've used the United Nations' *Demographic Yearbook*, you also might want to look into that. Very often, using earlier research as a resource in this way is all you'll need to find the data you'll want to use.

ADVANTAGES AND DISADVANTAGES OF USING AVAILABLE DATA

Available data offer many distinct advantages over data collected through questionnaires, observation, or experiments. Most striking among these are the savings in time, money, and effort they afford a researcher. The analysis of secondary data or existing statistics might cost little more than the price of your Internet connection (though depending on your institutional affiliation, there can be a charge of several hundred dollars—still relatively little—for access to and the right to use such data) and the time it takes to figure out how you want to analyze them. Countries might spend billions on their national censuses, but comparing 150 nations based on the results of those collections might cost you only the price of a library card or a few keystrokes on your computer.

Martinez and Lee's work with police logs, however, provides a useful caveat to the general rule that available data are cheap and time-saving. In fact, Martinez estimates that he spent roughly five years, off and on, collecting data on 2,500 homicide cases that occurred from 1980 to 1995 and probably spent about $200,000, using grants, salary, and sabbatical supplements, to do so. Part of this expense was because he started his work while working in Delaware. He says, "Flying to Miami from Delaware sounded great at first but it was not cheap. You needed airfare, hotel and food monies, as well as money for supplies. You'd manually pull out homicide records from a file room, copy them, and ship them back to your home base. Once you were at home, you needed to store, read and code the data, then have them entered and readied for statistical analysis [usually by a hired student]" (personal communication). Martinez speaks of "maxing out two credit cards" while getting the project going and hoping that a "grant proposal would hit." Eventually, he did "hit" on grants, but, when he didn't, even later on, "it was depressing. At times I felt like quitting. But then I remembered my credit cards."

Matthew Lee suggests that we should think of available data as being on a "continuum of availability" that reinforces the point that, although much available data is, in fact, easily available, some can come at considerable cost to the researcher (personal communication). He points out, however, that sometimes the less available data are the only means to true conceptual breakthroughs. He points out that he got interested in the migrant/crime question because he'd been so impressed by Cliff Shaw and Henry McKay's "idea that social processes like immigration and residential turnover (rather than 'bad' individuals) weaken community levels of social control and thereby facilitate crime" (1931). Yet Lee and Martinez found little support for that notion. In fact, immigrants appear to be stabilizing urban communities and suppressing levels of violent crime. Lee believes that one reason why many people see newcomers as a disorganizing influence on "institutions and sensibilities" is that the most easily available data, such as the FBI's UCR, cannot be used to address the validity of conventional wisdom on immigration and crime. And it's hard work to get alternative data. But, Lee points out, "Our experiences suggest that taking a gamble on collecting secondary data that are not easily available can lead to a bigger theoretical payoff" (personal communication).

A related advantage of using available data is the possibility of studying social change over time and social differences over space. Robert Putnam, in his national bestseller, *Bowling Alone: The Collapse and Revival of American Community* (2000), analyzes available data from many sources (election data, survey data, organizational data) to substantiate his argument that political, civic, religious, and work-related participation have substantially declined in America during the past quarter century, as have informal social connections and volunteer and philanthropic activities. Again, using data from a variety of sources, Frank et al. (2009) found that the reform of rape laws was strongly and positively associated with elevated police reporting of rapes between 1945 and 2005. And using U.S. Census data collected since 1790, gerontologists have been able to discern that the proportion of the American population that is 65 years or older has grown from 2 percent in 1790 to 4 percent in 1890 to 12.4 percent in 2000. Most individual researchers don't live long enough to observe such trends directly, but, because of available data and the imagination to use them, they don't have to.

 **STOP & THINK** *Imagine you want to find out how much the size of the "average" American family household has changed over the last two centuries. Would you be more inclined to use a survey, participant observation, an experiment, or U.S. Census reports as your data source? Why?*

Available data also mean that retiring and shy people, people who don't like the prospect of persuading others to answer a questionnaire or to participate in a study, can make substantial research contributions. Moreover, the researcher can work pretty much where he or she wants and at whatever speed is convenient. Roger once coded data from a source he'd borrowed from the library while waiting (along with 150 other people who could do no more than knit or play cards or read) in a courthouse to be impaneled on a jury. (At other times, when unable to sleep, he's collected data in the wee hours of the morning—a time when researchers committed to other forms of data collection pretty much have to shut down.) Again, this advantage doesn't always exist. Martinez found it useful, for instance, to cultivate many MPD officers, if only to provide details about cases that were left out of the logs he studied (personal communication).

A final, and most important, advantage of using available data is that the collection itself is unobtrusive and extremely unlikely to affect the nature of the data collected. Unlike researchers who use questionnaires, interviews, observations, or experiments, those of us who use available data rarely have to worry that the process of data collection will affect our findings—only that the original collection by primary researchers may affect those findings.

On the other hand, the unobtrusiveness of using available data is bought at the cost of knowing less about the limitations of one's data than if one collects the data firsthand. One major disadvantage of available data is that one is never quite sure about its reliability. Martinez and Lee, for instance, were evidently concerned that the "homicide logs" they investigated provided adequate definitions of victim and offender ethnicity. Martinez observes, "If a homicide record written in 1980 was not clear on certain things, you were stuck with it." Eventually, though, he had access to an atypical resource for researchers relying on available data. "With time," he reports, "I became acquainted with detectives, especially the 'old-timers' and asked them about certain cases.... If I had not had direct access, it would have been very difficult. I was very fortunate" (personal communication).

Most researchers, using available data, are not so fortunate. Cross-national researchers (like Arthur and Clark, forthcoming) using data on female labor force participation, for instance, have been plagued by their awareness that, in many countries, especially in those where women's work outside the home is frowned upon, there is considerable undercounting of that work.

 **STOP & THINK** *What kinds of nations, do you think, would be most likely to undercount women's work outside the home? Why?*

Sometimes the best you can do is to be explicit about your suspicion that some of your data are not reliable. You can at least try to clarify the kinds of biases the sponsors of the original data collection might have had, as we did when we used available data on female labor force participation. This kind of explicitness provides readers with a chance to evaluate the research critically. At other times, checks on reliability are possible, even without the capacity to perform the kinds of formal tests of reliability (e.g., inter-rater tests—see Chapter 6) that you could use while collecting primary data, or access, such as Martinez's, to the collectors of the original data. Steven Messner (1999) found that—the conventional wisdom about watching violence on television and personal acts of violence to the contrary notwithstanding—cities where people watched lots of violent television programs were likely to have lower violent crime rates than other cities were. Messner was concerned, however, that the "underreporting" of violent crimes in cities where many people were desensitized to crime by the viewing of violent television programming might have been responsible for his unexpected finding that such viewing was negatively associated with the presence of such crimes. But he knew that one kind of violent crime, homicide, was very unlikely to be underreported under any circumstances. So, when he found that the unexpected negative association between viewing violent programs and violent crime rates persisted even for homicide rates, he concluded that it was less plausible that his other unexpected findings were only the result of systematic underreporting.

 **STOP & THINK** *Why is homicide less likely than, say, shoplifting to be underreported, whatever the television viewing habits of inhabitants?*

A second disadvantage of using available data is that you might not be able to find what you want. In some cases, there just might not have been anything collected that's remotely appropriate for your purposes. Modern censuses are designed to enumerate the entire population of a nation, so you can reasonably expect to be able to use their data to calculate things such as sex ratios (the number of males per 100 females, used, for instance, by Messner as a control variable in his analysis). But ancient Roman censuses focused on the population of adult male citizens (the only group, after all, owning property and eligible for military service and therefore worth knowing about for a government interested primarily in taxation and military recruitment [Willcox, 1930, cited in Starr, 1987]), and so the data necessary for the computation of, say, sex ratios, not to mention population density or the percentage of the population that's of a certain age, might simply not be available. Also, available data might not present data the way you want. Martinez notes, for instance, that crime data from the FBI are presented in racial categories ("black" and "white") and not ethnic ones (e.g., "Mariel," "Haitian," and "Jamaican").

Sometimes, however, you can locate data that very nearly (if still imperfectly) meet your needs, so the issue of validity arises. Numerous cross-national studies (e.g., Arthur and Clark in their study of domestic violence mentioned earlier) have used the women's labor force participation rate as a rough measure of women's economic status in a nation. A women's labor force participation rate refers to the proportion of women in an area (in this case, a country) who are employed in the formal workforce. The rationale behind its use as a measure of women's economic status is the notion that, if women don't work for pay, they won't have the power and prestige within the family and in the community that access to an independent income can buy. But what if women and men don't always (as we know they don't) receive equal monetary returns for their work? Wouldn't the economic status of women then depend on, among other things, how closely their average income approximated that of men's—something that their labor force participation rate doesn't begin to measure? We're sure you can think of other problems with labor force participation as a valid measure of economic status, but you probably get the idea: Sometimes researchers who employ available data are stuck using imperfectly valid indicators of the concepts they want to measure because more perfect indicators are simply not available.

 **STOP & THINK** *Can you think of other problems with using women's labor force participation, or the proportion of women over 16 years old who are in the labor force, as an indicator of women's economic status in an area? (Hint: What other things, besides working at all, contribute to our sense that a person has a high economic status?)*

One peculiar danger of using available data, especially existing statistics, is the possibility of becoming confused about your unit of analysis (or subjects) and committing the ecological fallacy. Much of the research on the migrant-crime connection (e.g., Hagan and Palloni, 1998; U.S. Commission on Immigration Reform, 1994), before Martinez and Lee's study, had looked at the relationship between the proportion of migrants in a city and its crime rates. This research, however plausible, was nonetheless subject to the ecological fallacy because its units of analyses (cities) were not the same as units of analyses (people) researchers wanted to make inferences about. Martinez and Lee avoid the ecological fallacy by obtaining information that enables them to directly compare the homicide rates of various immigrant groups and of immigrants and non-immigrants.

SUMMARY

In contrast with primary data, or data that the researcher collects himself or herself, available data exist before you begin your research. Secondary data, or data that have been collected by someone else, have become increasingly popular in social science research. More and more social scientists are using survey data that have been collected by other researchers and have been made accessible for secondary analysis. Two major repositories for such data are the ICPSR at the University of Michigan and the Roper Center at the University of Connecticut.

A major advantage of using available data is that the data are usually cheap and convenient. Martinez and Lee's experience indicates, however, that there is a "continuum of availability" and that sometimes the least available of available data are well worth exploring. Available data also allow the study of social change over long periods and of social difference over considerable space. Finally, their collection is unobtrusive (or unlikely to affect the interactions, events, or behavior under consideration) even if their original collection might not have been.

Common sources of existing statistics are governmental agencies, like the U.S. Bureau of the Census, or supra-governmental agencies, like the United Nations. These statistics tend to be summaries at some aggregate level of information that has been collected at the individual level. These data have become increasingly available on the Internet.

The main disadvantage of using available data is that, because someone else has collected the data, their reliability might be unknown or unknowable by their user. Moreover, one frequently has to settle for indicators that are known to be imperfectly valid. Finally, especially when using existing statistics, the researcher has to be especially cognizant of the temptation to commit the ecological fallacy—of making inferences about individuals from data about aggregates.

EXERCISE 12.1

Urbanization and Violent Crime: A Cross-Sectional Analysis

This exercise gives you the opportunity to examine one assumption that seems to lie behind Martinez and Lee's analysis: Cities, or more particularly standard metropolitan statistical areas (SMSAs), are especially good places to study violent crime. We'd like you to see whether, in fact, population aggregates that are more highly urbanized (i.e., have a greater proportion of their population in urban areas) have higher violent crime rates than those that are less highly urbanized.

Let's examine a population aggregate that can be more or less urbanized: states in the United States. Find a recent resource, perhaps in your library, that has data on two variable characteristics of states: their levels of urbanization and their overall violent crime rate. Collect data on these variables for the first 10 states, listed alphabetically. (Hint: *The Statistical Abstract of the United States* [http://www.census.gov/compendia/statab/] is a

good source of the data you'll want. Most college libraries will have this in their reference section.)

1. Prepare 10 code sheets like this (actually, you can arrange all data on one page):
   State: Total Violent Crime Rate: Percent Urban:
2. Go to your library (or the website) and find the data on crime and urbanization. (If you use the website, you can just type in "crime rates" and "percent urban" in the "Search the Abstract" box. Using the code sheets you've prepared, collect data on 10 states of your choice.)
3. Once you've collected the data, analyze them in the following ways:
   a. Mark the five states with the lowest percent urban as **LU** (low urban) states. Mark the five states with the highest percent urban as **HU** (high urban) states.
   b. Mark the five states with the lowest total violent crime rates as **LC** (low crime) states. Mark the five states with the

highest total violent crime rates as **HC** (high crime) states.

c. Calculate the percentage of LU states that you have also labeled **HC**. Calculate the percentage of HU states that you have also labeled **HC**.

4. Compare your percentages. Which kinds of states tend to have the highest violent crime rates: highly urbanized or not-so-highly urbanized states? Interpret your comparison relative to the research question that sparked this exercise.

EXERCISE 12.2

Life Expectancy and Literacy: A Cross-National Investigation

This exercise asks you to investigate the connection between education and health by examining the research question: Do more educated populations have the longest life expectancies? Investigate this hypothesis using data from 10 countries (your choice) about which you obtain data on the percentage of the population that's literate and life expectancy. [You can find the data in most world almanacs and in a volume of Human Development Reports, any year, online. You can search for them by using key words "Human Development Reports" and, then, "Life Expectancy."]

1. Prepare 10 code sheets like this (again, you can arrange all data on one page):
   Country: Literacy: Life Expectancy:
2. Go to your library (or the website) and find the appropriate data. Using the code sheets you've prepared, collect data on 10 countries of your choice.
3. Once you've collected the data, analyze them using the technique introduced in Exercise 12.1.

a. Mark the five countries with the lowest literacy rate as **LL** (low literacy) nations. Mark the five countries with the highest literacy rates as **HL** (high literacy) nations.

b. Mark the five countries with the lowest life expectancy as **LLE** (low life expectancy) nations. Mark the five countries with the highest life expectancy as **HLE** (high life expectancy) nations.

c. Calculate the percentage of LL nations that you have also labeled **HLE**. Calculate the percentage of HL nations that you have also labeled **HLE**.

4. Compare your percentages. Which kinds of nations tend to have the highest life expectancy: those with high literacy rates or those with not-so-high literacy rates? Interpret your comparison relative to the research question that sparked this exercise.

5. What is the ecological fallacy? Do you think you succumbed to this fallacy in your interpretation of your analysis in this exercise? Why or why not?

EXERCISE 12.3

For Baseball Fanatics

This exercise gives you the chance to examine an age-old social science dispute: whether hitting or pitching wins ball games. Help shed light on this dispute by, again, turning to a recent almanac, this time to collect winning percentages, team batting averages, and team earned-run averages for all the teams in either the American or the National League for one year. Use data collection techniques similar to those you used for the last two exercises, this time collecting data on three variables, rather than two. Moreover, you should use the data analysis techniques of each exercise, this time examining the relationship between winning percentage and team batting average, on the one hand, and winning percentage and team earned-run average, on the other. Are both batting average and earned-run average associated with winning percentage? Which, would you say, is more strongly associated with it? Why?

Write a brief essay in which you state how your analysis bears on this research question.

EXERCISE 12.4

A Photo Finish

This exercise gives you a chance to exercise your skills as a visual sociologist through a comparison of two photos used to announce Barack Obama's win of the 2008 Presidential election. One of these will be the one that appeared on the first page of the *New York Times* the day after the election; the other, from the first page of the *Philippine Daily Inquirer* that day. Your job will be to compare the messages of these photos and to try to understand how the editors of these two newspapers were similar and how they differed in their take on the election.

1. Find Google Image and type in "Obama wins election New York Times." Find the largest picture you can of the picture the *New York Times* put on its front page and study the image. What, would you say, are three or four main messages that the editors of the *New York Times* may have tried to convey by using this image?

2. Using Google Image again, type "Obama wins election Philippines Daily Inquirer." Find the largest picture you can of the picture the *Philippine Daily Inquirer* put on its front page and study the image. What, would you say, are three of four main messages that the editors of the *Philippine Daily Inquirer* may have tried to convey by using this image?

3. What are clearly some of the messages that both the *New York Times* and the *Philippine Daily Inquirer* editors tried to convey through their images?

4. What is one message conveyed by the *New York Times* image that is not conveyed by the *Philippines Daily Inquirer* image?

5. What is one message conveyed by the *Philippines Daily Inquirer* image that is not conveyed by the *New York Times* image?

6. How might you interpret the similarities and differences you've discovered?

CHAPTER **13**

Content Analysis

© Jonathan Littlejohn / Alamy

INTRODUCTION

Have you ever taken an American literature course? If you'd done so in the 1960s, you probably wouldn't have read the works of very many women, African American, or Native American authors. They just didn't appear in the American literature anthologies of the time. If you've taken such a course in recent years, your chances of reading the works of women, African Americans, and Native Americans would have been much better, because representative authors of each group are in recent anthologies. This observation is what motivated Roger, his son, Adam, a student, Jennifer Racine, and a faculty colleague, Mikaila Arthur, to pursue the focal research of this chapter, research that examines the changing visibility of women, African American, Native American, Hispanic American, and Asian American authors in American literature anthologies over the last half century. The piece illustrates the technique of **content analysis**, a technique that is particularly useful for doing historical investigations. Content analysis is a method of data collection in which some form of communication (speeches, television programs, newspaper articles, films, advertisements, even literature anthologies) is studied systematically. Content analysis is one form of available data analysis.

content analysis, a method of data collection in which some form of communication is studied systematically.

FOCAL | RESEARCH

Reloading the Canon: The Timing of Diversification in American Literature Anthologies

By Mikaila Mariel Lemonik Arthur, Adam Clark, Roger Clark, and Jennifer Racine*

In *Loose Canons: Notes on the Culture Wars*, Henry Louis Gates (1992) noted that opening up American literature anthologies to previously marginalized voices has been fraught with controversy. At issue is what should be included in the canon, or "authors and works included in basic American literature college courses and textbooks" (Lauter, 1983a: 23). At times, the argument has been nothing short of a battle between academic innovators, on the one hand, and religious and political conservatives, on the other, with the stakes being, at least in part, about what it means to be American (Jay, 1997).

But the question addressed here is more mundane than why the re-shaping of the canon has been controversial. It's about the general outline of that re-shaping and, in particular, about the timing of the "arrival" of previously marginal voices—i.e., women, African Americans, Native Americans, and Asian Americans.

*This research was prepared especially for this text.

"Rampart Stormers" vs. "Gate Keepers" Perspectives

While the controversy over who should be included in American literature texts has implicated the larger political forces of the time, it has also pitted factions within the academic community against one another. The initial push for change in the "traditional" canon, which typically included mainly white male authors, came from the academic arms of major political movements of the 1960s and 1970s—the civil rights movement and the women's movement. Arthur (2009) calls these social movements within the academic community New Knowledge Movements and has examined the contentious relationships between political and intellectual insiders and outsiders in the academy. African American studies programs emerged between 1965 and 1968 (Rojas, 2007), the first women's studies program in 1971 (Boxer, 1998), and the first programs in Asian American, Native American, and Hispanic studies in 1969 (Barlow & Shapiro, 1971). One of the goals of the New Knowledge Movements was to develop knowledge about the literary experiences of the groups in question. Rediscovery of lost "classics," like the work of Harriet Beecher Stowe, William Wells Brown, and Rebecca Harding Davis, was a first order of business. But, recovering previously under-appreciated writing, like that of William Apess, Sui-Sin Far, Juan Nepomuceno Seguin, and Carlos Bulosan, became an important second-order task. Anthologies devoted to the literatures of each specific group came quite soon after the founding of the first of these programs. We see the New Knowledge Movements of the 1960s and 1970s, having established various new programs within academia and having supplied crucial information about relevant literary figures, as "rampart stormers" with respect to more general anthologies of American literature. There is no doubt that the degree of institutionalization of women's studies and African American studies programs has outstripped those of Native American studies, Hispanic studies, and Asian American studies programs,** but if rampart storming were a sufficient condition for the inclusion of various groups in American literature anthologies, one would expect that all five "minority" groups considered here might have experienced substantial representation in the more general anthologies, if not by the late 1970s, then surely by the 1980s.

"Rampart stormers" often encounter resistance, however. As Lauter notes (1983a), the editors of earlier anthologies did not value

---

**The U.S. College Search (2009) website, for instance, suggests there were 169 women's studies, 124 African American studies, 39 Native American studies, 36 Hispanic American studies, and 9 Asian American studies at "Colleges and Universities" in the United States in 2009. The College Board (2009) reports the number of women's studies programs at 320; of black studies programs, 225; of Chicano/Latino studies, 70; Asian American studies, 24; and Native American studies, 78. Similar patterns are also noted by the Integrated Post-Secondary Data System of the U.S. Department of Education (2009).

diversity in their offerings nor were they sensitive to cultural history that might have fostered such diversity. The dominant critical paradigm in the 1960s and 1970s, that of the New Criticism, valued examining works in isolation and studying how their parts worked together to form a complex whole. The New Criticism valued complexity and nuance; many of the works sponsored by the New Knowledge Movements were simpler and more direct. Something like a paradigm shift in criteria for inclusion needed to occur before such inclusion could take place. Consequently, and to the extent that this paradigm shift will have involved a time lag and may have been differentially resisted by different gate keepers, our expectations are 1) that the five "minority" groups we examine here will not have experienced substantial representation in general American literature anthologies until after the 1970s, and perhaps after the 1980s and 2) that the increased representation will have occurred at substantially different rates in different anthology series.

Methodology

Using books obtained from many libraries and sources, we examined a large convenience sample of 27 general American literature anthologies published since 1950: three from the 1950s; four from the 1960s; three from the 1970s; six from the 1980s; six from the 1990s; and five from the 2000s. A Worldcat search informs us that we haven't completely exhausted the population of general American literature anthologies published after 1950, but we've come pretty close.

Our key variable was the visibility of various "minority," or at least previously under-represented, groups (women, Native Americans, African Americans, Hispanics, and Asian Americans) in the anthologies. We conceptualized "visibility" as the degree to which works by members of these groups were present in an anthology. We operationalized visibility in terms of the proportion of pages devoted to authors recognized by the editors as belonging to a particular minority group. "Recognition" may be implicit, as in the use of personal pronouns (to distinguish women and men authors), or explicit. We are aware of objections that may be raised to this measurement scheme. First, it clearly ignores certain minority groups that have New Knowledge Movement counterparts today (e.g., queer studies) simply because appropriately identifying authors within the anthologies is impossible. Second, it clearly ignores groups that would have been considered minority groups in the past (e.g., Irish Americans, Italian Americans), privileging those with New Knowledge Movement counterparts today. Third, it implies that there are more similarities among members of the groups defined (e.g., women, Native Americans, Hispanics) than there are differences. Another concern about our measurement scheme, of course, is that it privileges what editors say about an author's minority status, possibly over the author's self-definition.

Our strategy permits relatively reliable measurements. Our final measure was the average of two raters' assessments of the corresponding proportions. Inter-rater tests show high levels of agreement (correlations between two raters' use of the percentage of all pages measure, for instance, range between .91 [for Native and African Americans] and .98 [for Asian Americans] for all "minority" groups).

Findings

There has indeed been a revolution in the presentation of previously ignored groups in American literature anthologies. As Table 13.1 shows, only very small portions of general American literature anthologies were devoted to the works of minorities in the 1950s and 1960s anthologies but the 2000 anthologies tell a different story. While women's works commanded, on average, only 3.9% of pages in 1950s anthologies, about 25.7% of pages in 2000s anthologies were been devoted to women's works. We see similar changes for the other groups as well.

Inspection of Table 13.1 suggests some support for both the "rampart stormer" and the "gate keeper" hypotheses, but more for the latter than the former. The average percentage of pages devoted to women's work did grow from 3.3% in the 1960s anthologies to 9.1% in the 1970s and then to 11.5% in the 1980s. And the average percentage devoted to works by African Americans grew from 0.8% in the 1960s to 2.9% in the 1970s and 3.4% in the 1980s. But these decades saw little growth in the representation of Native Americans and Hispanics, and no growth at all in the representation of Asian Americans in the anthologies. On average, 0.4% of pages were devoted to works by Native Americans, 1.0% to works by Hispanics, and 0% to works by Asian Americans in 1980s anthologies.

Indicative of greater support for the "gate keepers" hypothesis is that all previously under-represented groups enjoyed much greater visibility in 1990s anthologies than they had in 1980s anthologies, suggesting a delay of more than 20 years between the founding of the New Knowledge Movements and their now remarkable impact on the canon. The average percentage of pages devoted to women's works rose dramatically from 11.5% in the 1980s to 27.6% in the 1990s; to Native Americans' works, from 0.4% to 4.3%; to Hispanics' works, from 1.0% to 3.2%; to African Americans' works, from 3.4% to 10.9%; and to Asian Americans' works, from 0% to 2.1%. Surely, something had changed for the "gate keepers" by the 1990s that permitted them to welcome in previously excluded groups *en masse*. Inspection of editors' prefaces suggests that at least part of this change involved the kind of paradigm shift our "gate keeper"

TABLE **13.1**

Percent of Pages in Text Devoted to Authors with Certain Characteristics

Text	Edition	Year	Males	Females	Whites	Native Americans	Hispanics	African Americans	Asian Americans
Cady	1	1956	94.4	5.3*	100	0	0	0	0
Bradley	1	1956	92.1	5.1	98.6	0	1.2	0.2	0
Ray	1	1959	98.6	1.4	100	0	0	0	0
50s Average			**95.0**	**3.9**	**99.5**	**0**	**0.4**	**0.1**	**0**
Bradley	2	1962	91.5	5.7	95.5	0	1.1	2.1	0
Anderson	1	1965	98.7	1.3	100	0	0	0	0
Bode	1	1966	97.4	2.0	100	0	0	0	0
Bradley	3	1967	95.2	4.3	98.5	0	0.8	0.6	0
60s Average			**95.7**	**3.3**	**98.5**	**0**	**0.5**	**0.8**	**0**
McMichael	1	1974	93.6	6.4	96.3	0	0	3.7	0
Bradley	4	1974	91.4	7.7	97.6	1.0	0	1.5	0
Norton	1	1979	86.6	13.3	95.2	0	1.2	3.6	0
70s Average			**90.5**	**9.1**	**96.7**	**0.3**	**0.4**	**2.9**	**0**
McMichael	2	1980	90.6	9.2	96.4	0	0.9	2.8	0
Bradley	5	1981	91.9	7.7	96.5	0.9	0.6	2.0	0
Norton	2	1985	84.6	15.3	94.4	0	0.5	5.2	0
Bradley	6	1985	83.9	12.5	95.8	0.7	0.9	2.6	0
McMichael	4	1989	87.5	12.2	95.7	0	0.9	3.0	0
Norton	3	1989	87.8	12.0	92.3	0.6	2.1	4.8	0.2
80s Average			**87.7**	**11.5**	**95.2**	**0.4**	**1.0**	**3.4**	**0.0**
Heath	1	1990	63.4	31.3	76.0	6.0	4.4	12.0	2.5
Norton	4	1994	79.1	20.2	76.2	3.1	1.9	7.4	0.3
Heath	2	1994	63.4	31.0	69.4	5.8	4.4	15.3	5.0
McMichael	6	1997	74.6	24.1	90.0	1.4	0.8	7.9	0.5
Norton	5	1998	73.6	25.1	86.7	3.6	1.1	8.0	0.5
Heath	3	1998	61.3	34.4	68.5	6.0	6.8	14.9	3.8
90s Average			**69.2**	**27.6**	**77.8**	**4.3**	**3.2**	**10.9**	**2.1**
Heath	4	2002	65.5	30.5	72.4	4.9	6.0	13.9	2.5
Bradley	10	2003	80.4	17.1	94.1	1.0	1.2	3.6	0.5
Norton	6	2003	70.8	27.2	84.2	2.7	2.7	8.7	1.2
McMichael	8	2004	74.0	24.0	87.2	1.6	2.6	8.1	0.9
Heath	5	2006	65.6	29.6	77.0	4.9	4.9	12.0	0.8
00s Average			**71.3**	**25.7**	**83.0**	**3.0**	**3.5**	**9.3**	**1.2**

*The percent of anthologies devoted to male and female writers rarely adds up to 100% because of the presence of works by anonymous authors.

hypothesis was based on. Ray wrote, for instance, in the "prefatory note" to his *Masters of American Literature* that he and his co-editors were guided by the principle "that the student ... will profit more from regarding the works he reads to be studied and enjoyed in their own terms than he will from viewing them as illustrations of the course of literary or cultural history..." (1959: v). "Literary merit" was also the criterion for including authors in the long-running *The American Tradition in Literature* (by Bradley et al., beginning in 1956) and McMichael et al.'s *Anthology of American Literature* (beginning in 1974). This is not to say that the paradigm of pure "literary merit" did not undergo some decay even during the 1970s. The 1st edition of McMichael's et al.'s anthology did add a principle of "inclusion" to its primary concern for "scholarly responsibility" (1974: xv). The 4th (1974) edition of *The American Tradition in Literature* (what we have called "Bradley" in the table), did not, in its preface, explicitly claim to be revising the canon, but it did, in fact, include more women and African Americans than any anthology of the 1950s and 1960s, and was the first anthology, as far as we can tell, to include unattributed selections of Native American "tales, speeches, and poems." (It did so while deleting, for instance, previously included materials from, say, James Fenimore Cooper, which presumably represented "[t]he Indian heritage" [1974: xvii].) But the 1st edition of *The Norton Anthology of American Literature* (by Gottesman et al. in 1979) introduced a completely new principle for selection: "copiousness." This principle contrasts markedly with Ray's, and was "designed to allow teachers to set up their own reading lists, without the need to ask students to buy extra books" (Gottesman, 1979: xxxi). The 1985 (second) edition of *The Norton Anthology* recognized Lauter's (1983b) *Reconstructing American Literature* for providing an additional push to "open up" the traditional canon, a push that the *Norton Anthology* editors were responsive to.

It wasn't until after the publication of Lauter's first *Heath Anthology* in 1990 that the editorial tide seems to have turned completely. Here, the first principle was to provide access to a "*range* of writers" so that students could draw "stimulating comparisons and contrasts between canonical and non-canonical figures, between female and male, between one ethnic writer and another" (1990: xxxv). This approach clearly affected the filter used by other editors of 1990s anthologies. The preface of the sixth (1997) edition of McMichael et al.'s anthology still selected works "primarily" for their "literary significance," but now there were secondary criteria at work as well. None of the three anthologies dating back to at least the 1970s, however, came close to matching the *Heath Anthology* in recognition of previously marginalized groups. However, in the 10th edition of *The American Tradition*, the editors (Perkins and Perkins) simultaneously took pride in the "greatly enlarged attention to African Americans, Native Americans, and women that distinguished our fourth [1974] edition," while expressing some hesitation about the "industry of much longer books so focused on the margins that they engulfed the center" (2002: xxxiii).

There is still the question of why it was mainly women and African American authors who made it, however tentatively, into the 1970s and 1980s editions of general anthologies. Lauter's preface to the 1st *Heath Anthology* provides a clue, again one that supports the "gate keepers" more than the "rampart stormers" hypothesis. As Lauter describes it, his (1983a) *Reconstructing American Literature* became a repository for syllabi and course materials developed by professors interested in changing the canon, but it also alerted its readers (and editors, presumably) to the absence, in these syllabi, of works by Native American, Hispanic, and Asian-American writers (Lauter, 1990: xxxiv). This recognition may, in large measure, account for why it was women and African American authors who broke into the anthologies of the 1970s and 1980s and why it wasn't until the 1990s that Hispanic, Asian American, and Native American writers were introduced. The omission of these other groups simply hadn't been noticed even by sympathetic gate keepers yet.

Conclusion

Our analysis of the content of American literature anthologies in the last half century demonstrates the substantial re-shaping of American literary canon with all of the "minority" groups making enormous inroads. Coverage of works authored by males still outstrips that of works by females, but no longer in a ratio of 32 to 1. Now the ratio is more like 2.75 to 1. Some other "minority" groups are slightly over-represented, most still under-represented, in comparison with their share of the American population. Native Americans, with 3% of the pages in 2000s anthologies devoted to their work, are over-represented as their share of the population measured by the 2000 census was about 1%. African Americans, Hispanics, and Asian Americans, with about 9.3, 3.5, and 1.2% of the pages, respectively, are still under-represented (with population percentages respectively of 12.3, 12.5, and 3.6).

We entertained two hypotheses for explaining the timing of the change we expected to see: a "rampart stormers" hypothesis, which predicted that the bulk of the change would have occurred during the 1970s or the 1980s, and a "gate keepers" hypothesis, which predicted that the bulk of the change would have occurred later and that it would have occurred differentially in different anthology series. We find some support for the "rampart stormers" hypothesis as— representation of women, African Americans, and Native Americans did increase during the 1970s and the 1980s. Thus, consistent with the "rampart stormers" hypothesis, even the anthologies that were still, according to editors' prefaces, holding primarily to the standard of "literary merit" for including previously excluded works, found some such works that merited inclusion. Thus, the 4th (1974) edition of *The American Tradition in Literature* introduced 23 women artists, four African Americans, and some material by Native Americans. But, as we said, the 1st (1979) edition of the *Norton Anthology* introduced a new standard for inclusion: the standard of "copiousness." And it included 29 women writers and 14 African American authors. So, even the addition of women and African American authors in the 1970s and 1980s anthologies reflected a transition in traditional standards for inclusion, an incipient paradigm shift.

By the 1990s, though, the keepers had well and truly opened the gates. The average 1990s anthology devoted almost 20% less coverage to males and to whites than the average 1980s anthology. Women's share of the available space increased by 150%; Hispanic Americans, by 220%; African Americans, by about 220%; Native Americans, by almost 1000%; and Asian Americans, almost infinitely. Importantly, though, this "opening up" did not occur uniformly across all anthology series. Heath clearly led the way, followed by Norton, McMichael's (Pearson's), and, finally, Bradley's (now Perkins and Perkins's) anthology. The precise politics involved in the gate opening are beyond the scope of this paper. That it took a while for those politics to yield dramatic change in the anthologies and that the politics played out at different speeds in different anthology series, however, is now beyond dispute. But the canon has indeed been reloaded in a way that defines America as multivocal.

This multivocality has implications, of course, for the whole of the English literature curriculum. The anthologies are, if you will, the gate keepers to the curriculum. Multicultural curricular content remains much debated in upper-level courses as well as in anthologies (see Bryson, 2005). These upper-level courses have the capacity to forge ahead of what the anthologies provide—when and if faculty members are interested in such change. But in departments lacking progressive literature faculty, students will at least learn from their literature anthologies that diverse authors are a significant part of the canon.

References

Arthur, Mikaila Mariel Lemonik. 2009. "Thinking Outside the Mater's House: New Knowledge Movements and the Emergence of Academic Disciplines." *Social Movement Studies* 8(1):73–87.

Barlow, William, & Peter Shapiro. 1971. *An End to Silence: The San Francisco State College Movement in the '60s*. New York: Pegasus.

Baym, Nina, general editor. 1985. 1989. 1994. 1998. 2003. *Norton Anthology of American Literature*. 2nd, 3rd, 4th, 5th, and 6th editions. New York: Norton & Company. From 2 to 5 volumes.

Boxer, Marilyn Jacoby. 1998. *When Women Ask Questions: Creating Women's Studies in America*. Baltimore: Johns Hopkins University Press.

Bradley, Sculley et al. 1956. 1962. 1967. 1974 *The American Tradition in Literature.* 1st, 2nd, 3rd, and 4th editions. New York: Norton & Co. (Until, 1974, when Grosset & Dunlap became the publisher.) 2 volumes each year.

Bryson, Bethany Paige. 2005. *Making Multiculturalism: Boundaries and Meaning in U.S. English Departments.* Stanford: Stanford University Press.

College Board. 2009. College Board Site. Retrieved 5/14/09. http://college-search.collegeboard.com/search/adv_typeofschool.jsp

Efros, Susan. 1974. *This Is Women's Work.* San Francisco: Panjandrum Press.

Gates, Henry Louis. 1992. *Loose Canons: Notes on the Culture Wars.* New York: Oxford University Press.

Gottesman, Ronald, general editor. 1979. *The Norton Anthology of American Literature.* New York: Norton & Company.

Integrated Post-Secondary Education Data System. 2009. Integrated Post-Secondary Education Data System. Retrieved 5/14/09. http://nces.ed.gov/IPEDS/

Jay, Gregory. 1997. *American Literature & the Culture Wars.* Ithaca, NY: Cornell University Press.

Lauter, Paul. 1983a. "Race and Gender in the Shaping of the American Literary Canon: A Case Study from the Twenties." *Feminist Studies* 9(3): 22–47.

Lauter, Paul. 1983b. *Reconstructing American Literature: Course, Syllabi, Issues.* Old Westbury, NY: The Feminist Press.

Lauter, Paul, general editor. 1990. 1994. 1998. 2002. 2006. *The Heath Anthology of American Literature.* 1st, 2nd, 3rd, 4th, and 5th editions. Boston: Houghton-Mifflin. 2 to 5 volumes.

McMichael, George, general editor. 1974. 1980. 1989. 1997. 2004. *Anthology of American Literature.* 1st, 2nd, 4th, 6th, and 8th editions. New York: MacMillan (for 1st, 2nd, and 4th editions). Upper Saddle River, NJ: Prentice Hall (for 6th and 8th editions). 2 volumes each.

National Center for Education Statistics. 2009. Retrieved 5/14/09. http://www.nces.ed.gov/fastfacts/display.asp?id=98

Perkins, George, Sculley Bradley et al. 1985. *The American Tradition in Literature.* New York: Random House. 2 volumes.

Perkins, George, & Barbara Perkins. 2002. *The American Tradition in Literature.* Boston: McGraw-Hill. 2 volumes.

Ray, Gordon. 1959. *Masters of American Literature.* Cambridge, MA: The Riverside Press.

Rojas, F. 2007. *From Black Power to Black Studies: How a Radical Social Movement Became an Academic Discipline.* Baltimore: Johns Hopkins University Press.

U.S. College Search. 2009. Retrieved 5/14/09. http://www.uscollegesearch.org/

 **THINKING ABOUT ETHICS** *In this research, Arthur, Clark, Clark, and Racine are studying published materials rather than human subjects, so their main ethical responsibility is to report their findings honestly.*

APPROPRIATE TOPICS FOR CONTENT ANALYSIS

Although Arthur, Clark, Clark, and Racine applied content analysis to American literature anthologies, content analytic methods can be applied to any form of communication,[1] including movies, television shows, speeches, letters, obituaries, editorials, and song lyrics. Gilbert Shapiro and Philip Dawson (1998) looked at the notebooks of revolutionaries to gain insight into the causes of the French Revolution; Hale, Olsen, and Fowler (2009) compared the election coverage of Spanish- and English-language news stations in the United States; Loe (2004) examined newspaper advice columns to make sense of sexuality in the Viagra era; and Zhao, Grasmuck, and Martin (2008) studied Facebook accounts to examine how users create digital identities. Classic content analyses on racism include one by the freed slave Ida Wells (1892, cited in Reinharz [1992]) and G. J. Speed (1893, cited in Krippendorff [2003]) who analyzed newspaper articles to show, respectively, the extent to which black men were being lynched in the South and how gossipy newspapers in New York were becoming during the 1880s. More recently, Hazell and Clarke (2008) examined recent advertisements in two black-oriented magazines (*Essence* and *Jet*), finding that ideologies of racism persist in them.

Materials Appropriate for Content Analysis

Studies have focused on suicide notes (Bourgoin, 1995, to amend Durkheimian suicide theory), letters (Kenschaft, to examine nineteenth-century efforts to reinvent marriage, 2005), magazines (Friedan, to define the *Feminine Mystique*, 1963, and Taylor, to see how they construct the sexuality of young men, 2005), blogs (Attwood, 2009), chat rooms (Donelle and Hoffman-Goetz, 2008), wills (Finch and Wallis, 1993), radio programs (Albig, 1938), personal advertisements (Jaggar, 2005), speeches (Rutherford, 2004), verbal exchanges (Bales, 1950), diaries (Ulrich, 1990), parenting-advice books (Krafchick, Zimmerman, Haddock, and Banning, 2005), films—both conventional (Leslie, 2005) and pornographic (Cowan and Campbell, 1994), and websites (Holster, 2004). The list is endless. Systematic content analysis seems to date back to the late 1600s, when Swedish authorities and dissidents counted the words in religious hymns and sermons to prove and disprove heresy (Dovring, 1973; Krippendorff, 2003; Rosengren, 1981).

Questions Asked by Content Analyzers

Generally the questions asked by content analyzers are the same as those asked in all communications research: "Who says what, to whom, why, how, and with what effect?" (Lasswell, 1965: 12). Arthur, Clark, Clark, and Racine were interested in the degree to which American literature anthologies included the works of women, African Americans, Native Americans, Hispanic Americans, and Asian Americans from the 1950s to the 2000s and

---

[1]In fact, although content analysis is most often applied to communications, it may, in principle, be used to analyze any *content*. Thus, for instance, Rathje and Murphy's (1992) "garbology" project mentioned in Chapter 12 employed content analysis to study the contents of garbage cans.

held that content up against the proportion of such "minority groups" in the larger population and in the population of college students. They found that, during that period, the degree of "minority" visibility in anthologies increased enormously, but that the increase was timed in such a way as to suggest that vital role played by "gate keepers" (editors and publishers) and the mediated role played by "rampart stormers" (activists in what Arthur has called New Knowledge Movements). Consequently, they found more support for their "gate keepers" hypothesis than the "rampart stormers" hypothesis. Evaluating communications against hypotheses, or some other standards, and describing trends in communication content are two general purposes of content analysis (see Holsti, 1969: 43).

 STOP & THINK *Do you think local, regional, or national newspapers are more likely to "headline" international stories on the top of the front page? What is your hypothesis? Why? (You might test this hypothesis in Exercise 13.1.)*

QUANTITATIVE CONTENT ANALYSIS

quantitative content analysis, analysis focused on the variable characteristics of communication.

When researchers want to test hypotheses or to describe trends in communication, they are apt to engage in quantitative content analysis, in the way Arthur, Clark, Clark, and Racine did. **Quantitative content analysis**, like all other quantitative research, focuses on variable characteristics (in the case of the focal research, the degree of minority visibility, the year of publication, etc.) of units of analysis (in this case, of American literature anthologies).

Units of Analysis

units of analysis, the units about which information is collected.

Sampling for quantitative content analysis involves many of the same issues that sampling for any type of social research involves. First, you identify the **units of analysis** (or elements), the kinds of subjects you want to focus on. In the focal research on the American literary canon, this task was relatively simple. The population could, of course, have been all the books used in American literature courses or those that have ever been highly recommended by some set of reviewers. But once the researchers ran across Paul Lauter's definition of the canon as "authors and works included in basic American literature college courses and textbooks" (1983a: 23), it seemed appropriate to use the American literature anthologies that tend to be used in introductory American literature courses.

 **STOP & THINK** *Suppose you wanted to examine the question of whether recent heroines in Disney animated films, like Pocahontas and Mulan, were more or less stereotypically feminine than heroines of earlier Disney films, like Sleeping Beauty and Snow White. What would your unit of analysis be?*

Deciding on the appropriate unit of analysis is not always so simple as it was in the anthology study. One of Roger's students, Katie Grandchamp (2009), for instance, examined the question mentioned in the last "Stop and Think"; whether more recent heroines in Disney animated films were portrayed more or less stereotypically than heroines in earlier Disney films (she found that they were portrayed less stereotypically, by the way). In this case, the unit of analysis wasn't the film, but the heroine within the film. To use a distinction made by Earl Babbie (2009), the unit of analysis—the unit about which information is collected (heroines)—was imbedded within the **units of observation**—the units from which information is collected (films). The distinction between units of analysis and observation is one without a difference when the unit of analysis and observation are the same. For Arthur, Clark, Clark, and Racine the units of observation and the units of analysis were literature anthologies. But when the units of analysis and observation are different, it's useful to remember that the constituents of one's sample are the units of analysis, not the units of observation. Clark, Guilmain, Saucier, and Tavarez (2003), for instance, examined the presentation of gender in children's picture books from the 1930s to the 1960s. At times, they, like Grandchamp (2009), focused on characters, but sometimes they focused on pictures and sometimes the books as a whole. Thus, while their unit of observation was always children's books, their units of analysis were sometimes characters, sometimes pictures, and sometimes the books themselves.

If you were interested in the sexual content of television shows, as the Kaiser Foundation (2005) was, television programs would be both your unit of analysis and your unit of observation. In Box 13.1, some of the highlights of this interesting study are presented.

units of observation, the units from which information is collected.

BOX **13.1**

Major Findings of the Kaiser Family Foundation Biennial Report (2005) of "Sex on TV"

- Seventy percent of all shows have sexual content, up from 56 percent in 1998 and 64 percent in 2002.
- In shows that include sexual content, the number of sexual scenes is also up, to an average of 5.0 scenes an hour, compared to 3.2 scenes in 1998 and 4.4 scenes in 2002.
- Sexual intercourse is depicted or strongly implied in 11 percent of television shows, up from 7 percent in 1998, but down from 14 percent in 2002.
- Reality shows are the only genre of programming studied in which less than two-thirds of shows include sexual content. In 92 percent of movies shown on television, 87 percent of sitcoms, 87 percent of drama series, and 85 percent of soap operas, there is sexual content.
- About half of all scenes with intercourse (53 percent) involve characters who have an established relationship.
- Fifteen percent of scenes with intercourse present characters having sex when they have just met, compared to 7 percent in 2002.

SAMPLING

In content analysis, units of analysis can be words, phrases, sentences, paragraphs, themes, photographs, illustrations, chapters, books, characters, authors, audiences, or almost anything you want to study. Once you've chosen the units of analysis, you can sample them with any conventional sampling technique (see the various probability and nonprobability sampling techniques of Chapter 5) to save time or effort. In the analysis of American literature anthologies, for instance, Arthur, Clark, Clark, and Racine might have used the list they found from Worldcat and done a stratified random sample of, say, three anthologies from each decade between the 1950s and the 2000s. Instead, they chose to take data from as many of the anthologies from each decade as they could locate through their local college libraries, through faculty in English departments they knew, and through InterLibrary loan. Thus, theirs was a convenience sample. By contrast, Hazell and Clarke (2008) examined the total population of advertisements in *Essence* and *Jet* for 2003 and 2004. In fact, nonprobability sampling and the study of whole populations is typical even of quantitative content analysis, though probability sampling does occur. For example, the Kaiser Foundation (2005), for its study of sex on television reported in Box 13.1, random sampled television programs from October 2003 to March 2004.

CREATING MEANINGFUL VARIABLES

Quantitative content analysis is a technique that depends on the researcher's capacity to create and record meaningful variables for classifying units of analysis. Thus, for instance, in classifying literature anthologies, Arthur, Clark, Clark, and Racine were interested in the percentage of pages devoted to all works that were devoted to works by women, works by Native Americans, works by African Americans, works by Hispanic Americans, and works by Asian Americans. In each of these cases, variables were *interval-ratio* scale. Their coding sheet, a portion of which is depicted in Figure 13.1, shows the information they collected to determine the percentage of pages devoted to Native Americans. (Information about works by non-Hispanic European Americans, Hispanic Americans, African Americans, and Asian Americans were also included on the actual coding sheets.)

To calculate, for instance, the percentage of pages devoted to works by Native Americans, the researchers divided the total number of pages attributed to Native Americans (the sum of responses to items 9, 10, and 11) and divided that by the total number of pages attributed to authors of all sorts (the sum responses to items 6, 7, and 8) and multiplied the result by 100, obtaining an interval-ratio scale variable.

Hazell and Clarke (2008), in their study of racism in *Essence* and *Jet* advertisements, coded each advertisement for things like the product advertised, the audience of the advertisement, the physical characteristics of the model (i.e., gender, race, skin tone, hair, nose, etc.), and the interaction between models (i.e., formal or informal). Each of these variables is a nominal scale.

1. Name of anthology and edition number: *American Tradition in Literature* 10th ed.
2. Editor: George Perkins and Barbara Perkins
3. Number of volumes: 2
4. Year of publication: 2002
5. Publisher and place of publication: McGraw-Hill Co.: New York
6. Number of pages on which materials by or about male authors appear: 3,095
7. Number of pages on which materials by or about female authors appear: 619
8. Number of pages on which materials by or about an author of unknown gender appear: 43
9. Number of pages on which materials by or about male Native American authors appear: 0
10. Number of pages on which materials by or about female Native American authors appear: 5
11. Number of pages on which materials by or about Native American authors of unknown gender appear: 27

FIGURE **13.1** Partial Sample Coding Sheet from "Reloading the Canon" Study

The Kaiser Foundation (2005) "Sex on TV" study used both scenes and television programs as units of analysis. Scenes were coded as containing sexual material, or not, according to some fairly clear standards (whether it depicted sexual activity, clearly spelled out, contained sexually suggestive behavior, again clearly spelled out, or talked about sexuality). Thus, like Hazell and Clarke (2008), the Kaiser study created a nominal-scale variable at the scene level: whether or not it contained sexual material.

QUALITATIVE CONTENT ANALYSIS

qualitative content analysis, content analysis designed for verbal analysis.

We have discussed the piece by Arthur, Clark, Clark, and Racine, as well as those by Hazell and Clarke (2008) and the Kaiser Foundation (2005), to give you some appreciation for the potential of a relatively quantitative kind of content analysis, a kind that was designed with the statistical analysis of variables in mind. Quantitative content analysis is the most common kind in the social sciences. But we wouldn't want to leave you with the impression that content analysis must be quantitative to be effective (see also Reinharz, 1992). Actually, we think content analysis is well employed when qualitative social scientists aim for primarily verbal, rather than statistical, analysis of various kinds of communication. Roger did—with two students, Joel McCoy and Pamela J. Keller—a **qualitative content analysis** of 20 recent award-winning books for young adults, 10 Newbery medalists or runners-up written by white women about white female protagonists, and 10 Coretta Scott King award winners or runners-up written by black women about black female protagonists (Clark, McCoy, and Keller, 2008). We came to these books with few preconceived ideas but "listened" to them as if we were eavesdropping on white and black mothers talking to their adolescent daughters about life's struggles. We did so not to try to find ways to "quantify the data," as the previous section suggests we might have. Instead, we found ourselves "listening" to themes of struggle and noting patterns.

Compared with the relatively well-accepted techniques (e.g., for quantifying and ensuring the reliability of data) for doing quantitative content analysis, however, guidelines for doing qualitative content analysis are few and far between. (For those interested in such guidelines, however, we recommend Reinharz, 1992; Strauss, 1987: 28ff.) Roger, Joel, and Pamela tried to discern theme patterns by a process of comparison and contrast that would be congenial to the advocates of grounded theory (see also Glaser, 2005; Glaser and Strauss, 1967). Thus, we noticed black literary mothers were much more likely than white literary mothers to bring up the topic of oppression and much more likely, when "talking" about oppression, to locate it at the present, rather than in the past.

Generally speaking, then, quantitative content analyses tend to be of the deductive sort described in Chapter 2—analyses that begin, perhaps, with explicit or implicit hypotheses that the researcher wants to test with data. Explicit in the focal research of this chapter, for instance, is a hypothesis that minority authors will not have achieved substantial representation in American literature anthologies until at least two decades after the New Knowledge Movements that may be said to have encouraged that representation began. Qualitative content analyses, on the other hand, tend to be of the inductive sort, analyses that might begin with research questions, but are then likely to involve observations about individual texts (or portions of texts) and build to empirical generalizations about texts (or portions of texts) in general. Clark, McCoy, and Keller, for instance, began with the general research question "What differences are there in the messages about struggle presented by white and black authors?" and then analyzed the kinds of presentations they discovered as they read, and viewed, such books.

VISUAL ANALYSIS

visual analysis, a set of techniques used to analyze images.

A relatively new set of techniques, closely related to those of content analysis, is **visual analysis.** Visual analysis, a branch of the general field of visual sociology, we would argue, is actually a kind of content analysis. Visual analysts have studied photographs, video images, paintings, drawings, maps, and other kinds of images using a variety of analytic approaches. The earliest sociological research based on visual analysis was probably Erving Goffman's *Gender Advertisements* (1979). Here Goffman argues that advertisers, to create a kind of comfort with their product, will often depict them in social situations that we all might try to emulate in our day-to-day lives:

> Thus, just as a Coca-Cola ad might feature a well-dressed, happy looking family at a posh beach resort, so a real family of modest means and plain dress might step up their level of spending during ten days of summer vacation, indeed, confirming that a self-realizing display is involved by making sure to photograph themselves onstage as a well-dressed family at a posh summer resort. (1979: 27)

Goffman examined magazine advertisements to see what they implied about gender relations in society. A famous example is his assertion that relative size, especially height, in pictures is one way to mark relative power or authority and his observation that, typically, in images associated with products, men are depicted in ways that exaggerate their height advantage over women, much as powerful men are depicted in ways that make them look taller than other men (Goffman, 1979: 28).

This kind of close analysis of images in advertisements also occurs in John Grady's (2007a) study of advertisements depicting blacks in *Life* magazine, from 1936 to 2000. Following procedures typical of qualitative analysis, largely developed after Goffman's work, Grady coded the images, identifying variables as he noted that some photos were easily grouped together and others weren't. He, for instance, found that some advertisements caricatured blacks stereotypically and some did not. And then he discovered that virtually all of the caricaturing advertisements occurred in editions of *Life* published before 1968 and that virtually none of them occurred afterward. This kind of finding led him to the conclusion that white commitment to racial integration, implicit in the advertisements approved by *Life*'s editors and used by companies that advertised in *Life*, grew in the period from 1936 to 2000.

 **STOP & DO**

Why don't you use Google Images to look at the first few advertisements that pop up when you type in "Aunt Jemima pancakes"? Would you be inclined to code these advertisements, using Grady's scheme, as ones that caricature blacks or ones that don't? Why or why not?

Some visual analysts advocate the kind of qualitative approach to images employed by Goffman and, at least in the analysis described in the previous paragraph, by Grady, one that explicitly involves observing the "overtones and subtleties" of images (e.g., Collier, 2001: 39). Some (e.g., Bell, 2001) advocate the value of a kind of quantitative content analysis of pictures. Grady did this kind of analysis when he looked at all the advertisements that appeared in *Life* between 1936 and 2000 and calculated the percentage of all racially identifiable advertisements that contained blacks. (He found that this percentage grew from a very low level, never more than about 2 percent before 1961, to about 12 percent between 1998 and 2000.)

STOP & DO

Why don't you use Google Images again, but this time see what pops up when you type in "Sports Illustrated Ads"? Examine the first page and identify the advertisements that were indeed from "Sports Illustrated." Then count the number of these advertisements that include women or girls and calculate the percentage of all the advertisements on the page that include women or girls. Does this percentage surprise you at all? Why or why not?

Van Leeuwen and Jewitt's *Handbook of Visual Analysis* (2001) is perhaps the best guide to the variety of approaches taken by visual analysts; *Visual Studies* is the journal in which one can find the best examples of its uses by sociologists.

ADVANTAGES AND DISADVANTAGES OF CONTENT ANALYSIS

One great advantage of content analysis is that it can be done with relatively little expenditure of time, money, or person power. Arthur, Clark, Clark, and Racine worked as a team of four on their examination of literature anthologies, all of which they borrowed from local libraries or college professors. They did so partly for the fun of working with one another and partly for the enhanced data reliability that comes from cross-checking one another's observations in various ways. However, their project could have been done by one of them, with no special equipment other than a library card.

Another related plus is that content analysis doesn't require that you get everything right the first time. Frankly, despite appearances to the contrary, the partial coding sheet shown in Figure 13.1 is really a compilation (and distillation) of several separate efforts. Indeed, the authors had made a first pass at several anthologies *before* realizing they wanted to record the total number of pages devoted to male and female authors. They then returned to their subjects (the anthologies) to record that total. This kind of recovery from early oversights is much less feasible in survey or experimental research.

 **STOP & THINK** *Compare this to having surveyed 500 people and then deciding you had another question you wanted to ask them.*

Yet another advantage is that content analysis is unobtrusive; our content analysis itself can hardly be accused of having affected the content of the anthologies we read. Nor can the Kaiser Foundation's (2005) study of "Sex on TV" be criticized for having affected the sexual content of the television programming it examined. This is not to say that content analysis might not affect the content of subsequent communications. In fact, we believe that something like a content analysis of available anthologies did propel Paul Lauter and his colleagues (e.g. Lauter, 1983a, 1983b) to envision the more inclusive anthology that became their first *Heath Anthology* in 1990. And, when the Kaiser Foundation introduces its study by referring to the "a third (34 percent) of young women still become pregnant at least once before they reach the age of 20—about 820,000 a year" (2005: 2), it seems to be encouraging a diminution of the number of sexual messages on television.

 **STOP & THINK** *Can you imagine whose attitudes would need to be affected so that the "Sex on TV" study would have its desired effect?*

Like using other kinds of available data, content analysis permits the examination of times (and peoples) past—something surveys, experiments, and interviews don't do so easily. Arthur, Clark, Clark, and Racine's analysis of anthologies from the 1950s to the 2000s is a good example, as are Shapiro and Dawson's (1998) examination of French revolutionaries' notebooks and

Sheldon Olson's (1976) use of the Code of Hammurabi, a list of laws and punishments from ancient Babylonia.

 **STOP & THINK** *The Code of Hammurabi consists of 282 paragraphs, each specifying an offense and its punishment. One could code each of the paragraphs by the severity of the punishment associated with an offense, for instance, to gain a sense of what kinds of behavior the Babylonians most abhorred. Based on the punishments associated with stealing from a temple and striking another man, mentioned in the following paragraphs, which of these offenses was held in greater contempt?*

Paragraph 6. If a man steal the property of a god (temple) or palace, that man shall be put to death.

Paragraph 204. If a freeman strike a freeman, he shall pay ten shekels of silver.

One of Emily's future projects is an analysis of life expectancy, family structure, and gender roles in previous centuries by doing content analysis of the inscriptions on head stones in New England cemeteries.

Yet another advantage of content analysis is that, because it typically permits one to avoid interaction with human subjects, it also permits the avoidance of the usual ethical dilemmas associated with research involving human subjects and with the time involved in submitting a proposal to an Institutional Review Board, unless one plans to use private documents.

Some disadvantages of content analysis might have occurred to you by now. Because content analysis is usually only applied to recorded communication, it can't very well be used to study communities that don't leave (or haven't left) records. Thus, we can study the legal codes of ancient Babylonia because we've discovered the Code of Hammurabi, but we can't study the codes of societies for which no such discoveries have been made. And Arthur, Clark, Clark, and Racine can content-analyze literature anthologies with some sense that what college students read might have predictable consequences for their attitudes. But to tap those attitudes directly, it would be more valid to question or observe such students (perhaps as Lang and Podeschi in Chapter 9's focal research and Mueller, Dirks, and Houts Picca did for the focal research in Chapter 11). Content analysis can also involve social class biases because communications are more likely to be formulated by educated persons.

There are also questions about the validity of content-analytic measures. For example, is the invisibility of female and other minority authors in 1950s and 1960s anthologies really a valid indicator of attitudes about women and minorities? Perhaps, but maybe not. And even if one were pretty sure that the invisibility was a valid measure, whose attitudes are being measured: those of the editors and publishers, or the faculty users and student readers?

 **STOP & THINK** *Emily and a colleague thought they'd found evidence of significant social change when they content-analyzed 20 years of the "Confidential Chat" column (that published reader contributions) in the Boston Globe and*

found the column increasingly dealing with women's work outside the home between 1950 and 1970. They were discouraged when a reviewer of their research report for a scholarly journal suggested that their results might reflect changes in the editorial climate at the newspaper, rather than changes in the attitudes of the readership. When they interviewed the editor of the column, they found, indeed, he and the former editor published only about 10 percent of the letters he received and that changes in content were likely to reflect his sense of what his readership (mostly women) wanted to read. Emily and her co-author never resubmitted their paper because they now felt it failed to say anything significant about social change. What do you think of this decision?

Finally, content analysis doesn't always encourage a sensitivity to context that, say, literary criticism often does. The percentage of pages devoted to females isn't necessarily the best measure of the visibility of female authors in anthologies if, for instance, female authors do more with genres, like poetry, whose products tend to be shorter than the genres (e.g., novels) men engage. Questions of validity plague those who do content analysis just as much as they plague those who employ other methods of data collection.

 **STOP & THINK** *In the larger study from which the focal research of this chapter was taken, Arthur, Clark, Clark, and Racine used an additional measure of minority visibility: the percentage of all authors that were from a particular minority group (say, women). By this measure, all minority groups appeared to be considerably more visible in the 1990s and 2000s anthologies than they did when the percentage of pages measure was used. Do you think the percentage of all authors or the percentage of all pages is a more valid measure of minority visibility?*

Validity concerns are less likely to be legitimate in the case of qualitative content, and visual analyses, where the researchers focus on things like context and author (artist) intention and the verbal analysis is more likely to take such things into account. Thus, a project that focused on the research question, "What kinds of struggle does the author emphasize?" is more likely to identify such struggles validly than one that begins with the hypothesis "Books about male characters are more likely to be about personal struggle than books about female characters."

SUMMARY

Content analysis (as well as its offspring, visual analysis) is an unobtrusive, inexpensive, relatively easy, and flexible method of data collection that involves the systematic and—typically, but not necessarily—quantitative study of one kind of available data—some form of communication. It, like other forms of available data analysis, provides a way to study the past. Content analysis can be used to test hypotheses about communication, to compare the content of such communication with some standard, or to describe trends in communication, among other things. Arthur, Clark, Clark, and Racine simultaneously looked at trends in the presentation of minority authors in American literature anthologies and used these trends to evaluate competing hypotheses.

An early step in quantitative content analysis is to identify a unit of analysis. This is the unit about which information is collected and can be words, phrases, or paragraphs; stories or chapters; themes; photographs; illustrations; authors; audiences; or almost anything associated with communication. In Arthur, Clark, Clark, and Racine's study, the unit of analysis was the literature anthology.

The quality of a content analysis depends on the creation and meaningful usage of variables for classifying units of analysis. Arthur, Clark, Clark, and Racine used the percentage of pages devoted to all authors devoted to minority authors of different sorts as their main way to determine the degree to which anthologies made minority authors visible. Explicit and reliable coding schemes for recoding variation are most important. Checks on measurement reliability can be pretty simple but are always appropriate. Arthur, Clark, Clark, and Racine used statistics to check the reliability of their classification efforts.

On the down side, content analysis cannot tell us much about communities that have left no recorded communications. Moreover, content analysis is frequently criticized on validity grounds, mainly because those criticizing aren't sure whether the content analyzers are measuring what they think they are. Validity concerns are less well founded, however, in the case of qualitative content and qualitative visual analyses.

We've now covered the major methods of data collection in the social sciences: questionnaires, interviews, observations, and available data (including content analysis). For a summary of the advantages and disadvantages of each method, see Appendix A. For a discussion of how using more than one method can enhance the validity of findings, see Appendix B.

EXERCISE 13.1

How Newsworthy Is the News?

This exercise offers a chance to observe, and speculate about, differences in the treatment of a given news story in daily newspapers. Your research questions in this exercise is this: Does the treatment of a given news story vary from one newspaper to another and, if so, what might account for this variation?

1. Select a current national or international news story. What story did you choose?
2. Select four national, regional, or local newspapers (from either a newsstand or your library). Get copies of these papers on the same day. List the names of the newspapers and the date.

Date:
Newspaper 1:
Newspaper 2:
Newspaper 3:
Newspaper 4:

3. Decide how you will measure the importance given a story in a newspaper. (Hint: You might consider one of the following options or combinations thereof: What page does the story appear on? Where does the story appear on the page? How big is the typeface used in the headline? How many words or paragraphs are devoted to the story? Is it illustrated? What percentage of the total words or paragraphs in the paper is devoted

to it? Does the newspaper devote an editorial to the story?) Describe the measure you will use to determine the relative importance given your story in each of the papers.

4. Report your findings. Which paper gave the story the greatest play, by your measures?

Which paper, the least? Interpret your findings by trying to account for differences in terms of characteristics of the papers you've examined. Please submit copies of the stories with the assignment.

EXERCISE 13.2

Analyzing Website Commentaries on School Shootings

This exercise adapts an idea from Professor Molly Dragiewicz of George Mason University to give you the chance to analyze website commentaries on school shootings in the United States.

1. Do a web search using a search engine or a subject guide such as Google or Yahoo, using the terms "school shootings" and "commentary." Find four commentaries about school shootings that have taken place in the past decade and print them out.

2. Your research questions are the following:
   a. Do website commentaries tend to focus on the perpetrators or the victims of school shootings?
   b. Do these commentaries tend to focus most on gender, race, or class in their interpretations of shootings in American suburban schools?

3. Prepare four code sheets like this:
   Commentary: Author: Publication date:

   a. Would you say that the commentary focuses more on the perpetrators or the victims of school shootings?
   b. If the commentary focuses on the perpetrators, in your view, does it focus on the race, gender, or class of these perpetrators? (Yes or no)
   c. If so, which of these characteristics seem most important to the author?
   d. If commentary focuses on the victims, in your view, does it focus on the race, gender, or class of these victims? (Yes or no)
   e. If so, which of these characteristics seem most important to the author?

4. Write a paragraph summarizing your findings. (You might summarize, for instance, whether the commentaries seem more concerned with the perpetrators or the victims of the shootings. Then you might summarize whether race, gender, or class seems to be the "lens" through which the commentaries tend to be made.)

5. Include the copies of the commentaries as an appendix to the exercise.

EXERCISE 13.3

The Utopianism of Television Commercials

It has been said that commercials for children's television programs are utopian—that they tap into utopian sentiments for something different, something better; that they express utopian values of energy, abundance, and community (see, for example, Ellen Seiter, 1993: 133)—rather than extolling the virtues of things as they are. But are commercials for children's programs really all that different in this regard from those for adult

programs? This is your research question for this exercise.

1. Select a children's television show and an adults' television show on commercial stations. (Hint: Saturday mornings are particularly fruitful times to look for children's commercial television programs.) Pick shows of at least 30 minutes in duration (preferably 1 hour).

2. Indicate the shows you've chosen and the times you watched them.

   Children's show:
   Time watched:
   Adults' show:
   Time watched:

3. Describe the measures you will use to determine whether a commercial appealed to utopian sentiments or values. (For example, you might ask whether the commercial makes an explicit appeal to the viewer's desire for a better life [for instance, by showing an initially depressed-looking character fulfilled by the discovery of a soft drink or an initially bored breakfaster enlivened by the acquisition of a new cereal] or if it makes implicit appeals to, say, the viewer's wish for energy [is there dancing or singing associated with a product?], abundance [is anyone shown to be constrained by considerations of price?], or community [do folks seem to get along better after acquiring brand X?].) Show an example of the code sheet you will use, including a space for the product advertised in the commercial and a place for you to check if you observe the presence of what it is you're using to measure an appeal to utopian sentiments or values.

4. Use your coding scheme to analyze the first five commercials you see in both the children's and the adults' show.

5. Report your findings. In which kind of programming (for children or adults) did you find a higher percentage of commercials appealing to utopian sentiments or values? Interpret your findings.

EXERCISE 13.4

Gender among the Newberys: An Internet Exercise

The American Library Association (ALA) awards the Newbery medal for adolescent fiction. For this exercise, you'll want to use the ALA website (www.ala.org) to answer questions about the gender of characters in Newbery winners and honor books over the last decade. (Hint: One way of finding such a list, after getting on the ALA site, is to select the "search website" option, type "Newbery" at the request line, request information about one of the recent winners, then request the list of other winners, and honor books at the bottom of that page.)

1. Find a list of Newbery award winners and honor books for the last 10 years.

2. Begin your content analysis of the titles of these books by listing the titles that suggest to you that there is a female character in the book.

3. Now list those books whose titles suggest to you that there is a male character in the book. (The two lists might overlap. For instance, the title *Lizzie Bright and the Buckminster Boy*, an earlier title, suggests to us that there are at least two characters in the book, one male and one female. It would, therefore, make both your lists, if it fell into the period of your study.)

4. Does your brief content analysis of recent Newbery award and honor books suggest that more attention has been paid to males or females?

EXERCISE 13.5

Gender Advertisements

This exercise gives you an opportunity to practice visual analysis ala Erving Goffman's *Gender Advertisements* (1979). Your job will be to find four magazine advertisements that seem to tell a story using human models and to identify ways in which the advertisement displays "ideal" male and/or female behavior or traits.

1. Find four advertisements that seem to tell stories or portray relationships in magazines

using human models that you can get easily from magazines or from Google Images by using key words, "magazine ads." Cut out the advertisements or print them out and display them on a sheet or sheets of paper that you'll turn in with the exercise. Label the four advertisements with numbers: 1, 2, 3, and 4.

2. For each advertisement, briefly (in a sentence or two) tell the story of the advertisement as you see it.
3. Now, describe ways in which the model or models in each advertisement are posed to create an image of the "ideal" male and/or female.

CHAPTER 14

Applied Social Research

© Kosuke Okahara / Agence VU / Aurora Photos

INTRODUCTION

applied research,
research intended to be useful in the immediate future and to suggest action or increase effectiveness in some area.

By now you've read quite a bit of social research. Our hope is that you found at least some of the studies interesting. You also might have found some of them focused on topics of interest to you personally or professionally. Perhaps you've even thought that one or two of the studies could be of use to you in a current or future work setting. If so, then you've probably identified the kind of research we'll be considering in this chapter—**applied research**.

We've explored a variety of topics in this text: creating research questions and hypotheses, considering ethical and practical issues, selecting populations, sampling strategies and research designs, and choosing one or more methods of data collection. Before covering data analysis, we want to expand on our discussion of applied social science research.

Basic and Applied Research

basic research, research designed to add to our fundamental understanding and knowledge of the social world regardless of practical or immediate implications.

All social research is designed to increase our understanding of human behavior and can be useful to individuals, groups, or the whole society, but some work is more *immediately* useful than other research. As we noted in Chapter 1, **basic research** seeks to create new knowledge for the sake of that knowledge, whereas applied research is designed to help solve immediate social problems and answer broad research questions about settings in complex social and political environments. Although practical applications can usually be derived from basic research, those projects are designed primarily to provide greater understanding of our social world, to describe individuals or groups, to see relationships between variables, to develop or test theories, and the like. Applied work has a practical focus and is designed to provide organizations such as schools, legislatures, communities, social service agencies, health care institutions, and other organizations with information that can be used in the immediate future. Purity of theory is much less important in applied research; utility of theory is what counts as researchers look for concepts, variables, and theories that will help solve a problem (Bickman and Rog, 2009: xi). A great deal of research can be seen as falling somewhere between the poles of basic and applied research; identifying a study's goal or emphasis is helpful in determining where a project fits on the continuum.

Social Problems and Social Solutions

In the twentieth century in the United States, awareness of social problems and programs directed toward solving them became more common. The efforts before World War I focused on literacy, occupational training, and controlling infectious diseases, with projects designed to determine if funding initiatives in these areas were worthwhile (Rossi and Freeman, 1993: 9). Through World War II and into the 1960s, federal human service projects in health, education, housing, income maintenance, and criminal justice became widespread, as did the efforts to demonstrate their effectiveness.

By the end of the 1970s, many interested parties started questioning the continued expansion of government programs, not only because of changes in ideology, but also because of disenchantment with the outcomes and implementation of many of the programs advocated by public officials, planners, and politicians (Rossi and Freeman, 1993: 24). Since both old and newly identified social problems remained, despite disappointments with earlier efforts, programs continued to be created and implemented. With many millions of program dollars spent on local, state, and federal levels, policy makers wanted answers: Are the programs effective? Should they be continued? Modified or eliminated entirely? Applied research is a way to answer these questions.

 **STOP & THINK** *What do you think are the most serious social problems that we face as a society today? Unemployment? Violence in our schools and communities? Access to health care? Poverty? Racism? AIDS? Obesity? Teenage pregnancy? Homelessness? Drug and alcohol abuse? Are you aware of programs nationally or in your state or local community designed to work toward solving any of these problems? If you are familiar with a program, do you think it is effective? Are those who are served by such programs involved in deciding if they are useful?*

You may believe that all the problems mentioned warrant attention. In fact, there have been programs addressing each of them. Sometimes we read or hear about efforts to improve things, but many such efforts are ignored by the media. We rarely learn whether programs and policies designed to address social problems actually work; even more rarely do we hear how those affected by the programs and policies feel about the situation or what their concerns are. With pressure for *effective* and *accountable* use of public dollars, recent years have brought a growing interest in evidence-based programs, which are easily disseminated programs with rigorous evaluations that have demonstrated their efficacy (Small, Cooney, and O'Connor, 2009).

EVALUATION RESEARCH

evaluation research, research specifically designed to assess the impacts of programs, policies, or legal changes.

There are several kinds of applied research. One of the best known is **evaluation research**—research specifically designed to assess the impacts of programs, policies, or legal changes. Most often the focus of an evaluation is whether a program, policy, or law has succeeded in effecting intentional or planned change.

Evaluation research is not a unique research *method*. Rather, it is research with a specific *purpose*. Rossi and Freeman call evaluation research "a political and managerial activity, an input into the complex mosaic from which emerge policy decisions and allocations for the planning, design, implementation, and continuance of programs to better the human condition" (1993: 15).

Outcome Evaluation

outcome evaluation, research that is designed to "sum up" the effects of a program, policy, or law in accomplishing the goal or intent of the program, policy, or law.

The most common evaluation research is **outcome evaluation**, which is also called impact or summative analysis. This kind of evaluation seeks to estimate the effects of a treatment, program, law, or policy and thereby determine its utility. These research projects typically begin with the question "Does the program accomplish its goals?" or the hypothesis "The program or intervention (independent variable) has a positive effect on the program's objectives or planned outcome (dependent variable)." The researcher first determines *what* the goal of the program is and *how* the program was implemented and designs a project useful to practitioners and policy makers. At the end of an outcome evaluation, the researcher should be able to answer questions about a program's success and to speculate about how to improve it.

A wide variety of programs, policies, or laws can be evaluated—ones that serve many or few people; last for days, weeks, or years; have a large or limited geographic target area; and have broad or narrow goals. Examples of the many hundreds of research questions that have guided outcome evaluations in the past decade include:

1. Do batterer programs lead to a decrease in domestic violence? (Labriola, Rempel, and Davis, 2008)
2. Does the Experience Corp program, which brings older adults into public elementary schools to tutor and mentor at-risk children, actually improve student academic outcomes? (Morrow-Howell, Jonson-Reid, McCrary, Lee, and Spitznagel, 2009)
3. Do marriage education programs originally designed for white, middle-class, well-educated couples strengthen couple relationships in disadvantaged populations? (Dion, 2005)

While an outcome evaluation can tell us what works, it is as important to know what *doesn't* work. Reynolds, Mathieson, and Topitzes (2009) looked at 15 studies of 14 programs designed to reduce rates of child abuse and neglect to determine the extent to which early childhood interventions actually prevent child maltreatment. In each of the evaluations, the program was the independent variable and child abuse or maltreatment was the dependent variable. Most used an experimental design with a control group and an experimental group that took part in the program. Official reports of maltreatment from child protective services or the courts were used as indicators of the dependent variable. Review of the outcome evaluations leads Reynolds et al. (2009: 200) to conclude program success is not typical. "Of the 12 intervention models investigated, only 4 or one third reported that program participants had significantly lower rates of maltreatment than comparison groups.... Although 5 studies reported reductions in either substantiated or parent reported maltreatment, only for 3 programs was there consistent evidence of enduring effects." An outcome evaluation of one of the successful programs is discussed in Box 14.1.

Cost-Benefit Analysis

cost-benefit analysis, research that compares a program's costs to its benefits.

cost-effectiveness analysis, comparisons of program costs in delivering desired benefits based on the assumption that the outcome is desirable.

Even if a program has accomplished its goals, there is always the issue of cost and whether it might be better to expend the resources consumed by the program in other ways. An evaluation which considers that issue is a **cost-benefit analysis**, a study designed to weigh all expenses of a program (its costs) against the monetary estimates of the program's benefits (even putting dollar values on intangible benefits), or a **cost-effectiveness analysis**, which estimates the approach that will deliver a desired benefit most effectively (at the lowest

BOX **14.1**

Outcome Evaluation of a School-Based Intervention Program

The Chicago public schools administer Child Parent Centers (CPCs), a center-based early intervention program. Since 1967, this program has provided comprehensive educational and family support services for low income children beginning at age three through early elementary school. There have been many evaluations of the program that have measured success on different outcomes. One question evaluators addressed was whether CPC program participation was associated with lower rates of reported child maltreatment (Reynolds and Robertson, 2003). The evaluation followed a cohort of 1,539 low income minority children born in 1980 who were in either the intervention program or an alternative program. The intervention group began with 989 children who were in a CPC program (preschool, kindergarten, and child support services) in one of the 20 CPCs in the city while the matched comparison group was composed of 550 children who participated in a full-day kindergarten program for low income families, most of whom lived in areas of the city that were not served by a CPC program (Reynolds and Robertson, 2003). The evaluation researchers found that the program and comparison groups were similar on many child and family characteristics in the beginning of the project but differed in the outcome variable of levels of reported child maltreatment through age 17, using child protective services records and court records as data. After controlling for the influences of preprogram maltreatment, family risk, and other explanatory factors, the researchers found CPC participants had

"a 5.5% lower rate of child maltreatment than the comparison group (5% versus 10.5%).... This translates into a 52% reduction in the rate of child maltreatment associated with [program] participation" (Reynolds and Robertson, 2003: 13–14).

If we draw a diagram of the evaluation of the CPC program using the study design models presented in Chapters 7 and 8, you can see that the researchers are using a quasi-experimental design. Not being able to control who enters the intervention program and who attends an alternative program, they are unable to have a true experimental and control group. However, they used comparison group that matches the experimental group in most ways and they followed the two groups over time. The results give support to the explanatory hypothesis that the Chicago CPC program decreases child maltreatment.

Other evaluations of the program have investigated different outcomes by specifying different dependent variables. These evaluations indicate that the program has had that other important effects for the children who participated when compared to those in the comparison group. Among these outcomes for the intervention group are significantly lower rates of juvenile arrest for both violent and nonviolent offenses, lower rates of needing school remedial services, significantly higher rates of high school completion and four-year college attendance and healthier adult lifestyles (Reynolds et al., 2007; Topitzes, Godes, Mersky, Ceglarek, and Reynolds, 2009).

Groups	Time 1 (1985–1986)	Introduction of independent variable	Time 2 (14 years later)
Experimental group	Measures family and social characteristics	Children attend a CPC program for preschool and kindergarten that includes family support services	5% parental maltreatment rate
Comparison group	Measures family and social characteristics	Children attend other programs	10.5% parental maltreatment rate

cost) without considering the outcome in economic terms. Comparing program costs and benefits can be helpful to policy makers, funding agencies, and program administrators in allocating resources. With large public expenditures on health care, correctional facilities, social welfare programs, and education in the United States, saving even small percentages of costs add up.

Typical questions asked in cost-benefit and cost-effectiveness analyses are these: How effective is the program? How expensive is it? Is it worth doing? How does this program compare with alternative programs? These analyses can compare outcomes for program participants with those not in programs or with those in alternative programs. For example, a cost-benefit analysis of Minnesota's Challenge Incarceration Program (CIP), which is an intensive, structured, and disciplined program for offenders (commonly called "boot camp"), examined whether the program lowered recidivism (effectiveness) and saved money (cost) (Duwe and Kerschner, 2008). The outcome or dependent variable, recidivism, was operationalized as a rearrest, a felony reconviction, and a return to prison for a new criminal offense or a technical violation. The rearrest, felony reconviction, and reimprisonment rates were lower for CIP offenders compared to those in the control group, although the two groups returned to prison for all reasons (including technical violations) at about the same rate (Duwe and Kerschner, 2008: 627). The evaluators judge this boot camp program to be cost effective because when the program's offenders returned to prison, they stayed 40 fewer days than control group offenders because they were less likely to return for a new crime. Overall, considering the per diem costs of prison, the evaluators estimate that over the life of the program, it saved Minnesota at least $6.2 million by providing early release to program graduates and reducing the time the participants later spend in prison (Duwe and Kerschner, 2008).

In another cost-benefit analysis, evaluators of the Chicago Child-Parent Center (CPC) programs looked at the possible outcomes to low income youth who did not attend the program. Considering the cost of attending the Center programs versus the expenditures for things like remedial education, criminal justice costs, economic well-being, and tax revenues, they conclude that being in the program at least from preschool through kindergarten provides a return to society of $7.14 per dollar invested (Reynolds, Temple, Robertson, and Mann, 2002).

 **STOP & THINK** *Once a program is in place, it can be evaluated. But what about situations that might be problematic but for which there are no programs? What might be done to identify a problem and suggest solutions?*

stakeholders, people or groups that participate in or are affected by a program or its evaluation, such as funding agencies, policy makers, sponsors, program staff, and program participants.

Needs Assessments and Other Kinds of Evaluations

The tasks an evaluator sets out to accomplish are determined by the stage of the program (whether it is a new, innovative program or an ongoing one) and the needs and interests of the **stakeholders** (those involved in or affected by the program in some way). Before a program or project is designed, it's important to determine the needs of a geographic area or a specific population. This kind of

needs assessment, an analysis of whether a problem exists, its severity, and an estimate of essential services.

evaluation is a **needs assessment**, an analysis of the existence and severity of a problem and the number of people affected. It may also include an appraisal of existing resources and suggestions for new services or programs that could ameliorate the problem. Needs assessments can focus on a wide variety of concerns—educational, health, housing, and the like—and a variety of populations—students, the elderly, the incarcerated, and the like. Assessments can be designed by those within a community or by those who are outside of it.

After Hurricane Katrina, for example, researchers based in Mississippi did a needs assessment in East Biloxi, Mississippi, a community that had been heavily damaged by the hurricane. Based on their survey, Kleiner, Green, and Nylander (2007) concluded that repair to damaged housing and the building of new housing were among the most important needs of the community as few of the respondents had flood insurance and only about half had home owner's or renter's insurance at the time of the hurricane. The researchers were able to report on the needs of residents, facilitating the provision of services by community agencies (Kleiner et al., 2007).

Two researchers, Jill Harrison and Maureen Norton-Hawk, conducted a needs assessment far from home. In the focal research piece that follows, you'll read about their study of incarcerated women in Ecuador and the many inmate needs that their research documented. Jill Harrison had become familiar with Ecuador's prisons when, living and working in Bogota, Colombia, she visited Ecuador and was arrested under the ruse of being an illegal alien. She remembers having no money in her pocket to buy a container of water, no food to eat, and having to read by candlelight even during the day. Her personal history was one of the reasons she was interested in studying the current conditions in the women's prisons (Harrison, personal communication).

FOCAL RESEARCH

A Needs Assessment of Incarcerated Women in Ecuador

By Jill Harrison and Maureen Norton-Hawk*

The relationships between abuse, poverty, drugs, incarceration, and the inequitable impact on women have been well documented. We conducted a needs assessment of the inmates in Ecuador's largest women's prison. This exploratory research indicates that gendered injustice extends well beyond American borders.

Incarceration of women in Ecuador has increased dramatically since this small South American country adopted U.S. foreign policy whereby favorable trade relations and military support increases as Ecuador cracks down on the distribution and possession of drugs

*Jill Harrison is an Assistant Professor of Sociology at Rhode Island College. Maureen Norton-Hawk is an Associate Professor of Sociology at Suffolk University. This article is published with permission.

(Nuñez, 2006; Merolla, 2008). Most favored nation trading status with the United States is an incentive for a partnership in the War on Drugs, and such laws do not differentiate between large scale traffickers and low level carriers or "mules," who are mostly women (Preston & Roots, 2004; Merolla, 2008). In 2001, 2,310 people were incarcerated for drug violations in Ecuador, and women's incarceration doubled for drug offenses from 376 in 2001 to 840 in 2004 (Boletin Estadistico, 2001–2004). The women are relatively easy to convict because they are generally caught "red-handed" and have no money to mount a legal defense. Whether the carrier has two ounces of marijuana or two kilos of unprocessed coca, the minimum mandatory sentence is between 6 and 8 years (Associated Press International, 2008).

Method

El Centro de Rehabilitación Social, colloquially called el Inca, the largest women's prison in Ecuador was selected as the primary research site. Maureen Norton-Hawk had visited incarcerated women at el Inca frequently over the past 5 years and both researchers visited several times in 2007 and 2008. Formal IRB approval was secured from Suffolk University in Boston, Massachusetts. As there is no formal review process for researchers at el Inca, informal approval was granted by the Comite de las Mujeres, an inmate group that negotiates donations, and the prison director.

The Comite de las Mujeres "advertised" our research the day before we arrived for data collection. Inmates are generally free to move about the prison between 7:00 AM and 10:00 PM and were invited to come to a classroom in the morning to be part of the study. We talked about the purpose of the research and explained the concept of informed consent. As this was a normal visiting day, those without visitors were more likely to hear about the study. Each woman was provided with a small gift: a plastic bag filled with a roll of toilet paper, tampons, shampoo, conditioner, a bar of soap, and 10 hard candies. They were free to leave at that point, although none did. We went over informed consent with each woman individually and offered the option of not signing the form in order to protect confidentiality if papers were seized by guards, committee members or the director. Most of the women chose the option of signing with an X rather than a signature while a researcher served as the witness.

A convenience sample of 50 female inmates, or approximately 15% of the total prison population in December 2008, completed a needs assessment questionnaire anonymously. Ten women had the questions read to them by a researcher who also recorded their answers because they could not read, had poor eyesight or did not have corrective eye glasses. All the women surveyed were incarcerated on drug charges and were serving an average sentence of 6.5 years. Most were under age 30, mothers of small children, with limited access to education and were unemployed when arrested.

The women were asked to assess their housing, educational needs, medical treatment, access to legal counsel, visitation rights, and nutritional needs during incarceration. The needs assessment survey also asked them about their children and how they were being cared for either inside or outside the prison.

Results

Our findings indicate that the women experience hardships during their incarceration that would be unconstitutional in the United States. As Table 14.1 indicates, on almost every measure, from visitation to violence, medical care, legal assistance to the quantity and quality of food, the vast majority of women reported that their basic needs were not being met.

TABLE **14.1**
Distribution of Selected Questions from Needs Assessment
Survey (n = 50)

Received some legal assistance	64%
Thought assistance was helpful	50%
No lawyer	40%
Family Visitation: Yes	69%
No family visitation	31%
Average frequency of visits	Weekly
Procedures for visitation satisfactory	40%
Quantity of prison food satisfactory	50%
Quality of prison food satisfactory	16%
Job training in prison satisfactory	48%
Health care received in prison	
Medical	20%
Obstetric/gynecological	12%
Other	10%
Dental	10%
No medical assistance reported	22%
No response	26%
Reporting fear for personal safety in prison	76%

Many women complained that the one meal per day supplied by the prison was simply not enough to eat. Most women worked inside the prison and/or relied on family members or charitable organizations to get supplemental food. Supplementing prison rations is expected and encouraged by prison administration, so typically bags of groceries arrive with visitors. It is not uncommon for illegal drugs or other contraband, such as cell phones, to be smuggled in with the food. Some women prostituted themselves during visiting days while others sold their crafts, such as dolls, cards, and jewelry.

The women reported that they provide their own clothes, and if family members or other inmates cannot assist them, the poorest of women wear the clothes they had on their backs at the time of their arrest. A hierarchy develops among the inmates based on quantity and quality of supplies each woman is able to obtain from the outside. Power is a function of those who have enough or more than enough to get by.

Medical care is the number one service the women want to see improved. Just over 50% of the women said they received any prison medical services at all, and, for those that reported receiving medical services, the most common was for either obstetric or gynecological care. No preventative care is available and mental health services do not exist. Even inmates with severe health problems, such as HIV-AIDS and advanced diabetes, receive no treatment.

Seventy-six percent of women often felt afraid for their safety while in prison as inmates and visitors have free access within the prison and prison guards do not monitor inmate interactions or patrol cell blocks. Drugs are reported to be widely available inside the prison and are used by visitors and inmates alike. Over 50% of the women reported that they wished the guards had more training in order to deal with the violence they are confronted with periodically from guests or other inmates.

We find the conditions in the prison problematic for the overall custody and care of inmates that typically spans 6 to 8 years. Every woman we encountered at el Inca recounted very difficult experiences before and during incarceration, including poverty, isolation, abuse, sexual assault, rape, and in some cases, torture. Both long standing trauma and chronic anxiety fuel women's inability to resist similar mistreatment by others and may be a dominant factor in recidivism.

Conclusion

The U.S. "War on Drugs" has an impact on Ecuador. Curbing illegal employment in the drug trade is a hard sell where poverty, corruption, and deep political, economic, and social unrest leave many women struggling for survival (Sherret, 2005). Patriarchal structures within the culture exacerbate this problem in which dependent, naïve, poor, young women often feel as though they have little choice but participate and know that they must endure or suffer the consequences (Alvirez, Bean, & Williams, 1981). At the same time, their participation is vilified by the police and they easily become targets for mistreatment, neglect, and violence at all levels of the drug trade (Epele,

2003). The women feel they are easy prey for the police who use the arrest data to argue that they are doing their part to stop the flow of drugs into the United States.

Our assessment points to the need for large scale policy change as well as for more immediate programs for the incarcerated women of el Inca. We will be sharing our assessment with the prison director and Marcha Blanca, a leading Ecuadorian quasi-governmental organization that works with national security issues that include a focus on incarcerated women and their children. We will seek external funding to develop cottage industries for incarcerated and paroled women, as well as to facilitate other programs related to health care and employment. For example, a group of volunteers has organized a women's cooperative of recently released inmates to make reusable grocery bags. We will also help those in Ecuador to evaluate the long term impact of the U.S. War on Drugs, especially as it affects the women and the children of the incarcerated. Our efforts along with the work of local organizations may begin the process of long term social change.

References

Associated Press International, July 7, 2008.

Alvirez, D., Bean, F., & Williams, D. 1981. The Mexican American Family. In C. Mindel and R. Habenstein, eds., *Ethnic Families in America*. New York: Elsevier.

Boletin Estadistico El Sistem Penitenciario Ecuatoriano. 2001–2004. Direccion Nacional de Rehabilitacion Sociale (DNRS).

Epele, Maria E. 2003. Changing Cocaine-Use Practices. *Substance Use and Misuse*. 38:9, 1189–2126.

Merolla, D. 2008. The War on Drugs and the Gender Gap in Arrests: A Critical Perspective. *Critical Sociology*. 34:2, 255–270.

Nuñez, Jorge. 2006. Caceria de Brujos Drogas Ilegales y Sistema de Carceles en Ecuador Quito Ecuador: Digitale Abya-Yala.

Preston, F.W. & Roots, R.I. 2004. Law & Its Unintended Consequences. *American Behavioral Scientist*. 47:11, 1371–1376.

Sherret, Laurel. 2005. Futility in Action: Coca Fumigation in Colombia. *Journal of Drug Issues*. 35:1.

 **THINKING ABOUT ETHICS**

While prisons in the United States have their own internal review boards to protect inmates against ethical violations, no such review boards exist in Ecuador. For this reason, the researchers sought and received Institutional Review Board (IRB) approval from Suffolk University, Norton-Hawk's home institution. Of special concern to the IRB was that the gift the researchers intended to give each inmate could be construed as coercive in nature if the gift was too large to refuse and would make the inmate feel obligated to participate. The researchers satisfied the IRB concern by giving a small bag of necessities valued at under $3.00 as a token of appreciation for participation.

After collecting and analyzing the data, Harrison and Norton-Hawk wrote their needs assessment report describing the current situation and proposing some changes in the area of health care and employment. In June 2009, Jill Harrison met with the director of the prison and also with the director of Marcha Blanca, the quasi-governmental organization that works with national issues including the concerns of incarcerated women. Both were very pleased to have the data identifying the more immediate and long-term needs of the inmates (Harrison, personal communication). Identifying medical care as a primary issue led Harrison to contact Brown University's Alpert School of Medicine. As a result of her efforts, Brown University is now planning a conference on HIV prevention in Ecuador and a program for their medical students to do clerkships in el Inca and other public health facilities (Harrison, personal communication). This needs assessment is likely to lead to some very successful outcomes in the future.

While sociologists can work far from home, they can do important work much closer to home. Box 14.2 describes Kristy Maher's view of practicing applied sociology in her local community of Greenville, North Carolina.

formative analysis, evaluation research focused on the design or early implementation stages of a program or policy.

If a new program is funded and there is time in the early stages of design or implementation to make improvements, program staff and developers can benefit from a **formative analysis**. This kind of analysis occurs when an evaluator involved in the beginning stages of program development and implementation provides program staff with an evaluation of the current program and suggestions for improving the program design and service delivery. Suppose, for example, a high school principal sought assistance with a new anti-bullying program. A formative analysis would carefully review the program's goals and its current instructional materials (including comparisons of the program with programs and materials used in other school systems) and collect data on the ongoing program. Administrators, teachers, and students might be interviewed, teacher training and program sessions might be observed, and

---

BOX **14.2**

Practicing Sociology as a Vocation

Kristy Maher believes in practicing sociology in a way that uses her knowledge and skills to effect a positive change in her local community of Greenville, North Carolina. Her first project was a needs assessment of the health care needs of the local Hispanic population and her second was an outcome evaluation of a prenatal outreach program for high-risk pregnancies. The first resulted in the hospital system's hiring full-time translators for the Hispanic community and the second was helpful in securing continuing funding for the successful program. Since then she has worked on a variety of projects, including serving as the medical sociologist in a community health care assessment and as an evaluator for a program aimed at providing quality of health care access for children in low income schools. Maher (2005) says that she believes in practicing sociology as a *vocation*, which means using her expertise as a medical sociologist, a researcher, and a statistician to better her community.

the first students involved in the program might be asked to complete a questionnaire or to participate in focus groups. After analysis of the data, evaluators should be able to offer advice, including suggestions for staff development, for modifying instructional materials or techniques, and for making adjustments in program management.

process evaluation, research that monitors a program or policy to determine if it is implemented as designed.

A related kind of evaluation is **process evaluation,** which is research to determine if a program is implemented as designed and if modifications, including alternative procedures, would be useful. If there are variations in program delivery in different sites or for different recipients, then the program and its outcome can vary. It's especially important for the program staff responsible for delivering subsequent versions of a program to know exactly how the initial version was implemented and what parts are working. A process evaluation of a community-based volunteer-led health awareness program with peer educators focused on the implementation and ways to make improvements (Karwalajtys et al., 2009). In this program in two Canadian communities, peer volunteers were recruited, trained, and supported by a local organization, and then helped the attendees in a cardiovascular health program by giving information, explaining automated blood pressure instruments, recording readings, and the like. The process evaluation focused on the experiences of the volunteers and found that while there were differences between their experiences in the two program sites, overall, the volunteers reported positive experiences and rewarding aspects of their involvement as well as a desire for additional, ongoing training. The process evaluation helped the program planners strengthen support for and expand the role of the peer educator (Karwalajtys et al., 2009).

Designing Evaluation Research

As in all kinds of research, the evaluation researcher needs to select research strategies. Making specific choices for study design, measurement, data analysis, and the like depends on the specific evaluation, including intended audience, resources available, ethical concerns, and the project's time frame.

Typically, many parties will have an interest in an evaluation, including program administrators, legislators, client groups, interest groups, and the general public. Different constituencies or stakeholders may want different information and have different expectations about the initiative. Scientists are most often concerned with conducting high quality research and obtaining knowledge; the public and legislators may focus on using tax dollars wisely and improving the well-being of individuals and society; program administrators want to manage resources effectively and efficiently (Trochim, Marcus, Mâsse, Moser, and Weld, 2008).

As the objective of most evaluation research is to determine if a program, policy, or law accomplished what it set out to do, one of the study's concepts or variables must be the goal or desired outcome of the program, policy, or law. A typical evaluation research question asks whether the program accomplished its goal(s). If a researcher constructs a hypothesis, typically the program, policy, or law will be the independent variable and the goal(s) or outcome(s) will be the dependent variable(s). In the evaluation of the CPC program presented in Box 14.1, the program offering comprehensive education to preschool and kindergarten children and services for their families for between four and six years was the independent variable and a variety of outcomes were the dependent variables. One hypothesis tested by evaluation research was that CPC program participation was associated with lower rates of reported parental maltreatment of children. Another was that CPC program participation was associated with better educational outcomes for the participants through high school and college years.

If a program has broad or amorphous goals, it can be difficult to select one or more specific dependent variables. Therefore, a critical challenge for evaluators is to articulate appropriate dependent variables and select valid measurement techniques. Even when program personnel help define a specific goal, it still might be hard to determine the desired outcome precisely and how to measure it. Some of the issues the evaluation researchers must confront are specified in Box 14.3.

 **STOP & THINK** *Evaluation researchers almost always have an explanatory purpose and typically test a causal hypothesis about the effect of an independent variable such as a program, law, or policy on a desired outcome. Can you remember some of the study designs discussed in Chapters 7 and 8? If you have a study with an explanatory purpose, what is a particularly useful study design?*

causal hypothesis, a testable expectation about an independent variable's effect on a dependent variable.

internal validity, agreement between a study's conclusions about causal connections and what is actually true.

We talked about experiments in Chapter 8 as useful for testing **causal hypotheses.** Built into the experimental design are ways to decrease systematic differences between the experimental and control groups before introducing the program or policy. The pretest–posttest experimental design can handle many of the possible challenges to **internal validity** because there are two measurements (before and after) and the two groups are exposed to the same conditions *except* for the program, policy, or law. Therefore, any preexisting differences before the program can be determined and the impact of the independent variable (program, policy, or law) on the dependent variable (goal of the program, policy, or law) can be determined.

BOX **14.3**

Evaluation Issues

When evaluating a program, policy, or law, the following must be decided:

- What is the desired outcome?
- Is the outcome to be short or long term?
- Are attitudinal changes sufficient, or is it also essential to study behavioral change?
- How should change be determined?
- Should all aspects of the program be studied or only certain parts?
- Should all the targets of an intervention be studied or only some of them?

 **STOP & THINK** *The traditional experimental design is possible only when participants can be assigned to experimental and control groups. Can you think of evaluation research situations where a researcher either couldn't or wouldn't want to put people in a control group?*

Not all situations lend themselves to using the true experimental model. Outside a laboratory, the evaluator might not be able to control all aspects of the design. A well-known field study of the impact of police intervention on wife abuse in Minneapolis (Berk, Smyth, and Sherman, 1988) was supposed to be a true experiment. Police were to randomly assign domestic violence calls to one of three treatments (mediating the dispute, ordering the offender to leave the premises for several hours, or arresting the abuser). Despite the intended design, however, in about one third of the calls, officers used individual discretion, not random assignment, to determine treatment. Without a pretest measurement or random assignment, there was no way to estimate prior differences between the experimental and control groups.

Sometimes practical concerns of time or money limit design choices. Evaluation research can take five or more years from design to the final report—often too much time for administrators and policy makers (Rossi and Wright, 1984). In other situations, researchers know from the beginning that assigning participants to experimental or control groups is not possible or desirable. The Chicago CPC evaluation, for example, could not control who enrolled in this program versus other programs and therefore used a quasi-experimental design with a comparison rather than a control group. In studying the impact of "no fault" divorce laws on husbands' and wives' post-divorce economic statuses, Lenore Weitzman (1985) could not regulate where a couple would live when filing for divorce nor when couples would want to end their marriages. And clearly, in situations where there is an ethical responsibility to offer services to the most needy—such as AIDS treatment studies and child abuse programs—using random selection as the criterion for assignment to treatment or no-treatment groups is not acceptable. Instead, it is ethical to give treatment and services to the whole sample and have no control group.

If no control group is possible, a useful design for evaluation is the panel study—a design that uses before and after comparisons of a group that receives program services or that is affected by policy changes. If, for example, a school intended to change a discipline code by increasing the use of in-school suspension and decreasing out-of-school suspension, student behavior for the months preceding the change could be compared with behavior afterward. If student behavior improved after the policy was introduced, school personnel might decide to keep the new policy, even if they weren't sure about the extent to which other factors (time of year, a new principal, a new course scheduling pattern, and so on) contributed to the differences in behavior. A panel study can be useful if a comparison group is not available, even though this design is better at documenting changes in the group under study than in identifying the causes of changes.

Cross-sectional studies are sometimes used in evaluating programs or policies, especially if other designs are problematic. In a cross-sectional design, data about the independent and dependent variables are collected at the same time. Say, for example, that a city has been using a series of public service announcements about AIDS prevention in all buses for the past six months. Officials might commission a survey of city residents to determine the effectiveness of the advertisements. Interviewers could survey a random sample of respondents asking if they had seen the advertisements, their opinions of the advertisements, and then include a series of questions about AIDS and AIDS prevention. If the survey's data show that a higher percentage of those who remember seeing and liking the advertisements are more knowledgeable about AIDS and its prevention than those who either don't remember or didn't like the advertisements, the city officials could conclude that the advertisement campaign was successful.

 **STOP & THINK** *Why would we need to be cautious in concluding that the advertisement campaign was successful?*

In our hypothetical city with our imagined advertisement campaign, with this relatively inexpensive and easy-to-do study, we could determine that a *correlation* exists between two variables. Although the connection between noticing advertisements and knowledge is suggestive of causation, we don't have evidence of the time order. For example, it's possible that those already most interested in and best informed about AIDS prevention would be the people to pay the most attention to the advertisements. And those who are least informed about AIDS to begin with could be the ones who didn't read the public service announcements on the topic. In other words, the proposed independent variable could actually be the dependent variable!

Evaluation research can be judged by many standards: usefulness for intended users, degree of practicality, meeting ethical responsibilities, balancing client needs against responsibilities for public welfare, and issues of validity and reliability. For a statement of program evaluation standards, see the American Evaluation Association's Guiding Principles for Evaluators posted on their website at http://www.eval.org/Publications/GuidingPrinciples.asp.

In most evaluation projects, like the one by Jill Harrison and Maureen Norton-Hawk on the Ecuadorian prison, the researchers are typically "outsiders" to the organization under study. Although Harrison and Norton-Hawk were originally viewed with suspicion by the local prison authorities, the researcher from outside the setting is, in many situations, seen as a credible professional, detached, and objective. Such researchers can use what William Foote Whyte called the "professional expert model," where the researcher is "called upon by a client organization (or gets himself or herself invited) to evaluate existing policies and to suggest policy changes to meet new objectives" (1997: 110). When outside evaluators are hired, they are expected to give priority to the perspectives and needs of their clients or the funding agency. One of the limitations of traditional evaluation research is that the research participants and other stakeholders are given very limited opportunities to get involved and, as a result, they might not be invested in the outcome.

Some researchers want more equity for research participants. For example, psychologist Anne Brodsky (2001) and a team of students were invited to evaluate job training and education programs for low income women at a community center. During the two-year project, the research team learned a great deal about the center and the research process and provided feedback about the programs to the center. However, Brodsky (2001: 332) was troubled by the difficulty in balancing the needs of the center and those of the researchers, noting that the team was asking research questions that were more theoretical in nature and probably more important in the long run to her career and her students' education than to the center's effectiveness. Moreover, in conducting the process evaluation of the center, the research team was aware of power differences among its own members based on things like position (faculty/student) and prior experiences and the need to make it clear to the those attending the program that, as researchers, they were outsiders and distinct from program staff (Brodsky et al., 2004). While priorities, insider/outsider status, power differences, access, and the like affect all kinds of applied research, they may be less of a concern in more activist and participatory kinds of applied research to which we now turn.

PARTICIPATORY ACTION RESEARCH

In the past few decades, a growing body of activist research from different disciplines, historical traditions, and geopolitical environments has emerged with an explicit value system and view of stakeholders. Some of the names for the recent kinds of activist research are empowerment evaluation, empowerment research, participatory research, collaborative research, action research, critical action research, and participatory action research (PAR).

Explicit Goals

participatory action research or PAR, research done by community members and researchers working as co-participants, most typically within a social justice framework to empower people and improve their lives.

In our discussion, we'll concentrate on PAR, although much of the commentary applies to the other forms as well. **Participatory action research or PAR** differs from more conventional sociological research in terms of the purpose of the work and the roles and views of study participants, including the researcher. PAR is defined as research in which "some of the people in the organization or community under study participate actively with the professional researcher throughout the research process from the initial design to the final presentation of results and discussion of action implications.... The social purpose underlying PAR is to empower low status people in the organization or community to make decisions and take actions that were previously foreclosed to them" (Whyte, 1997: 111–112). Such research may involve collaborations among professional researchers and local stakeholders or it may be completed entirely by those who share the lived experience of the problem (Boser, 2006: 11).

An essential aspect of this kind of applied research is that those who would normally be the subjects of a study are now also the researchers, actively engaged in all stages of the research process. The other essential part of this work is the *action* that comes from the research—action that can help the community that is participating, rather than the researcher and the research community.

Some argue that PAR can have conservative goals, as when corporations, foundations, and government bureaucracies use participation in research to get people at the grassroots to comply with the organization's goals, such as profit-making (Feagin and Vera, 2001: 165). However, PAR usually aims to have an emancipatory cast. Feagin and Vera (2001) argue, in fact, that the roots of PAR are in the very beginnings of U.S. sociology—in the efforts of female sociologists, such as Jane Addams, working out of and in the Hull House settlement in Chicago in the 1890s. Addams sought the advice of women at all class levels, brought them onto the stage at gatherings, and provided them with research data to help local residents "understand community patterns in order to make better decisions" (Feagin and Vera, 2001: 66).[1]

The practical significance of research and its potential impact are critical to this growing activist tradition. Rejecting the view of research as a value-free endeavor, participant action researchers see themselves as combining popular education, community organizing, and issue-based research in the service of the community (Brydon-Miller, 2002: 2). Researchers in this tradition argue that human beings can't be value neutral because *not* stating values is another way of supporting the status quo (Serrano-Garcia, 1990: 172). They argue that all action requires research, and that research is impossible without action because once we start uncovering the social construction of reality we affect it (Serrano-Garcia, 1990: 172). Themes in participatory work are a focus on social justice and pragmatism centered on what works. Peter Park (1993: 2) describes PAR as seeking to "bring about a more just society in which no groups or classes of people suffer from the deprivation of life's essentials, such as food, clothing, shelter, and health, and in which all enjoy basic human freedoms and dignity.... Its aim is to help the downtrodden be self-reliant, self-assertive, and self-determinative, as well as self-sufficient." Unlike mainstream social science research, PAR has traditionally been a methodology of the margins as it works with subordinate or oppressed groups to better their circumstances within society, rejects research objectivity, and acknowledges that research is inherently political (Jordan, 2003).

Participation and the Researcher's Role

Doing participatory research means working in partnership with those in the community being studied to obtain and use the knowledge that is generated to empower the community. These research projects are "done *with* the community and not *to* the community" (Nyden, Figert, Shibley, and Burrows, 1997: 7). John Gaventa, a practitioner of PAR, says these researchers consider the questions of "Knowledge for whom?" and "Knowledge for what?" but also ask "Knowledge by whom?" as they understand that knowledge is a form of power (O'Neill, 2008).

---

[1]Feagin and Vera indicate that Addams very clearly saw herself as a sociologist and that her "demotion" to mere reformer and social worker, by subsequent generations of sociologists, is more a reflection of the loss of the original emancipatory vision of sociology than of Addams's lack of concern with empirical research.

In the more traditional research perspectives, the researchers are sometimes seen by community organizations and social service agencies as coming in to judge how well they have been performing their jobs. Often in nonparticipatory evaluation research, the researcher treats the members of organizations and communities being studied as passive subjects, and the relationship between researcher and researched is hierarchical, with the researcher being above others. But in participatory action work, the relationship is more equal as the researcher acts collaboratively with community and organization members throughout the research process.

In PAR, the researcher must be involved intimately in the life of a community and its problems and be in dialogue with other researcher collaborators while eschewing hierarchy. The research groups are usually composed of professionals and ordinary people, with all viewed as authoritative sources of knowledge and experience. The success of the project is not seen as depending on expertise related to technical skill but, rather, on the ability to communicate and be responsive (Esposito and Murphy, 2000: 181).

Researchers in activist traditions can select from all methods of social inquiry and both qualitative and quantitative approaches, including questionnaires, interviews, focus groups, observation, archival material, and visual methods. The more naturalistic methods of inquiry, such as participant observation and informal interviewing (perhaps in a more reciprocal way), are more typically selected because the research focus is usually on the participants and their interpretations of the social context. Kemmis and McTaggart (2003: 375) argue that in participatory action and related kinds of research, the researchers may make sacrifices in methodological and technical rigor in exchange for better immediate face validity so that the evidence makes sense within its context. They argue that this tradeoff is acceptable because a goal of the research is transformative social action, where participants live with the consequences of the transformations while others (Cook, 2006) worry that by using participatory methods to both develop something and evaluate it, there is a danger of losing critical perspective.

The critical issue in PAR is that members of the community are involved in deciding which methods to use, developing or adapting the research instruments, and carrying out the research and analysis. Every aspect of the research is seen as an opportunity for interaction and discussion. In this perspective, information is not transmitted from others to researchers but is co-created (Esposito and Murphy, 2000: 182).

Two recent studies illustrate the goals of PAR, its methodology, the participants, and the way authority is distributed in the research process. The first is a project by more traditional research partners, a project that was designed to understand the cultural context of domestic violence in a community with multiple minority groups and to examine access to and satisfaction with services for battered women. The participants included community-based organizations, government agencies, advocates, survivors of domestic violence, and activists working to end domestic violence (Sullivan, Bhuyan, Senturia, Shiu-Thornton, and Ciske, 2005).

A research team composed of university researchers, Health Department personnel, and advocates from the partner agencies was formed; an advisory group with members from the larger domestic violence communities was selected; and a community activist who was also a survivor of domestic violence was hired to help coordinate the work (Sullivan et al., 2005: 980). Decision making was shared by researchers and community partners at each step of the process. For example, while the original plan was to conduct focus groups and individual interviews, some community advocates argued that women in their communities would be reluctant to talk about domestic violence in groups, but others disagreed. The resolution was suggested by one of the advocates: let the women themselves chose the kind of interview for themselves. Recruitment of a sample of women aged 18 and older who had experienced domestic violence by an intimate partner or family member living in their households was also a joint process, with advocates in some communities spending a great deal of time finding participants. Similarly, the data collection methodology, including the specific questions asked of participants, was first designed by brainstorming in team meetings.

The researchers report that those with the most limited participation in the research process were the immigrant and ethnic minority women participants: "Although these informants reported having enjoyed taking part in the focus groups and interviews, the structure of the project did not facilitate them being further involved. In particular, constraints arose because of the demands on informants' lives as related to possible ongoing abuse, parenting responsibilities, and employment" (Sullivan et al., 2005: 990). The community-based agencies participated in the research process and then took a leadership role in implementing the kinds of projects that the research suggested would work best. The power of the women's stories galvanized agency members. "Even advocates who had worked in the field for years learned new information from listening to the women's stories. Hearing, reading, and rereading the stories also developed a strong sense of responsibility to the participants" (Sullivan et al., 2005: 991).

A less traditional research team was in charge of another recent research project. Youth in Focus (YIF) is a nonprofit organization in northern California that fosters youth development and organizations by supporting youth-led research, using interviews, focus groups, surveys, observation, photography, and video as methods of data collection in efforts to investigate disparities and to advocate for justice in fields such as health, juvenile justice, education, and community planning (Sanchez, Lomeli-Loibl, and Nelson, 2009). A recent YIF Health Justice Initiative project was conducted in partnership with the Sacramento Gay and Lesbian Center and adult allies in local gay-straight alliances. Over the course of the project, the research team included 25 young people ages 13 to 22 and 3 adults. The broad research question tackled was "What are the needs of lesbian, gay, bisexual, transgender, and questioning (LGBTQ) youth in Sacramento?" The initial effort was a needs assessment survey with an emphasis on mental health using a sample of students and counselors at local high schools. Based on the results, the research team made a series of recommendations that included increases in affordable and accessible mental health services for these LGBTQ youth, greater visibility of and training for providers who work with this population, and strong outreach to community centers and schools. The team also filmed a documentary focused on the narratives and

BOX **14.4**

Planning PAR

- Find a community and one or more agencies or organizations to be your partner(s).
- With your community partners, make joint decisions on a research topic or question and the time frame you have in mind.
- Find out about the requirements for approval of your project from the IRB at your college or university and apply for approval in a timely manner.
- Think about your stake in the research, your expertise, and what you want to contribute.
- With your community partners, estimate the time and the costs of doing the project, locate possible resources, and put together an action plan.

needs of LGBTQ youth throughout Sacramento. The PAR project has been well received by local agencies and has attracted attention from local service providers. "With an overwhelming sense of both need and optimism, the youth research team developed its strategy to be not only an informative process but also an exercise in advocacy for the many who are silenced by homophobia" (Sanchez et al., 2009: 7–8).

The steps in doing PAR are described in Box 14.4. If you'd like to read about some new work in the field, check out the website of the Institute for Participatory Action Research at the City University of New York (CUNY). This collective *dreams wildly* about critical inquiry, social theory and the politics of social justice for youth. With the craft of PAR, our projects seek to reveal theoretically and empirically the contours of injustice and resistance while we challenge the very bases upon which traditional conceptions of 'expert knowledge' sit" (Fine, 2009).

FINAL CONSIDERATIONS

A Middle Ground

Some argue that while "flattening" the research hierarchy may be an ideal in community research, the roles of researcher and participant are not equal. They advocate acknowledging the "power and resource differences that typically exist and, thereby, accept additional responsibilities as researchers that are not necessarily balanced by parallel or complementary responsibilities on the part of participants" (Bond, 1990: 183). Although at first glance it appears that researchers must choose between the more conventional and the more activist kinds of applied work, because projects fall along the continuum of community involvement and subject participation, there is a middle ground. Many organizations have barely sufficient staff and volunteers to offer their programs and services and have no additional resources to commit to evaluation or action research. As a result, groups might be willing to work in collaboration with researchers to obtain information about effectiveness of programs for their own internal use or to use in seeking recognition or funding for future projects. In addition, PAR and its approach are gaining acceptance by traditional evaluators. For example, they are increasingly given credence by researchers and used by international organizations such as the World Bank and government agencies (Jordan, 2003).

Leonard Krimerman (2001) argues that PAR will not replace mainstream or expert-directed work but that it is possible and sensible for the practitioners from both traditions to collaborate. He believes that social scientists can play a major role in the invention and extension of key concepts: initiating identification of and response to oppressive conditions; contributing macro-level policy analyses; providing the cross-national analysis. On the other hand, he notes that any research focused on the poor, the disenfranchised, or the marginalized that does not include their voices and contributions will be flawed or incomplete (Krimerman, 2001).

Using a combined approach, described by some as engaged evaluation or community-based research, the researcher and the organization or group work together and share responsibilities to assess a program or project and chart future directions. Evaluation is typically different when done for internal rather than external purposes. In internal evaluations, questions are asked about "our program," whereas external evaluation asks about "your program" (Fetterman and Eiler, 2000). The researcher working *with* the community can bridge the gap between these approaches.

The researcher and community group can split the work. One study that gathered information from 20 teams of evaluators and project directors found that most had shared many of their roles (Cartland, Ruch-Ross, Mason, and Donohue, 2008). While some reported tension and confusion in role sharing, most resolved the problems early and willingly adopted collaborative stances. Few evaluators avoided participation in project development and almost all invited stakeholders to make some of the decisions about the evaluation itself (Cartland et al., 2008). This cooperative model of evaluation may become more prevalent in the future, with greater discussion and communication among the various stakeholders and more self-evaluation of programs with discussion, data collection, and analysis conducted through the sharing of roles.

A model of work that balances conventional evaluation research perspectives with more participatory viewpoints can be found at the Center for AIDS Prevention Studies (CAPS) and its efforts to stimulate collaboration among academic researchers, public health professionals, and community-based organizations. CAPS is committed to working with communities in the areas of information dissemination, by offering technical assistance to community-based organizations and scientists and by doing community collaborative research (Fernández-Peña et al., 2008).

One CAPS project was a study of 11 partnerships between community-based organizations and university-based researchers. The study found that the community-based organizations took the lead in developing the research question, delivering the program, and collecting the data; the academic researchers took the lead in developing the instrument, consent procedures, and data analysis; together the groups trained the staff, interpreted the data, and disseminated the findings (CAPS, 2000). CAPS (2006) has supported many collaborative research and evaluation projects using their model of community collaborative research and has produced *Working Together* (2001), a manual on doing collaborative research with community organizations.

 **STOP & THINK** *Applied research is usually field research. What do you think are some of the consequence of the real-world setting for the research and its outcome?*

Politics and Applied Research

Regardless of approach, all research is political in nature and is affected, to some degree, by the social, political, and economic climates that surround the research community. Knowledge is socially constructed, so the choice of research topics and questions and all methodological decisions are related to the social and political context. Participatory action researchers are aware of the politics of setting and, even though they have professional expertise, usually work as non-privileged members of a team. Communities make an initial judgment about researchers and choose whether or not to invite them into the setting. Because of shared responsibility, the working group will determine the form and interpretation of the research project.

In evaluation research, the specific choice of research project is affected not only by societal values and the priorities of funding agencies, but also by the perspectives of various constituencies and program stakeholders. Evaluations are typically conducted in field settings, within ongoing organizations such as school systems, police departments, health care facilities, and social service agencies. Within these organizations and the world beyond them, politics is a major factor. There is ideology to consider, "turf" to protect, loyalties and long-standing personal relationships to look after, and careers and economic survival to think about. Organizational participants and their allies can either help or hinder evaluation research efforts depending on their assessment of the politics of the research and its social context.

Sara McClelland and Michelle Fine (2008) demonstrate the way that politics and evaluation interact in their fascinating discussion of the federally sponsored abstinence-only-until-marriage (AOUM) programs. As the country increased spending on these programs between 2001 and 2006, during the presidency of George W. Bush, with a total of almost $800 million spent on them by the end of 2006, there was great interest in evaluations of these programs. One way that politics affected evaluation was in the choice of the outcome to be evaluated. One group of evaluations that assessed various AOUM programs focused on *attitudes not behavior*. The evaluation studies asked young people about their attitudes and intentions to abstain from sex until marriage and found evidence of pre- and post-attitudinal shifts among preteens, but most did not measure behavior as an outcome variable (McClelland and Fine, 2008). It wasn't until 2007 that Mathematica Policy Research, Inc., a highly regarded, nonpartisan social science research institute, released the final results of its quasi-experimental, longitudinal evaluation of abstinence-only education. After nine years of evaluation, the reports were released and presented convincing evidence that these federally funded abstinence-only education programs do not delay sexual activity, increase abstinence rates, or change rates of condom use (McClelland and Fine, 2008).

Beyond Our Control

Organizations are not "neutral territory," and in most cases, the researcher is an outsider, working in someone else's sphere. The results of any assessment will have the potential to affect the organization and the individuals under study. As a result, a special challenge in evaluation research can be obtaining the cooperation of program staff for access to data. Sometimes this is especially challenging. When one of us, Emily, was hired by a special education professor and researcher to help with the evaluation of special needs educational policies in several communities, feasibility became a concern. The research plan was for Emily to observe school meetings at which individual education plans for special-needs students were designed by school personnel in consultation with parents. There was no difficulty in getting cooperation at three of the schools selected, but, at the fourth school, the principal kept "forgetting" to communicate when meetings were canceled or had been rescheduled because of "unavoidable problems and crises." Several times, on arriving for a meeting at this particular school, Emily was told that the meeting had already been held—the day or even the hour before. The message was clear: The principal did not want an observer. Because the schools selected were a purposive, rather than random, sample, the meetings at another school were substituted and observed.

Harrison and Norton-Hawk also had problems obtaining data in their needs assessment of the prison in Ecuador. Even with letters from their home institutions, access to inmates for the research was difficult. The researchers were not initially viewed as credible or as having any authority (Harrison, personal communication). The Comite de las Mujeres, the inmate group they had worked with before, was helpful in gaining access to the prison director as there is no formal review process for researchers. The director was concerned that they would exploit the inmates by using the information in inappropriate ways. After they explained the IRB approval process at Suffolk University and talked about their experience studying female incarceration in prisons in the United States, he allowed the research. However, they were not able to use a random selection process and were told that they were lucky to be permitted access at all (Harrison, personal communication).

The sentiment that evaluators aren't welcome isn't unusual. Evaluators and their work are often distrusted by program personnel who often feel that the time and money spent on evaluation would be better spent on programs. Data collection for an evaluation can make it harder to run to the program; evaluators must be sensitive to the fact that the program staff has a job to do that might be made more difficult by ongoing evaluation.

Participant action researchers have less control over (and responsibility for) the research than evaluation researchers. They must be willing to contribute their expertise and work as team members rather than in a traditional hierarchy. They may find that while they can be invited in, they can also be asked to leave.

Having an Impact

As education and action are explicit parts of their work, participatory action researchers can usually see the impact of the work. For example, at the end of the PAR on domestic violence described by Sullivan et al. (2005), culturally appropriate programs were developed to provide education and skill building through support groups. At the end of the Youth in Action needs assessment project, a documentary film was produced and recommendations for services were made to a variety of local organizations (Sanchez et al., 2009). Working with local groups and agencies, participatory action researchers know that their work will have an impact because a group of knowledgeable and committed community participants remain when the research is completed.

More conventional evaluation faces the challenge of implementing change after the research is completed. This is not a new story. In 1845, for example, the Board of Education of the city of Boston initiated the use of printed tests to assess student achievement. The resulting evaluation of test scores showed them to be low, but rather than analyze the causes of the poor performance (Traver, cited in Chambers, Wedel, and Rodwell, 1992: 2), the board decided to discontinue the test! Unfortunately, this example, more than a century and a half old, still has relevance.

Evaluation work can influence policy and programs in several ways. It can be used to give direction to policy and practice (i.e., close down or modify programs); to justify preexisting preference and actions (i.e., provide legitimation); to provide new generalizations and ideas for making sense of the situation even when the decision makers don't act on them immediately or directly (Weiss, Murphy-Graham, and Birkeland, 2005). But, even if evaluators are trusted, allowed to collect data, and do a credible job of evaluating a program or policy, their conclusions might have little impact on the decision to keep a program, modify it, or eliminate it. Ideological and political interests can sometimes have more influence on decisions about the future of social interventions than evaluative feedback. Even if a program is shown to be ineffective, it might be kept if it fits with prevailing values, satisfies voters, or pays off political debts. Often changes in social programs are very gradual because frequently no single authority can institute radical change (Shadish, Cook, and Leviton, 1991: 39).

A long-time and very popular program designed to reduce drug use among youth provides an interesting example of how programs that receive negative evaluations can continue to flourish. Drug Abuse Resistance Education (DARE) originated on a small scale in the early 1980s and over the next decade spread around the country. By 2001, estimates of its costs varied from about $200 million to around $1 billion annually, with the DARE organization estimating the annual costs of the officer services to be about $215 million (Shepard, 2001).

Over the years, evaluations indicated that DARE programs were not successful (Ennett, Tobler, Ringwalt, and Flewelling, 1994; Perry, Komro, Veblen-Mortenson, and Bosma, 2000; West and O'Neal, 2004; Wysong, Aniskiewicz, and Wright, 1994). The studies concluded that any positive effects on students' knowledge, attitudes, and behavior (often observed right after the program) were temporary so that by late adolescence students exposed and not exposed to the program were indistinguishable (Birkeland, Murphy-Graham, and Weiss, 2005). The negative findings received a great deal of attention in the media, spreading the message that DARE doesn't work (Weiss et al., 2005). However, some stakeholders—in this case, organizations with direct and indirect involvement— accumulated sufficient resources and legitimacy to protect the program.

School districts have had mixed responses to DARE in the decade since negative evaluations became widespread and public. Some have ignored the evidence and continue to offer DARE for a variety of reasons, such as believing that perhaps there is some long-term effect not measured by short-term evaluations or seeing a positive effect in the relationships between the police officers who teach the program and the students who take them (Birkeland et al., 2005). Other systems were persuaded by the evaluations or dropped the program for other reasons (Weiss et al., 2005). An analysis of how decisions were made in 16 communities in 4 states that had DARE programs finds that there was no single way that communities made the decision to keep or cut the program (Weiss et al., 2005).

Even when evaluations are very good and the results are indisputable, programs can be cut due to economic and political factors. Box 14.5 tells the sad story of the demise of a very successful educational program for parents and children that we described earlier in the chapter.

It's important to remember that outside evaluators usually have an advisory role, "closest to that of an expert witness, furnishing the best information possible under the circumstances; it is not the role of judge and jury"

BOX **14.5**

Successful Programs and Political Realities

A blog entry on the website of New American Foundation, a nonprofit, nonpartisan public policy institute, tells a depressing story.

Consider an education program so effective that its impact can be measured 19 years later, so well-studied that it can be backed up with decades of scientific evidence on children's improved skills in math and reading, and so impressive to policy makers that it continues to be championed around the country 40 years after its launch. These are the superlatives that come with Chicago's Child Parent Centers. So you might figure they're flourishing as part of the Chicago Public Schools' early childhood programs, right? Not so. Their numbers are dwindling. In the mid-1980s, there were at least 25 CPCs serving more than 1,500

children. By 2006, there were 13. Today, 11 are still open, according to the Promising Practices Network. Enrollment in 2009, as reported by the Chicago Public Schools, is down to 670, less than half of what it once was. It now represents just 2 percent of the system's total preschool enrollment ... Shifts in budget priorities within the school system and across the state played a big role, along with changing neighborhood demographics that have complicated the CPC's mission of helping low-income children ... CPCs offer some of the strongest lessons of all, and their closures send a warning about how difficult it can be to sustain the programs that have been shown to do the most good. (Guernsey, 2009)

(Rossi and Freeman, 1993: 454). Evaluators can provide reports with significance for policy making, discuss the larger context of their work, argue forcefully for their positions, and work toward disseminating their findings widely, but they rarely have the power to save, close, or institute changes in programs or policies.

SUMMARY

Applied research is research with a practical purpose. It can be undertaken by a researcher with the participation of one or more stakeholders. In evaluation research, the researcher or a specific stakeholder develops a research question or hypothesis, most typically a causal hypothesis with a program or its absence as the independent variable and the goal of the program as the dependent variable. Evaluation research can provide reasonably reliable and valid information about the merits and results of particular programs that operate in particular circumstances. Necessary compromises in programs and research methodologies can mean that the users of the information will be less than fully certain of the validity of the findings. Because it is better to be approximately accurate than have no information at all, researchers might have to settle for practical rather than ideal evaluations.

Participatory action researchers and others in the activist tradition work collaboratively with participants. With the goals of education and social justice, outside professionals and those who are participants in the setting decide collaboratively on the research questions, the study's methodologies, and ways to apply the findings.

Applied work is done in real-world settings with political and social contexts. In each project, there will be multiple stakeholders. At the end of each study, the analyses and their practical implications can be valuable to program staff, clients, other stakeholders, other researchers, and the general public. If used wisely, applied research can help to create, modify, and implement programs and activities that make a difference in people's lives.

EXERCISE 14.1

Designing Evaluation Research

Find a description of a social program in a daily newspaper or use the following article of a hypothetical program. (If you select a newspaper article, attach it to the exercise.)

"Pets Are Welcome Guests"

Residents of the Pondview Nursing and Rehabilitation Center have a series of unusual guests once a week. The VIPs (Volunteers Interested in Pondview) have organized a "Meet the Pets Day" at the local facility. Each week, one or more volunteers bring a friendly pet for short one-to-one visits with residents. On a typical day, a dozen or so owners will bring dogs, cats, and bunnies to Pondview, but sometimes companions include hamsters and gerbils.

Last week, Buffy, a spirited golden retriever with a wildly wagging tail, made her debut at Pondview. In a 15-minute visit with Mrs. Rita Williams, an 85-year-old widow recovering from a broken hip, Buffy managed to bestow at least several "kisses" on the woman's face and hands. Mrs. Williams said she has seen more sedate pets, but wasn't at all displeased with today's visit.

Margaret Collins, facility administrator, said that the program had been adapted from one she had read about in a nearby city. She was glad that the VIPs had organized the new program. "It gives the residents something to look forward to," she said. "I think it makes them more alert and attentive. If it really does aid the residents' recoveries and results in their improved mental health, we'll expand it to several days a week next year."

Design a research project to evaluate either "Meet the Pets Day" or the program described in your local newspaper by answering the questions that follow.

1. What is an appropriate research question that your evaluation should seek to answer or hypothesis that you would test?
2. Describe the social program that is being offered.
3. Who are the program's participants?

4. In addition to the program's participants, who are the *other* stakeholders?
5. What is the goal or intended outcome of the program?
6. Describe how you would decide if the program was successful in meeting its goal by designing an *outcome evaluation study*. Be sure to describe the study design you would use, who your sample would include, and how you would measure the dependent variable.
7. Comment on the internal validity of the study you designed.
8. What are the ethical considerations you would need to consider if you were interested in doing this study?
9. What are the practical issues (time, money, and access) that you would need to consider if you were interested in doing this study?

EXERCISE 14.2

Participatory Action Research

Imagine that you would like to do a piece of applied research that focuses on a group in your community, such as a shelter for homeless people, a literacy program, a neighborhood preservation association, a local boys or girls club, a volunteer services for animals organization, or any group of interest.

Visit one such group or organization in your community. See if you can meet either formally or informally with members of the staff and with clients. Ask them what they feel their most pressing needs are and what problems the organization or group has had in meeting these needs.

Based on what you find out, write a short paper describing a possible PAR project, which includes the following:

1. A description of the organization, agency, or group.
2. A listing and description of the major stakeholders.
3. A possible research question that could be answered if you were to work collaboratively with the major stakeholders.
4. The benefits of doing the work you're proposing.
5. Any practical or ethical concerns you think you'd run into in doing the project.

EXERCISE 14.3

Evaluating Your Course

The course you are taking this semester is an educational program designed to help students understand and use social research methods. As such, it can be evaluated like any other program. For this exercise, select one aspect of your course

(such as the textbook, the instructional style of the teacher, the frequency of class meetings per week, the length of each class session, the lectures, or the exercises) as an independent variable. Design an evaluation research project that could be conducted to test the effectiveness of the aspect of the course you've selected on the dependent variable of student learning.

Answer the following questions.

1. What aspect of the course are you focusing on as your independent variable?
2. Who are the significant stakeholders in this setting?
3. How would you define and measure the dependent variable "student learning of social research methods?"

4. What would you do to determine the effectiveness of the aspect of the course that you are interested in?
5. Using your imagination, what are some results that you think you'd get from your evaluation?
6. Comment on the practical and ethical considerations you would need to consider if you wanted to do this study.

EXERCISE 14.4

Comparing Evaluation Research Projects

The website of the Office of Juvenile Justice and Delinquency Prevention has a large database of programs archived by approach (prevention, sanction, residential, and re-entry) and by target population (age, gender, ethnicity, first-time offenders, etc.). The archive is available at http://www2.dsgonline.com/mpg/mpg_search_factors.aspx.

Think about your interests and issues your community is facing and find at least two programs for children and youth in the database that have addressed this concern with a program. Read about the evaluation research that was done of the programs. Then describe and analyze the evaluations using the following format:

1. Describe and compare the programs in terms of cost, populations served, approach, and so on, based on the information you can find online about them.
2. Discuss the evaluations of the programs. Do they have different outcome variables? How are the variables measured? Does one program seem to be more effective than the other? Do you think one or more of the programs should be modified? Expanded? Eliminated?
3. Print out the listings of the programs from the database and attach them to your exercise.

CHAPTER 15

Quantitative and Qualitative Data Analysis

© Royalty-Free/CORBIS

INTRODUCTION

Perhaps you've asked yourself, as we've discussed all the ways of collecting data, "Once I've collected all the data, how do I organize all these facts so that I can share what I've found?"

We'd now like to introduce you to the most rewarding part of the research process: analyzing the data. After all the work you've done collecting your information, there's nothing quite like putting it together in preparation for sharing with others—a little like finally cooking the bouillabaisse after shopping for all the individual ingredients. But like the excitement of cooking bouillabaisse, the excitement of data analysis needs to be balanced by having a few guidelines (or recipes) and principles to follow. In this chapter, we'll introduce you to quantitative and qualitative data analyses. Rather like an elementary cookbook, this chapter offers rudimentary data-analytic recipes, and some insights into the basic delights and principles of data analysis. Bon appetit!

Quantitative or Qualitative?

What is the difference between quantitative and qualitative data? One answer is that quantitative data results from quantitative research and qualitative data results from qualitative research. This sounds like question begging, and of course it is, but it carries an important implication: The distinction drawn between quantitative and qualitative data isn't as important (for data-analytic purposes) as the distinction between the strategies driving their collections. Sure, you can observe, as James Dabbs (1982: 32) and Bruce Berg (1989: 2) have, that the notion of qualitative refers to the essential nature of things and the notion of quantitative refers to their amounts. But, as we've hinted before (e.g., in chapters on qualitative interviewing and observation techniques), there's really nothing about qualitative research that precludes quantitative representations (see Morrow, 1994: 207). Enos (see Chapter 10), of course, did count the number of inmates' children who had been placed in foster care. Similarly, research that might seem primarily quantitative never totally ignores essences. Martinez and Lee (in Chapter 12) used qualitative judgments about which murders were committed, for instance, by non-intimates before they used quantitative techniques to decide what percent of murders were committed by non-intimates. Indeed, although the mix of quantitative and qualitative approaches does vary in social science research, almost all published work contains elements of each. Consequently, the distinction we draw (and have used to organize this text) is not one of mutually exclusive kinds of analysis but, rather, of kinds that, in the real world, stand side by side and, in an ideal world, would always be used to complement one another.

An Overview of Quantitative Data Analysis

quantitative data analysis, analysis based on the statistical summary of data.

However artificial the distinction between quantitative and qualitative research and data might be, there's no denying that the motives driving quantitative and qualitative researchers are distinguishable. Although there are exceptions, **quantitative** researchers normally focus on the relationships between or among variables, with a natural science-like view of social science in the backs of their minds. Thus, in the fashion of the physicist who asserts a relationship between the gravitational attraction and distance between two bodies, Ferrari (in Chapter 6) examined whether impostors were more or less likely to engage in academic dishonesty than non-impostors and Ramiro Martinez and Matthew Lee (in Chapter 12) studied whether migrant communities were more likely to have high crime rates than nonmigrant communities. Projects like these engaged in other natural science-like activities as well: for example, looking at aggregates of units (whether these were characters in picture books or communities), more or less representative samples in some studies (see Chapter 5 on sampling) of these units, and the measurement of key variables (see Chapter 6 on measurement). For data-analysis purposes, the distinguishing characteristic of this type of study is attention to whether there are associations among the variables of concern: Whether, in general, as one of the variables (say, whether one is an immigrant or not) changes, the other one (whether one is likely to commit homicide or not) also changes. To demonstrate such an association, social science researchers generally employ *statistical* analyses. Thus, we will aim toward a basic understanding of statistical analyses in the part of the chapter on quantitative data analysis.

 **STOP & THINK** *Recall Adler and Clark's study (in Chapter 7) of retirement. Did people who said they expected to work after retirement actually end up being more likely to work after retirement than people who didn't expect to work after retirement? Do you see that the focus of this article, like that of others using quantitative data, is on variables (in this case whether the expectation to work after retirement is associated with actually working after retirement) and not really on people, groups, or social organizations?*

An Overview of Qualitative Analyses

qualitative data analysis, analysis that results in the interpretation of action or representations of meanings in the researcher's own words.

The strategic concerns of qualitative researchers differ from those of quantitative researchers and so do their characteristic forms of data analysis. Generally, **qualitative** researchers tend to be concerned with the interpretation of action and the representation of meanings (Morrow, 1994: 206). Rather than pursuing natural science-like hypotheses (as Durkheim advocated in his *The Rules of Sociological Method* [1938]), qualitative researchers are moved by the pursuit of empathic understanding, or Max Weber's *Verstehen* (as described in his *The Theory of Social and Economic Organization* [1947]), or an in-depth description (e.g., Gertz's [1983] *thick description*). Sandra Enos (in the focal research in Chapter 10) aspired to an empathic understanding of the meaning of motherhood for women inmates and of the differences in those meanings for different inmates. Mueller, Dirks, and Houts Picca (in Chapter 11) provide a thick description of cross-racial Halloween costuming. Qualitative researchers pursue single cases or a limited number of cases, with much less concern than quantitative researchers have about how well those cases "represent" some larger universe of cases. Qualitative researchers look for interpretations that can be captured in words rather than in variables and statistical language. Later in the chapter, we'll seek a basic understanding of how qualitative researchers reach their linguistic interpretations of data.

QUANTITATIVE DATA ANALYSIS

Quantitative data analysis presumes one has collected data about a reasonably large, and sometimes representative, group of subjects, whether these subjects (or units of analysis) are individuals, groups, organizations, social artifacts, and so on. Ironically, the data themselves don't always come in numerical form (as one might expect from the term *quantitative* data).

A brief word to clarify this irony: You'll recall that the focus of quantitative data analysts isn't really the individual subjects being studied (e.g., school children or communities) but, rather, the variable characteristics of those subjects (e.g., whether they were exposed to caring literature or whether they have high crime rates). Remember that variables come in essentially three levels of measurement: nominal, ordinal, and interval-ratio. The most common level of measurement is the nominal level (all variables are at least nominal), because all variables (even marital status) have categories with names (like "married" and "single"). Thus, an individual adult might be married or single, and the information you'd obtain about that person's marital status might, for instance, be "single." But, "single" is obviously not a number. Nonetheless, it is information about marital status that when, taken together with data about marital status from other subjects, can be subjected to numerical or quantitative manipulation, as when we count the number of single people in our sample.

This is just a long way of saying that the descriptor "quantitative" in "quantitative data analysis" should really be thought of as an adjective describing the kind of analysis one plans to do, rather than as an adverb modifying the word "data." When you do a quantitative analysis, the data might or might not be in the form of numbers.

Sources of Data for Quantitative Analysis

coding, assigning observations to categories.

Data appropriate for quantitative analysis can come from many sources, including a survey conducted by the researcher herself. When the data are collected by the researcher, an important first step often is **coding**. In quantitative analyses, this frequently means making data computer usable. In qualitative analyses, this means associating words or labels with passages and then collecting similarly labeled passages into files. Coding is the process by which raw data are given a standardized form. In most cases, quantitative coding involves assigning a number to each observation or datum. Thus, when coding gender, you might decide to assign a value of "1" for each "female" and "2" for each "male." The assignment of a number is often pretty arbitrary (there is no reason, for instance, that you couldn't assign "1" for females and "0" for males), but it should be consistent within your data set.

As we suggested in Chapter 12, more and more published research in the social sciences is now produced by researchers who have analyzed secondary survey data—data collected by large research organizations and made available to other researchers, sometimes for a fee. According to one study (Clark, 1999), about 40 percent of all articles published in sociology toward the end of the 1990s were based on analyses of secondary survey data. You might recall that the General Social Survey (GSS) is the most popular of these sets of survey data. One of the great advantages of secondary survey data is that they usually come pre-coded and ready for analysis via one of the software packages social scientists use for statistical analyses. The GSS comes, in an abridged version, for instance, along with abridged versions of a number of other data sets, in *The SPSS Guide to Data Analysis* (Norusis, 2009). This guide introduces students to statistical analysis using the software package, the Statistical Package of the Social Sciences (SPSS). GSS data are so widely available that you can even visit a website, the Social Documentation and Analysis (SDA) Archive (http://sda.berkeley.edu/archive.htm), and do your own analysis of GSS data from 1972 to 2008, and other data sets. In some of the embedded "Stop and Do" sections that follow, we'll be asking you to do so, so you can see what can be done with GSS data online.

In any case, we will be using GSS data to illustrate some of the kinds of quantitative analysis most frequently used by social scientists in our discussion of elementary quantitative analyses.

Elementary Quantitative Analyses

Perhaps the defining characteristic of quantitative analyses is their effort to summarize data by using statistics. Statistics, being a branch of mathematics, is an area of research methods that some social science students would just as soon avoid. But the advantages of a little bit of statistical knowledge are enormous for those who would practice or read social science research, and the costs of acquiring that knowledge, we think you'll agree, aren't terribly formidable, after all. Our primary goal in the section on quantitative analysis is to provide you with the kind of overview that will give students who have never had a course in statistics, and those with more experience, a feeling for how statistics advance the practice of social research.

descriptive statistics, statistics used to describe and interpret sample data.

inferential statistics, statistics used to make inferences about the population from which the sample was drawn.

To organize and limit our discussion of social statistics, however, we'd like to begin with two basic sets of distinctions: one distinction, between **descriptive** and **inferential** statistics, relates to the generalizability of one's results; the other distinction among **univariate**, **bivariate**, and **multivariate** statistics relates to how many variables you focus on at a time. Descriptive statistics are used to describe and interpret data from a sample. We find that about 62 percent of the people sampled in the GSS from 1972 to 2008 were married, so that modal (the most common) marital status of the GSS sample is "married" (data and analysis done using GSS data from the SDA Archive, 2009). The mode is a descriptive statistic because it describes only the data one has in hand (in this case, data on 53,070 respondents). Inferential statistics, on the other hand, are used to make estimates about characteristics of a larger body of data (population data—see Chapter 5 on sampling). We find, using the GSS data again, that men are significantly more likely than women to have been employed full time (SDA Archive, 2009). The word "significantly" in this context means that there's a good chance that, in the larger population from which the GSS sample has been drawn (basically adult Americans), men are more likely to be employed full time than women. We used an inferential statistic to draw this conclusion. In the rest of this chapter, we'll focus on descriptive statistics, but we'll briefly introduce inferential statistics, too.

Much powerful quantitative research can be done with an understanding of relatively simple univariate (one variable) or bivariate (two variable) statistics. Univariate statistics, like the mode (the most frequently occurring category), tell us about one characteristic. The fact that the modal marital status is "married" in the GSS sample tells us about the typical marital status of that sample. Bivariate statistics, on the other hand, tell us something about the association between two variables. When we wanted to find out whether men or women have been more likely to be employed full time—that is, that there is a relationship between "gender" and "work status"—we used bivariate statistics. Multivariate statistics permit us to examine associations while we investigate the role of additional variables. We hypothesized that the association between people's gender and their likelihood of being employed full time might have been accounted for in terms of their level of education. We used GSS data (from the SDA Archive, 2009) and found that the association between gender and employment status could not be accounted for in terms of education by using multivariate statistical methods.

Univariate Analyses
Measures of Central Tendency

Perhaps the most common statistics that we use and encounter in everyday life are ones that tell us the *average* occurrence of something we're interested in. In 2008 Dustin Pedroia, the Red Sox second baseball and the American League Most Valuable Player, sometimes got a hit and sometimes didn't, but his batting average for the year was .326, meaning that he got a hit just under 33 out of every 100 times he came to bat. In most distributions, values hover around an average, or central tendency. The most common of these are the **mode**, the **median**, and the **mean**.

In Chapter 6, we observed that the mode, median, and mean are designed for nominal, ordinal, and interval level variables, respectively. The mode, designed for nominal level variables, is that value or category that occurs most frequently. Have a look at the following data about five students' gender, age, and height. Because more students are female than male, the modal gender for the sample is "female."

Student	A	B	C	D	E
Gender	Male	Female	Female	Female	Female
Age	18	18	17	19	20
Height	Tall	Tall	Short	Short	Medium

 **STOP & THINK** *Look at those data again to see if you can decide what the mode is for "age" in this sample.*

Although the mode has been designed for nominal level variables, like gender, it can be calculated for ordinal and interval level variables (like age) because ordinal and interval level variables, whatever else they are, are also nominal. As a result, the mode is an extremely versatile measure of central tendency: It can be computed for any variable.

univariate analyses, analyses that tell us something about one variable.

bivariate analyses, data analyses that focus on the association between two variables.

multivariate analyses, analyses that permit researchers to examine the relationship between variables while investigating the role of other variables.

mode, the measure of central tendency designed for nominal level variables. The value that occurs most frequently.

median, the measure of central tendency designed for ordinal level variables. The middle value when all values are arranged in order.

mean, the measure of central tendency designed for interval level variables. The sum of all values divided by the number of values.

STOP & THINK *We say a variable is unimodal when, like "gender" and "age" in the sample of students mentioned earlier, it has only one category that occurs most frequently. We say it is bimodal when it has two categories that occur most frequently. Check out the data for the variable "height." Is it unimodal or bimodal?*

 STOP & DO *The mode is easy enough to calculate when, as in our example of students, there are five cases. But let's see if you can apply the principles to a much larger sample? Let's look at data from the GSS on the SDA Archive. Access the Archive by going to http://sda.berkeley.edu/archive.htm. Now hit on the most recent GSS Cumulative Data File, and see if you can figure out what the modal highest educational degree has been for respondents to this survey. (Hint: You'll want to browse the codebook for "education" and then look at the information for "Rs highest degree," a variable called DEGREE. Hit on DEGREE and see what you can see.)*

The median, the measure of central tendency designed for ordinal level variables, is the "middle" case in a rank-ordered set of cases. The variable "height," when measured as it is for our students mentioned earlier in categories like "short," "medium," and "tall," is an ordinal level variable. The height for the Students A through E is reported to be "Tall," "Tall," "Short," "Short," and "Medium," respectively. If you arrange these data from shortest to tallest, they become "Short," "Short," "Medium," "Tall," and "Tall." The third, or middle case, in this series is "Medium," so the median is "Medium." Though it is designed for ordinal level variables, the median also can be used with interval variables, because interval variables, whatever else they are, are also ordinal variables.

 STOP & THINK *Can you calculate the median age for the sample mentioned earlier?*

The five students mentioned earlier have ages of 18, 18, 17, 19, and 20. Arranging these in order, from lowest to highest, they become 17, 18, 18, 19, and 20. The third, or middle, case in this series is 18, so the median is 18 years of age.

The median is pretty easy to calculate (especially for a computer) when you have an odd number of cases (provided, of course, you have at least ordinal level information). What you do with an even number of cases is locate the median halfway between the two middle cases. If, for instance, the "number of years of education" for eight cases were 9, 9, 2, 9, 9, 7, 9, and 12, you'd want to arrange them in order so that they would become 2, 7, 9, 9, 9, 9, 9, and 12. Then, you'd take the two middle cases (9 and 9) and find that the median for these cases is $(9 + 9)/2 = 9$.

The mean, the measure of central tendency designed for interval level variables, is the sum of all values divided by the number of values. The mean years of education for the first seven cases mentioned in the previous paragraph is

$$\frac{9 + 9 + 2 + 9 + 9 + 7 + 9}{7} = 7.7 \text{ years.}$$

How does a researcher know which measure of central tendency (mode, median, or mean) to use to describe a given variable? Beyond advising you not to use a measure that is inappropriate for a given level of measurement (such as a mean or a median for a nominal level variable like gender), we can't give you hard and fast rules. In general, though, when you are dealing with interval level variables (such as age, years in school, or number of confidantes), variables that *could* be described by all three averages, the relative familiarity of reading audiences with the mean makes it a pretty sound choice. Thus, although you could report that the modal number of years of education among the first seven cases listed above is 9 or that their median is 9, we'd be inclined to report that the mean is 7.7 years, other things being equal. See Box 15.1 for an example of a time when things are not equal.

Variation

Measures of central tendency can be very helpful at summarizing information, but as Stephen Jay Gould (1997) eloquently argued, they can also be misleading. Gould gave his own bout with a rare form of cancer, abdominal mesothelioma, as an example. He'd been diagnosed with this disease as a 40-year-old and learned, almost immediately, that mesothelioma was incurable and that people who had it had a median life expectancy of eight months. Gould reported that his understanding of statistics provided him almost immediate consolation (talk about knowing about statistics being personally useful!) because he knew that if all the journals were reporting life expectancy in terms of the median, then it was very likely that life expectancy for individuals was strongly skewed. Otherwise, the journals would be reporting a mean life expectancy. Moreover, he reckoned, the extreme values of life expectancy after diagnosis were almost surely on the high end, perhaps even many, many years after diagnosis. (One or two life expectancies of zero, after all, couldn't skew a distribution with a median of eight months that much.) Gould also realized that his own personal life expectancy was almost surely going to be on the high end of the scale. He was, after all, young, eager to fight the disease, living in a city (Boston) that offered excellent medical treatment, gifted with a supportive family, and "lucky that my disease had been discovered relatively early in its course" (Gould, 1997: 50).

Gould's larger point is a crucial one: human beings—indeed, anything we're likely to study as social scientists—are not measures of central tendency, not means, nor medians, nor modes. All things worth studying entail variation. In his case, he needed to be able to place himself in a region of variation based on particulars of his own case. Similarly, if we're interested in the educational level, income, or wealth of a group of people, averages will only get us so far: We need to know something about the variation of these variables.

We can discover a lot of important information about variation by describing the distribution of a sample over various categories of a variable. One of the most commonly used techniques for such descriptions is the **frequency distribution**, which shows the frequency (or number) of cases in each category of a variable. To display such a distribution, the categories of the variable are

frequency distribution, a way of showing the number of times each category of a variable occurs in a sample.

BOX **15.1**

Another Factor in Choosing an Average: Skewness

skewed variable, an interval level variable that has one or a few cases that fall into extreme categories.

Although our suggestion that you use the mean as your measure of central tendency for describing interval level variables works some of the time, it can yield pretty misleading "averages" for those many real-world interval level variables that are skewed. A **skewed variable** is an interval level variable that has one or a few cases that fall into extreme categories. Although we made up the following distribution of incomes for 11 people, it is not completely unlike the real distribution of incomes (and wealth) in the United States and the world, inasmuch as it depicts few very high income people and many low income people.

$20 million

$1,926,363.60 The Mean Income

$1 million

$100,000

$50,000

Median (the one in the middle; five above, five below)

$10,000

Mode (occurs most frequently)

This income variable is skewed because it contains at least one extreme case (the one falling into the $20 million category. The one falling in the $1 million category is "out there" too.). Because of this case, the mean of $1,926,363.60 provides a sense of average that somehow seems to misrepresent the central tendency of the distribution. In this case, as in the case of most skewed variables, we would advise reporting an average, like the median, that is less sensitive to extreme cases.

Do you agree that the median provides a better sense of this distribution's central tendency than the mean?

---

listed, and then the number of times each category occurs is counted and recorded. Box 15.1, with its display of hypothetical data about the income of 11 people, provides an example of a frequency distribution.

Let's imagine how we might create a frequency distribution for data about the gender of a hypothetical sample of respondents. Suppose we had 20 people in our sample and we found that there were 17 females and 3 males. The frequency distribution for this variable, then, is displayed in Table 15.1.

TABLE **15.1**
Frequency Distribution of Gender of Respondent

Gender	Frequency (f)	Percentages
Female	17	85
Male	3	15
Total	N = 20	100

One thing this frequency distribution demonstrates is that the overwhelming majority of our respondents were female. The table has three rows. The first two rows display categories of the variable (gender). The third row displays the total number of cases appearing in the table.

The middle column shows the number of cases in each category (17 and 3). The number is called the *frequency* and is often referred to by the letter *f*. The total of all frequencies, often referred to by the letter N, is equal to the total number of cases in the sample examined (in this case, 20). The third column shows the percentage of the total number of cases that appears in each category of the variable. Because 17 of the 20 cases are female, the percentage of cases in this category is 85 percent.

The variable in this example, gender, is a nominal level variable, but frequency distributions can be created for variables of ordinal and interval level as well. Let's confirm this by working with an interval level variable: the number of close friends our respondents have. Suppose our 20 respondents told us that they had 1, 0, 0, 0, 2, 1, 1, 1, 3, 0, 2, 2, 2, 1, 1, 1, 0, 0, 1, and 1 close friends, respectively.

Create the outline of an appropriate table on a separate piece of paper, using Table 15.2 as a guide.

Now complete the distribution. First, count the number of respondents who report having "0" close friends and put that number (frequency) in the appropriate spot in the table. Do the same for categories "1," "2," and "3." Do the frequencies in each of these categories add up to 20, as they should? If not, try again. Once they do, you can calculate the percentage of cases that fall into each category. (Calculate percentages by dividing each *f* by the N and multiplying by 100. We count six respondents with 0 close friends, so the percentage of

TABLE **15.2**
Frequency Distribution of Number of Close Friends

Number of Close Friends	Frequency (f)	Percentages
0		
1		
2		
3		
Total	N = 20	100

respondents with 0 close friends is $6/20 \times 100 = 30$.) Into which category does the greatest number of respondents fall? (We think it's category "1," with nine respondents.) In which category does the smallest number of respondents fall? What percentage of the respondents said they had 3 close friends?

Frequency distributions, simple as they are, can provide tremendous amounts of useful information about variation. One of our students, Angela Lang (2001), did her senior thesis on Americans' attitudes toward income inequality. Some of her sources had suggested that the American public is basically apathetic toward income inequality in the country and that a majority, in fact, felt that income inequality provided an "opportunity ... for those who are willing to work hard" (Ladd and Bowman, 1998: 36). Lang was therefore surprised to find, upon examining GSS data from 1996, that, when asked whether they agreed or disagreed with the statement "Differences in income in America are too large," Americans gave responses that are summarized in the frequency distribution depicted in Table 15.3.

Most striking to Lang was the percentage of the 1,468 respondents who answered the question and either strongly agreed or agreed with the statement "Differences in income in America are too large:" 32.3 percent and 32.5 percent, respectively. You can spot these percentages in the column labeled "percent." Using the cumulative percentage column, which presents the percentage of the total sample that falls into each category and every category above it in the table, Lang was also able to discern that 64.8 percent of this sample either strongly agreed or agreed with the statement. This was a much higher level of agreement than she expected based on her reading. Notice that the median and the mode for this variable were both "agree." The point, though, is that reporting that the median response was "agree" doesn't convey nearly as much information about the level of agreement as saying "About 65 percent of respondents either agreed or strongly agreed with the statement, whereas only about 20 percent disagreed or strongly disagreed."

 **STOP & THINK** *Can you see where the figure of 20 percent comes from?*

TABLE **15.3**

Frequency Distribution of Responses to the Statement "Differences in Income in America Are Too Large"

Response	Frequency	Percent	Cumulative Percent
Strongly agree	464	32.3	32.3
Agree	467	32.5	64.8
Neither	178	12.4	77.2
Disagree	169	11.8	89.0
Strongly disagree	117	8.1	97.1
Can't choose	41	2.9	100.0
Total	1,468	100.0	

Source: General Social Survey, 1996 (see Davis and Smith, 1998).

STOP & DO

Over the years the GSS has asked about 25,474 respondents whether, if their party nominated a woman for president of the United States and she were qualified for the job, they would vote for her. What percentage, would you guess, has said, "Yes"? What percentage has said, "No"? Check your answers against the actual percentage by going to http://sda.berkeley.edu/ archive.htm, and finding the frequency distribution for the variable FEPRES.

Measures of Dispersion or Variation for Interval Scale Variables

Examining frequency distributions, and their associated percentage distributions, is a pretty good way of getting a feel for dispersion or variation in nominal or ordinal level variables. If, for instance, you're looking at gender and discern that 100 percent of your sample is female and 0 percent is male, you know that there is no variation in gender in your sample. If, on the other hand, you find that 50 percent of your sample is female and 50 percent is male, you know that there is as much variation as there could be over two categories. Statisticians have actually given us more elegant ways of describing variation for interval level variables. Again, though, we start with the observation that the mean, perhaps the most frequently used measure of central tendency for interval level variables, can hide a great deal about the variable's *spread* or *dispersion*. Thus, a mean of 3 could describe both of the following samples:

measures of dispersion, measures that provide a sense of how spread out cases are over categories of a variable.

Sample A: 1, 1, 5, 5

Sample B: 3, 3, 3, 3

range, a measure of dispersion or spread designed for interval level variables. The difference between the highest and the lowest values.

Inspection shows, however, that Sample A's values (varying between 1 and 5) are more spread out or dispersed than those of Sample B (*all* of which are 3). To alleviate this problem, researchers sometimes report **measures of dispersion** for individual variables. The simplest of these is the **range**: the difference between the highest and the lowest value. The range of Sample A would be $5 - 1 = 4$, while that of Sample B is $3 - 3 = 0$. The two ranges tell us that the spread of Sample A is larger than the spread of Sample B.

STOP & THINK

Calculate the range of Samples C and D:

Sample C: 1, 1, 5, 5; Sample D: 1, 3, 3, 5

Having calculated the range, can you think of any disadvantage of the range as a measure of spread or dispersion?

standard deviation, a measure of dispersion designed for interval level variables that accounts for every value's distance from the sample mean.

There are several measures of spread that, like the range, require interval scale variables. The most commonly used is the **standard deviation**. The major disadvantage of the range is that it is sensitive only to extreme values (the highest and the lowest). Samples C and D in the last "Stop and Think" exercise have ranges of 4, but the spreads of these two samples are obviously not the same. In Sample C, each value is two "units" away from the sample mean of 3. In Sample D, two values (1 and 5) are two "units" away, but two values (3) are zero "units" away. In other words, the average "distance" or "variation" from the mean is greater in Sample C than in Sample D. The average variation is 2 in Sample C, but less than 2 in Sample D. The standard deviation is meant to capture this difference (one that isn't caught by the range) and to assign higher measures of spread to samples like Sample C than to those like Sample D. And, in fact, it does.

It does so by employing a computational formula that, in essence, adds up the "distances" of all individual values from the mean and divides by the number of values—a little like the computation of the mean in the first place. That's the essence. In fact, the computational formula is

$$s = \sqrt{\frac{\sum (X - \overline{X})^2}{N}}$$

where s stands for the standard deviation

$\overline{X}$ stands for the sample mean

X stands for each value

N stands for the number of sample cases

Although this formula might look a bit formidable, it's not very difficult to use. We'll show you how it works for Sample D.

First notice that the computational formula requires that you compute the sample mean ($\overline{X}$). For Sample D, the mean is 3. Then subtract this mean from each of the individual values, in turn ($X - \overline{X}$). For Sample D (whose values are 1, 3, 3, and 5), these differences are −2, 0, 0, and 2. Then square each of those differences [$(X - \overline{X})^2$]. For Sample D, this results in four terms: 4, 0, 0, and 4 (−2 squared = 4). Then sum these terms [$\sum(X - \overline{X})^2$]. For Sample D, this sum is 8 (4 + 0 + 0 + 4 = 8). Then the formula asks you to divide the sum by the number of cases [$\sum (X - \overline{X})^2/N$]. For Sample D, this quotient is 2 (8/4 = 2). Then the formula asks you to take the square root of this quotient. For Sample D, this is the square root of 2, or about 1.4. So the standard deviation of Sample D is about 1.4.

STOP & THINK *Now try to calculate the standard deviation of Sample C. Is the standard deviation of Sample C greater than (as we expected it would be) or less than the standard deviation of Sample D?*

We hope you've found that the standard deviation of Sample C (we calculate it to be 2) is greater than the standard deviation of Sample D. In any case, we're pretty sure you'll see by now that the standard deviation does require variables of interval or ratio scale. (Otherwise, for instance, how could you calculate a mean?) You might also see why a computer is helpful when you are computing many statistics.

The standard deviation has properties that make it a very useful measure of variation, especially when a variable is normally distributed. The graph of a **normal distribution** looks like a bell, with its "hump" in the middle and cases diminishing on both sides of the hump (see Figure 15.1). A normal distribution is symmetric. If you folded it in half at its center, one half would lie perfectly on the other half. One of the nice mathematical properties of a variable that has a normal distribution is that about 68 percent of the cases would fall between one standard deviation above the mean (the center of the distribution) and one standard deviation below the mean.

normal distribution, a distribution that is symmetrical and bell-shaped.

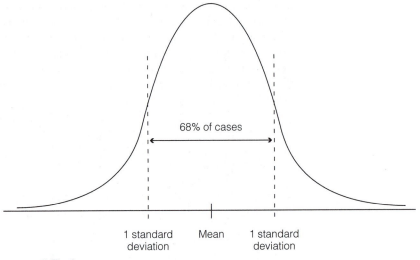

FIGURE **15.1** Normal Distribution

The standard deviation becomes less intuitively useful when a variable's distribution does not conform to the normal distribution, however. Many socially interesting variables do not so conform. Most of these nonconformers, like income in Box 15.1, are skewed. Figure 15.2 shows another skewed distribution: the distribution of the number of children respondents said they had in the 1996 GSS. As you might expect, most of the 2,889 who answered this question said they had zero, one, or two children. But just enough reported that they had five, six, seven, and eight (or more), that the distribution has a

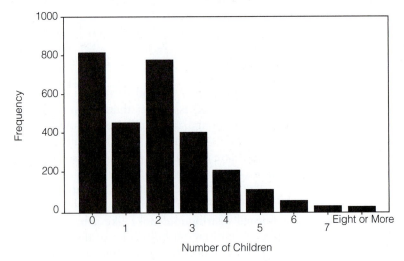

FIGURE **15.2** Number of Children Reported by General Social Survey Respondents. Mean = 1.8. Skewness = 1.0.

Source: The General Social Survey, 1996 (see Davis and Smith, 1998). Graph was produced using SPSS.

"tail" … this time to the right. Statisticians have provided another very useful statistic that we won't try to show you how to compute: skewness. We discussed the concept of skewness earlier, but skewness is so important that it also merits its own statistic. If the computed skewness statistic is zero, the variable is very likely to be nearly normally distributed. The skewness statistic for number of children, as depicted in Figure 15.2, is actually about 1, indicating that the "tail" of the distribution is to the right. Skewness values can be both positive and negative (when the "tail" is to the left) and can vary from positive to negative infinity. But when skewness gets near 1 (or −1), as it does in the case of number of children, or in the case of life expectancy at diagnosis, as it did for mesothelioma when Gould was diagnosed with the disease, one can no longer pretend that one is dealing with a normal distribution. As Gould's case indicates, however, distributions don't have to be normal to be interesting, enlightening, and even life-affirming … especially if one has some appreciation of statistical variation.

Bivariate Analyses

Analyzing single variables can be very revealing and provocative. (Think of Gould's musing about the "life expectancy at diagnosis" variable.) But sometimes the real fun begins only when you examine the *relationship* between variables. A relationship exists between two variables when categories of one variable tend to "go together" with categories of another. You might have found, if you tried to do the last "Stop and Do" exercise, that about 86.5 percent of adult Americans queried by the GSS from 1972 to 2008 have said that they would vote for a woman if she were nominated by their party and if she were qualified. But do you think this percentage has gone up, down, or remained the same over time since 1972? Let's suppose you think it might have generally gone up since 1972, perhaps because the women's movement since the late 1960s has made it more acceptable to Americans to envision women in leadership positions. If you think of this hypothesis in the variable language introduced in Chapter 2 and the passage of time and people's willingness to vote for a woman for president as variables, then you're saying that you expect the passage of time and people's willingness to vote for a woman to "go together." One way of depicting this expectation is shown in Figure 15.3.

crosstabulation, the process of making a bivariate table to examine a relationship between two variables.

One way of showing such a relationship is to crosstabulate, or create a bivariate table for, the two variables. **Crosstabulation** is a technique designed for examining the association between two nominal level variables and therefore, in principle, applies to variables of any level of measurement (because all ordinal and interval-ratio variables can also be treated as nominal level variables).

FIGURE **15.3** Our Hypothesis About the Relationship Between Time and the Willingness to Vote for a Woman President

Bivariate tables (sometimes called contingency tables) provide answers to questions such as "Has the percentage of adult Americans who are willing to vote for a woman president increased since 1972?" Or, more generally to questions such as "Is there a difference between sample members that fall into one category of an independent variable and their counterparts in other categories in their experience of another characteristic?"

 **STOP & DO**

Let's see what's actually happened to the percentage of adult Americans who say they'd be willing to vote for a woman president over time by going back to http://sda.berkeley.edu/archive.htm. This time, after you've gotten into the most recent GSS Cumulative Data File, and "started" the "Frequencies or crosstabulation (with charts)," put FEPRES in the "row" box and YEAR in the "column" box. You'll have crosstabulated the variable "Would You Vote for a Female President?" with the variable "Year of Survey." Examine the resulting table and see whether the percentage of adult Americans who say they would vote for a woman president has gone up, down, or remained the same since 1972.

Isn't it interesting how the percentage of American adults who say they would vote for a qualified woman for president inched up after 1972? At 73.6 percent in 1972, it rose, with minor exceptions, pretty steadily until it reached 94.0 percent in 2008. In fact, if we just focused on the years 1972 and 2008, as categories of the independent variable, "Year of Survey," and on "Yes" and "No," as categories of the dependent variable, "Would Respondent Vote for a Qualified Woman for President?" the crosstabulation of these two variables could be reduced to the bivariate (or contingency) table, Table 15.4. (We've calculated a couple of other statistics, "Cramer's V" and "p," and put them in Table 15.4 as well. We'll be getting to them soon.)

This table demonstrates that, as we hypothesized, the year of the survey is related to or associated with whether respondents answered they would vote for a qualified woman for president; it makes a difference what year it was.

Let's create a contingency table using data that Roger collected with students and colleagues (Clark et al., 2006) for a recent study of children's picture books. In this study, the researchers were interested to see, among other things, whether award-winning picture books created by black illustrators are more likely to make female characters central than award-winning picture books created by

TABLE **15.4**

Whether Respondent Would Vote for a Woman for President by Year of Survey, 1972 or 2008

Would Respondent Vote for Woman?	Yes	No	Total
1972	1,129 (73.6%)	404 (26.4%)	1,533 (100.0%)
2008	1,238 (94.0%)	80 (6.0%)	1,318 (100.0%)
	Cramer's V = .27	p < .05	

non-black illustrators. They collected data on, among other things, Caldecott award and honor books from 2000 to 2004 (created, as it turns out, by non-black illustrators and authors) and Coretta Scott King award and honor books from 2000 to 2004 (created by black illustrators and authors). The researchers hypothesized that the King books would make females more central to their books than the Caldecotts, because the King committee, unlike the Caldecott committee, is enjoined to pick "educational" books and because women have been central to the survival of the African American community in the United States.

Here are data on two variables for the 36 relevant award-winning books from 2000 to 2004: the type of award (Caldecott or King) and whether there is a female central character:

Award	Central Female?	Award	Central Female?
1. King	Yes	19. Caldecott	No
2. Caldecott	Yes	20. Caldecott	No
3. King	Yes	21. King	No
4. Caldecott	No	22. Caldecott	No
5. Caldecott	Yes	23. Caldecott	No
6. King	Yes	24. King	No
7. King	Yes	25. King	Yes
8. Caldecott	No	26. Caldecott	No
9. Caldecott	No	27. King	No
10. King	Yes	28. Caldecott	No
11. Caldecott	No	29. King	No
12. Caldecott	Yes	30. Caldecott	Yes
13. Caldecott	No	31. King	No
14. King	No	32. Caldecott	No
15. King	No	33. Caldecott	Yes
16. Caldecott	No	34. King	No
17. Caldecott	No	35. Caldecott	No
18. King	Yes	36. Caldecott	No

Note that 15 of these 36 books were Coretta Scott King award or honor books and that 21 were Caldecott award or honor books. Of the 15 King books, 7 (46.7 percent) have female central characters and 8 (53.3 percent) do not. Of the 21 Caldecott books, 5 (23.8 percent) have female central characters and 16 (76.2 percent) do not. We'll put this information into a bivariate (or contingency) table, Table 15.5.

 **STOP & THINK**   *What percentages would you compare from the table to show that King books were more likely to have female central characters than Caldecott books?*

TABLE **15.5**

Do Caldecotts or Kings from 2000 to 2004 Have More Female Central Characters?

Award	Does Book Have Female Central Character?		
	No	Yes	Total
Caldecott	16 (76.2%)	5 (23.8%)	21 (100.0%)
King	8 (53.3%)	7 (46.7%)	15 (100.0%)
	Cramer's V = .24	p > .05	

Note that the way we've formulated our hypothesis forces us to look at the relationship between type of book award and presence of central female character in a certain way. Using the variable language introduced in Chapter 2, we expected "type of book award" to affect "presence of central female character," rather than "presence of central female character" to affect "type of book award." In other words, "type of book award" is our independent variable and "presence of central female character" is our dependent variable. Given the formulation of our hypothesis and given the analysis of Table 15.5, we're in a good position to say that King books are more likely than Caldecott books to have central female characters.

Measures of Association

Before we leave the topic of bivariate analyses, let's note a few more important points. Table 15.4 contains some curious new symbols that are worth mentioning, mainly because they illustrate a whole class of others. One such class is called **measures of association**, of which "Cramer's V" is a specific example. Measures of association give a sense of the strength of a relationship between two variables, of how strongly two variables "go together" in the sample. Cramer's V can vary between 0 and 1, with 0 indicating that there is absolutely no relationship between the two variables and 1 indicating that there is a perfect relationship. A perfect relationship exists between two variables when change in one variable is always associated with a predictable change in the other variable. The closer Cramer's V is to 0, the weaker the relationship is; the farther from 0 (closer to 1), the stronger the relationship.

measures of association, measures that give a sense of the strength of a relationship between two variables.

 **STOP & THINK** *You'll note, from Table 15.5, that Cramer's V for the relationship type of award and the presence of female main characters is .24 and, from Table 15.4, that Cramer's V for year of survey and the willingness of respondents to vote for a qualified woman for president is .27. Which relationship is stronger?*

Measures of Correlation

Cramer's V is a measure of association that can be used when both variables are nominal level variables (and have at least two categories each). Statisticians have cooked up literally scores of measures of association, many of which can be distinguished from others by the levels of measurement for which they're designed: some for nominal, some for ordinal, and some for

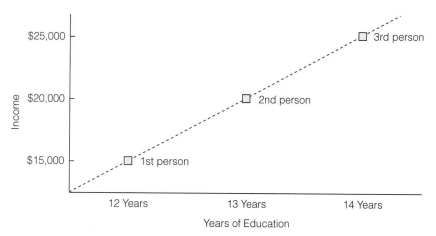

FIGURE **15.4** A Graph of the Relationship Between Education and Income for Our Hypothetical Sample

measures of correlation, measures that provide a sense not only of the strength of the relationship between two variables but also of its direction.

interval level variables. A particular favorite of social science researchers is one called Pearson's r, designed for examining relationships between interval level variables. Pearson's r falls in a special class of measures of associations: it's a **measure of correlation.** As such, it not only provides a sense of the strength of the relationship between two variables, it also provides a sense of the *direction* of the association. When variables are intervally (or ordinally) scaled, it is meaningful to say that they can go "up" and "down," as well as to say that they vary from one category to another.

Suppose, for instance, that you had data about the education and annual income of three people. Suppose further that the first of your persons had 12 years of education and made $15,000 in income, the second had 13 years of education and made $20,000, and the third had 14 years of education and made $25,000. In this case, not only could you say that the two variables (education and income) were related to each other (that every time education changed, income changed), but you could also say that they were *positively* or *directly* related to each other—that as education rises so does income. You might even be tempted to graph such a relationship, as we have done in Figure 15.4.

One feature of our graphical representation of the relationship between education and income is particularly striking: All the points fall on a straight line. One striking feature of this line is that it rises from the bottom left of the graph to the top right. When such a situation exists (i.e., when the data points from two variables fall on a straight line that rises from bottom left to top right), the correlation between two variables is said to be perfect (because all points fall on a line) and positive (because it goes from the bottom left up to the top right), and the Pearson's r associated with their relationship is 1.

STOP & THINK *Suppose your sample included three individuals, one of whom had 14 years of education and made $15,000 in income, one of whom had 13 years of education and made $20,000, and one of whom had 12 years of education* *and made $25,000. What would a graphical representation of this data look like? Can you imagine what the Pearson's r associated with this relationship would be?*

If, on the other hand, the data points for two variables fell on a line that went from the "top left" of a graph to the "bottom right" (as they would for the data in the "Stop and Think" exercise), the relationship would be perfect and negative and the Pearson's r associated with their relationship would be −1. Thus, Pearson's r, unlike Cramer's V but like all other measures of correlation, can take on positive *and* negative values, between 1 and −1.

In general, negative values of Pearson's r (less than 0 to −1) indicate that the two variables are *negatively* or *indirectly* related, that is, as one variable goes up in values, the other goes down. Positive values of Pearson's r (greater than 0 to 1) indicate that the two variables are directly related—as one goes up, the other goes up. In both cases, r's "distance" from 0 indicates the strength of the relationship: the farther from 0, the stronger the relationship. Thus, a Pearson's r of .70 and a Pearson's r of −.70 indicate relationships of equal strength (quite strong!) but opposite directions. The first (r = .70) indicates that as one variable's values go up, the other's also go up (pretty consistently); the second (r = −.70) indicates that as one variable's values go up, the other's go down (again pretty consistently). Figure 15.5 provides a graphical representation (known as a *scattergram*) of what relationships (with many data points) with r's of .70 and −.70 might look like. Notice that, although in neither case do the points perfectly conform to a straight line, in both cases they all hover around a line. r's of .30 and −.30 would, if you can imagine them, conform even less well to a straight line.

STOP & THINK *Would you expect the association between education and income for adults in the United States to be positive or negative?*

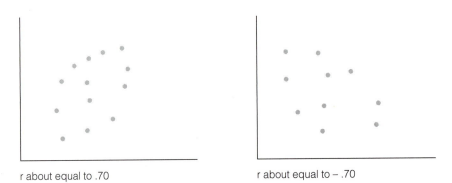

r about equal to .70   r about equal to − .70

FIGURE **15.5** Idealized Scattergrams of Relationships of r = .70 and −.70

STOP & DO   *Try doing a correlation analysis for the association between years of education and income for adults in the United States in 2006. Go back to http://sda.berkeley.edu/archive.htm, enter the most recent GSS Cumulative Data File, start the "correlation matrix" action, enter the variable names "educ" and "rincom06" into the first two input boxes before hitting "run correlations." Check the correlation in the middle of the page. Look at the intersection of a row headed by "educ" and a column headed by "rincom06" (or a row headed by "rincom06" and a column headed by "educ") and read the correlation coefficient (the Pearson's r).*

We did the analysis in the "Stop and Do" and found what we hope you did as well, that the Pearson's r for the association between education and income in the 2006 GSS sample was .27. As it turns out, the association is a positive one (the positive "sign" of .27 indicates a direct relationship), but the relationship is nowhere near a perfect one (.27 is closer to 0 than to 1)—so one would not expect the data points to conform very well to a straight line.

STOP & THINK   *Suppose the Pearson's r for the relationship between years of education and incomes had proven to be −.67, instead of .27. How would you describe the relationship?*

We could tell you more about correlation analysis, but any standard statistics book could do so in more detail. We think you'll find correlation analysis a particularly handy form of analysis if you ever want to study the associations among interval level variables.

Inferential Statistics

We draw your attention to a final detail about Table 15.4 before we conclude our discussion of bivariate relationships. On the same row as Cramer's V in the table is a funny-looking set of symbol, "p < .05." This symbol, standing for "probability is less than .05," enables the reader to make an inference about whether a relationship, like the one shown in the table for the sample (of 2,854 respondents), exists in the larger population from which the sample is drawn. This "probability" is estimated using what we've called inferential statistics, in this case chi-square, perhaps the most popular inferential statistic used in crosstabulation analysis. There are almost as many inferential statistics as there are measures of association and correlation, and a discussion of any one of them stands outside the scope of this chapter. Inferential statistics are properly computed when one has a probability sample (see Chapter 5). They are designed to permit inferences about the larger populations from which probability samples are drawn.

You might recall from Chapter 5, however, that, even with probability sampling, it is possible to draw samples whose statistics misrepresent (sometimes even badly misrepresent) population parameters. Thus, for instance, it is possible to draw a sample in which a strong relationship exists between two variables even when no such relationship exists in the larger population. Barring any other information, you'd probably be inclined to infer that a relationship also exists in the larger population. This kind of error—the one where you infer that a relationship exists in a population when it really doesn't—is called a Type I error. Such an error could cause serious problems, practically and scientifically.

Social scientists are a conservative lot and want to keep the chances, or probability, of making a Type I error pretty small—generally lower than 5 times in 100, or "$p < .05$." When the chances are greater than 5 in 100, or "$p > .05$," social scientists generally decline to take the risk of inferring that a relationship exists in the larger population.

 **STOP & THINK** *When "$p < .05$," social scientists generally take the plunge and infer that a relationship, like the one in the sample, exists in the larger population. Would social scientists take such a plunge with the relationship between year of survey and respondents' willingness to vote for women for president as displayed in Table 15.4?*

The probability of making a Type 1 error, the "p," is related to the size of the sample. The larger the sample, the easier it is to achieve a p of less than .05, and therefore it is easier to feel comfortable with an inference that a relationship exists in the larger population. Consequently, it is not surprising that the association between year of survey and respondents' willingness to vote for women for president as displayed in Table 15.4 is associated with a p less than .05, even though the strength of the association (as measured by a Cramer's V of .27) is not terribly strong. The number of cases in the sample is 2,854, after all.

Measures of association, measures of correlation, and inferential statistics are not an exotic branch of statistical cookery. They are the "meat and potatoes" of quantitative data analysis. We hope we've presented enough about them to tempt you into learning more about them. We'd particularly recommend Marija Norusis's *The SPSS Guide to Data Analysis* (2009), which, in addition to offering a pretty good guide to SPSS (the statistical package we used to generate several tables and figures), offers a good guide to the kinds of statistics we've referred to here.

Multivariate Analysis and the Elaboration Model

Examination of bivariate relationships is the core activity of quantitative data analysis. It establishes the plausibility of the hypothesized relationships that inspired analysis in the first place. Suppose we expected that the happiness of a person's marriage would be positively associated with the person's overall, or general, happiness. To use the variable language introduced in earlier chapters, we'd expect that an independent variable, marital happiness, would be associated with the dependent variables, general happiness. We could then turn to GSS data to test our proposition. Table 15.6 uses such data (from 2008) and demonstrates that our guess was right: 58.9 percent of people who reported being very happy with their marriages reported being very happy generally, while only 11.5 percent of people who reported being less

TABLE **15.6**

Relationship between Marital and General Happiness (2008)

| | | General Happiness | | |
		Very Happy	Less than Very Happy	Total
	Very Happy	58.9%	41.1%	100%
Marital		415	290	705
Happiness	Less than Very Happy	11.5%	88.5%	100%
		50	382	432
		Cramer's V = .47	p < .05	

Source: General Social Survey, 2008 (http://sda.berkeley.edu/archive.htm).

than very happy with their marriages reported being very happy generally. Our bivariate analysis would demonstrate what we expected: that happily married people are more likely to be happy than other people. But the examination of bivariate relationships is frequently just the beginning of quantitative data analyses.

You might ask: Why would a researcher want to examine more than two variables at a time? One answer is that we might be worried that there's some obvious challenge to the thesis we're proposing: in this case, that marital happiness causes general happiness. You may recall, from Chapter 2, that one such challenge is the possible presence of a third variable, antecedent to the independent and dependent variables, which is responsible for its association. Hans Zeisel, in his classic *Say It with Figures* (1968: 108), reminds us of an old Chinese statistical joke that people who visit a doctor have considerably higher mortality rates than people who don't. Obviously, in this case, we should look at the association between doctor visits and mortality separately for those who were sick and those who were not. This, essentially, is the point of multivariate (more-than-two variable) analyses.

As the doctor/mortality example suggests, a researcher will often want to pursue questions like "Is there a third variable (say, sickness), associated with both the independent (say, doctor visiting) and the dependent variable (say, mortality), that can account for the association between the two?" or "Are there certain conditions under which the association between the independent and dependent variables is particularly strong or weak?" or "Are there any other variables, affected by the independent variable and affecting the dependent variable, that can help me understand how the independent variable affects the dependent variable?" The pursuit of any such question can lead researchers to elaborate on the original bivariate relationship by introducing additional variables. In doing so, they will employ the **elaboration** model. A complete description of Paul Lazarsfeld's (see Rosenberg, 1968) elaboration model is beyond the scope of this book. But, we think you deserve at least a taste of it.

elaboration, the process of examining the relationship between two variables by introducing the control for another variable or variables.

Recall our finding, from Table 15.6, that people who are happy with their marriages are more likely to be generally happy than others. We may want to figure out why. Perhaps being happy in one area of life makes people happier in general. But before we jump to this conclusion we might want to consider that marital and general happiness are *not* causally related at all. We might fear that others will argue that they are not really related and their association reflects their common association with a third variable, say gender, that forces them to "go together," as it were. The explanation might go something like this: Women tend to be happier with their marriages than men and women tend to be happier in general than men, therefore the relative grumpiness of men (both with their marriages and in general) and the relative happiness of women (with both) accounts for the apparent association between marital and general happiness. In this case, gender is hypothesized to be an antecedent **control variable** because it is expected to affect both marital happiness and general happiness. (In the variable language of earlier chapters, it is a possible independent variable for both marital and general happiness.) This possibility is depicted in Figure 15.6.

control variable, a variable that is held constant to examine the relationship between two other variables.

Let's be clear. We've entertained the possibility that gender might prove the reason why marital and general happiness are related, but we're doing it partly because we think our critics might think it's the reason. We don't really want gender to prove to be the reason. (That would undermine our argument that marital happiness itself has a significant impact on general happiness.) To show that it wasn't the reason, we need to understand how one would go about showing that it was (or, at least might be). To do that, one would have to do three things: (1) show that gender and marital happiness are related, (2) show that gender and general happiness are related, and (3) show that when gender is *controlled*, the association between marital and general happiness "disappears." Thanks to Lazarsfeld's elaboration model, we know that all three of these conditions have to exist to conclude that gender "explains" the relationship between

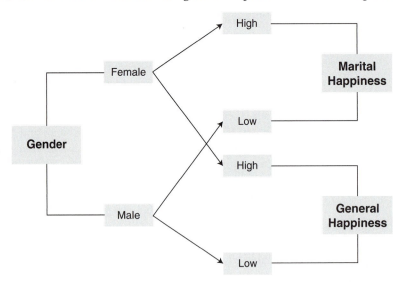

FIGURE **15.6** Gender as an Antecedent Control Variable of Marital Happiness and General Happiness

marital and general happiness. Actually, explanation is the term Lazarsfeld gave to that form of elaboration that shows a spurious relationship: a relationship that is explained away by an antecedent control variable.

Can you imagine what it would mean to show that when gender is *controlled*, the association between marital and general happiness "disappears"? Understanding the introduction of a control variable is really the key to understanding the elaboration model, so we'd like to show you how it's done.

In the context of crosstabulation, the concept of *controlling* a relationship (e.g., between marital and general happiness) for a third variable (e.g., gender) is reasonably straightforward. It means looking at the original association separately for cases that fall into each category of the control variable. For the control variable, gender, there are two categories: female and male. Controlling the relationship between marital and general happiness for gender means, then, examining that relationship only for females, on the one hand, and only for males, on the other.

But what does it mean to say that a relationship "disappears" when you control for a third variable? Table 15.7 illustrates with a hypothetical example.

TABLE **15.7**

Hypothetical Example Demonstrating How the Relationship between Marital and General Happiness Could "Disappear" with Gender Controlled

		General Happiness		
		Very Happy	Less than Very Happy	Total
Full Sample				
Marital Happiness	Very Happy	58.9%	41.1%	100%
		415	290	705
	Less than Very Happy	11.5%	88.5%	100%
		50	382	432
		Cramer's V = .47	p < .05	
Females				
Marital Happiness	Very Happy	100%	0%	100%
		415	0	415
	Less than Very Happy	100%	0%	100%
		50	0	50
		Cramer's V = .00	p > .05	
Males				
Marital Happiness	Very Happy	0%	100%	100%
		0	290	290
	Less than Very Happy	0%	100%	100%
		0	382	382
		Cramer's V = .00	p > .05	

The first segment of this table simply replicates the information about the association between marital and general happiness from Table 15.6 and shows that, for the full sample of 1,137 respondents, a substantial relationship exists. We could emphasize the strength of this relationship by comparing the percentages (e.g., 58.9 percent of very happily married people reported being generally very happy, whereas only 11.5 percent of people reporting themselves less than happily married reported being generally very happy) or by looking at the measure of association (a Cramer's V of .47 indicates a strong association).

The second and third segments of this table reflect the introduction of the control variable, gender. The second segment presents only hypothetical information about females. Here we examine what is sometimes called a **partial relationship** (or partial, for short) because we're only examining the original relationship for part of the sample—for females. Among females, the table suggests, all 415 respondents who were reported being happily married were also reported being happy generally, and all 50 respondents who were reported being less than very happily married were also reported being happy generally. In other words, there is no relationship between marital and general happiness for women—a fact that is underscored by a Cramer's V of .00. Similarly, the third segment of the table presents hypothetical information about males. Within this group too, there is no relationship between marital and general happiness.

partial relationship, the relationship between an independent and a dependent variable for that part of a sample defined by one category of a control variable.

 **STOP & THINK** *How can we tell there's no relationship in the partial involving males?*

Consequently, because there's no relationship between marital and general happiness for females and males, the relationship that existed with no control has "disappeared" in each of the controlled situations. Table 15.7, then, makes the kind of demonstration that would lead researchers to believe that gender "causes" the association between marital and general happiness. It shows a situation where the control for an antecedent variable makes the relationship between an independent and a dependent variable "disappear."

This table was based on hypothetical data. As it turns out, nothing like what the table shows occurs in reality. Table 15.8 shows what really occurs when we control the relationship between marital and general happiness for gender, using data collected in the 2008 GSS. The real data indicate that the partial relationship between marital and general happiness is almost exactly as strong for females (Cramer's V = .48), on the one hand, and for males (Cramer's V = .46), on the other, as it had been in the larger sample (Cramer's V = .47). Certainly, the relationship doesn't "disappear" for either females or males.

Thus, the introduction of a control for gender does not do what we feared it might: "explain away" the original relationship. As a result, we're in a stronger position to assert that the relationship might be "causal" (rather than spurious, or caused by the action of an antecedent variable that makes the variables in the relationship appear to vary together).

TABLE **15.8**

What Really Happened to the Relationship between Marital and General Happiness with Gender Controlled

		General Happiness		
		Very Happy	Less than Very Happy	Total
Females				
Marital Happiness	Very Happy	62.8%	37.2%	100%
		222	132	353
	Less than Very Happy	14.5%	85.5%	100%
		34	200	234
		Cramer's V = .48	p < .05	
Males				
Marital Happiness	Very Happy	55.1%	44.9%	100%
		194	158	352
	Less than Very Happy	7.9%	92.1%	100%
		16	182	198
		Cramer's V = .46	p < .05	

Source: General Social Survey, 2008 (http://sda.berkeley.edu/archive.htm).

 STOP & THINK *Of course, you cannot entirely eliminate the possibility that a relationship is spurious. In Chapter 2, we suggested that such an elimination would require showing that there is no antecedent variable whose control makes the original relationship disappear. Can you see why this is not possible?*

To summarize and, well, elaborate, elaboration permits us to better understand the meaning of a relationship (or lack of relationship) between two variables. We've actually discussed two types of elaboration earlier, but there are others as well. Let us briefly introduce you to four kinds of elaboration: replication, explanation, specification, and interpretation.

replication, a kind of elaboration in which the original relationship is replicated by all the partial relationships.

1. **Replication.** The original relationship can be replicated in each of the partial relationships. This was essentially true when we controlled the relationship between marital and general happiness by gender (see Table 15.8). The partials (with Cramer's Vs of .48 and .46) were just about the same as the original (Cramer's V = .47).

explanation, a kind of elaboration in which the original relationship is explained away as spurious by a control for an antecedent variable.

2. **Explanation.** The original relationship can be *explained away* (or found to be *spurious*) through the introduction of the control variable. Two variables that were related in the bivariate situation might no longer be related when an *antecedent* variable is controlled. This happened in the hypothetical (but unreal) situation, shown in Table 15.7, when the relationship between marital and general happiness "disappeared" both for females and for males.

specification, a kind of elaboration that permits the researcher to specify conditions under which the original relationship is particularly strong or weak.

3. **Specification.** Introducing a control variable might permit the researcher to *specify* conditions under which the original relationship is particularly strong or weak. Table 15.9 provides an example of specification, using data from the 2008 GSS. Here, again, the major concern is with the relationship between marital and general happiness, but now we're controlling for self-reported health—whether a respondent feels in excellent health or not. When we separate the original sample into those reporting excellent health and those reporting less than excellent health (i.e., when we control for health), we find that the relationship between marital and general happiness differs for the two groups. Thus, the association between marital and general happiness is stronger for people in excellent health (Cramer's V = .50) than it is for people in less than excellent health (Cramer's V = .44). In other words, marital happiness seems to do more for the happiness of people in excellent health than for people in less than excellent health. In still other words, we can specify a condition—that is, being in excellent health—under which marital happiness does more for general happiness than under other conditions (in this case, being in less than excellent health).

TABLE **15.9**
Relationship between Marital and General Happiness with State of Health Controlled

| | | General Happiness | | |
		Very Happy	Less than Very Happy	Total
People in Excellent Health				
Marital Happiness	Very Happy	72.1%	27.9%	100%
		93	53	130
	Less than Very	19.2%	80.8%	100%
	Happy	13	53	66
		Cramer's V = .50	p < .05	
People in Less than Excellent Health				
Marital Happiness	Very Happy	55.4%	44.6%	100%
		198	159	357
	Less than Very	11.0%	89.0%	100%
	Happy	23	185	205
		Cramer's V = .44	p < .05	

Source: General Social Survey, 2008 (http://sda.berkeley.edu/archive.htm).

**STOP & THINK** *Can you imagine why people in less than excellent health get less of a boost in their general happiness by being in a happy marriage than people in excellent health?*

interpretation, a kind of elaboration that provides an idea of the reasons why an original relationship exist without challenging the belief that the original relationship is causal.

4. **Interpretation.** A control variable can give the researcher an idea about why the independent variable is related to the dependent variable *without challenging the belief that the one causes the other*, as it does in explanation. In this case, the control is said to *interpret* rather than *explain* the relationship. Interpretation occurs when, in the mind of the researcher, the control isn't antecedent to both the independent and the dependent variable (as it would be in explanation) but, rather, intervenes between them. An *intervening* control variable is one that "comes between" the independent and dependent variable and, therefore, at least partially accounts for their association. In such a situation, Lazarsfeld tells us, the partials (as measured by, say, Cramer's V) will be lower than the original. We provide an example of what we believe to be interpretation in Table 15.10, which is based on 2006 GSS data. Here the major concern is with whether a person's income is associated with whether he or she voted for John Kerry, the Democratic candidate, or George Bush, the Republican candidate, in the 2004 presidential election (people who voted for other candidates were eliminated). What the table shows is that people whose family income was less than $35,000 were more likely to have voted for Kerry than Bush and those whose family income were $35,000 or more were more likely to have voted for Bush than for Kerry.

STOP & THINK *What percentage of the people with family incomes of less than $35,000 voted for Kerry? What percentage of these people voted for Bush? Can you imagine why higher income people might have voted for Bush, while lower income people might have voted for Kerry?*

The Cramer's V for the association, .10, is modest at best, but noticeable. We then asked ourselves "What variable might intervene between people's income and their vote for president?" and considered their political party affiliation as a possibility. In particular, we thought it possible that people who had lower incomes might well affiliate themselves with the Democratic party and, then, in turn, vote for the Democratic candidate, in this case John Kerry. We also thought it possible that people who had higher incomes might affiliate with the Republican party and, in turn, vote for the Republican candidate, in this case George Bush. We depict this expectation, that party affiliation might be an intervening variable between income and presidential vote, in Figure 15.7. Note that, in this figure, unlike the situation in Figure 15.6, the control variable is envisioned as falling between the independent and dependent variables.

We coded three categories of political party affiliation: Democrat, Independent, and Republican. Table 15.10 shows the association between income and presidential vote for each of these categories. Given our expectation that

TABLE **15.10**

An Example of Interpretation: The Association between Family Income and Presidential Vote for Democrats, Independents, and Republicans

	Voted for Kerry	Voted for Bush
Full Sample		
Income less than $35,000	57.5%	42.6%
	614	453
Income $35,000 or more	43.1%	56.9%
	1,034	1,364
	Cramer's V = .10	p < .05
Democrats		
Income less than $35,000	85.7%	14.4%
	527	88
Income $35,000 or more	86.0%	14.0%
	868	141
	Cramer's V = .01	p > .05
Independents		
Income less than $35,000	50.0%	50.0%
	52	52
Income $35,000 or more	40.0%	60.0%
	87	125
	Cramer's V = .08	p > .05
Republicans		
Income less than $35,000	7.8%	92.2%
	26	307
Income $35,000 or more	6.3%	93.7%
	71	1,061
	Cramer's V = .03	p > .05

Source: General Social Survey, 2006 (http://sda.berkeley.edu/archive.htm).

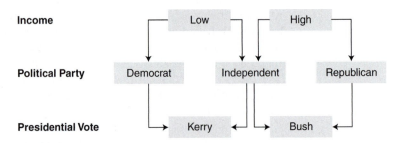

FIGURE **15.7** Political Party as an Intervening Variable between Income and Presidential Vote

party affiliation would help "interpret" the relationship between income and vote, we expected each of the partials to be lower than the original (Cramer's $V = .10$). And they were. The partial for Democrats, as you can see from the table, was .01. The partial for Independents was .08, and the partial for Republicans was .03. Crucially, however, the diminution in this case has a very different meaning than it does when the control variable is antecedent to both the original variables. Now it suggests that income affects political party affiliation, which in turn affects presidential vote. Income's causal effect on presidential vote is not questioned; it's simply interpreted as income's effect on party affiliation and party affiliation's effect on voting.

The great advantage of elaboration is that it gives us the tools to think beyond mere two-variable relations: to question, and test, whether they might not be causal (as in the case of explanation); to wonder, and see, whether they might be stronger under some conditions and weaker under others (as in the case of specification); and to think about, and assess, possible mechanisms through which the effects of an independent variable work on a dependent variable (as in the case of interpretation). The model even permits us to anticipate why others might believe a favored two-variable relationship is not causal and show that their objection does not withstand the test of empirical investigation (as in the case of replication).

QUALITATIVE DATA ANALYSIS

> Qualitative data are sexy. They are a source of well-grounded, rich descriptions and explanations of processes in identifiable local contexts.... Then, too, good qualitative data are more likely to lead to serendipitous findings and to new integrations.... Finally, the findings from qualitative studies have a quality of "undeniability." Words, especially organized into incidents or stories, have a concrete, vivid, meaningful flavor that often proves far more convincing to a reader ... than pages of summarized numbers. (Miles and Huberman, 1994: 1)

As we indicated at the beginning of this chapter and as the quotation from Matthew Miles and Michael Huberman (1994) suggests, a major distinction between qualitative and quantitative data analyses lies in their products: words, "especially organized into incidents or stories," on the one hand, and "pages of summarized numbers," on the other. Even though, as Miles and Huberman argue, the former are frequently more tasty to readers than the latter, the canons, or recipes, that guide qualitative data analysis are less well defined than those for its quantitative counterpart. As Miles and Huberman elsewhere observe, "qualitative researchers ... are in a more fluid—and a more pioneering— position" (1994: 12). There is, however, a growing body of widely accepted principles used in qualitative analysis. We turn to these in this chapter.

If the outputs of qualitative data analyses are usually words, the inputs (the data themselves) are also usually words—typically in the form of extended texts. These data (the words) are almost always derived from what the researcher has observed (e.g., through the observation techniques discussed in Chapter 11), heard in interviews (discussed in Chapter 10), or found in documents (discussed in Chapter 13). Thus, the data analyzed by Mueller, Dirks, and Houts Picca were their students' observations and interviews of other students dressing for Halloween, and the data analyzed by Enos were the words of incarcerated women. The outputs of both data analyses were the essays themselves (see Chapters 10 and 11).

Social Anthropological versus Interpretivist Approaches

Although there might be general agreement about the nature of the inputs and outputs of qualitative analyses (words), there are nonetheless some fundamental disagreements among qualitative data analysts. One basic division involves theoretical perspectives about the nature of social life. To illustrate, and following Miles and Huberman (1994: 8ff.), look at two general approaches: a social anthropological and an interpretative approach. Social anthropologists (and others, like grounded theorists and life historians) believe that there exist behavioral regularities (e.g., rules, rituals, relationships) that affect everyday life and that it should be the goal of researchers to uncover and explain those regularities. Interpretivists (including phenomenologists and symbolic interactionists) believe that actors, including researchers themselves, are forever interpreting situations and that these, often quite unpredictable, interpretations largely affect what goes on. As a result, interpretivists see the goal of research to be their own (self-conscious) "accounts" of how others (and they themselves) have come to these interpretations (or "accounts"). The resulting concerns with law-like relationships, by social anthropologists, and with the creation of meaning, by interpretivists, can lead to distinct data-analytic approaches. Given this difference, our discussion in this chapter will generally be more in line with the beliefs and approaches of the social anthropologists (and those with similar perspectives) than it is with those of the interpretivists. Nevertheless, we believe that much of what we say applies to either approach and that their differences can be overstated. Thus, although Enos clearly sought to discover law-like patterns in the experience of inmate mothering (remember how white mothers were more likely to have their children placed in foster care than black mothers), she was nonetheless sensitive to variation in the interpretations or accounts given by all mothers.

 **STOP & THINK** *When Enos pursued law-like patterns in placement of children, she was doing what the social anthropologists advocate. What about when she looked to the accounts given by the mothers?*

Does Qualitative Data Analysis Emerge from or Generate the Data Collected?

In qualitative field studies, analysis is conceived as an *emergent* product of a process of gradual induction. Guided by the data being gathered and the topics, questions, and evaluative criteria that provide focus, analysis is the fieldworker's *derivative ordering* of the data. (Lofland and Lofland, 1995: 181)

In ethnography the analysis of data is not a distinct stage of the research. It begins in the pre-fieldwork phase, in the formulation and clarification of research problems, and continues into the process of writing up. (Hammersley and Atkinson, 1983: 174)

These quotations from John Lofland and Lyn Lofland (1995) and Martyn Hammersley and Paul Atkinson (1983) seem to point to another debate concerning qualitative analysis: whether qualitative analysis *emerges from* or *generates* the data collected. At issue seems to be, to paraphrase the more familiar chicken-or-egg dilemma, the question of which comes first: data or ideas about data (e.g., theory). Lofland and Lofland stress the creative, after-the-fact nature of the endeavor by describing it as a "process of gradual induction"—of building theory from data. Hammersley and Atkinson emphasize how data can be affected by conceptions of the research topic, even before one enters the field. These positions, however different in appearance, are, in our opinion, both true. Rather than being in conflict, they are two sides of the same coin. Patricia Adler and Peter Adler (1994b), in discussing their own practice in studying afterschool activities, have told us, in personal communication, about their efforts to immerse themselves in their setting first, *before* examining whatever literature is available on their topic, to better "grasp the field as members do" without bias. But *something* compelled their interest in afterschool activities, and whatever that was surely entailed certain preconceptions. Sandra Enos (the author of the focal research report in Chapter 10) is much more explicit about her preconceptions and, in conversation, has confided that one of her expectations before observing mother–child visits was that she'd see mothers attempting to create "quality time" with their children during the visits. Still, she remained open to the "serendipitous findings" and "new integrations" mentioned by Miles and Huberman. When she found mothers casually greeting their kids with "high-fives," rather than warmly embracing them, as she'd expected, she dutifully took notes. Because Enos has been explicit about various stages of her data analysis, and because she's been willing to share her field notes with us, we use her data to illustrate our discussion of qualitative data analysis in this chapter.

The Strengths and Weaknesses of Qualitative Data Analysis Revisited

Before we actually introduce you to some of the techniques that are increasingly associated with qualitative data analysis, we want, once again, to suggest one of its great strengths. As Gary Fine and Kimberly Elsbach suggest, qualitative data can produce theories that "more accurately describe real-world issues and processes than do quantitative data" (2000). Because qualitative data usually come from some combination of the observation or interview of real-world participants, they are more likely to be grounded in the immediate experiences of those participants than in the speculations of researchers. Thus, rather than simply handing out a set of closed-ended questions to students who participated in Halloween activities, Mueller, Dirks, and Houts Picca (see Chapter 11) asked other students to observe them as they participated in those activities. Consequently, they didn't need to worry, quite as much, about whether they were asking the right questions (and supplying the right answers, as they would if they'd used closed-ended questions). They received observations as well as accounts of what the observations meant. The resulting capacity of qualitative data to yield insights into the longer-term dynamics of behavior is also a great advantage in producing theories that "describe real-world ... processes." We've mentioned earlier (see, for instance, Chapter 11) that one of the disadvantages of qualitative data is the questionable generalizability of the theories they lead to. Because Enos's study was based on a convenience sample of incarcerated mothers, for instance, the extent to which her findings may be typical is unclear. In general, though, we feel that the richness and accuracy of theories generated from qualitative data analysis are well worth the trouble, especially if their limitations are kept in mind and, perhaps, investigated further through future research.

Are There Predictable Steps in Qualitative Data Analysis?

Quantitative data analysis often follows a fairly predictable pattern: Researchers, after they have coded their own data or acquired computer-ready available data, almost always begin by doing some sort of univariate analyses, perhaps followed by bivariate and multivariate analyses. The "steps" involved in executing good qualitative analyses, however, are much less predictable, and a lot "more fluid," as Miles and Huberman suggest. Miles and Huberman prefer to see qualitative analysis as consisting of "concurrent flows of activity," rather than as "steps." They dub these flows "data reduction," "data display," and "conclusion drawing/verification" and depict the interactions among these flows as we do in Figure 15.8.

This depiction of qualitative data analysis involves several notable assumptions. First, even data collection is not immune to the "flows" of data analysis, being itself subject to the (perhaps tentative) conclusions a researcher entertains at various stages of a project (note the arrow from "conclusions" to "data collection"). Enos had expected inmate mothers to greet their children with warm embraces (even before she entered the field) and so focused attention on the gestures of greeting and, as you'll see, collected data about them. Second, Figure 15.8 suggests that data reduction and data display (both discussed later) are not only products of data collection but of one another and of conclusions, even if such conclusions are not yet fully developed. Thus, Enos's coding of her data (a data reduction process described later) not only contributed to her final conclusions about mothering in prison, but might have been itself a product of earlier, perhaps more tentative, conclusions she'd reached. Finally, the figure emphasizes that the conclusions a

FIGURE **15.8** Components of Data Analysis: Interactive Model from Miles and Huberman: 12, 1994. Used with permission

researcher draws are not just products of earlier flows (data reduction and data display) but continually inform both of the other "flows," as well as the data collected in the first place.

We'd like to emphasize that one danger of the model depicted in Figure 15.8 is that it could give the impression that qualitative data analysis invariably follows a predictable pattern of data collection and transcription, followed by data reduction (particularly coding and memoing), followed by data displaying, followed by conclusion drawing and verification. It's not that simple. We think, however, that the advantages of such a model outweigh its dangers. One important advantage is that it gives new (and even experienced) practitioners a sense of order in the face of what otherwise might feel like terribly uncharted and, therefore, frightening terrain. Angela Lang (in Clark and Lang, 2001: 1) speaks of the anxiety she experienced when confronted with the demands of generating theory from qualitative data for an undergraduate project: "I soon realized I had to reinvent my creative side. I was nervous that I would discover that I wasn't creative at all." But, once she'd finished the project, she reflected on the model implicit in Figure 15.8 as a source of inspiration and comfort. She said that, by doing qualitative data analysis, she "had gained a tremendous amount of confidence in [her]self." And that, "If I were to do a qualitative project again, I'd probably try to organize my work in much … [the same way]. I'd probably plan to create fieldnotes, code data, memo, do data displays, review literature, write drafts of the final paper … and hope that the ideas and research questions began to float into my head, as they did this time" (Clark and Lang, 2001: 5–6).

Data Collection and Transcription

We've discussed at some length the process of collecting qualitative data in earlier chapters: through qualitative interviewing (Chapter 10), observation techniques (Chapter 11), studying the content of texts (Chapter 13), and studying the results of various forms of visual analysis (Chapters 10, 11, and 13). So far, however, we haven't shown you any actual data. We'd like to present a small portion of Enos's field notes, as she composed them on the evening after an interaction with one mother in the prison parenting program. These notes were based on observation and conversation in the off-site parenting program, where women got together with their children. The excerpt in Figure 15.9 is from nearly 150 pages of field notes Enos took before she engaged in the qualitative interviewing she reports on in the essay within Chapter 10.

 **STOP & THINK** *One of Enos's expectations had been that women would greet their children with hugs. How did this woman greet her son? How did he greet her?*

The physical appearance of these field notes brings up the issue of computer usage in qualitative data analysis. Enos did all her written work, note collection, and analysis with an elementary word-processing program, a practice that is extremely common these days. Enos did not use one of the increasingly available software packages (e.g., NUD*IST, NVIVO, Atlas.ti, Kwalitan, MAX, and QUALPRO) designed to facilitate the processing of qualitative

I spoke to a woman who had remained
silent in the group. She was
SITTING ALONE AT A TABLE. I ASKED
if I could talk with her. She has
six children, two of her children
will be visiting today. Her mother
has always had her kids. Last year
her mother passed away and now her
sister is taking care of the kids.
When she's released, she'll have
responsibility for the whole bunch.
She says it was hard to see your
kids while you were in jail but that
if you were DCYF[1] involved sometimes
it was easier. (This has to do with
termination of rights.) She
remarked that it was important not
to be a "stranger" to your kids
while you were in jail. It made
everything worse. I asked about
telling your kids where you are.
She said she just straight out told
them and they accepted. "Mommy's in
jail. That's it." She thought
having such direct experience with
jail that maybe her kids would
"avoid" it when they got older.
I asked her when she'd be released.

[1]Department of Children, Youth & families.

FIGURE **15.9** A Segment from Enos' Field Notes

data, until she entered the qualitative interview phase of her work when she used a package called HyperResearch. If you think you'd like to try one of these packages, you might find Lewins and Silver's (2007) guide helpful for getting started. We believe that much good qualitative analysis can be done without the use of anything more sophisticated than a word processor—and are inclined to point out that much has been achieved with nothing more sophisticated than pen and paper. (See Box 15.2 for some sense of the advantages and disadvantages of computer-assisted qualitative data analysis software [CAQDAS].)

Before we return to Enos's transcribed field notes, however, we'd like to point to one trend in CAQDAS that shows great promise: the direct analysis of audio-recorded data using computer software. The transcription of audio-tapes, as mentioned in Chapter 10, is enormously time consuming, sometimes taking, in the case of interviews, up to four hours for an hour's worth of interview. Gibson, Callery, Campbell, Hall, and Richards (2005) and Hutchinson (2005) have used Atlas.ti and GoldWave software, respectively, to analyze sound files and found them useful. But Gibson et al. (2005), in particular,

In a few months, but she'd be
going to a drug treatment program.
She's been in prison many times
before and this time she's going to
do things right. She's made a
decision to change, to do something
new.
She is very happy with this program.
She likes the fact that there are
no officers around, that kids can
run and scream, that there is an
extended time out of the joint and
that she can spend more time with
her kids.
Her children arrived, a boy five
years old or so and his older
sister. They walked over to us.
The mother greeted her son, "Hey,
Dude! How you doing?" He didn't say
anything. The woman asked for the
children's coats and hung them on the
chair backs. She introduced me to
the kids, who were quiet and
beautiful, very well cared for. I
complimented her on these kids and
excused myself so she could be alone
with them.

FIGURE **15.9** Continued

claim that there remain substantial obstacles to the direct integration of audio materials into qualitative analysis: for example, absence of good enough
recording technologies and the insufficient development of interfaces between recording devices and computers. An even more recent test of the relative efficiency of current speech recognition software finds that it is still
not much quicker than traditional methods of transcription (Dresing, Pehl,
and Lombardo, 2008). But this is one technology that, in the near future,
could make the analysis of audio data faster and cheaper than before. For
now, though, back to Enos's field notes.

The physical appearance of these field notes might be the first thing that
strikes you. Enos uses only half a page, the left column, for her field notes in
anticipation of a need to make notes about the notes later on (see "Data Reduction"). Leaving room for such notes about notes is excellent practice.
Enos didn't have an interview guide at this stage of her investigation, but she
evidently had certain questions in mind as she approached this woman. Her
notes indicate that she asked about how the woman explained her incarceration to her children and about her expectations for release and her current
child-care arrangements.

BOX **15.2**

The Debate about CAQDAS

Advocates of CAQDAS claim that it will

- Speed up and enliven coding (see "Coding")
- Offer a clear-cut structure for writing and storing memos (see "Memoing")
- Provide a "more complex way of looking at the relationships in the data" (Barry, 1998: 2.1)

Opponents see the possibility that CAQDAS will

- Distance analysts from their data
- Lead to qualitative data being used quantitatively

- Generate an unhealthy sameness in the methods used by analysts
- Continue to favor word-processed text over other forms of data such as sketches, maps, photos, video images, or recorded sound (Dohan and Sanchez-Jankowski, 1998: 483)

Barry (1998) gives a relatively balanced appraisal of these positions and offers specific insights into two of the most popular varieties of CAQDAS, NUDIST, and Atlas.ti. You can read it on the web at www. socresonline.org.uk/socresonline/3/3/4.html.

---

anticipatory data reduction, decision making about what kinds of data are desirable that is done in advance of collecting data.

Enos's interest in the aforementioned questions indicates that she's probably engaged in a fair amount of what Miles and Huberman (1994: 16) call **anticipatory data reduction**—that is, decision making about what kinds of data she'd like to collect. Admittedly, her research strategy (more or less "hanging out" at an off-site parenting program for female prisoners) is pretty loose at this stage.

There is quite a bit of debate about the relative merits of "loose" and "tight" (those that entail explicit conceptual frameworks and standard instruments) research strategies in field research. Those advocating relatively loose strategies or "designs" (e.g., Wolcott, 1982) point to the number of questions and answers that can be generated through them. Susan Chase, for instance, in the study mentioned in Chapter 10, was primarily interested in getting female school superintendents to tell their life stories and was willing to use any question that stimulated such stories. She was willing to "hang out" during these interviews, though she does confess that she eventually found asking female superintendents to describe, generally, their work histories, was a pretty good way to induce "lively, lengthy, and engrossing stor[ies]" (Chase, 1995: 8).

But the advantages of relatively tight strategies, especially for saving time spent in the field, are worth noting. Enos developed an appreciation of tight strategies during the course of her fieldwork. She's indicated how tempting it is to pursue every idea that occurs in a setting:

> So many questions occurred to me after I left the site. I would consider what women had said to me and then try to figure out if that was related to race/ethnicity, to class, to previous commitments to prison, to number of children, to street smarts, to sophistication about the system, and so on and so on. (Enos, personal communication)

In the end, she realized that she couldn't pursue all those questions that she needed, in her words, "to keep on track." She claims that the most important task was to continually sharpen her focus as the fieldwork went on and that this meant "finally coming to the conclusion that you can't look at everything" (personal communication).

By the time Enos entered the interview stage of her project, she'd developed a very tight design indeed, having, among other things, composed the interview guide shown at the end of her report in Chapter 10. This guide led Enos to ask very specific questions of her respondents, about their age of first incarceration, and so on.

Data Reduction

data reduction, the various ways in which a researcher orders collected and transcribed data.

Qualitative **data reduction** refers to the various ways a researcher orders collected data. Just as questioning 1,000 respondents in a closed-ended questionnaire can generate more data on most variables than the researcher can summarize without statistics, field notes or transcripts also can generate volumes of information that need reduction. Some researchers can do a reasonably good job of culling out essential materials, or at least can tell compelling stories, without resorting to systematic data reduction procedures in advance of report writing. But most can't, or don't. Especially in long-term projects with ongoing data collection, most researchers engage in frequent data reduction. In the following section, we'll focus on two common data reduction procedures: coding and memoing. In doing so, we'll be emphasizing one thing implicit in Figure 15.8: Data reduction is affected by the kinds of data you collect but is also affected by, and affects, the conclusions you draw.

Coding

Coding, in both quantitative and qualitative data analysis, refers to assigning observations, or data, to categories. In quantitative analysis, the researcher usually assigns the data to categories of variables he or she has already created, categories that he or she frequently gives numbers to in preparation for computer processing. In qualitative analysis, coding is more open-ended because both the relevant variables and their significant categories are apt to remain in question longer. In any case, coding usually refers to two distinct, but related, processes: The first is associating words or labels with passages in one's field notes or transcripts; the second is collecting similarly labeled passages into files. The goal of coding is to create categories (words or labels) that can be used to organize information about different cases (whether situations, individuals, or so on). When you don't have much of a framework for analysis, as Enos didn't initially, the codes (words or labels), at least initially, can feel a little haphazard and arbitrary. They can nonetheless be useful guides to future analysis. Thus, Enos's own coding of the initial paragraph of her notes (from Figure 15.9) looked something like what appears in Figure 15.10.

Here we see Enos, in her first informal interview, developing codes that will appear again and again in the margins of her field notes: codes that permit her to characterize how social mothering is managed while a biological mother is in prison. The codes "permanent substitute" and "temporary substitute," referring to this particular woman's mother and sister, respectively, become categories for classifying the kinds of mothering that all other children receive.

 **STOP & THINK** *Suppose you wanted to compare all the data that Enos coded with the words "temporary substitute." What might you do to facilitate such a comparison?*

Fieldnotes	Codes
I spoke to a woman who had remained silent in the group. She was sitting alone at a table. I asked if I could talk with her. She has six children, two of her children will be visiting today. Her mother has always had her kids. Last year her mother passed away and now her sister is taking care of the kids. When she's released, she'll have responsibility for the whole bunch. She says it was hard to see your kids while you were in jail but that if you were DCYF involved sometimes it was easier. (This has to do with termination of rights.) She remarked that it was important not to be a "stranger" to your kids while you were in jail. It made everything worse. I asked about telling your kids where you are. She said she just straight out told them and they accepted. "Mommy's in jail. That's it." She thought having such direct experience with jail that maybe her kids would "avoid" it when they got older.	women as birthgivers, not caretakers motherhood management permanent substitute temporary substitute

FIGURE **15.10** Enos' Coding of Fieldnotes

Assigning a code to a piece of data (i.e., assigning "temporary substitute" to this woman's sister or "permanent substitute" to her mother) is just the first step in coding. The second step is physically putting the coded data together with other data you've coded in the same way. There are basically two ways of doing this: the old-fashioned way, which is to put similar data into a file folder, and the new-fashioned way, which is to do essentially the same thing using a computer program that stores the "files" in the computer itself. Old-fashioned manual filing is simple enough: you create file folders with code names on their tabs, then place a copy of every note you have coded in a certain way into the folder with the appropriate label. In these days of relatively cheap photocopying and the capacity of computers to make multiple copies, you can make as many copies of each page as there are codes on it and slip a copy of the page into each file with the corresponding code. (In the bad old days, this process required relatively cumbersome manipulations of carbon paper and scissors.) Thus, Enos might have placed all pages on which she placed the codes "temporary substitute" into one folder and all pages on which she wrote the code "permanent substitute" into another. Later in her analysis, when she wanted to see all the data she'd coded in the same way, she'd just have to pull out the file with the appropriate label.

The logic of the new-fashioned, computer-based filing is the same as that of the old-fashioned one. The newer way requires less paper and no actual file folders (and can therefore result in a great savings of money and trees), but it does generally require special computer software (so the initial outlay can be substantial). "Code-and-retrieve programs," for instance, like Kwalitan, MAX, and NUDIST, help you divide notes into smaller parts, assign codes to the parts, and find and display all similarly coded parts (Miles and Huberman, 1994: 312). Computer-based filing also makes revising codes simpler. For those who are interested, we recommend Weitzman and Miles (1995), Richards and Richards (1994), Barry (1998), and Dohan and Sanchez-Jankowski (1998) for discussions of what's available.

Types of Coding

One purpose of coding is to keep facts straight. When you code for this purpose alone, as you might be doing if you were to assign a person's name to all paragraphs about that person, you'd be engaged in purely *descriptive coding*. Descriptive codes can emerge from the data and usually do from when "loose" research designs are employed. Initially Lang (in Clark and Lang, 2001: 11), who looked at respondents' open-ended reflections to an image of Sojourner Truth, found herself coding the reflections as "complex" or "simple," with no particular idea where she might be going with those codes. Sometimes, and especially when research designs are of moderate "tightness," as in Levendosky, Lynch, and Graham-Bermann's (2000) interview study of women's perceptions of abuse, the codes can be suggested by participants themselves. Responses, in this study, to the question, "How do you think that violence you have experienced from your partner has affected your parenting of your child?" led naturally to seven categories, such as "no impact on parenting" and "reducing the amount of emotional energy and time available," into which all accounts could be placed. Sometimes, especially when a very "tight" research design is employed, descriptive codes actually exist before the data are collected. Thus, Dixon, Roscigno, and Hodson (2004: 3–33) report that even before their coders read the 133 book-length ethnographies of workplaces in the United States and England, they'd been prepared to look for and code evidence of strikes, union organization, internal solidarity, and a legacy of workplace activism. The coders found 21 workplaces (out of the 133) that had strikes during the period of observation, 67 that were unionized, 78 that exhibited average to strong levels of solidarity, and 16 that had a legacy of frequent strike activity.

Because the basic goal of coding is to advance your analysis, you'd eventually want to start doing a more *analytic* type of coding. Enos's simultaneous development of the two codes "permanent substitute" and "temporary substitute" is an example of analytic coding. Kathy Charmaz calls this preliminary phase of analytic coding *initial coding* and defines it as the kind through which "researchers look for what they can define and discover in the data" (1983: 113). This is also a point when one's early, provisional conclusions (see Figure 15.8) can inform one's choice about codes.

Eventually, according to Charmaz, initial coding gives way to *focused coding*, during which the initial codes themselves become the subject of one's coding exercise. At this stage, the researcher engages in a selective process, discarding, perhaps, those codes that haven't been terribly productive (e.g., Enos never used the "women as birthgivers" code after its first appearance) and concentrating or elaborating on the more enlightening ones. After Enos saw all the ways in which women found temporary substitutes, for instance, she began to use variations in those ways—for example, relatives, foster care, friends—as codes, or categories, of their own.

Memoing

memos, more or less extended notes that the researcher writes to help herself or himself understand the meaning of codes.

All data analyses move beyond coding once the researcher tries to make sense of the coded data. In quantitative analyses, the researcher typically uses statistical analyses to make such sense. In qualitative analyses, other techniques, such as memoing and data displaying, are used. **Memos** are more or less extended notes that the researcher writes to help understand the meaning of coding categories. Memos can vary greatly in length (from a sentence to a few pages) and their purpose, again, is to advance the analysis of one's data (notes). An example (labeled distinctly as a theoretical note) of a memo appears about 100 pages after Enos's write-up of the encounter with the woman whose children had been (permanently) cared for by her mother until the mother died, but were now being (temporarily) cared for by her sister. It appears to be a reflection on all the material that's been coded "temporary substitute" and "permanent substitute" throughout the field notes. This memo is shown in Figure 15.11.

 **STOP & THINK** *Enos tells herself to think about the literature on family breakdown, something that the mother's imprisonment can also lead to. What do you think are some of the more familiar causes of family breakdown?*

Notice that Enos makes meaning by drawing an analogy between family separations brought about by imprisonment and those brought about by divorce, separation, death, and so on. She's focusing her attention on what happens to the care of children when they lose daily access to a parent (in this case, the mother). This has clearly become a central concern for future consideration. With this memo as a springboard, it's not surprising that child care and relationships with family members are a key issue in the research she reports in Chapter 10.

> Theoretical note
> Consider the work on the breakdown
> of the family and compare that to
> the efforts of these families to
> keep their children in their midsts.
> These children are being raised by
> persons other than their birth
> mothers but they are well dressed,
> taken care of, appear healthy, and
> so forth. If they are on welfare,
> there appears to be another source
> of support.

FIGURE **15.11** Illustration of Memo from Enos

Because fieldwork and coding can be very absorbing, it's easy to lose track of the forest for the trees. The practice of memoing, of regularly drawing attention to the task of analysis through writing, is one way that many qualitative researchers force themselves to "look for the big picture." Memos can be thought of as the building blocks of the larger story one eventually tells. Some memos, such as Enos's, still need some refinement (e.g., through the reading of the literature on family breakdown), but others qualify for inclusion in the final report without much alteration. Some, especially those that are memos about linkages among memos, can provide the organizing ideas for the final report. Memos are a form of writing and have the potential for creative insight and discovery that writing always brings. One can easily imagine that Hannah Frith and Celia Kitzinger's (1998) key theoretical question came from a memo about whether qualitative data are transparent reflections of reality or context-dependent constructions of participants. In their own analysis of young women in focus groups talking about how they refuse men sex, Frith and Kitzinger decided that the talk needed to be seen as generative and perhaps reflective of group constructions. The women spoke of their efforts to protect the feelings of the man they'd refused. Frith and Kitzinger saw the women's talk as "constructing" men as needy, and perhaps weak, and themselves as knowledgeable agents, in opposition to traditional stereotypes of men as sexual predators and women as vulnerable prey. The talk was not a "transparent" window on what went on during the refusals, Fritz and Kitzinger conclude. Roger (Clark, 2001) was also made aware of such group constructions when he asked faculty at a particularly successful urban middle school to talk, in focus groups, about how they'd managed to make differences in test performance between high income and low income students "disappear." His memos led to this awareness as he tried to interpret the focus-group interactions in which one faculty member would come up with an explanation and others would agree vigorously. Thus, for instance, one faculty member mentioned at one point that he or she maintained high standards for all of his or her students, both high and low income. When others voiced immediate agreement (and still others independently wrote of high standards in response to an open-ended question), Roger didn't necessarily feel he'd "seen" what went on in individual classrooms, but he did feel he had evidence of a common definition of faculty values at this school.

Memos can be potent theoretical building blocks, but they can also be vital methodological building blocks. Enos tells us, for instance, that memos were a particularly useful way for her to figure out what she needed to "learn next" during her study and how to actually learn it (personal communication). One memo (presented in Figure 15.12) shows Enos deciding to adopt a more formal interview strategy than she'd used before. This memo constituted a transition point between the fieldwork and the interview stages of Enos's study.

Data Displays

If memos, in their relative simplicity, are analogous to the relatively simple univariate examinations of quantitative analyses, data displays can be seen as a rough analogue to the more complex bivariate and multivariate analyses

Methodological note
While it has been easy to use my role as researcher and "playmate" to gain some information about mothers and children in prison, it has also been frustrating from a data collection perspective. I am involved in some long and wide-ranging conversations with the women and trying to remember the tone, the pacing, and the content of discussion while engaging in these as an active member. I can disappear for a few minutes to an adjoining room to scribble notes but this is difficult. I focus on the points that are most salient to my interests but can get sidetracked because so much is of interest to me. A more regimented conversation in an atmosphere that is more controlled than this one would probably yield "better" notes.

FIGURE **15.12** Methodological Memo from Enos

used there. Miles and Huberman (1994) and Williamson and Long (2005) are particularly adamant about the value of data displays for enhancing both the processes of data reduction, admittedly a means to an end, and conclusion making and verifying, the ultimate goals of data analysis.

data display, a visual image that summarizes information.

A **data display** is a visual image that summarizes a good deal of information. Its aim is to present the researcher himself or herself with a visual image of what's going on and help him or her to discern patterns. Miles and Huberman's figure (Figure 15.8) itself can be seen as an example of such a display, summarizing the interaction among various processes in qualitative analysis.

The crucial feature of data displays is that they are tentative summaries of what the researcher thinks he or she knows about his or her subject. Therefore data displays permit him or her to examine what he or she thinks by engaging in more data examination (and reduction) and, eventually, to posit conclusions. Enos, for instance, tends to use the space on the right half of her field note pages not only for coding but also for data displays, often in the form of exploratory typologies about the kinds of things she's seeing. Thus, next to notes she's taken about one mother, "Pam," and her interaction with her children, "Tom" (a baby) and "Joy" (a 3-year-old), we find a display (in Figure 15.13) about how mothers seem to manage their weekly interactions with their children in the parenting program.

Enos's coding of the first paragraph (referring to Pam's attempt to normalize her interaction with Tom and Joy) with a label that she's used before ("attempt at normalization") clearly has pushed her, by the time she rereads the second paragraph, to play with a list of ways she's seen mothers interact with

There were a number of infants visiting today. One of them, Tom, [2], is the son of Pam who also has a three year old girl. Pam plays actively with her children, is with them on the carpeted area, holding the baby and roughhousing with her daughter. Joy is an active and very outgoing kid who asks to have her hair fixed by her mom who is struggling to feed the baby and pay attention to her daughter.

attempt at normalization

The baby is tiny for his age and pale. These children are being taken care of by Pam's boyfriend's mother. Pam has a problem with cocaine and got high the day before Tom was born. She says that before he was born all she was interested in was getting high and the "chase." During her pregnancy with Joy, she quit smoking and was drug free.

Setting management
Mother Behavior
1) no playing with child
2) active playing
3) sporadic conversation
4) steady conversation
5) steady interaction
6) kid in other's care

2 All names are pseudonyms.

FIGURE **15.13** Enos' Emergent Typology of Mothering Behavior in the Parenting Program

their children at the parenting program. (Nothing in the second paragraph, in fact, is likely to have sparked this reflection.) This typology is the kind of data display that Miles and Huberman advocate. It pictorially summarizes observations Enos has made about the varieties of mothering behavior she's seen, and its presence facilitates her evaluation of its adequacy for subsequent presentation.

The forms in which data displays occur are limited only by your imagination. Figure 15.8 is a flowchart; Enos's display is a more conventional-looking typology. Miles and Huberman (1994) also advocate using various kinds of matrices, something that Roger and a student co-author, Heidi Kulkin, used in preparing their findings about 16 young-adult novels about non-white, non-American, or non-heterosexual main characters. Heidi and Roger found themselves most interested in themes of oppression and resistance in such novels and needed a quick way of summarizing the variation they'd discovered. The matrix shown in Table 15.11 worked. Preparing this matrix was helpful in two ways. First, it revealed some things they definitely wanted to say in their final analysis. It gave them the idea, for instance, of organizing their written presentation into four parts: stories about contemporary American characters, historical American characters, non-American characters, and historical non-American characters. Reviewing the table also led them to valid observations

TABLE **15.11**

Content Analytic Summary Matrix Used in Preparation of Clark and Kulkin's (1996) "Toward Multicultural Feminist Perspective on Fiction for Young Adults"

Place	Contemporary	Historical
United States	No. of books: 4	No. of books: 3
	Varieties of oppression: racism, classism, cultural imperialism, heterosexism	Varieties of oppression: sexism, racism
	Strategies of resistance: escape, resort to traditional custom, communion with similarly oppressed others	Strategies of resistance: escape, cooperation, extraordinary self-sacrifice, learning, writing
Outside United States	No. of books: 4	No. of books: 5
	Varieties of oppression: sexism, classism, racism, imperialism	Varieties of oppression: all the usual suspects
	Strategies of resistance: education, communion with similarly oppressed persons, creative expression, self-reliance, self-sacrifice	Strategies of resistance: escape, communion with similarly oppressed others, writing, self-definition, education, self-sacrifice

of between-group differences for their sample. Thus, for instance, they observed that authors writing about contemporary American characters were less likely to emphasize the importance of formal education and extraordinary self-sacrifice than those writing about other subjects.

Preparing the matrix, then, led Heidi and Roger to some observations they directly translated into their final report (Clark and Kulkin, 1996). It also helped them in another, perhaps more important, way: It warned them against saying certain things in that report. Thus, for instance, although each mode of oppression and resistance listed in each "cell" of the matrix was prominent in some novel that fell within that cell, each was not equally evident in all the novels of the cell. Heidi and Roger knew, then, that their final written presentation would have to be much more finely attuned to variation within each of their "cells" than the matrix permitted.

Sometimes, after coding, a researcher will have essentially comparable and simple information about cases. Such was Lang's case after she'd coded respondents' reflections on images of Sojourner Truth. Initial coding had led her to observe that some respondents wrote relatively complex reflections. A memo, which consciously located all reflections in the early morning class in which they'd been received, led Lang to ask respondents whether they considered themselves "morning," "afternoon," or "night" people. Lang then organized the two sets of information (about complexity and whether people considered themselves "morning" people, etc.) using a kind of display that Miles and Huberman call a meta-matrix or "monster-dog." Table 15.12 is one of the monster-dogs that Lang created, albeit a relatively small and tame one. Lang's monster-dog, like all good examples of its breed, is a "stacking up" of "single-case displays" (Miles and Huberman, 1994: 178), one on top of the other, so that the researcher can make comparisons. As Lang reports, "I looked at several Monster Dogs, each displaying different sets of variables. I was soon able to see helpful patterns. For instance, self-described morning people looked as though they gave more elaborate responses to the photograph of Sojourner Truth than [others] did. When I crosstabulated these two variables, there was in fact clearly a relationship" (Clark and Lang, 2001: 14).

Data displays of various kinds, then, can be useful tools when you analyze qualitative data. Miles and Huberman would take this as too weak an assertion. They claim

> You know what you display. Valid analysis requires, and is driven by, displays that are focused enough to permit a viewing of a full data set in the same location, and are arranged systematically to answer the research questions at hand. (Miles and Huberman, 1994: 91–92)

TABLE **15.12**

Example of Lang's Monster-Dog Data Displays: Elaborateness of Response versus When Respondent Feels Most Awake

Respondent	Elaborate	Morning	Afternoon	Night
A				
B		X		
C	X		X	
D			X	
E	X	X		
F		X		
G	X			
H	X	X		
I			X	
J	X			X
K				X
L		X		
M	X	X		
N	X	X		
O				X
P	X	X		

Note: X indicates presence of a trait; no X indicates absence.

Source: Adapted from Clark and Lang, 2001.

Although we think this position is overstated, if only because we know of good qualitative analyses that make no explicit reference to data displays, we think you could do worse than try to create data displays of your own while you're engaged in qualitative data analysis. When you do, you might well look to Miles and Huberman (1994) for advice.

Conclusion Drawing and Verification

The most important phase of qualitative data analysis is simultaneously its most exciting and potentially frightening: the phase during which you draw and verify conclusions. Data displays are one excellent means to this end because they force you to engage in two essential activities: pattern recognition and synthesis. Coding, of course, is a small step in the ongoing effort to do these things, and memoing is a larger one. But, you can hardly avoid these activities when you're engaged in creating a data display.

Enos emphasizes that one very important resource for verifying conclusions, especially for researchers who are doing announced field observations or qualitative interviews, is the people being observed or interviewed. These people, she reminds us, can quickly tell you "how off-base you are:" "This was very helpful to me, as was asking questions they could theorize about, like 'It seems that white women use foster care a lot more than black women do. Is that right?' [And, if they agree, 'Why is that the case?']" (personal communication).

Elliot Liebow (1993) advocates, however, an "Ask me no questions, I'll tell you no lies" principle, especially relative to participant observation. In his study of homeless women, he consciously decided not to bring the kind of theoretical questions to the women that Enos recommends. In doing so, he claimed that he was less concerned with the "lies" he might receive in return for his questions than with "contaminating the situation with questions dragged in from the outside" (Liebow, 1993: 321).

Whatever you decide to do about the issue of "bouncing ideas" off those you're studying, we think you'll find that a vital phase of conclusion drawing and verification comes when you compose your final report. The very act of writing, we (and others [e.g., Lofland and Lofland, 1995]) find, is itself inseparable from the analysis process. As a result, we divide the remainder of the section on drawing and verifying conclusions into a discussion of two phases: the pre-report-writing and the report-writing phase.

Pre-Report Writing

Lofland and Lofland (1995) point out that the growing use of computers for qualitative data analysis has helped reinforce two points. The first is that qualitative analysis can be facilitated by the kinds of things computers do, like storing and retrieving information quickly. The new programs that facilitate data coding testify to these strengths. The second is that good qualitative analysis requires what computers aren't really much good at. One of these tasks, and something that really distinguishes qualitative research from other brands of social research, is deciding what problems are important. Enos began her study of women in prison with an open mind about what the real problems were. The same was true of Mueller, Dirks, and Houts Picca (see Chapter 11) as they approached cross-racial Halloween costuming, and Clark and Kulkin (1996) as they began to read nontraditional novels for young adults. These authors all came to their settings with open eyes, open ears, and, particularly, open minds about what they'd find to be the most important problems.

 **STOP & THINK** *Contrast this openness with the hypothesis-driven approach of Martinez and Lee (see Chapter 12).*

Finding out just what problem you should address requires something more than what computers are good at: the imagination, creativity, and moral insight of the human brain.

Imagination and creativity are necessary even after you've chosen your topic and are well into the data analysis. One way to stimulate such imagination and creativity is to use the kinds of data displays advocated by Miles and Huberman. But, as Lofland, Snow, Anderson, and Lofland (2005) point out, other practices can create fertile conditions for qualitative analysis. One practice is to *alter data displays*. If you've created a data display that, somehow, isn't useful, try a different form of display. We can imagine, for instance, Miles and Huberman building the flowchart that we've reproduced in Figure 15.8, indicating the various interrelationships that might exist among data collection, data reduction, data display, and conclusions, out of a simple typology of (with no indication of relationships among) activities involved in qualitative analysis: data collection, data reduction, data display, and conclusions. Note that altering data displays involves conclusion verification and dismissal, both important goals of data analysis.

Another pre-report-writing activity, advocated by Lofland and Lofland, and before them by Glaser and Strauss (1967), is that of *constant comparison*. The key question here is this: How is this item (e.g., person, place, or thing) similar to or different from itself at a different time or comparable items in other settings? One can, again, easily imagine how such comparisons led Enos to her eventual interest (see Chapter 10) in how substitute mothering differs for women from different racial backgrounds, or Mueller, Dirks, and Houts Picca (Chapter 11) to their focus the various meanings of cross-racial Halloween dressing, or to Clark and Kulkin's interest in how themes of oppression and resistance vary by geographical or historical context, or to Lang's (see Clark and Lang, 2001) belief that "morning" and other people might differ in their capacity to give complex responses early in the morning. Constant comparison is also implicit in Janet Mancini Billson's (1991) progressive verification method for analyzing individual and joint interviews (see Chapter 10).

Lofland et al. (2005) actually recommend several other pre-report-writing techniques for developing creative approaches to your data, but we'd like to highlight just one more of these: *speaking and listening to fellow analysts*. Conversations, particularly with friendly fellow analysts (and these needn't, of course, be limited to people who are actually doing qualitative data analysis but can include other intelligent friends) can not only spark new ideas (remember the adage "two heads are better than one?") but also provide the kind of moral support we all need when ideas aren't coming easily. Friendly conversation is famous not only for showing people what's "right in front of their noses," but also for providing totally new insights. Lang is explicit about the value of sharing her ideas with other students in a data-analysis class devoted to this purpose: "The discussion was similar to having someone proofread a paper for you. Sometimes outsiders help you find better ways to phrase sentences, etc. In this case, they helped me articulate to myself what I was most interested in. The class felt like one big peer review session" (Clark and Lang, 2001: 11–12).

Qualitative Report Writing

A qualitative report writer on report writing had this to say:

> When I began to write, I still had no order, outline, or structure in mind ...
> I plunged ahead for many months without knowing where I was going, hoping and believing that some kind of structure and story line would eventually jump out. My original intention was to write a flat descriptive study and let the women speak for themselves—that is, to let the descriptive data and anecdotes drive the writing. In retrospect, that was a mistake. Ideas drive a study, not observations or unadorned facts ... I shuffled my remaining cards again and discovered that many cards wanted to be grouped under Problems in Day-to-Day Living, so I undertook to write that chunk. The first paragraph asked the question, How do the women survive the inhuman conditions that confront them? After I wrote that paragraph it occurred to me that most situations, experiences, and processes could be seen as working for or against the women's survival and their humanity—that the question of survival might serve as the opening and organizing theme of the book as a whole and give the enterprise some kind of structure. (Liebow, 1993: 325–326)

Lang's reflections on her writing of an undergraduate paper, based on qualitative data, made related, but slightly different points:

> I finally sat down and began writing the paper. It was now that the project actually began to make sense and fit together. One of my biggest concerns going into this step was how I would incorporate the literature reviewed with the rest of my findings. I'd found literature, I knew, that was somewhat relevant, but nothing that spoke directly to my findings that morning people would write more elaborately in the morning. As I wrote, I discovered that my worries were misplaced. Things did flow. The fact that the literature I reviewed didn't speak directly to what I'd found became a virtue; I could simply say that I was addressing a hole in what the literature already said about sleep patterns. Once I began to put it all into words, the paper "practically wrote itself." That was neat. (Clark and Lang, 2001: 17)

Most authorities on qualitative data analysis attest to the difficulty of distinguishing between report writing and data analysis (see Hammersley and Atkinson, 1983: 208; and Lofland and Lofland, 1995: 203). Data collection, data reduction, data display, and other pre-report-writing processes all entail an enormous amount of writing, but, whatever else it is, *report writing is a form of data analysis*. Liebow's confession that he came to the report-writing phase of *Tell Them Who I Am*, his touching portrayal of the lives of homeless women, with "no order, outline, or structure in mind" could probably have honestly been made by many authors of qualitative social science—and, frankly, by many authors of quantitative social science. Lang might have had a better sense of "order" going into her short essay than Liebow did going into his book, but writing clearly helped her to refine her ideas, too. The discovery, through writing, of one vital organizing question or another is probably the rule rather than the exception for qualitative analyses.

The writing involved in pre-report phases and the report-writing phase of qualitative analysis can be distinguished, notably by their audiences. Early phases involve writing in which the researcher engages in conversations with himself or herself. We're privy to Enos's codes, her memos, and her typologies, for instance, only because she has generously permitted us to peer over her shoulder, as it were. (Lang's [in Clark and Lang, 2001] self-conscious renderings are exceedingly unusual.) Enos's intended audience was Enos and Enos alone. Formal report writing, on the other hand, is much more clearly a form of communication between author and some other audience. This is not to say that the author himself or herself won't gain new insights from the process. We've already said that he or she will. But it does imply that at least a nodding acquaintance with certain conventions of social science writing is useful, even if you only intend to "play" with those conventions for effect. As a result, we'd like to introduce you to a few such conventions and do so in our Appendix C on report writing.

Liebow's confession suggests a few guidelines we don't mention in Appendix C. One is that, even if you're not sure just where a particular segment of a report will fall within the whole, as long as you've got something to say, write. Don't insist on seeing the forest, before you write, especially if you've got a few good trees. Liebow wrote and wrote, long before he'd gotten a sense of his main theme. He wrote with the kind of hope and belief that an experienced writer brings: that a story line will emerge. But in writing what he took to be a minor theme, the "Problem of Day-to-Day Living," he first glimpsed what would be the organizing principle of his whole book, "the question of survival." Once he discovered this principle, he found that much of what he'd written before was indeed useful.

Another principle, implicit in Liebow's confession, is that you needn't look for perfection as you write. Especially if you have access to a word processor and time, you can write relatively freely, in the certain knowledge that whatever you write can be revised. Don't assume you need to say what you want to say in the best (the most grammatical, the most concise, the most stylish) way possible the first time around. Assure yourself that you can always go back, revise, and rethink as necessary. Such assurances are terribly liberating. Just remember to keep your implicit promise to yourself and revise whenever you can.

SUMMARY

The most common goal of quantitative data analysis is to demonstrate the presence or absence of association between variables. To demonstrate such associations, and other characteristics of variables, quantitative data analysts typically compute statistics, often with the aid of computers.

Statistics can be distinguished by whether they focus on describing a single sample, in which case they are descriptive statistics, or on speaking to the generalizability of sample characteristics to a larger population, in which case they are inferential statistics. Statistics can also be distinguished by the number of variables they deal with at a time: one (as in univariate statistics), two (bivariate statistics), or more than two (multivariate statistics).

The most commonly used univariate statistical procedures in social data analysis are measures of central tendency, frequency distributions, and measures of dispersion. Measures of central tendency, or average, are the mode, the median, and the mean. Measures of dispersion include the range and standard deviation. All the univariate statistical procedures described in this chapter focus on describing a sample and are therefore descriptive, rather than inferential, in nature.

We demonstrated bivariate and multivariate statistical procedures by focusing on crosstabulation, one of many types of procedures that allow you to examine relationships between two variables. In passing, we discussed descriptive statistics, like Cramer's V and Pearson's r, used to describe sample relationships, and inferential statistics, like chi-square, used to make inferences about population relationships. We also illustrated ways of controlling the relationship between two variables through the introduction of control variables.

The canons for analyzing qualitative data are much less widely accepted than those for analyzing quantitative data. Nonetheless, the various processes involved in such analysis are broadly accepted in general terms: data collection itself, data reduction, data display, and conclusion drawing and verification. These processes are interactive in the sense that they do not necessarily constitute well-defined and sequential steps and insofar as all the individual processes can affect or be affected by, directly or indirectly, any of the others.

Word processors are very useful tools in data collection. Soon after each field or interview session, it is useful to word-process field notes or transcripts. It's often a good idea to leave plenty of room in your notes for coding, comments, memos, and displays.

Data reduction refers to various ways a researcher orders collected and transcribed data. We've talked about two sub-processes of this larger process: coding and memoing. Coding is the process of associating words or labels with passages in your field notes and then collecting all passages that are similarly coded into appropriate "files." These files can be traditional file folders or computer files. Codes can be created using predefined criteria, based on a conceptual framework the researcher brings to his or her data, or using postdefined criteria, that "emerge," as it were, as the researcher reviews his or her field notes. Codes can be merely descriptive of what the researcher's seen or analytic, in which case the codes actually advance the analysis in some way. Memos, unlike codes, are more or less extended notes that the researcher writes to help create meaning out of coding categories. Memos are the building blocks of the larger story the researcher eventually tells about his or her data.

Data displays are visual images that present information systematically and in a way that allows the researcher to see what's going on. The forms in which data can be displayed are limitless, but we've discussed four types in this chapter: charts, typologies, matrices, and "monster-dogs."

Data displays are particularly useful for drawing conclusions that are then subject to verification or presentation. We discussed a few other pre-report techniques for developing such conclusions: altering data displays, making comparisons, and talking with and listening to fellow analysts. We also described report writing itself as an important stage in the analysis process.

EXERCISE 15.1

Analyzing Data on the Web

Visit the GSS Cumulative Data file by going to http://sda.berkeley.edu/archive.htm and hitting on the most recent GSS Cumulative Data File. Do your own univariate, bivariate, and multivariate analyses of marital status (called "marital" in the codebook) and labor force status ("wrkstat"), controlling for gender ("sex").

1. Do you remember what we said the modal marital status (married, divorced, separated, or never married) is among adults in the United States? Why don't you check for yourself? Start the "Browse codebook" action, hit the "standard codebook," hit the "sequential variable list," and click on "Personal and Family Information." Find the variable named "marital" and hit on it. Which, in fact, has been the modal marital status among adults in the United States?

2. Now, consider the variable "labor force status" ("wrkstat" in the codebook). Which labor force status would you expect to be the most common among adults in the United States: working full time, working part time, temporarily not working, retired, in school, keeping house, or something else? State your guess. Now find the variable named "wrkstat." Which, in fact, is the modal labor force status among adults in the United States? Were you right?

3. Now consider both marital status and labor force status: Which marital status would you expect to have the greatest representation in the full-time labor force? State your guess in the form of a hypothesis. Now return to the original GSS menu (where you found the "browse codebook" option before). This time, click "frequencies and crosstabulation." Where the menu asks for "row" variable, type "wrkstat." Where it asks for "column" variable, type "marital." Then "run the table." Which marital status does, in fact, have the greatest percentage of its members employed "full time"? Were you right?

4. Now consider the relationship between marital status and labor force participation, controlling for gender. Which marital status would you expect to have the greatest representation in the full-time labor force **for men?** Which marital status would you expect to have the greatest representation in the full-time labor force **for women?** Now return to the original GSS menu. This time, click "frequencies and crosstabulation." Where the menu asks for "row" variable, type "wrkstat." Where it asks for "column" variable, type "marital." Where it asks for "control" variable, type "sex." Which marital status does, in fact, have the greatest percentage of members employed "full- time" **among men? Among women?** Were you right?

EXERCISE 15.2

Interpreting a Bivariate Analysis

After Lang (2001) had analyzed Americans' GSS responses to the question of how they felt about income differences in America, she decided to look at how this feeling was related to other variable characteristics of the respondents.

Look at the following table from her paper and write a paragraph about what it shows about the association between respondents' satisfaction with their own financial situation and their sense of whether income differences in America are too large.

TABLE

Crosstabulation of Satisfaction with Present Financial Situation and Response to Whether Differences in Income in America Are Too Large

Satisfaction with Financial Situation	Differences in Income in America Are Too Large					
	Strongly Agree	Agree	Neither	Disagree	Strongly Disagree	Total
Pretty well satisfied	27.6%	29.7%	16.3%	13.7%	12.6%	100%
More or less satisfied	30.0%	37.5%	13.2%	13.0%	6.3%	100%
Not satisfied at all	44.7%	30.7%	8.5%	8.5%	7.7%	100%

Source: Lang (2001: 25).

EXERCISE 15.3

Reconstructing a Qualitative Data Analysis

Return to either Mueller, Dirks, and Houts Picca's essay (in Chapter 11) or Enos's essay (in Chapter 10). See if you can analyze one of these articles in terms of concepts developed in this chapter.

1. See if you can find two portions of transcribed field notes or interviews that made it into the report. (Hint: Your best bet here is to locate quotations that might have been taken from those notes.) Write these out here.
2. See if you can imagine two codes or categories that the author or authors may have used to organize his or her/their field notes. (There are no wrong answers here. We aren't actually told, in either article, what

the codes were. You're looking here for categories that come up in the analysis and that seem to have linked two or more segments of the field notes.) Briefly discuss why you think these codes might have been used.
3. See if you can imagine a memo that the author or authors might have written to organize his or her/their memos. (Again, there are no wrong answers here.) Briefly discuss why you think this memo might have been used.
4. See if you can imagine and draw a data display that the author or authors might have used to organize his or her/their essay. (Again, there are no wrong answers, though a very good answer might be the presentation of a typology.)

EXERCISE 15.4

Data Collection and Coding

For this exercise, you'll need to collect data through either observation or qualitative interview. If you prefer to do observations, you'll need to locate a social setting to which you can gain access twice for at least 20 minutes each time. You can pick some easily observable activity (e.g., a condiment table at a cafeteria, an elevator on campus, a table at a library where several people

sit together). You can choose to tell others of your intention to do research on them or not. You can work out some research questions before you go to the setting the first time or not. You can take notes at the setting, tape-record what's going on, or choose to do all your note-taking later.

If you prefer to do two interviews, we recommend that you do something like the work suggested in Exercise 10.1, pick an occupation you know something, but not a great deal, about (such as police officer, waiter, veterinarian, letter carrier, and high school teacher) and find two people who are currently employed in that occupation and willing to be interviewed about their work. Conduct 20-minute semi-structured interviews with each of these people, finding out how they are trained or prepared for the work, the kinds of activities they do on the job, about how much time is spent on each, which activities are enjoyed and which are not, what their satisfactions and dissatisfactions with the job are, and whether or not they would like to continue doing their work for the next 10 years.

Whatever the setting or whoever the interviewee, write down your notes afterward, preferably using a word processor and preferably leaving room for coding and memoing (i.e., use wide right margins).

After writing down your notes, spend some time coding your data by hand. Remember: This means not only labeling your data but also filing away similarly labeled data (in this case, in file folders you make or purchase).

EXERCISE 15.5

Memoing and Data Displaying

Some time shortly after you've recorded your notes (for Exercise 15.4) and finished coding them (at least for the first time), reread both the notes and your coding files.

Spend about 20 minutes or so writing memos (either word processing them or hand-writing them) at the end of your notes. You can, of course, spend more than 20 minutes at this

activity. The important thing is to develop some facility with the process of memoing.

Once you've written your memos, spend some time trying to create a data display that helps you understand patterns in the observations or interviews you've done. This display can be in the form of a chart, a typology, a graph, a matrix, or a mini-monster-dog. The purpose is to help you "see" patterns in the data that you might not have already seen.

EXERCISE 15.6

Report Writing

Having completed Exercises 15.4 and 15.5, write a brief report (no more than two or so typed pages) about your project. You should consult Appendix C on report writing for guidance here. You can omit a literature review (though, of course, if you found a couple of relevant articles,

they might help), but you might otherwise be guided by the list (of items to include in a report) provided in Appendix C. (In particular, consider giving your report a title; providing a brief abstract, an overview, and a sense of your methods; and highlighting your major findings and your conclusions.) Good luck!

Comparing Methods

The text covers the major methods of data collection in the social sciences: questionnaires, interviews, observations, and available data (including content analysis). We believe that no one method is inherently better than others. Instead, we believe that some methods are more or less useful for studying specific topics and in specific situations.

Researchers should consider the following issues when deciding whether to observe something, to ask questions about it, or to use some available data as the data collection method:

1. Which method(s) will allow the collection of valid information about the social world?
2. Which method(s) will allow the collection of reliable information about the social world?
3. To what extent will the results be generalizable?
4. To what extent are these data collection methods ethical?
5. To what extent are these data collection methods practical?

Each method of data collection has advantages and disadvantages. A brief overview of these is found in the following table:

TABLE
Advantages and Disadvantages of Methods of Data Collection

Method	Advantages	Disadvantages
Asking Questions		
Group-administered questionnaire	Inexpensive Time efficient Can be anonymous, which is good for "sensitive" topics Good response rate	Not all populations possible Must have literate sample Must be fairly brief Needs permission from setting or organization
Mailed questionnaire	Fairly inexpensive Completed at respondent's leisure Time efficient Can be anonymous, which is good for "sensitive" topics	Response rate may be relatively low Needs names and addresses Must have literate sample Can't explain questions
Individually administered questionnaire	Good response rate Completed at respondent's leisure	Most expensive questionnaire Needs names and addresses Must have literate sample
Web or Internet-based questionnaire	Very inexpensive and time efficient Can be completed at respondent's leisure	Not all populations possible Random samples might not be possible Low response rate
Structured phone interview	Good response rate Little missing data Can explain questions and clarify answers Can use "probes" Don't need names or addresses Good for populations with difficulty reading/writing	Expensive Time consuming Respondents might not feel anonymous Some interviewer effect Lower response rate than in-person
Structured in-person interview	Good response rate Little missing data Good for populations with difficulty reading/writing Can assess validity of data Can explain questions and answers and use "probes"	Expensive Time consuming Needs names and addresses Interviewer effect Confidential, not anonymous

Semi-structured interview	Allows interviewer to develop rapport with study participants Good response rate Questions can be explained and modified for each participant Useful for discussing complex topics because it is flexible and allows follow-up questions Can be used for long interviews Useful when the topics to be discussed are known	Time consuming Expensive Requires highly skilled interviewer Interviewer effect Data analysis is time consuming
Unstructured interview	Allows interviewer to develop rapport with study participants Good response rate Questions can be explained and modified for each participant Participant can take lead in telling his or her own story Useful for exploratory research on new topics Useful for discussing complex topics and seeing participant's perspective Can be used for long interviews	Time consuming Expensive unless small sample Requires highly skilled interviewer Interviewer effect Data analysis is time consuming May not have comparable data for all participants
Focus group interview	Allows participants to have minimal interaction with interviewer and keep the emphasis on participants' points of view Typically generates a great deal of response to questions Often precedes or supplements other methods of data collection	Attitudes can become more extreme in discussion Possibility of "group think" outcome Requires highly skilled interviewer Data less useful for analysis of individuals

(Continued)

Method	Advantages	Disadvantages
Observation		
Participant and nonparticipant observation	Permits researcher to get handle on the relatively unknown Permits researcher to obtain understanding of how others experience life Permits the study of quickly changing situations Relatively inexpensive Permits valid measurement especially of behavior Relatively flexible, at least insofar as it permits researcher to change focus of research	Results are often not generalizable Being observed sometimes elicits characteristics (demand characteristics) in observed simply because of observation Especially in unannounced observations, requires deceit Time consuming
Available Data and Content Analysis		
Available data	Savings in time, money, and effort Possibility of studying change and differences over space Good for shy and retiring people and for those who like to work at their own convenience Data collection is unobtrusive	Inability to know limitations of data Might not be able to find information about topic Sometimes one has to sacrifice measurement validity Possibility of confusion about units of analysis Danger of ecological fallacy
Content analysis	Savings of time, money, and effort Don't need to get everything right the first time Good for shy and retiring people and for those who like to work at their own convenience Data collection is unobtrusive	Can't study communities that have left no records Validity issues, especially resulting from insensitivity to context of message examined

APPENDIX B

Multiple Methods

Each individual method has advantages and disadvantages, so it is always useful to employ more than one method of data collection whenever it is practical. Using multiple methods, or what has been called triangulation (Denzin, 1997), allows the strengths of one method to compensate for the weaknesses of another and increases the validity of the study's conclusions. For example, in a study focusing on the connection between violence, masculinity, and football (or soccer), Ramon Spaaij (2008) collected data on football fan culture in several countries. He was able to understand the creation of the "hooligan" identity within social and historical contexts by using multiple methods, including semi-structured interviews, participant observation, and available data such as websites, media reports, and fan magazines.

Of course, using additional methods means increases in time and cost, the need to obtain access more than once, and the obligation to reconcile discrepant findings. These considerations must be weighed against the increases in validity or generalizability that the use of multiple methods brings.

EXAMPLES OF MULTIPLE METHODS

Research questions: Do nurses treat terminally ill patients differently from those who are expected to recover?

Possible Methods: Being a participant observer in a hospital for at least several weeks coupled with semi-structured, in-person interviews with samples of nurses, patients, and patients' families and an analysis of a sample of patient records.

Research questions: How do two communities in the same urban area differ? Do they receive different levels of services? Do residents in one community feel more socially isolated than the other? Does one community experience higher levels of crime than the other?

Possible Methods: Available data (such as the annual reports from local agencies and organizations, editorials, news stories in local newspapers, information on websites and blogs, National Crime Victim Survey data, and the Federal Bureau of Investigation (FBI)'s *Uniform Crime Reports*), along with observations in public places taking the role of a complete observer, and structured phone interviews conducted with a random sample of residents.

Research questions: What do high school students think about careers in law enforcement? Do attitudes differ by race?

Possible Methods: Interview high school guidance counselors concerning student views, do an Internet-based survey of a sample of high school students, be a complete observer at career fairs that include law enforcement opportunities, and use available data from law enforcement recruitment efforts.

Writing the Research Report

The research report can be communicated in a book, a written paper, or an oral presentation. The intended audience(s) can be the teacher of a class, the supervisor at work, an agency that funded the research, an organization that gave access to the data, or the general public. Although the report might differ for each audience, each version should describe the research process and the research conclusions in an orderly way. Students wanting a detailed discussion of the "nuts and bolts" of technical writing, including coverage of footnoting, referencing, and writing style, should consult one of the many excellent guides, such as Turabian's *A Manual for Writers of Term Papers, Theses, and Dissertations* (1996), Pyrczak and Bruce's *Writing Empirical Research Reports: A Basic Guide for Students of the Social and Behavioral Sciences* (2007), or Lester and Lester, Jr.'s *Writing Research Papers* (2009).

We'd like to tell you about some organizational conventions of report writing, even though we feel a bit presumptuous in doing so. After all, what will work for book-length manuscripts probably won't work equally well for article-length reports. Moreover, generally speaking, research reports involving qualitative data are less formulaic than those involving quantitative data. Nonetheless, because report writing does involve an effort to reach out to an audience, and because social science report writing frequently involves an effort to reach out to a social science audience, it's not a bad idea to have some idea of the forms expected by such audiences.

We'd like to suggest, then, a checklist of features that often appear in article-length research reports. (Books tend to be much less predictable in their formats.) Audiences can include policy makers or practitioners, who typically will expect practical information about either policy or practice, and the general public, who are interested in facts, new ideas, or information on their particular interests (Silverman, 2004: 369). Many of you, however, will write for an academic audience, so we'll focus on what that audience will expect most of the time. Those features are the following:

1. *A title* that tells what your research is about. We think the title should give the reader some idea of what you have to say about your subject. Four good examples are from reports presented in Chapters 7, 9, 10, and 11:
   - "Moving On? Continuity and Change after Retirement" (Adler and Clark, Chapter 7)
   - "Environmentalism among College Students: A Sociological Investigation" (Lang and Podeschi, Chapter 9)
   - "Managing Motherhood in Prison" (Enos, Chapter 10)
   - "Unmasking Racism: Halloween Costuming and Engagement of the Racial Other" (Mueller, Dirks, and Houts Picca, Chapter 11)

2. *An abstract* or a short (usually more than 100 but less than 150 words) account of what the paper does. Such accounts should state the paper's major argument and describe its methods. We haven't included abstracts for any of our focal research articles because we wanted to keep the length of the text manageable. But we've been very self-conscious about not following this convention and think we should give you at least one good example of an abstract, this one from Mueller, Dirks, and Houts Picca's original version of "Unmasking Racism: Halloween Costuming and Engagement of the Racial Other:"

   > We explore Halloween as a uniquely constructive space for engaging racial concepts and identities, particularly through ritual costuming. Data were collected using 663 participant observation journals from college students across the United States. During Halloween, many individuals actively engage the racial other in costuming across racial/ethnic lines. Although some recognize the significance of racial stereotyping in costuming, it is often dismissed as being part of the holiday's social context. We explore the costumes worn, as well as responses to cross-racial costuming, analyzing how "playing" with racialized concepts and making light of them in the "safe" context of Halloween allows students to trivialize and reproduce racial stereotypes while supporting the racial hierarchy. We argue that unlike traditional "rituals of rebellion," wherein subjugated groups temporarily assume powerful roles, whites contemporarily engage Halloween as a sort of "ritual of rebellion" in response to the seemingly restrictive social context of the post-Civil Rights era, and in a way that ultimately reinforces white dominance. (2007:315)

   We think this abstract does a wonderful job of encapsulating key points of the article while also noting the method and the context of the research. Abstracts frequently come before the main body of a report—just after the title—but are generally (and ironically) written *after* the rest of the report's been put together, so the author knows what points to summarize. Here again, however, there are no rules: Drafting an abstract can also be a useful heuristic earlier in the process—to help the writer focus (in a process sometimes called "nutshelling"). Notice, however, that in this abstract, the authors have actually engaged in a final act of conclusion drawing—the ultimate goal, after all, of data analysis itself.

3. *Overview of the problem or Introduction* and *literature review*. These two features are sometimes treated distinctly but can also be combined. The basic function of these features is to justify your research, although they're also good places to give credit to those who have contributed ideas, like theories, conceptual definitions, or measurement strategies, which you're using in some way. The literature review, in particular, should justify the current research through what's already been written or not written about the topic. In Chapter 10, Enos does this with two kinds of "literatures:" literature that has discussed the "imprisonment boom" in the United States and has pointed to the special problems of women in prison and literature that implies that family forms should have certain "normative" aspects, such as mothers who engage in full-time mothering. If you've never written a research report before, we think you could do worse than study the ways in which the authors of articles in this book have used literature reviews to justify their projects. Literature reviews are an art form, and social science literature reviews are a specialized genre of the larger one. As an art form, there are few hard-and-fast rules, but there are some. One is this: Never simply list the articles and books you've read in preparation for your report, even if you include some of what they say while you're listing them. Always use your literature review to build the case that your research has been worth doing.

    Literature reviews themselves offer an excellent chance to engage in analysis. Whether your review occurs before, after, or throughout your field research, it will compel you to make decisions about what your study's really about. (See Chapter 4 for further tips on preparing your literature review.)

4. *Data and methods.* This section involves the description and explanation of research strategies. Authors generally lay out the study design, population, sample, sampling technique, method of data collection, and any measurement techniques that they used. They also sometimes discuss practical and ethical considerations that influenced their choice of method. Authors also sometimes justify methodological choices by discussing things such as the validity and reliability of their measures in this section.

5. *Findings.* Being able to present findings is usually the point of your analysis. You might want to subdivide your findings section. In the findings section, researchers always incorporate data or summaries of data. If the project involved quantitative data analyses, this is where researchers display their statistical results, tables, or graphs (see, e.g., Martinez and Lee, Chapter 12). If the research involved qualitative data analyses, this is where the authors present summary statements or generalizations, usually supported by anecdotes and illustrations from their interview or field notes. Qualitative researchers usually spice up and document their accounts with well-selected quotations in this section (see, e.g., Enos,

Chapter 10). In both kinds of studies, writers generally offer interpretations of their data in this section.

6. *Summary and conclusions.* This section should permit you to distill, yet again, your major findings and to say something about how those findings are related to the literature or literatures you've reviewed earlier. It's very easy to forget to draw such relationships (i.e., between your work and what's been written before), but once you get started, you might find you have a lot more to say that could interest other social scientists than you anticipated.

7. *References.* Don't forget to list the references you've referred to in the paper! Here you'll want to be guided by certain stylistic conventions of, perhaps, a formal guide (e.g., American Sociological Association, 1997). If you're preparing a paper for a course or other academic setting, you might even ask the relevant faculty member what style he or she prefers.

8. *Appendices.* It's sometimes useful to include a copy of a questionnaire, an interview schedule, or a coding scheme if one was used. Sometimes appendices are used to display elements of a sample (e.g., particular states, countries, or organizations) that were used in the research and sometimes even the raw data (especially when the data were difficult to come by and other researchers might be interested in using them for other purposes). In general, researchers tend to use appendices to display information that is germane, but not central, to the development of the paper's main argument or that would require too much space in the article.

There are few absolutes here, however. Perhaps the most important rule is to give yourself oodles of time to write. If you think it might take about 25 hours to write a report, give yourself 50. If you think it might take 50, give yourself 100. Writing is hard work, so give yourself plenty of time to do it.

Report writing is an art form. Although guidance, such as what we've presented in this section, can be useful, this is one of those human activities where experience is your best guide. Good luck as you gather such experience!

Checklist for Preparing a Research Proposal

A good research proposal should impress someone (e.g., a teacher, a faculty committee, an Institutional Review Board [IRB], or a funding agency) with the worthiness and feasibility of a project and with the thought that has gone into its design. The following items usually, but not always, appear in a research proposal.

I. **A Title** A title that captures the theme or thesis of the proposed project in a nutshell.

II. **A Statement of the Project's Problem or Objective** In this section, you should answer questions such as What exactly will you study? Why is it worth studying? Does the proposed study have practical significance?

III. **Literature Review** In general, a good literature review justifies the proposed research. (See Appendix C for a discussion of literature reviews in research reports. Literature reviews in research proposals should do the same things that literature reviews in research reports do.) The literature review normally accomplishes this goal (of justification) by addressing some of the following questions: What have others said about this topic and related topics? What research, if any, has been done previously on the topic? Have other researchers used techniques that can be adapted for the purposes of the proposed study? In a literature review, one normally cites references that appear in the proposal's reference section (see later) using a style that is appropriate to one's discipline (e.g., American Sociological Association style for sociology, American Psychological Association style for psychology and education). It is often appropriate to end the literature review with a statement of a research question (or research questions) or of a hypothesis (or hypotheses) that will guide the research.

IV. **Methods** In a methods section, you should answer questions such as Whom or what will you study to collect data? How will you select your sample? What, if any, ethical considerations are relevant? What method(s) of data collection will you use—a questionnaire, an interview, an observation, and/or an available data? You might also, depending on the nature of the study (e.g., whether it is quantitative or qualitative), want to answer questions such as What are the key variables in your study? How will you define and measure them? Will you be borrowing someone else's measures or using a modified form of measures that have been used before? What kind of data analysis, or comparisons, do you intend to do, or make, with the data you collect?

V. **Plan for Sharing Your Findings** In this section, you will want to answer questions like these: Will you write up your results in the form of a paper (or book) to be shared with others? What kinds of reporting outlets might you use: a journal or magazine, a conference, or other audience? (If you might prepare a book manuscript, what publisher might be interested in the material?) In general, you want to address the question: With whom will you share your findings?

VI. **Budget** What are the major foreseeable costs of your project and how will you cover them?

VII. **References** In this section, you want to provide a list of the references (books, articles, websites) already cited in the proposal, using some standard style format appropriate for one's discipline (e.g., American Sociological Association style for sociology).

Random Number Table

3250	2881	2326	9108	3702	6844	9827	3210	4895
7961	7952	3660	1503	4644	7656	3856	1672	0695
7739	5850	5998	9821	9476	2165	7662	5742	5048
0417	0262	7442	0873	6101	8592	2862	5346	7284
5816	4578	4061	7262	6328	1355	9386	5446	0666
7843	1922	3038	8442	5367	5952	6782	4206	4630
4912	7287	8526	9602	4545	3755	4159	3288	2900
2105	4791	5097	7386	6040	3575	4143	4820	5547
3263	3102	3555	4204	9411	0088	4873	1407	0858
4619	0105	9946	3234	1265	9509	0227	6671	3992
4460	7800	4476	3613	1252	8579	0611	6919	4208
6041	6103	3043	4634	1191	7886	4579	6301	4488
6718	8181	1425	6396	5312	4700	8077	1604	0870
8253	0766	2292	0146	6302	4413	6170	9764	3377
2061	5261	7862	8368	6533	9481	4098	1313	8527
4712	2096	6000	4173	8920	6913	3092	2028	2678
3344	6985	0656	2416	8367	5673	3272	8878	8714
5965	7229	0064	8356	7767	0960	5365	4980	9568
1203	8216	0354	1005	8063	3853	6732	0012	5391
6619	7927	0067	5559	2929	8706	6366	1111	9997

(Continued)

5545	8811	0833	1259	1344	0579	2735	9419	8533
8446	2198	8304	0803	5947	6322	9623	4419	8007
6017	2981	0880	6879	0193	6721	9130	2341	4528
7582	0139	5100	9315	5571	4508	1341	3450	1307
4118	7411	9556	9907	0612	0601	9545	8084	7117
8443	2486	1121	6714	8133	9113	4613	2633	6828
0231	2253	0931	9757	3471	3080	6369	6270	6296
8393	2801	8322	2871	6504	4972	1608	4997	2211
6092	5922	5587	8472	4256	2888	1344	4681	0332
8728	4779	8851	7466	9308	3590	3115	0478	6956
8585	4266	8533	9121	1988	4965	1951	4395	3267
9593	6571	7914	1242	8474	1856	0098	2973	2863
9356	6955	4768	1561	6065	3683	6459	4200	2177
0542	9565	7330	0903	8971	5815	2763	3714	0247
6468	1257	2423	0973	3172	8380	2826	0880	4270
1201	1959	9569	0073	8362	6587	7942	8557	4586
3616	5075	8836	1704	9143	6974	5971	0154	5509
4241	9365	6648	1050	2605	0031	3136	5878	9501
1679	5768	2801	5266	9165	5101	7455	9522	4199
5731	9611	0523	7579	0136	5655	6933	7668	8344
1554	7718	9450	9736	4703	9408	4062	7203	8950
1471	3213	7164	3788	4167	4063	7881	5348	4486
9563	8925	5479	0340	6959	3054	9435	6400	7989
3722	8984	2755	2560	1753	1209	2805	2763	8189
4237	1818	5266	2276	1877	4423	7318	1483	3153
2859	9216	5303	2154	3243	7359	1393	9909	2482
8303	7527	4566	8713	0538	7340	6224	3135	8367
4150	4056	1579	9286	4353	3413	6618	9524	9284
1507	1745	7221	9009	4769	9333	6803	0198	7879
1365	3544	3753	1530	7384	9357	2231	6166	3217
7414	5143	8912	6611	3396	7969	6559	9463	9114
9658	7935	7850	4297	4908	9653	9540	8753	3920
3169	0990	6016	0631	8451	2488	8724	3625	1805
7294	6007	0887	9667	5211	0082	2522	7055	4431
0514	8862	2713	9944	7880	5101	7535	1200	2583

7016	0451	2039	5827	2820	4979	1997	0920	0988
0389	0734	3322	0822	7189	7229	1031	2444	5827
4903	9574	0313	6611	6589	4437	8070	4480	6481
3746	6799	3082	6987	3766	8230	4376	7150	6963
7695	8918	3141	8417	7200	6066	4097	8883	5254
8423	2024	8911	9430	1711	6618	6768	0251	8926
5009	0421	9412	0227	4881	2652	8863	2900	1878
3265	7124	8628	0386	7584	3412	9002	7013	2898
2263	6500	7451	5922	8289	3552	5546	7389	1959
4629	5201	3214	6479	0313	4100	0596	9274	0415
6833	0952	8529	2108	2219	8249	2184	6631	7768
1991	4762	7349	3282	6394	3786	5792	0337	7781
9628	1897	4009	1946	2765	3552	9865	0398	4612
5860	8540	7938	2799	6545	0922	9124	4352	2057
3394	0297	8151	6445	6994	4566	0323	6185	3117
6976	4713	7478	9092	7028	8140	2490	9508	9481
5359	7743	8669	8469	5126	1434	7695	8129	3184
7127	6156	8500	7593	0670	9534	3945	9718	0834
8690	6983	6359	3205	6167	4362	5340	4104	3004
5705	2941	2505	3360	5976	2070	8450	6761	7404
7210	4415	4744	2061	5102	7796	2714	1876	3398
0777	2055	8932	0542	1427	8487	3761	3793	0954
0270	9605	7099	9063	2584	5289	0040	2135	6934
7194	7521	3770	6017	4393	3340	8210	6468	6144
6029	0732	9672	6507	1422	3483	2387	2472	6951
6631	3518	5642	2250	3187	8553	6747	6029	1871
6958	4528	6053	9004	6039	5841	8685	9100	4097
1589	6697	9722	9343	3227	3063	6808	2900	6970
2004	8823	1038	1638	9430	7533	8538	6167	1593
2615	5006	1524	4747	4902	7975	9782	0937	5157
6487	6682	2964	0882	2028	2948	6354	3894	2360
7238	7432	4922	9697	7764	7714	4511	3252	6864
0382	2511	7244	7271	4804	9564	8572	6952	6155
8918	3452	3044	4414	1695	2297	6360	2520	5814
9331	0179	6815	5701	0711	5408	0322	8085	0080

(Continued)

1065	9702	9944	9575	6714	9382	8324	3008	9373
6663	5272	3139	4020	5158	7146	4100	3578	0683
7045	5075	6195	5021	1443	2881	9995	8579	7923
6483	7316	5893	9422	0550	3273	5488	2061	1532
4396	9028	4536	8606	4508	3752	9183	5669	5927
5766	7632	9132	7778	6887	8203	6678	8792	3159
2743	8858	5094	2285	2824	9996	7384	4069	5442
3330	8618	1974	2441	0457	6561	7390	4102	2311
4791	2725	4085	3190	9098	2300	6241	7903	9150
6833	9145	2459	2393	7831	2279	9658	5115	6906
9474	9070	4964	2035	7954	7001	8688	7040	6924
8576	0380	5722	6149	4086	4437	9871	0850	3248

GLOSSARY

access the ability to obtain the information needed to answer a research question.

account a plausible and appealing explanation of the research that the researcher gives to prospective participants.

accretion measures indicators of a population's activities created by its deposits of materials.

anonymity when no one, including the researcher, knows the identities of research participants.

antecedent variable a variable that comes before both an independent variable and a dependent variable.

anticipatory data reduction decision making about what kinds of data are desirable that is done in advance of collecting data.

applied research research intended to be useful in the immediate future and to suggest action or increase effectiveness in some area.

authorities socially defined sources of knowledge.

available data data that are easily accessible to the researcher.

basic research research designed to add to our fundamental understanding and knowledge of the social world regardless of practical or immediate implications.

biased samples samples that are unrepresentative of the population from which they've been drawn.

bivariate analyses data analyses that focus on the association between two variables.

case study a research strategy that focuses on one case (an individual, a group, an organization, and so on) within its social context at one point in time, even if that one time spans months or years.

causal hypothesis a testable expectation about an independent variable's effect on a dependent variable.

causal relationship a nonspurious relationship between an independent and a dependent variable with the independent variable occurring before the dependent variable.

closed-ended questions questions that include a list of predetermined answers.

cluster sampling a probability sampling procedure that involves randomly selecting clusters of elements from a population and subsequently selecting every element in each selected cluster for inclusion in the sample.

coding assigning observations to categories. In quantitative analyses, this frequently means making data computer usable. In qualitative analyses, this means associating words or labels with

passages and then collecting similarly labeled passages into files.

cohort a group of people born within a given time frame or experiencing a life event, such as marriage or graduation from high school, in the same time period.

cohort study a study that follows a cohort over time.

complete observer role being an observer of a situation without becoming part of it.

complete participant role being, or pretending to be, a genuine participant in a situation one observes.

composite measures measures with more than one indicator.

concepts words or signs that refer to phenomena that share common characteristics.

conceptual definition a definition of a concept through other concepts. Also called a theoretical definition.

conceptualization the process of clarifying what we mean by a concept.

concurrent criterion validity how well a measure is associated with behaviors it should be associated with at the present time.

confidentiality also called privacy, is when no third party knows the identities of the research participants.

construct validity how well a measure of a concept is associated with a measure of another concept that some theory says the first concept should be associated with.

content analysis a method of data collection in which some form of communication is studied systematically.

content validity a test for validity that involves the judgment of experts in a field.

contingency questions questions that depend on the answers to previous questions.

control variable a variable that is held constant to examine the relationship between two other variables.

controlled (or systematic) observations observations that involve clear decisions about what is to be observed.

convenience sample a group of elements that are readily accessible to the researcher.

cost-benefit analysis research that compares a program's costs to its benefits.

cost-effectiveness analysis comparisons of program costs in delivering desired benefits based on the assumption that the outcome is desirable.

cover letter the letter accompanying a questionnaire that explains the research and invites participation.

coverage error an error that results from differences between the sampling frame and the target population.

cross-sectional study a study design in which data are collected for all the variables of interest using one sample at one time.

crosstabulation the process of making a bivariate table to examine a relationship between two variables.

data display a visual image that summarizes information.

data reduction the various ways in which a researcher orders collected and transcribed data.

deductive process process that moves from more general to less general statements.

demand characteristics characteristics that the observed take on simply as a result of being observed.

dependent variable a variable that is seen as being affected or influenced by another variable.

descriptive statistics statistics used to describe and interpret sample data.

descriptive study research designed to describe groups, activities, situations, or events.

dimensions aspects or parts of a larger concept.

double-blind experiment an experiment in which neither the subjects nor the research staff who interact with them knows the memberships of the experimental or control groups.

ecological fallacy the fallacy of making inferences about certain types of individuals from information about groups that might not be exclusively composed of those individuals.

elaboration the process of examining the relationship between two variables by introducing the control for another variable or variables.

element a kind of thing a researcher wants to sample. Also called a sampling unit.

empirical generalizations statements that summarize a set of individual observations.

erosion measures indicators of a population's activities created by its selective wear on its physical environment.

ethical principles in research the set of values, standards, and principles used to determine appropriate and acceptable conduct at all stages of the research process.

evaluation research research specifically designed to assess the impacts of programs, policies, or legal changes.

exhaustive the capacity of a variable's categories to permit the classification of every unit of analysis.

existing statistics summaries of data collected by large organizations.

experimental design a study design in which the independent variable is controlled, manipulated, or introduced in some way by the researcher.

experimenter expectations when expected behaviors or outcomes are communicated to subjects by the researcher.

explanation a kind of elaboration in which the original relationship is explained away as spurious by a control for an antecedent variable.

explanatory research research designed to explain the cause of a phenomenon and that typically asks "What causes what?" or "Why is it this way?"

exploratory research groundbreaking research on a relatively unstudied topic or in a new area.

extraneous variable a variable that has an effect on the dependent variable in addition to the effect of the independent variable.

face validity a test for validity that involves the judgment of everyday people, like you and me.

feasibility whether it is practical to complete a study in terms of access, time, and money.

field experiment an experiment done in the "real world" of classrooms, offices, factories, homes, playgrounds, and the like.

focus group interview a type of group interview where participants converse with each other and have minimal interaction with a moderator.

formative analysis evaluation research focused on the design or early implementation stages of a program or policy.

frequency distribution a way of showing the number of times each category of a variable occurs in a sample.

gatekeeper someone who can get a researcher into a setting or facilitate access to participants.

generalizability the ability to apply the results of a study to groups or situations beyond those actually studied.

grounded theory theory derived from data in the course of a particular study.

group-administered questionnaire questionnaire administered to respondents in a group setting.

group interview a data collection method with one interviewer and two or more interviewees.

history the effects of general historical events on study participants.

honest reporting the ethical responsibility to produce and report accurate data.

hypothesis a testable statement about how two or more variables are expected to relate to one another.

in-person interview an interview conducted face-to-face.

independent variable a variable that is seen as affecting or influencing another variable.

index a composite measure that is constructed by adding scores from several indicators.

indicators observations that we think reflect the presence or absence of the phenomenon to which a concept refers.

individually administered questionnaires questionnaires that are hand delivered to a respondent and picked up after completion.

inductive reasoning reasoning that moves from less general to more general statements.

inferential statistics statistics used to make inferences about the population from which the sample was drawn.

informants participants in a study situation who are interviewed for an in-depth understanding of the situation.

informed consent the principle that potential participants are given adequate and accurate information about a study before they are asked to agree to participate.

informed consent form a statement that describes the study and the researcher and formally requests participation.

institutional review board (IRB) the committee at a college, university, or research center responsible for evaluating the ethics of proposed research.

internal consistency method a method that relies on making more than one measure of a phenomenon at essentially the same time.

internal validity agreement between a study's conclusions about causal connections and what is actually true.

interobserver (or interrater) reliability method a way of checking the reliability of a measurement strategy by comparing results obtained by one observer with results obtained by another using exactly the same method.

interpretation a kind of elaboration that provides an idea of the reasons why an original relationship exists without challenging the belief that the original relationship is causal.

intersubjectivity agreements about reality that result from comparing the observations of more than one observer.

interval level variables variables whose categories have names, whose categories can be rank-ordered in some sensible way, and whose adjacent categories are a standard distance from one another.

intervening variable a variable that comes between an independent and a dependent variable.

interview a data collection method in which respondents answer questions asked by an interviewer.

interview guide the list of topics to cover and the order in which to cover them that can be used to guide less structured interviews.

interview schedule the list of questions and answer categories read to a respondent in a structured or semi-structured interview.

interviewer effect the change in a respondent's behavior or answers that is the result of being interviewed by a specific interviewer.

keywords the terms used to search for sources in a literature review.

laboratory research research done in settings that allows the researcher control over the conditions, such as in a university or medical setting.

life story interview a short account of an interviewee's life created in collaboration with an interviewer.

literature review the process of searching for, reading, summarizing, and synthesizing existing work on a topic or the resulting written summary of a search.

longitudinal research a research design in which data are collected at least two different times, such as a panel, trend, or cohort study.

mailed questionnaire questionnaire mailed to the respondent's residence or workplace.

margin of error a suggestion of how far away the actual population parameter is likely to be from the statistic.

matching assigning sample members to groups by matching the sample members on one or more characteristics and then separating the pairs into two groups with one group randomly selected to become the experimental group.

maturation the biological and psychological processes that cause people to change over time.

mean the measure of central tendency designed for interval level variables. The sum of all values divided by the number of values.

measure a specific way of sorting units of analysis into categories.

measurement the process of devising strategies for classifying subjects by categories to represent variable concepts.

measurement error the kind of error that occurs when the measurement we obtain is not an accurate portrayal of what we tried to measure.

measures of association measures that give a sense of the strength of a relationship between two variables.

measures of correlation measures that provide a sense not only of the strength of the relationship between two variables but also of its direction.

measures of dispersion measures that provide a sense of how spread out cases are over categories of a variable.

median the measure of central tendency designed for ordinal level variables. The middle value when all values are arranged in order.

memos more or less extended notes that the researcher writes to help herself or himself understand the meaning of codes.

mode the measure of central tendency designed for nominal level variables. The value that occurs most frequently.

multidimensionality the degree to which a concept has more than one discernible aspect.

multistage sampling a probability sampling procedure that involves several stages, such as randomly selecting clusters from a population, then randomly selecting elements from each of the clusters.

multivariate analyses analyses that permit researchers to examine the relationship between variables while investigating the role of other variables.

mutually exclusive the capacity of a variable's categories to permit the classification of each unit of analysis into one and only one category.

natural experiment a study using real-world phenomena that approximates an experimental design even though the independent variable is not controlled, manipulated, or introduced by the researcher.

needs assessment an analysis of whether a problem exists, its severity, and an estimate of essential services.

nominal level variables variables whose categories have names.

nonparticipant observation observation made by an observer who remains as aloof as possible from those observed.

nonprobability samples samples that have been drawn in a way that doesn't give every member of the population a known chance of being selected.

nonresponse error an error that results from differences between non-responders and responders to a survey.

normal distribution a distribution that is symmetrical and bell-shaped.

objectivity the ability to see the world as it really is.

observational techniques methods of collecting data by observing people, most typically in their natural settings.

observer-as-participant role being primarily a self-professed observer, while occasionally participating in the situation.

open-ended questions questions that allow respondents to answer in their own words.

operationalization the process of specifying what particular indicator(s) one will use for a variable.

ordinal level variables variables whose categories have names and whose categories can be rank-ordered in some sensible way.

outcome evaluation research that is designed to "sum up" the effects of a program, policy, or law in accomplishing the goal or intent of the program, policy, or law.

panel attrition the loss of subjects from a study because of disinterest, death, illness, or inability to locate them.

panel conditioning the effect of repeatedly measuring variables on members of a panel study.

panel study a study design in which data are collected about one sample at least two times where the independent variable is not controlled by the researcher.

participatory action research (PAR) research done by community members and researchers working as co-participants, most typically within a social justice framework to empower people and improve their lives.

parameter a summary of a variable characteristic in a population.

partial relationship the relationship between an independent and a dependent variable for that part of a sample defined by one category of a control variable.

participant-as-observer role being primarily a participant, while admitting an observer status.

participant observation observation performed by observers who take part in the activities they observe.

passive consent when no response is considered an affirmative consent to participate in research; also called "opt out informed consent," this is sometimes used for parental consent for children's participation in school-based research.

personal inquiry inquiry that employs the senses' evidence.

personal records records of private lives, such as biographies, letters, diaries, and essays.

photo elicitation a technique used to encourage dialogue between interviewer and interviewee.

photo-interviewing a data collection technique using photographs to elicit information and encourage discussion usually in conjunction with qualitative interviewing.

physical traces physical evidence left by humans in the course of their everyday lives.

pilot test a preliminary draft of a set of questions that is tested before the actual data collection.

placebo a simulated treatment of the control group that is designed to appear authentic.

population the group of elements from which a researcher samples and to which she or he might like to generalize.

positivist view of science a view that human knowledge must be based on what can be perceived.

post-positivist view of science a view that knowledge is not based on irrefutable observable grounds, that it is always somewhat speculative, but that science can provide relatively solid grounds for that speculation.

posttest the measurement of the dependent variable that occurs after the introduction of the stimulus or the independent variable.

posttest-only control group experiment an experimental design with no pretest.

predictive criterion validity a method that involves establishing how well a measure predicts future behaviors you'd expect it to be associated with.

pretest the measurement of the dependent variable that occurs before the introduction of the stimulus or the independent variable.

pretest–posttest control group experiment an experimental design with two or more randomly selected groups (an experimental and a control group) in which the researcher controls or "introduces" the independent variable and measures the dependent variable at least two times (pretest and posttest measurement).

primary data data that the same researcher collects and uses.

probability samples samples drawn in a way to give every member of the population a known (nonzero) chance of inclusion.

process evaluation research that monitors a program or policy to determine if it is implemented as designed and if modifications, including alternative procedures, would be useful.

protecting study participants from harm the principle that participants in studies are not harmed physically, psychologically, emotionally, legally, socially, or financially as a result of their participation in a study.

purposive sampling a nonprobability sampling procedure that involves selecting elements based on the researcher's judgment about which elements will facilitate his or her investigation.

qualitative content analysis content analysis designed for verbal analysis.

qualitative data analysis analysis that results in the interpretation of action or representation of meanings in the researcher's own words.

qualitative interview a data collection method in which an interviewer adapts and modifies the interview for each interviewee.

qualitative research research focused on the interpretation of the action of, or representation of meaning created by, individual cases.

quantitative content analysis content analysis focused on the variable characteristics of communication.

quantitative data analysis analysis based on the statistical summary of data.

quantitative research research focused on variables, including their description and relationships.

quasi-experiment an experimental design that is missing one or more aspects of a true experiment, most frequently random assignment into experimental and control groups.

questionnaire a data collection instrument with questions and statements that are designed to solicit information from respondents.

quota sampling a nonprobability sampling procedure that involves describing the target population in terms of what are thought to be relevant criteria and then selecting sample elements to represent the "relevant" subgroups in proportion to their presence in the target population.

random assignment a technique for assigning members of the sample to experimental and control groups by chance to maximize the likelihood that the groups are similar at the beginning of the experiment.

random-digit dialing a method for selecting participants in a telephone survey that involves randomly generating telephone numbers.

range a measure of dispersion or spread designed for interval level variables. The difference between the highest and the lowest values.

rapport a sense of interpersonal harmony, connection, or compatibility between an interviewer and a respondent.

ratio level variables variables whose categories have names, whose categories may be rank-ordered in some sensible way, whose adjacent categories are a standard distance from one another, and one of whose categories is an absolute zero point—a point at which there is a complete absence of the phenomenon in question.

reliability the degree to which a measure yields consistent results.

replication a kind of elaboration in which the original relationship is replicated by all the partial relationships.

requirements for supporting causality the requirements needed to support a causal relationship include a pattern or relationship between the independent and dependent variables, determination that the independent variable occurs first, and support for the conclusion that the apparent relationship is not caused by the effect of one or more third variables.

research costs all monetary expenditures needed for planning, executing, and reporting research.

research question a question about one or more topics or concepts that can be answered through research.

research topic a concept, subject, or issue that can be studied through research.

researchable question a question that is feasible to answer through research.

respondents the participants in a survey who complete a questionnaire or interview.

response rate the percentage of the sample contacted that actually participates in a study.

sample a number of individual cases drawn from a larger population.

sampling the process of drawing a number of individual cases from a larger population.

sampling distribution the distribution of a sample statistic (such as the average) computed from many samples.

sampling error any difference between the characteristics of a sample and the characteristics of the population from which the sample is drawn.

sampling frame or study population the group of elements from which a sample is actually selected.

sampling variability the variability in sample statistics that can occur when different samples are drawn from the same population.

scale an index in which some items are given more weight than others in determining the final measure of a concept.

scientific method a way of conducting empirical research following rules that specify objectivity, logic, and communication among a community of knowledge seekers and the connection between research and theory.

screening question a question that asks for information before asking the question of interest.

secondary data research data that have been collected by someone else.

selection bias a bias in the way the experimental and control or comparison groups are selected that is responsible for preexisting differences between the groups.

self-administered questionnaires questionnaires that the respondent completes by himself or herself.

self-report method another name for questionnaires and interviews because respondents are most often asked to report their own characteristics, behaviors, and attitudes.

semi-structured interviews interviews with an interview guide containing primarily open-ended questions that can be modified for each interview.

simple random sample a probability sample in which every member of a study population has been given an equal chance of selection.

skewed variable an interval level variable that has one or a few cases that fall into extreme categories.

snowball sampling a nonprobability sampling procedure that involves using members of the group of interest to identify other members of the group.

Solomon four-group design a controlled experiment with an additional experimental and control group with each receiving a posttest only.

specification a kind of elaboration that permits the researcher to specify conditions under which the original relationship is particularly strong or weak.

spurious non-causal.

spurious relationship a non-causal relationship between two variables.

stakeholders people or groups that participate in or are affected by a program or its evaluation, such as funding agencies, policy makers, sponsors, program staff, and program participants.

standard deviation a measure of dispersion designed for interval level variables that accounts for every value's distance from the sample mean.

statistic a summary of a variable in a sample.

stimulus the experimental condition of the independent variable that is controlled or "introduced" by the researcher in an experiment.

stratified random sampling a probability sampling procedure that involves dividing the population into groups or strata defined by the presence of certain characteristics and then random sampling from each stratum.

structured interview a data collection method in which an interviewer reads a standardized list of questions to the respondent and records the respondent's answers.

study design a research strategy specifying the number of cases to be studied, the number of times data will be collected, the number of samples that will be used, and whether the researcher will try to control or manipulate the independent variable in some way.

survey a study in which the same data, usually in the form of answers to questions, are collected from one or more samples.

systematic sampling a probability sampling procedure that involves selecting every kth element from a list of population elements, after the first element has been randomly selected.

target population the population of theoretical interest.

telephone interview an interview conducted over the telephone.

test–retest method a method of checking the reliability of a test that involves comparing its results at one time with results, using the same subjects, at a later time.

testing effect the sensitizing effect on subjects of the pretest.

theoretical saturation the point where new interviewees or settings look a lot like interviewees or settings one has observed before.

theory an explanation about how and why something is as it is.

thick description reports about behavior that provide a sense of things like the intentions, motives, and meanings behind the behavior.

thin description bare-bone description of acts.

time expenditures the time it takes to complete all activities of a research project from the planning stage to the final report.

trend study a study design in which data are collected at least two times with a new sample selected from a population each time.

units of analysis the units about which information is collected, and the kind of thing a researcher wants to analyze. The term is used at the analysis stage, whereas the terms element or sampling unit can be used at the sampling stage.

units of observation the units from which information is collected.

univariate analyses analyses that tell us something about one variable.

unobtrusive measures indicators of interactions, events, or behaviors whose creation does not affect the actual data collected.

unstructured interview a data collection method in which the interviewer starts with only a general sense of the topics to be discussed and creates questions as the interaction proceeds.

validity the degree to which a measure taps what we think it's measuring.

variable a characteristic that can vary from one unit of analysis to another or for one unit of analysis over time.

video ethnography the video recording of participants and the reviewing of the resulting footage for insights into social life.

vignettes scenarios about people or situations that the researcher creates to use as part of the data collection method.

visual analysis a set of techniques used to analyze images.

visual sociology an approach to studying society and culture that employs images as a data source.

voluntary participation the principle that study participants choose to participate of their own free will.

web or Internet survey survey that is sent by e-mail or posted on a website.

REFERENCES

Acker, J., K. Barry, & J. Esseveld. 1991. Objectivity and the truth: Problems in doing feminist research. In *Beyond methodology: Feminist scholarship as lived research*, 133–153, edited by M. M. Fonow & J. A. Cook. Bloomington: Indiana University Press.

Adler, E. S. 1981. The underside of married life: Power, influence and violence. In *Women and crime in America*, 300–319, edited by L. H. Bowker. New York: Macmillan.

Adler, E. S., & J. S. Lemons. 1990. *The elect: Rhode Island's women legislators, 1922–1990*. Providence: League of Rhode Island Historical Societies.

Adler, P. A. 1985. *Wheeling and dealing*. New York: Columbia University Press.

Adler, P. A., & P. Adler. 1987. *Membership roles in field research*. Newbury Park, CA: Sage.

Adler, P. A., & P. Adler. 1994a. Observational techniques. In *Handbook of qualitative research*, 377–392, edited by N. K. Denzin & Y. S. Lincoln. Thousand Oaks, CA: Sage.

Adler, P. A., & P. Adler. 1994b. Social reproduction and the corporate other: The institutionalization of afterschool activities. *Sociological Quarterly* 35: 309–328.

Adler, P. A., & P. Adler. 2005. Foreword: Fighting the good fight. In *Contempt of court: A scholar's battle for free speech from behind bars*, edited by R. Scarce. Lanham, MD: AltaMira Press.

Albig, W. 1938. The content of radio programs—1925–1935. *Social Forces* 16: 338–349.

Allen, K. R., & D. H. Demo. 1995. The families of lesbians and gay men: A new frontier in family research. *Journal of Marriage and the Family* 57: 111–128.

Allen, S. 2006. Harvard to study Katrina's long-term psychological toll. *Boston Globe* 269(6): A4.

Altman, L. E. 2005. Nobel prize for discovering bacterium causes ulcers. *New York Times*, October 4, D3.

American Sociological Association (ASA). 1999. Code of ethics and policies and procedures of the ASA committee on professional ethics. Washington, DC: ASA. Retrieved on September 16, 2009, from http://www.asanet.org/galleries/default-file/Code%20of%20Ethics.pdf

American Sociological Association (ASA). 2005. NSF awarded $5.5 million to sociology in '04. *Footnotes*, January. Washington, DC: ASA. Retrieved on March 30, 2009, from http://www.asanet.org/footnotes/jan05/fn7.html

American Sociological Association (ASA). 2009. About ASA. Retrieved on June 23, 2009, from http://www.asanet.org/cs/root/about_asa/about_asa

Anden-Papadopoulos, K. 2009. US soldiers imaging the Iraq War on YouTube. *Popular Communication* 7(1): 17–27.

Anderson, E. 1999. *Code of the street*. New York: Norton.

Angrist, J. D., & J. H. Johnson IV. 2000. Effects of work-related absences on families: Evidence from the Gulf War. *Industrial & Labor Relations Review* 54: 41–58.

AoIR. 2002. Ethical decision-making and Internet research. Retrieved on July 21, 2009, from http://www.aoir.org/reports/ethics.pdf

Aquilino, W. S. 1993. Effects of spouse presence during the interview on survey responses concerning marriage. *Public Opinion Quarterly* 57: 358–376.

Aquilino, W. S. 1994. Interview mode effects in surveys of drug and alcohol use. *Public Opinion Quarterly* 58: 210–242.

Aronson, P. 2008. The markers and meanings of growing up: Contemporary young women's transition from adolescence to adulthood. *Gender & Society* 22(1): 56–82.

Arthur, C., & R. Clark. Forthcoming. Determinants of domestic violence: A cross-national study. *International*

Journal of Sociology of the Family 35.

Arthur, M. M. L. 2007. Getting away from the basics: Competing explanations of curricular change. Doctoral dissertation for New York University.

Asher, R. 1992. *Polling and the public.* Washington, DC: Congressional Quarterly.

Atchley, R. C. 2003. *Social forces and aging,* 10th edition. Belmont, CA: Wadsworth.

Atkinson, R. 2001. The life story interview. In *Handbook of interview research,* 121–140, edited by J. F. Gubrium & J. A. Holstein. Thousand Oaks, CA: Sage Publications.

Attwood, F. 2009. Intimate adventures: Sex blogs, sex 'blooks' and women's sexual narration. *European Journal of Cultural Studies* 12(1): 5–20.

Auerbach, J. 2000. Feminism and federally funded social science: Notes from inside. *Annals of the American Academy of Political and Social Science* 571: 30–41.

Babalola, S., L. Folda, & H. Babavora. 2008. The effects of a communication program on contraceptive ideation and use among young women in northern Nigeria. *Studies in Family Planning* 39(2): 211–220.

Babbie, E. 2009. *The practice of social research,* 12th edition. Belmont, CA: Wadsworth.

Bakalar, N. 2005. Ugly children may get parental short shrift. *New York Times,* May 3. Retrieved on September 16, 2009, from http://www.nytimes.com/2005/05/03/health/03ugly.html

Bales, K. 1999. Popular reactions to sociological research: The case of Charles Booth. *Sociology* 33: 153–168.

Bales, R. F. 1950. *Interaction process analysis.* Reading, MA: Addison-Wesley.

Barry, C. 1998. Choosing qualitative data analysis software: Atlas/ti and Nudist compared. *Sociological Research Online* 3. Retrieved on June 20, 2001, from www.socresonline.org.uk/socresonline/3/3/4.html

Barry, E. 2002. After day care controversy, psychologists play nice. *Boston Globe,* September 3, E1.

Bart, P. B., & P. H. O'Brien. 1985. *Stopping rape: Successful survival strategies.* New York: Pergamon.

Baruch, Y. 1999. Response rate in academic studies—a comparative analysis. *Human Relations* 52: 421–438.

Baruch, Y., & C. Holtom. 2008. Survey response rate levels and trends in organizational research. *Human Relations* 61: 1139–1160.

Bates, B., & M. Harmon. 1993. Do "instant polls" hit the spot? Phone-in versus random sampling of public opinion. *Journalism Quarterly* 70: 369–380.

Baumrind, D. 1964. Some thoughts on ethics of research: After reading "Milgram's behavioral study of obedience." *American Psychologist* 19: 421–423.

Becker, H. S. 1974. Photography and sociology. *Studies in the Anthropology of Visual Communication* 1: 3–26.

Becker, H. S., B. Geer, E. C. Hughes, & A. L. Strauss. 1961. *Boys in white: Student culture in medical school.* Chicago: University of Chicago Press.

Belinfante, A. 2005. Telephone subscribership in the United States (data through March 2005). Retrieved on September 16, 2009, from http://www.fcc.gov/Bureaus/Common_Carrier/Reports/FCC-State_Link/IAD/subs0305.pdf

Bell, P. 2001. Content analysis of visual images. In *Handbook of visual analysis,* 10–35, edited by T. van Leeuwan & C. Jewitt. Thousand Oaks, CA: Sage.

Belli, R. F., L. M. Smith, P. M. Andreski, & S. Agrawal. 2007. Methodological comparisons between CATI event history calendar and standardized conventional questionnaire instruments. *Public Opinion Quarterly* 71: 603–622.

Benney, M., & E. C. Hughes. 1956. Of sociology and the interview. *American Journal of Sociology* 62: 137–142.

Berg, B. L. 1989. *Qualitative research methods of the social sciences.* Boston: Allyn & Bacon.

Berg, J. A. 2009. White public opinion toward undocumented immigrants: Threat and interpersonal environment. *Sociological Perspectives* 52(1): 39–58.

Berinsky, A. 2006. American public opinion in the 1930s and 1940s: The analysis of quota-controlled sample survey data. *Public Opinion Quarterly* 70(4): 499–529.

Berk, R. A., G. K. Smyth, & L. W. Sherman. 1988. When random assignment fails: Some lessons from the Minneapolis spouse abuse experiment. *Journal of Quantitative Criminology* 4: 209–223.

Bernstein, R., A. Chadha, & R. Montjoy. 2001. Overreporting voting, why it happens and why it matters. *Public Opinion Quarterly* 65: 22–44.

Bickman, L., & D. J. Rog. 2009. *The SAGE handbook of applied social research,* 2nd edition. Thousand Oaks, CA: SAGE Publications.

Birkeland, S., E. Murphy-Graham, & C. H. Weiss. 2005. Good reasons for ignoring good evaluation: The case of the Drug Abuse Resistance Education (DARE) program. *Evaluation and Program Planning* 28: 247–256.

Bishop, G. F. 1987. Experiments with the middle response alternative in survey questions. *Public Opinion Quarterly* 51: 220–232.

Blumberg, S. J., & J. V. Luke. 2008. Wireless substitution: Early release of estimate from the National Health Interview Survey, July–December 2007. National Center for Health Statistics. Retrieved on March 5, 2009, from http://www.cdc.gov/nchs/data/nhis/earlyrelease/wireless200805.htm

Blumberg, S. J., & J. V. Luke. 2009. Wireless substitution: Early release of estimates from the National Health Interview Survey, July–December 2008. Center for Disease Control. Retrieved on June 23, 2009, from http://www.cdc.gov/nchs/data/nhis/earlyrelease/wireless200905.pdf

Bock, J. 2000. Doing the right thing? Single mothers by choice and the struggle for legitimacy. *Gender & Society* 14: 62–86.

Bogenschneider, K., & L. Pallock. 2008. Responsiveness in parent-adolescent relationships: Are influences conditional? Does the reporter matter? *Journal of Marriage and Family* 70: 1015–1029.

Bonanno, G. A., C. Rennicke, & S. Dekel. 2005. Self-enhancement among high-exposure survivors of the September 11th terrorist attack: Resilience or social maladjustment? *Journal of Personality and Social Psychology* 88: 984–998.

Bond, M. A. 1990. Defining the research relationship: Maximizing participation in an unequal world. In

Researching community psychology, 183–185, edited by P. Tolan, C. Keys, F. Chertock, & L. Jason. Washington, DC: American Psychological Association.

Boser, S. 2006. Ethics and power in community–campus partnerships for research. *Action Research* 4: 9–21.

Bostock, D., & J. Daley. 2007. Lifetime and current sexual assault and harassment victimization rates of active-duty United States air force women. *Violence Against Women* 13(9): 927–944.

Botha, C., & J. Pienaar. 2006. South African correctional officer occupational stress: The role of psychological strengths. *Journal of Criminal Justice* 34(1): 73–84.

Bourdieu, P. 1977. Cultural reproduction and social reproduction. In *Power and ideology in education*, edited by J. Karabel & A. H. Halsey. New York: Oxford University Press.

Bourgoin, N. 1995. Suicide in prison: Some elements of a strategic analysis. *Cahiers Internationaux de Sociologie* 98: 59–105.

Bradburn, N. M., & S. Sudman. 1988. *Polls and surveys*. San Francisco: Jossey-Bass.

Brick, J. M., W. S. Edwards, & S. Lee. 2007. Sampling telephone numbers and adults, interview length, and weighting in the California Health Interview Survey Cell Phone Pilot Study. *Public Opinion Quarterly* 71: 793–813.

British Sociological Association (BSA). 2002. Statement of ethical practice for the British Sociological Association. Retrieved on March 13, 2009, from http://www.britsoc.co.uk/equality/Statement+Ethical+Practice.htm

Brodsky, A. 2001. More than epistemology: Relationships in applied research with underserved communities. *Journal of Social Issues* 57(2): 323–335.

Brodsky, A. E., K. Rogers-Senuta, C. Leiss, C. M. Marx, C. Loomis, S. Arteaga, H. Moore, R. Benhorin, & A. Casteganera. 2004. When one plus one equals three: The role of relationships in community research. *American Journal of Community Psychology* 33: 229–241.

Brophy, B. 1999. Doing it for science. *U.S. News & World Report*, March 22. Retrieved on July 21, 2001, from http://www.usnews.

com/usnews/issue/990322/nycu/22nurs.htm

Browne, J. 1976. The used car game. In *The research experience*, 60–84, edited by M. P. Golden. Itasca, IL: F. E. Peacock.

Brydon-Miller, M. 2002. What is participatory action research and what's a nice person like me doing in a field like this? Paper presented at the 2002 SPSSI Convention, Toronto, Canada, June 28–30. Retrieved on September 16, 2009, from http://www.scu.edu.au/schools/gcm/ar/w/Brydon-Miller.pdf

Budiansky, S. 1994. Blinded by the cold-war light. *U.S. News & World Report* 116: 6–7.

Bulmer, M. 1982. *Social research ethics: An examination of the merits of covert participant observation*. New York: Holmes & Meier.

Bureau of Labor Statistics (BLS). 2001. *National longitudinal surveys: Mature women user's guide 2001*. http://www.bls.gov/nls/mwguide/2001/nlsmwg0.pdf

Burris, S., & K. Moss. 2006. U.S. health researchers review their ethics review boards: A qualitative study. *Journal of Empirical Research on Human Subjects Ethics* 1: 39–58.

Bushman, B. J., & C. A. Anderson. 2009. Comfortably numb: Desensitizing effects of violent media on helping others. *Psychological Science* 21: 273–277.

California Health Interview Survey. 2009. *CHIS 2007 methodology series: Report 4—response rates*. Los Angeles, CA: UCLA Center for Health Policy Research. Retrieved on May 31, 2009, from http://www.chis.ucla.edu/pdf/CHIS2007_method4.pdf

Calvey, D. 2008. The art and politics of covert research: Doing "situated ethics" in the field. *Sociology* 42: 905–918.

Cammorata, N. 2009. Down at the half. *Boston Globe*, March 29, C2. Retrieved on May 26, 2009, from http://www.boston.com/bostonglobe/ideas/articles/2009/03/29/down_at_the_half/

Campbell, D. T., & J. C. Stanley. 1963. *Experimental and quasi-experimental designs for research*. Chicago: Rand McNally.

Campbell, R., & A. E. Adams. 2009. Why do rape survivors volunteer for face-to-face interviews? A meta-study of victims' reasons for and

concerns about research participation. *Journal of Interpersonal Violence* 24: 395–405.

Canary, H. E. 2007. Teaching ethics in communication courses: An investigation of instructional methods, course foci, and student outcomes. *Communication Education* 56: 193–208.

CAPS. 2000. The legacy project: Lessons learned about conducting community-based research. Retrieved on March 11, 2006, from http://www.caps.ucsf.edu/capsweb/publications/LegacyS2C.pdf

CAPS. 2001. Working together: A guide to collaborative research in HIV prevention. San Francisco: University of California, Center for AIDS Prevention Studies. Retrieved on July 3, 2009, from http://www.caps.ucsf.edu/pubs/manuals/pdf/WorkingTogether.pdf

CAPS. 2006. Community collaborative research at CAPS. Retrieved on March 11, 2006, from http://www.caps.ucsf.edu/capsweb/commres.html

Carroll-Lind, J., J. W. Chapman, J. Gregory, & G. Maxwell. 2006. The key to the gatekeepers: Passive consent and other ethical issues surrounding the rights of children to speak on issues that concern them. *Child Abuse and Neglect* 30: 979–989.

Cartland, J., H. S. Ruch-Ross, M. Mason, & W. Donohue. 2008. Role sharing between evaluators and stakeholders in practice. *American Journal of Evaluation* 29: 460–477.

Chambers, D. E., K. R. Wedel, & M. K. Rodwell. 1992. *Evaluating social programs*. Boston: Allyn & Bacon.

Charmaz, K. 1983. The grounded theory method: An explication and interpretation. In *Contemporary field research: A collection of readings*, 109–126, edited by R. M. Emerson. Boston: Little, Brown.

Chase, B., T. Cornille, & R. English. 2000. Life satisfaction among persons with spinal cord injuries. *Journal of Rehabilitation* 66: 14–27.

Chase, S. E. 1995. *Ambiguous empowerment: The work narratives of women school superintendents*. Amherst: University of Massachusetts Press.

Childs, J. H., & A. Landreth. 2006. Analyzing interviewer/respondent interactions while using a mobile computer-assisted personal interview device. *Field Methods* 18: 335–351.

Chito Childs, E. 2005. Looking behind the stereotypes of the "angry black woman": An exploration of black women's responses to interracial relationships. *Gender & Society* 19(4): 544–561.

Chiu, C. 2009. Contestation and conformity: Street and park skateboarding in New York City public space. *Space and Culture* 12(1): 25–42.

Ciambrone, D. 2001. Illness and other assaults on self: the relative impact of HIV/AIDS on women's lives. *Sociology of Health and Illness* 23(4): 517–540.

Clance, P. R. 1985. *The impostor phenomenon: Overcoming the fear that haunts your success.* Atlanta, GA: Peachtree Publishers.

Clark, B. L. 2003. *Kiddie lit.* Baltimore: Johns Hopkins University Press.

Clark, M. C., & B. F. Sharf. 2007. The dark side of truth(s), ethical dilemmas in researching the personal. *Qualitative Inquiry* 13: 399–416.

Clark, R. 1982. Birth order and eminence: A study of elites in science, literature, sports, acting and business. *International Review of Modern Sociology* 12: 273–289.

Clark, R. 1999. Diversity in sociology: Problem or solution? *American Sociologist* 30: 22–43.

Clark, R. 2001. A curious incident: The disappearing class differences in academic achievement. *Middle School Journal* 32: 5–13.

Clark, R., J. Guilmain, P. K. Saucier, & J. Tavarez. 2003. Two steps forward, one step back: The presence of female characters and gender stereotyping in award-winning picture books between the 1930 and the 1960s. *Sex Roles* 49: 439–449.

Clark, R., P. J. Keller, A. Knights, J. A. Nabar, T. B. Ramsbey, & T. Ramsbey. 2006. Let me draw you a picture: Alternate and changing views of gender in award-winning picture books for children. Unpublished manuscript.

Clark, R., & H. Kulkin. 1996. Toward a multicultural feminist perspective on fiction for young adults. *Youth & Society* 27(3): 291–312.

Clark, R., & A. Lang. 2001. *Balancing yin and yang: Teaching and learn-ing qualitative data analysis within an undergraduate data analysis course.* Presented at the Eastern Sociological Society Meetings, March, Philadelphia.

Clark, R., J. McCoy, & P. Keller. 2008. Teach your children well: Reading lessons to and about black and white adolescent girls from black and white women authors. *International Review of Modern Sociology* 34(2): 211–228.

Clark, R., & A. Nunes. 2008. The face of society: Gender and race in introductory sociology books revisited. *Teaching Sociology* 36: 227–239.

Clark, R., & T. Ramsbey. 1986. Social interaction and literary creativity. *International Review of Modern Sociology* 16: 105–117.

Clark, R., & G. Rice. 1982. Family constellations and eminence: The birth orders of Nobel Prize winners. *Journal of Psychology* 110: 281–287.

Cohen, P. 2007. As ethics panels expand grip, no field is off limits. *New York Times*, February 28. Retrieved on February 26, 2009, from http:// www.nytimes.com/2007/02/28/arts/ 28board.html

Collier, M. 2001. Approaches to analysis in visual anthropology. In *Handbook of visual analysis*, 35–60, edited by T. van Leeuwan & C. Jewitt. Thousand Oaks, CA: Sage.

Connolly, K., & A. Reid. 2007. Ethics review for qualitative inquiry. *Qualitative Inquiry* 13: 1013–1047.

Converse, J. M., & H. Schuman. 1974. *Conversations at random: Survey research as interviewers see it.* New York: Wiley.

Converse, P. D., E. W. Wolfe, X. Huang, & F. L. Oswald. 2008. Response rates for mixed-mode surveys using mail and e-mail/web. *American Journal of Evaluation* 29: 99–107.

Cook, T. 2006. Collaborative action research within developmental evaluation: Learning to see or the road to myopia? *Evaluation* 12: 418–436.

Cooney, M., & C. H. Burt. 2008. Less crime, more punishment. *American Journal of Sociology* 114(2): 491–527.

COSSA. 2005. Budget issue: Proposed by 2006 budgets for social and behavioral science. *COSSA Washington Update* 24(4): 1–49. Retrieved on September 16, 2009, from http:// www.cossa.org/volume24/24.4.pdf

Couper, M. P., & S. E. Hansen. 2001. Computer-assisted interviewing. In *Handbook of interview research*, 557–575, edited by J. F. Gubrium & J. A. Holstein. Thousand Oaks, CA: Sage Publications.

Couper, M. P., & P. V. Miller. 2008. Web survey methods. *Public Opinion Quarterly* 72: 831–835.

Cowan, G., & R. R. Campbell. 1994. Racism and sexism in interracial pornography: A content analysis. *Psychology of Women Quarterly* 1918: 323–338.

Crawford, S., M. Couper, & M. Lamias. 2001. Web surveys: Perceptions of burden. *Social Science Computer Review* 19: 146–162.

Cronbach, L., & P. Meehl. 1955. Construct validity in psychological tests. *Psychological Bulletin* 52: 281–302.

Crow, G., R. Wiles, S. Heath, & V. Charles. 2006. Research methods and data quality: The implications of informed consent. *International Journal of Social Research Methodology* 9: 83–95.

Culver, L. 2004. The impact of new immigration patterns on the provision of policy services in Midwestern communities. *Journal of Criminal Justice* 32(4): 329–344.

Curtin, R., S. Presser, & E. Singer. 2000. The effects of response rate changes on the index of consumer sentiment. *Public Opinion Quarterly* 64(4): 413–428.

Curtin, R., S. Presser, & E. Singer. 2005. Changes in telephone survey nonresponse over the past quarter century. *Public Opinion Quarterly* 69: 87–98.

Davidson, J. O., & D. Layder. 1994. *Methods, sex and madness.* New York: Routledge.

Davis, D. W. 1997. Nonrandom measurement error and race of interviewer effects among African Americans. *Public Opinion Quarterly* 61: 183–206.

Davis, J. A., & T. W. Smith. 1998. *General social surveys, 1972–1998.* Chicago: National Opinion Research Center.

Davis, J. A., T. W. Smith, & P. V. Marsden. 2007. *General social surveys, 1972–2006.* Chicago: National Opinion Research Center.

De Meyrick, J. 2005. Approval procedures and passive consent considerations in research among young people. *Health Education* 105(4): 249–258.

Demant, J., & M. Jarvinen. 2006. Constructing maturity through alcohol experience—focus group interviews with teenagers. *Addiction Research and Theory* 14: 589–602.

Denscombe, M. 2006. Web-based questionnaires and the mode effect: An evaluation based on completion rates and data contents of near-identical questionnaires delivered in different modes. *Social Science Computer Review* 24: 246–254.

Denzin, N. 1989. *The research act: A theoretical introduction to sociological methods.* Englewood Cliffs, NJ: Prentice-Hall.

Dickert, N., & J. Sugarman. 2005. Ethical goals of community consultation in research. *American Journal of Public Health* 95(7): 1123–1127.

Diener, E. 2001. Satisfaction with life scale. Retrieved on August 12, 2001, from http://www.psych.uiuc.edu/~ediener/hottopic/hottopic.html

Diener, E., R. A. Emmons, R. J. Larsen, & S. Griffin. 1985. The satisfaction with life scale. *Journal of Personality Assessment* 49: 71–75.

DiLillo, D., S. DeGue, A. Kras, A. R. Di Loreto-Colgan, & C. Nash. 2006. Participant responses to retrospective surveys of child maltreatment: Does mode of assessment matter? *Violence and Victims* 21: 410–424.

Dillman, D., V. Lesser, R. Mason, J. Carlson, F. Milits, R. Robertson, & B. Burke. 2007. Personalization of mail surveys for general public and populations with a group identity: Results from nine studies. *Rural Sociology* 72: 632–646.

Dillman, D. A., M. D. Sinclair, & J. R. Clark. 1993. Effects of questionnaire length, respondent-friendly design and a difficult question on response rates for occupant-addressed census mail surveys. *Public Opinion Quarterly* 57: 289–304.

Diment, K., & S. Garrett-Jones. 2007. How demographic characteristics affect mode preference in a postal/web mixed-mode survey of Australian researchers. *Social Science Computer Review* 25: 410–417.

Dion, M. R. 2005. Healthy marriage programs: Learning what works. *Future of Children* 15(2): 139–156.

Divett, M., N. Crittenden, & R. Henderson. 2003. Actively influencing consumer loyalty. *Journal of Consumer Marketing* 20(2): 109–126.

Diviak, K. R., S. J. Curry, S. L. Emery, & R. Mermelstein. 2004. Human

participants challenges in youth tobacco cessation research: Researchers' perspectives. *Ethics & Behavior* 14(4): 321–334.

Dixon, M., V. J. Roscigno, & R. Hodson. 2004. Unions, solidarity, and striking. *Social Forces* 83: 3–33.

Dohan, D., & M. Sanchez-Jankowski. 1998. Using computers to analyze ethnographic field data: Theoretical and practical considerations. *Annual Review of Sociology* 24: 477–499.

Donelle, L., & L. Hoffman-Goetz. 2008. Health literacy and online health discussions of North American black women. *Women & Health* 47: 71–90.

Donohue, J. J., & S. D. Levitt. 2001. The impact of legalized abortion on crime. *Quarterly Journal of Economics* 116: 379–420.

Donohue, J. J., & S. D. Levitt. 2003. Further evidence that legalized abortion lowered crime: A reply to Joyce. NBER Working Paper No. 9532. February, 1–30.

Douglas, J. D. 1976. *Investigating social research: Individual and team field research.* Beverly Hill, CA: Sage.

Dovring, K. 1973. Communication, dissenters and popular culture in eighteenth-century Europe. *Journal of Popular Culture* 7: 559–568.

Dresing, T., T. Pehl, & C. Lombardo. 2008. Speech recognition software—an improvement to the transcription process. *Qualitative Social Research* 9(2). Retrieved on April 18, 2009, from http://www.qualitative-research.net/index.php/fqs/article/view/418

Du, P., & Y. Zhong. 2007. Effects and characteristics of senior students' demand for the internationalization of higher education in China. *Frontiers of Education in China* 2: 325–335.

Dunn, M. 1997. *Black Miami in the twentieth century.* Gainesville: University Press of Florida.

Dunning, T. 2008. Improving causal inference: Strengths and limitations of natural experiments. *Political Research Quarterly* 61: 282–293.

Durkheim, É. 1938. *The rules of sociological method.* New York: Free Press.

Durkheim, É. [1897] 1964. *Suicide.* Reprint. Glencoe, IL: Free Press.

Duwe, G., & D. Kerschner. 2008. Removing a nail from the boot camp coffin. *Crime and Delinquency* 54: 614–643.

The Economist. 1997. Don't let it happen again. 343: 27–28.

Ellis, C., A. Bochner, N. Denzin, Y. Lincoln, J. Morse, R. Pelias, & L. Richardson. 2008. Talking and thinking about qualitative research. *Qualitative Inquiry* 14: 254–284.

Ennett, S. T., N. S. Tobler, C. L. Ringwalt, & R. L. Flewelling. 1994. How effective is drug abuse resistance education? A meta-analysis of project DARE outcome evaluations. *American Journal of Public Health* 84: 1394–1401.

Erikson, K. T. 1976. *Everything in its path: Destruction of community in the Buffalo Creek flood.* New York: Simon & Schuster.

Esposito, L., & J. W. Murphy. 2000. Another step in the study of race relations. *Sociological Quarterly* 41: 171–187.

ESRC. 2009. ESRC facts and figures. Retrieved on August 28, 2009, from http://www.esrcsocietytoday.ac.uk/ESRCInfoCentre/PO/fast_facts_link/

Fairbrother, M. 2008. Why is trade policy so mercantilist? Neoliberal institutions, neoliberal ideas, and an enduring puzzle. Paper presented at the Annual Meeting of the American Sociological Association, Boston, August 1–4.

Farrell, G., & K. Clark. 2004. What does the world spend on criminal justice? HEUNI Paper No. 20. Helsinki, Finland: HEUNI. Retrieved on January 10, 2006, from http://www.heuni.fi/uploads/qjy0ay2w7l.pdf

Feagin, J. R., A. M. Orum, & G. Sjoberg. 1991. Conclusion: The present crisis in U.S. sociology. In *A case for the case study*, 269–278, edited by J. R. Feagin, A. M. Orum, & G. Sjoberg. Chapel Hill: University of North Carolina Press.

Feagin, J. R., & H. Vera. 2001. *Liberation sociology.* Boulder, CO: Westview.

Feeley, M. 2007. Legality, social research, and the challenge of institutional review boards. *Law & Society Review* 41: 757–776.

Felmlee, D., & A. Anna Muraco. 2009. Gender and friendship norms among older adults. *Research on Aging* 31: 318–344.

Fernández-Peña, J. R., L. Moore, E. Goldstein, P. DeCarlo, O. Grinstead, C. Hunt, D. Bao, & H. Wilson. 2008. Making sure research is used: Community-generated recommendations for disseminating research. *Progress in Community Health Partnerships: Research, Education, and Action* 2.2. Retrieved on July 22, 2009, from http://www.caps.ucsf.edu/CAPS/CAB/pdf/2.2.fernandez-pena.pdf

Ferraro, F. R., E. Szigeti, K. Dawes, & S. Pan. 1999. A survey regarding the University of North Dakota institutional review board: Data, attitudes and perceptions. *Journal of Psychology* 133: 272–381.

Fetterman, D., & M. Eiler. 2000. Empowerment evaluation: A model for building evaluation and program capacity. Paper presented at the American Evaluation Association Meetings, Honolulu, Hawaii, November 1–4. Retrieved on August 2, 2001, from http://www.stanford.edu/~davidf/Capacitybld.pdf

Finch, J., & L. Wallis. 1993. Death, inheritance and the life course. *Sociological Review Monograph* 50–68.

Fine, G. A., & K. D. Elsbach. 2000. Ethnography and experiment in social psychological theory building: Tactics for integrating qualitative field data with quantitative lab data. *Journal of Experimental Social Psychology* 36: 51–76.

Fine, M. 2009. A brief history of the participatory action research collective. Retrieved on September 16, 2009, from http://web.gc.cuny.edu/che/start.htm

Fine, M., N. Freudenberg, Y. Payne, T. Perkins, K. Smith, & K. Wanzer. 2003. "Anything can happen with police around": Urban youth evaluate strategies of surveillance in public places. *Journal of Social Issues* 59(1): 141–158.

Fine, M., & M. E. Torre. 2006. Intimate details, participatory action research in prison. *Action Research* 4(3): 253–269.

Fine, M., L. Weis, S. Wesson, & L. Wong. 2000. For whom? Qualitative research, representations, and social responsibilities. In *Handbook of qualitative research*, 2nd edition, 107–132, edited by N. K. Denzin & Y. S. Lincoln. Thousand Oaks, CA: Sage.

Fischman, W., B. Solomon, D. Greenspan, & H. Gardner. 2004. Making good: How young people cope with moral dilemmas at work. Cambridge, MA: Harvard University Press.

Flay, B. R., & L. M. Collins. 2005. Historical review of school-based randomized trials for evaluating problem behavior prevention programs. *Annals of the American Academy of Political and Social Science* 599(May): 115–146.

Fogel, C. 2007. Ethical issues in field-based criminological research in Canada. *International Journal of Criminal Justice Sciences* 2: 109–118.

Fontana, A., & J. H. Frey. 1994. Interviewing: The art of science. In *Handbook of qualitative research*, 361–376, edited by N. K. Denzin & Y. S. Lincoln. Thousand Oaks, CA: Sage.

Fontes, T. O., & M. O'Mahony. 2008. In-depth interviewing by instant messaging. *Social Research Update* 53: 1–4. Retrieved on September 16, 2009, from http://sru.soc.surrey.ac.uk/SRU53.pdf

Fox, J., C. Murray, & A. Warm. 2003. Conducting research using web-based questionnaires: Practical, methodological, and ethical considerations. *International Journal of Social Research Methodology* 6(2): 167–180.

Frank, D. J., B. Camp, & S. Boucher. 2008. Worldwide trends in the criminal regulation of sex, 1945–2005. Working Paper, Department of Sociology, University of California-Irvine.

Frank, D. J., T. Hardinge, & K. Wosick-Correa. 2009. The global dimensions of rape-law reform: A cross-national study of policy outcomes. *American Sociological Review* 74: 272–290.

Frank, K. 2005. Exploring the motivations and fantasies of strip club customers in relations to legal regulations. *Archives of Sexual Behavior* 34: 487–506.

Frankenberg, R. 1993. *White women, race matters: The social construction of whiteness*. Minneapolis: University of Minnesota Press.

Fricker, R. D., & M. Schonlau. 2002. Advantages and disadvantages of Internet research surveys: Evidence from the literature. *Field Methods* 14: 347–367.

Friedan, B. 1963. *The feminine mystique*. New York: Norton.

Friedman, J., & M. Alicea. 1995. Women and heroin: The path of resistance and its consequences. *Gender & Society* 9: 432–449.

Frohmann, L. 2005. The framing safety project: Photographs and narratives

by battered women. *Violence Against Women* 11: 1396–1419.

Gano-Phillips, S., & F. D. Fincham. 1992. Assessing marriage via telephone interviews and written questionnaires: A methodological note. *Journal of Marriage and the Family* 54: 630–635.

GAO. 2007. 2010 census: Design shows progress, but managing technology acquisitions, temporary field staff, and gulf region enumeration require attention. Retrieved on June 13, 2009, from http://www.gao.gov/new.items/d07779t.pdf

Garza, C., & M. Landeck. 2004. College freshman at risk—social problems at issue: An exploratory study of a Texas/Mexico border community college. *Social Science Quarterly* 85(5): 1390–1400.

Gauthier, J. B. 2002. *Measuring America: The decennial censuses from 1790–2000*. U.S. Census Bureau. Washington, DC: U.S. Government Printing Office.

Geer, J. G. 1988. What do open-ended questions measure? *Public Opinion Quarterly* 52: 365–371.

Geer, J. G. 1991. Do open-ended questions measure "salient" issues? *Public Opinion Quarterly* 55: 360–370.

General Social Survey. 2008. Cumulative data file. SDA Archive. Retrieved on April 2009 from http://sda.berkeley.edu/archive.htm

Gerich, J. 2008. Real or virtual? Response behavior in video-enhanced self-administered computer interviews. *Field Methods* 20: 356–376.

Gertz, C. 1983. *Local knowledge: Further essays in interpretative anthropology*. New York: Basic Books.

Ghose, T., D. Swendeman, G. Sheba, & D. Chowdhury. 2008. Mobilizing collective identify to reduce HIV risk among sex workers in Sonagachi, India: The boundaries, consciousness, and negotiation framework. *Social Science & Medicine* 67(2): 311–320.

Gibson, W., P. Callery, M. Campbell, A. Hall, & D. Richards. 2005. The digital revolution in qualitative research: Working with digital audio data through Atlas.ti. *Sociological Research Online* 10(1).

Gillespie, W. 2008. Thirty-five years after stonewall: An exploratory study of satisfaction with police among gay, lesbian, and bisexual persons at the 34th Annual Atlanta Pride

Festival. *Journal of Homosexuality* 55: 619–647.

Giordano, P., & S. Cernkovich. 2008. Juvenile delinquency, college attendance and the paradoxical role of higher education in crime and substance use. Paper presented at the Annual Meeting of the American Sociological Association, Boston, August 1–4.

Glaser, B. G., ed. 1993. *Examples of grounded theory: A reader.* Mill Valley, CA: Sociology Press.

Glaser, B. G. 2005. *The grounded theory perspective III: Theoretical coding.* Mill Valley, CA: Sociology Press.

Glaser, B. G., & A. L. Strauss. 1967. The discovery of grounded theory: Strategies for qualitative research. New York: Aldine.

Glazer, N. 1977. General perspectives. In *Woman in a man-made world,* 1–4, edited by N. Glazer & H. Youngelson. Chicago: Rand McNally.

Glenn, N. D. 1998. The course of marital success and failure in five American 10-year marriage cohorts. *Journal of Marriage and the Family* 60(3): 569–576.

Glover, T., W. Stewart, & K. Gladdys. 2008. Social ethics of landscape change: Toward community-based land-use planning. *Qualitative Inquiry* 14(3): 384–401.

Goffman, E. 1979. *Gender advertisements.* Cambridge, MA: Harvard University Press.

Gold, R. L. 1958. Roles in sociological field observations. *Social Forces* 36: 217–223.

Gold, R. L. 1997. The ethnographic method in sociology. *Qualitative Inquiry* 3: 388–402.

Goode, E. 1996. The ethics of deception in social research: A case study. *Qualitative Sociology* 19: 11–33.

Gorden, R. L. 1975. *Interviewing, strategy, techniques and tactics.* Homewood, IL: Dorsey.

Gorin, S., C. Hooper, C. Dyson, & C. Cabral. 2008. Ethical challenges in conducting research with hard to reach families. *Child Abuse Review* 17: 275–287.

Goritz, A. 2008. The long-term effect of material incentives on participation in online panels. *Field Methods* 20: 211–225.

Gould, S. J. 1995. *Dinosaur in a haystack: Reflections in natural history.* New York: Harmony.

Gould, S. J. 1997. *Full house: The spread of excellence from Plato to Darwin.* New York: Three Rivers Press.

Grady, J. 2001. Becoming a visual sociologist. *Sociological Imagination* 38(1/2): 83–119.

Grady, J. 2007a. Advertising images as social indicators: Depictions of blacks in *Life* magazine, 1936–2000. *Visual Studies* 22(3): 211–239.

Grady, J. 2007b. Visual sociology. In *21st century sociology: A reference handbook,* 63–70, edited by C. Bryant & D. Peck. Thousand Oaks, CA: Sage Publication.

Grady, J. 2007c. Visual methods. In *The encyclopedia of sociology,* 2286–2289, edited by G. Ritzer. London: Blackwell Publications.

Grandchamp, K. 2009. Disney animated movies: Friend or foe of femininity? Term paper for social research methods course. Rhode Island College.

Grenz, S. 2005. Intersections of sex and power in research on prostitution: A female researcher interviewing male heterosexual clients. *Signs: Journal of Women in Culture and Society* 30(4): 2091–2113.

Griesler, P. C., D. B. Kandel, C. Schaffran, M. Hu, & M. Davies. 2008. Adolescents' inconsistency in self-reported smoking: A comparison of reports in school and household settings. *Public Opinion Quarterly* 72: 260–290.

Groves, R. M. 1989. *Survey errors and survey costs.* New York: Wiley.

Groves, R. M. 2006. Nonresponse rates and nonresponse bias in households surveys. *Public Opinion Quarterly* 70(5): 646–675.

Groves, R. M., & E. Peytcheva. 2008. The impact of nonresponse rates on nonresponse bias: A meta-analysis. *Public Opinion Quarterly* 72(2): 167–189.

Groves, R. M., S. Presser, & A. Dipko. 2004. The role of topic interest in survey participation decisions. *Public Opinion Quarterly* 68(1): 2–31.

Gubrium, J. F., & J. A. Holstein. 2001. From the individual interview to the interview society. In *Handbook of Interview Research,* 3–32, edited by J. F. Gubrium & J. A. Holstein. Thousand Oaks, CA: Sage Publications.

Guernsey, L. 2009. What's been cut: The story of the child parent centers. Retrieved on February 26, 2009, from http://www.newamerica.net/blog/early-ed-watch/2009/

whats-been-cut-story-child-parent-centers-10341

Guillemin, M. 2004. Understanding illness: Using drawings as a research method. *Qualitative Health Research* 14: 272–289.

Guyll, M., R. Spoth, & C. Redmond. 2003. The effects of incentives and research requirements on participation rates for a community-based preventive intervention research study. *Journal of Primary Prevention* 24: 25–41.

Hadaway, C. K., & P. L. Marler. 1993. What the polls don't show: A closer look at U.S. church attendance. *American Sociological Review* 58: 741–752.

Hagan, J., & A. Palloni. 1998. Immigration and crime in the United States. In *The immigration debate,* 367–387, edited by J. P. Smith & B. Edmonton. Washington, DC: National Academy Press.

Hagestad, G. O., & P. Uhlenberg. 2005. The social separation of old and young: A root of ageism. *Journal of Social Issues* 61(2): 343–360.

Hale, M., T. Olsen, & E. F. Fowler. 2009. A matter of language or culture: Coverage of the 2004 U.S. elections on Spanish- and English language television. *Mass Communication & Society* 12(1): 26–51.

Hall, R. A. 2004. Inside out: Some notes on carrying out feminist research in cross-cultural interviews with South Asian women immigration applicants. *International Journal of Social Research Methodology* 7(2): 127–141.

Halse, C., & A. Honey. 2005. Unraveling ethics: Illuminating moral dilemmas of research ethics. *Signs: Journal of Women in Culture and Society* 30(4): 2149–2162.

Haluza-DeLay, R. 2008. Churches engaging the environment: An autoethnology of obstacles and opportunities. *Human Ecology Review* 15: 71–81.

Hamarat, E., D. Thompson, K. M. Sabrucky, D. Steele, K. B. Matheny, & F. Aysan. 2001. Perceived stress and coping resource availability as predictors of life satisfaction in young, middle-aged and older adults. *Experimental Aging Research* 27: 181–196.

Hamer, J. 2001. *What it means to be daddy: Fatherhood for black men living away from their children.*

New York: Columbia University Press.

Hammersley, M. 1995. *The politics of social research*. London: Sage.

Hammersley, M., & P. Atkinson. 1995. *Ethnography: Principles in practice*, 2nd edition. London: Routledge.

Hammond, P. E. 1964. *Sociologists at work*. New York: Basic Books.

Hamnett, M., D. Porter, A. Singh, & K. Kumar. 1984. *Ethics, politics, and international social science research: From critique to praxis*. Hawaii: University of Hawaii Press for East-West Center.

Hanks, R. S., & N. T. Carr. 2008. Lifelines of women in jail as self-constructed visual probes for life history research. *Marriage and Family Review* 42: 105–116.

Hara, S. 2007. Managing the dyad between independence and dependence: Case studies of the American elderly and their lives with pets. *International Journal of Japanese Sociology* 16: 100–114.

Hardie, J. H., & A. Lucas. 2008. Economic factors and relationship quality among young cohabitors. Paper presented at the Annual Meeting of the American Sociological Association, Boston, August 1–4.

Harding, S. 1993. Rethinking standpoint epistemology: What is "strong objectivity"? In *Feminist epistemologies*, 49–82, edited by L. Alcoff & E. Potter. New York: Routledge.

Harding, S., & K. Norberg. 2005. New feminist approaches to social science methodologies: An introduction. *Signs* 30: 2009–2015.

Harpel, T. S. 2008. Fear of the unknown: Ultrasound and anxiety about fetal health. *Health* 12: 295–312.

Harris, W., L. Skogrand, & D. Hatch. 2008. Role of friendship, trust, and love in strong Latino marriages. *Marriage & Family Review* 44: 455–488.

Hazell, V., & J. Clarke. 2008. Race and gender in media: A content analysis of two mainstream black magazines. *Journal of Black Studies* 39(1): 5–21.

Heinonen, T., & M. Cheung. 2007. Views from the village: Photonovella with women in rural China. *International Journal of Qualitative Methods* 6: 35–52.

Heise, L., M. Ellsberg, & M. Gottemoeller. 1999. Ending violence against women. *Population Reports*,

Series L, No. 11. Baltimore: Johns Hopkins University School of Public Health, Population Information Program, December.

Hennigan, K. M., C. L. Maxson, D. Sloane, & M. Ranney. 2002. Community views on crime and policing: Survey modes effects on bias in community surveys. *Justice Quarterly* 19(3): 565–587.

Herrera, C. D. 1997. A historical interpretation of deceptive experiments in American psychology. *History of the Human Sciences* 10: 23–36.

Herzog, A. R., & W. L. Rogers. 1988. Interviewing older adults. *Public Opinion Quarterly* 52: 84–99.

Hill, E. 1995. Labor market effects of women's post-school-age training. *Industrial and Labor Relations Review* 49: 138–149.

Hill, M. E. 2002. Race of the interviewer and perception of skin color: Evidence from the multi-city study of urban inequality. *American Sociological Review* 67: 99–108.

Hirsch, C. E. 2009. The strength of weak enforcement: The impact of discrimination charges, legal environments, and organization conditions on workplace segregation. *American Sociological Review* 74: 245–271.

Hirschi, T. 1969. *Causes of delinquency*. Berkeley: University of California Press.

Hobbs, F. 2005. *Examining American household composition: 1990 and 2000*. U.S. Census Bureau, Census 2000 Special Reports, CENSR-24. Washington, DC: I.S. Government Printing Office.

Hobza, C., & A. B. Rochlen. 2009. Gender role conflict, drive for muscularity and the impact of ideal media portrayals on men. *Psychology of Men & Masculinity* 12: 120–130.

Hochschild, A. R., with A. Machung. 1989. *The second shift*. New York: Avon Books.

Hofferth, S. L. 2005. Secondary data analysis in family research. *Journal of Marriage and Family* 67: 891–907.

Holbrook, A. L., M. C. Green, & J. A. Krosnick. 2003. Telephone versus face-to-face interviewing of national probability samples with long questionnaires. *Public Opinion Quarterly* 67(1): 79–125.

Holster, K. 2004. The business of Internet egg donation: An exploration of donor and recipient experiences.

Dissertation Abstracts International.

Holsti, O. R. 1969. *Content analysis for the social sciences and the humanities*. Reading, MA: Addison-Wesley.

Holyfield, L. 2002. *Moving up and out: Poverty, education and the single parent family*. Philadelphia: Temple University Press.

Hoover, E. 2005. The ethics of undercover research. *Chronicle of Higher Education* 51(47): A36. Retrieved on August 7, 2005, from http://chronicle.com/free/v51/i47/47a03601.htm

Howe, A., I. P. Cate, A. Brown, & J. Hadwin. 2008. Empathy in preschool children: The development of the Southampton test of empathy for preschoolers. *Psychological Assessment* 20(3): 305–309.

Hughes, E. C. 1960. Introduction: The place of field work in social science. In *Field work: An introduction to the social sciences*, iii–xiii, edited by B. H. Junker. Chicago: University of Chicago.

Huisman, K., & P. Hondagneu-Sotelo. 2005. Dress matters: Change and continuity in the dress practices of Bosnia Muslim refugee women. *Gender & Society* 19: 44–65.

Humphreys, L. 1975. *Tearoom trade: Impersonal sex in public places*. Chicago: Aldine.

Hurworth, R. 2003. Photo-interviewing for research. *Social Research Update* 40. Retrieved on June 21, 2009, from http://sru.soc.surrey.ac.uk/SRU40.pdf

Hutchinson, A. 2005. Analysing audio-recorded data: Using computer software applications. *Nurse Researcher* 12: 20–31.

Internet World Stats. 2008. Top 47 countries with the highest Internet penetration rates. Retrieved on March 6, 2009, from http://www.internetworldstats.com/top25.htm

Internet World Stats. 2009. The Internet big picture, world Internet users and population stats. Retrieved on June 23, 2009, from http://www.internetworldstats.com/stats.htm

Inter-Parliamentary Union. 2004. Women's suffrage. Retrieved on April 19, 2003, from http://www.ipu.org/wmn-e/suffrage.htm

Jacinto, C., M. Duterte, P. Sales, & S. Murphy. 2008. "I'm not a real dealer": The identify process of ecstasy dealers. *Journal of Drug Issues* 38(2): 419–444.

Jackall, R. 1988. *Moral mazes: The world of corporate managers.* New York: Oxford University Press.

Jaggar, E. 2005. Is thirty the new sixty? Dating, age and gender in postmodern, consumer society. *Sociology* 39: 89–106.

Johnson, T. P., J. Hougland, & R. Clayton. 1989. Obtaining reports of sensitive behavior: A comparison of telephone and face-to-face interviews. *Social Science Quarterly* 70: 174–183.

Jones, J. H. 1981. *Bad blood: The Tuskegee syphilis experiment.* New York: Free Press.

Jones, R. K. 2000. The unsolicited diary as a qualitative research tool for advanced research capacity in the field of health and illness. *Qualitative Health Research* 10: 555–567.

Jordan, S. 2003. Who stole my methodology? Co-opting PAR. *Globalisation, Societies and Education* 1: 185–200.

Jowell, R. 1998. How comparative is comparative research? *American Behavioral Scientist* 42: 168–177.

Joyce, T. 2003. Did legalized abortion lower crime? *Journal of Human Resources* 38: 1–37.

Kahneman, D., & A. Krueger. 2006. Developments in the measurement of subjective well-being. *Journal of Economic Perspectives* 20(1): 3–24.

Kahneman, D., A. B. Krueger, D. A. Schkade, N. Schwarz, & A. A. Stone. 2004. A survey method for characterizing daily life experience: The day reconstruction method. *Science* 306(5702): 1776–1780.

Kaiser Family Foundation. 2005. *Sex on TV: Executive summary.* Menlo Park, CA: Henry J. Kaiser Family Foundation. Retrieved on November 20, 2009, from http://www.kff.org/entmedia/upload/Sex-on-TV-4-Full-Report.pdf

Kane, E. W., & L. J. Macaulay. 1993. Interviewer gender and gender attitudes. *Public Opinion Quarterly* 57: 1–28.

Kane, E. W., & H. Schuman. 1991. Open survey questions as measures of personal concern with issues: A reanalysis of Stouffer's *Communism, conformity and civil liberties. Sociological methodology* 21: 81–96.

Kaplowitz, M. D., T. D. Hadlock, & R. Levine. 2004. A comparison of web and mail survey response rate.

Public Opinion Quarterly 68(1): 94–101.

Karney, B. R., J. Davila, C. L. Cohan, K. T. Sullivan, M. D. Johnson, & T. N. Bradbury. 1995. An empirical investigation of sampling strategies in marital research. *Journal of Marriage and the Family* 57: 909–920.

Karp, J. A., & D. Brockington. 2005. Social desirability and response validity: A comparative analysis of overreporting voter turnout in five countries. *Journal of Politics* 67(3): 825–840.

Karwalajtys, T., B. McDonough, H. Hall, M. Guirguis-Younger, L. W. Chambers, J. Kaczorowski, L. Lohfeld, & H. Brian. 2009. Development of the volunteer peer educator role in a community Cardiovascular Health Awareness Program (CHAP): A process evaluation in two communities. *Journal of Community Health* 34: 336–345.

Katz, J. 1972. *Experimentation with human beings.* New York: Russell Sage Foundation.

Keeter, S., C. Miller, A. Kohut, R. Groves, & S. Presser. 2000. Consequences of reducing nonresponse in a national telephone survey. *Public Opinion Quarterly* 64: 125–148.

Keleher, A., & E. R. A. N. Smith. 2008. Explaining the growing support for gay and lesbian equality since 1990. Paper presented at the Annual Meeting of the American Political Science Association, Boston, August 28–31.

Kemmis, S., & R. McTaggart. 2003. Participatory action research. In *Strategies of qualitative inquiry,* 2nd edition, 336–396, edited by N. K. Denzin & Y. S. Lincoln. Thousand Oaks, CA: Sage.

Kempner, J. 2008. The chilling effect: How do researchers react to controversy? *PloS Medicine* 5: 1571–1578.

Kenschaft, L. 2005. *Reinventing marriage: The love and work of Alice Freeman Palmer and George Herbert Palmer.* Urbana and Chicago: University of Illinois Press.

Kessler, R. C., S. Galea, M. J. Gruber, N. A. Sampson, R. J. Ursano, & S. Wessely. 2008. Trends in mental illness and suicidality after Hurricane Katrina. *Molecular Psychiatry* 13(4): 374–384.

Kim, J., & E. Hatfield. 2004. Love types and subjective well-being:

A cross-cultural study. *Social Behavior and Personality* 32(2): 173–182.

Kim, M. S., H. W. Kim, K. H. Cha, & J. Lim. 2007. What makes Koreans happy? Exploration on the structure of happy life among Korean adults. *Social Indicators Research* 82: 265–286.

King, A. C. 1994. Enhancing the self-report of alcohol consumption in the community: Two questionnaire formats. *American Journal of Public Health* 84: 294–296.

Kinsey, A. C., W. B. Pomeroy, & C. E. Martin. 1953. *Sexual behavior in the human female.* Philadelphia: Saunders.

Kirp, D. L. 1997. Blood, sweat and tears: The Tuskegee experiment and the era of AIDS. *Tikkun* 10: 50–55.

Kleiner, A. M., J. J. Green, & A. B. Nylander III. 2007. A community study of disaster impacts and redevelopment issues facing East Biloxi Mississippi. In *The sociology of Katrina: Perspectives on a modern catastrophe,* 155–172, edited by D. L. Brunsma, D. Overfelt, & J. S. Picou. Lanham, MD: Rowman & Littlefield.

Klofstad, C. A., S. Boulianne, & D. Basson. 2008. Matching the message to the medium: Results from an experiment on Internet survey email contacts. *Social Science Computer Review* 26: 498–509.

Kneipp, S., J. Castleman, & N. Gailor. 2004. Informal caregiving burden: An overlooked aspect of the lives and health of women transitioning from welfare to employment. *Public Health Nursing* 21(1): 24–31.

Knickman, J. R. 2005. When health policy is the problem. *Journal of Health Politics, Policy and Law* 30: 367–374.

Kolar, T., & I. Kolar. 2008. What respondents really expect from researchers. *Evaluation Research* 32: 363–391.

Korn, J. 1997. *Illusions of reality: A history of deception in social psychology.* Albany: State University of New York Press.

Koropeckyj-Cox, T., & G. Pendell. 2007. The gender gap in attitudes about childlessness in the United States. *Journal of Marriage and Family* 69: 899–915.

Kotlowitz, A. 1991. *There are no children here.* New York: Doubleday.

Krafchick, J. L., T. S. Zimmerman, S. A. Haddock, & J. H. Banning. 2005.

Best-selling books advising parents about gender: A feminist analysis. *Family Relations* 54: 84–100.

Kraszewski, J. 2008. Pittsburgh in fort worth football bars, sports television, sports fandom, and the management of home. *Journal of Sport & Social Issues* 32: 139–157.

Kravdal, O., & R. R. Rindfuss. 2008. Changing relationships between education and fertility: A study of women and men born 1940 to 1964. *American Sociological Review* 73(5): 854–874.

Kreuter, F., S. Presser, & R. Tourangeau. 2008. Social desirability bias in CATI, IVR, and web surveys: The effects of mode and question sensitivity. *Public Opinion Quarterly* 72: 847–865.

Krimerman, L. 2001. Participatory action research: Should social inquiry be conducted democratically? *Philosophy of Social Sciences* 31: 60–82.

Krippendorff, K. 2003. *Content analysis: An introduction to its methodology*, 2nd edition. Beverly Hills, CA: Sage.

Krysan, M., & M. P. Couper. 2003. Race in the live and the virtual interview: Racial deference, social desirability, and activation effects in attitude surveys. *Social Psychology Quarterly* 66: 364–383.

Krysan, M., H. Schuman, L. J. Scott, & P. Beatty. 1994. Response rates and response content in mail versus face-to-face surveys. *Public Opinion Quarterly* 58: 381–400.

Labriola, M., M. Rempel, & R. C. Davis. 2008. Do batterer programs reduce recidivism? Results from a randomized trial in the Bronx. *Justice Quarterly* 25: 256–282.

Ladd, E. C., & K. Bowman. 1998. *What's wrong: A survey of American satisfaction and complaint*. Washington, DC: AEI Press.

Landra, K., & T. Sutton. 2008. You think you know ghetto? Contemporizing the Dove "Black IQ Test." *Teaching Sociology* 36(4): 366–377.

Lang, A. 2001. Economic inequality: A quantitative study of opinions related to the income gap and government involvement in its reduction. Honors thesis at Rhode Island College.

Lasswell, H. D. 1965. Detection: Propaganda detection and the courts. In *The language of politics: Studies in quantitative semantics*, edited by

H. Lasswell et al. Cambridge, MA: MIT Press.

Lauder, M. A. 2003. Covert participation observation of a deviant community: Justifying the use of deception. *Journal of Contemporary Religion* 18(2): 185–196.

Laumann, E. O., J. H. Gagnon, R. T. Michael, & S. Michaels. 1994. *The social organization of sexuality*. Chicago: University of Chicago Press.

Lavrakas, P. J. 1987. *Telephone survey methods sampling, selection and supervision*. Newbury Park, CA: Sage.

Lawrence, S., R. Mukal, & J. Atlenza. 2009. *Foundation giving trends: Update on funding priorities*. New York: Foundation Center.

Lazarsfeld, P. F. 1944. The controversy over detailed interviews: An offer for negotiation. *Public Opinion Quarterly* 8: 38–60.

Lee, S., N. A. Mathiowetz, & R. Tourangeau. 2007. Measuring disability in surveys: Consistency over time and across respondents. *Journal of Official Statistics* 23: 163–184.

Lemert, C. 1993. *Social theory: The multicultural and classic readings*. Boulder, CO: Westview.

Leo, R. A. 1995. Trial and tribulations: Courts, ethnography, and the need for an evidentiary privilege for academic researchers. *American Sociologist* 26: 113–134.

Lester, D., & D. Lester, Jr. 2009. *Writing research papers*. New York: Longman.

Lever, J. 1981. Multiple methods of data collection. *Urban Life* 10: 199–213.

Levitt, S. D. 2004. Understanding why crime fell in the 1990s: Four factors that explain the decline and six that do not. *Journal of Economic Perspectives* 18: 163–190.

Lewins, A., & C. Silver. 2007. *Using software in qualitative research: A step-by-step guide*. London: Sage Publication.

Lewis, C. A., M. J. Dorahy, & J. F. Schumaker. 1999. Depression and life satisfaction among northern Irish adults. *Journal of Social Psychology* 139: 533–535.

Liebow, E. 1967. *Tally's corner: A study of Negro streetcorner men*. Boston: Little, Brown.

Liebow, E. 1993. *Tell them who I am: The lives of homeless women*. New York: Free Press.

Lin, J. Y., L. S. Chen, E. S. Wang, & J. M. Cheng. 2007. The relationship between extroversion and leisure

motivation: Evidence from fitness center participation. *Social Behavior and Personality* 35(10): 1317–1322.

Lindsay, J. 2005. Getting the numbers: The unacknowledged work in recruiting for survey research. *Field Methods* 17(1): 119–128.

Lindsey, E. W. 1997. Feminist issues in qualitative research with formerly homeless mothers. *Affilia: Journal of Women and Social Work* 12: 57–76.

Link, M. W., M. P. Battaglia, M. R. Frankel, L. Osborn, & A. H. Mokdad. 2008. A comparison of address-based sampling (ABS) versus random-digit dialing (RDD) for general population surveys. *Public Opinion Quarterly* 72: 6–27.

Link, M. L., & A. Mokdad. 2005. Advance letters as a means of improving respondent cooperation in random digit dial studies: A multistate experiment. *Public Opinion Quarterly* 69(4): 572–587.

Lo, C. 2000. Timing of drinking initiation: A trend study predicting drug use among high school seniors. *Journal of Drug Issues* 30: 525–534.

Loe, M. 2004. Sex and the senior woman: Pleasure and danger in the Viagra era. *Sexualities* 7: 303–326.

Lofland, J., & L. H. Lofland. 1995. *Analyzing social settings: A guide to qualitative observation and analysis*, 3rd edition. Belmont, CA: Wadsworth.

Lofland, J., D. Snow, L. Anderson, & L. H. Lofland. 2005. *Analyzing social settings: A guide to qualitative observation and analysis*, 4th edition. Belmont, CA: Wadsworth.

Logan, T. K., R. Walker, L. Shannon, & J. Cole. 2008. Combining ethical considerations with recruitment and follow-up strategies for partner violence and victimization research. *Violence Against Women* 14: 1226–1251.

Long, J. S., & E. K. Pavalko. 2004. The life course of activity limitations: Exploring indicators of functional limitations over time. *Journal of Aging and Health* 16(4): 490–516.

Lopata, H. Z. 1980. Interviewing American widows. In *Fieldwork experience qualitative approaches to social research*, edited by W. B. Shaffir, R. A. Stubbing, & A. Turowetz. New York: St. Martin's 68–81.

Louie, M. C. Y. 2001. *Sweatshop warriors: Immigrant women workers take on the global factory.* Cambridge, MA: South End Press.

Lu, Y., & D. J. Treiman. 2008. The effect of sibship size on educational attainment in China: Period variations. *American Sociological Review* 73(5): 813–835.

Lundquist, J. H. 2008. Ethnic and gender satisfaction in the military: The effect of a meritocratic institution. *American Sociological Review* 73(3): 477–497.

Lystra, K. 1989. *Searching the heart: Women, men and romantic love in nineteenth-century America.* New York: Oxford University Press.

Madge, C., & H. O'Connor. 2006. Parenting gone wired: empowerment of new mothers on the Internet? *Social & Cultural Geography* 7: 199–220.

Madriz, E. 2000. Focus groups in feminist research. In *Handbook of qualitative research*, 2nd edition, 835–850, edited by N. K. Denzin & Y. S. Lincoln. Thousand Oaks, CA: Sage.

Maher, F. A., & M. K. Tetreault. 1994. *The feminist classroom.* New York: Basic.

Maher, K. 2005. Practicing sociology as a vocation. In *Explorations in theology and vocation*, 31–40, edited by W. E. Rogers. Greenville, NC: Furman University's Center for Theological Explorations of Vocation.

Mair, M., & C. Kierans. 2007. Descriptions as data: Developing techniques to elicit descriptive materials in social research. *Visual Studies* 22: 120–136.

Majumdar, A. 2007. Researching South Asian women's experiences of marriage: Resisting stereotypes through an exploration of "space" and "embodiment". *Feminism & Psychology* 17: 316–322.

Malinowski, B. 1922. *Argonauts of the western pacific.* New York: Dutton.

Malter, S., L. Simich, N. Jacobson, & J. Wise. 2009. Reciprocity: An ethic for community-based participatory action research. *Action Research* 6: 305–325.

Mann, S. L., D. J. Lynn, & A. V. Peterson Jr. 2008. The "downstream" effect of token prepaid cash incentives to parents on their young adult children's survey participation. *Public Opinion Quarterly* 72: 487–501.

Manning, P. K. 2002. FATHETHICS: Response to Erich Goode. *Qualitative Sociology* 25(4): 541–547.

Manza, J., & C. Brooks. 1998. The gender gap in U.S. presidential elections: When? Why? Implications? *American Journal of Sociology* 103: 1235–1266.

Marecek, J., M. Fine, & L. Kidder. 1997. Working between worlds: Qualitative methods and social psychology. *Journal of Social Issues* 53: 631–645.

Marshall, B. 2002. *Heliobactor pioneers: Firsthand accounts from the scientists who discovered heliobactors 1892–1982.* Perth: University of Western Australia.

Martin, L. L., T. Abend, C. Sedikides, & J. D. Green. 1997. How would I feel if…? Mood as input to a role fulfillment evaluation process. *Journal of Personality and Social Psychology* 73: 242–254.

Martineau, H. 1962. *Society in America.* New York: Anchor.

Martinez, R., & M. T. Lee. 2000. Comparing the context of immigrant homicides in Miami: Haitians, Jamaicans and Mariel. *International Migration Review* 34: 794–812.

Marx, K. [1867] 1967. *Das Kapital.* Reprint. New York: International.

McCall, G. J. 1984. Systematic field observation. *Annual Review of Sociology* 10: 263–282.

McClelland, S. I., & M. Fine. 2008. Embedded science: Critical analysis of abstinence-only evaluation research. *Cultural Studies, Critical Methodologies* 8: 50–81.

McCorkel, J., & K. Myers. 2003. What difference does difference make? Position and privilege in the field. *Qualitative Sociology* 26(2): 199–231.

McGraw, L., A. Zvonkovic, & A. Walker. 2000. Studying postmodern families: A feminist analysis of ethical tensions in work and family research. *Journal of Marriage and the Family* 62(February): 68–77.

McLaren, C. 2009. Analyzing qualitative data about hospitalized children: Reflections on bodily expressions. *Qualitative Report* 14, 140–154. Retrieved on June 23, 2009, from http://www.nova.edu/ssss/QR/QR14-1/mclaren.pdf

McNabb, S. 1995. Social research and litigation: Good intentions versus good ethics. *Human Organization* 54: 331–335.

McNamara, J. M. 2007. Long-term disadvantage among elderly women: The effects of work history. *Social Service Review* 81: 423–452.

Mead, M. 1933. *Sex and temperament in three primitive societies.* New York: Morrow.

Mears, A., & W. Finlay. 2005. Not just a paper doll: How models manage bodily capital and why they perform emotional labor. *Journal of Contemporary Ethnography* 34: 317–343.

Melrose, M. 2002. Labour pains: Some considerations on the difficulties of researching juvenile prostitution. *International Journal of Social Research Methodology* 5(4): 333–351.

Menard, S. 1991. *Longitudinal research.* Newbury Park, CA: Sage.

Merton, R. K., M. Fiske, & P. L. Kendall. 1956. *The focused interview.* Glencoe, IL: Free Press.

Messner, S. F. 1999. Television violence and violent crime: An aggregate analysis. In *How it's done: An invitation to social research*, 308–316, edited by E. S. Adler & R. Clark. Belmont, CA: Wadsworth.

Michael, R. T., J. H. Gagnon, E. O. Laumann, & G. Kolata. 1994. *Sex in America.* Boston: Little, Brown.

Michell, L. 1998. Combining focus groups and interviews: Telling how it is; telling how it feels. In *Developing focus group research*, 36–46, edited by R. S. Barbour & J. Kitzinger. Thousand Oaks, CA: Sage.

Midanik, L. T., T. K. Greenfield, & J. D. Rogers. 2001. Reports of alcohol-related harm: Telephone versus face to face interviews. *Journal of Studies on Alcohol* 62: 74–78.

Mies, M. 1991. Women's research or feminist research? In *Beyond methodology: Feminist scholarship as lived research*, 60–84, edited by M. M. Fonow & J. A. Cook. Bloomington: Indiana University Press.

Miles, M. B., & A. M. Huberman. 1994. *Qualitative data analysis*, 2nd edition. Thousand Oaks, CA: Sage.

Milgram, S. 1974. *Obedience to authority: An experimental view.* New York: Harper & Row.

Miller, D. C., & N. J. Salkind. 2002. *Handbook of research design and social measurement.* Thousand Oaks, CA: Sage.

Miller, E. M. 1986. *Street woman.* Philadelphia: Temple University Press.

Miller, W. E., & S. Traugott. 1989. *American national election studies data sourcebook, 1952–1986.* Cambridge, MA: Harvard University Press.

Mitchell, A. 1997. Survivors of Tuskegee study get apology from Clinton. *New York Times* 146: 9–10.

Moller, V., P. Theuns, I. Erstad, & J. Bernheim. 2008. The best and worst times of life: Narratives and assessments of subjective well-being by anamnestic comparative self assessment (ACSA) in the Easter Cape, South Africa. *Social Indicators Research* 89(1): 1–22.

Moret, M., R. Reuzel, G. van der Wilt, & J. Grin. 2007. Validity and reliability of qualitative data analysis: Interobserver agreement in reconstructing interpretative frames. *Field Methods* 19: 24–39.

Morgan, D. L. 2001. Focus group interviewing. In *Handbook of interview research*, 141–159, edited by J. F. Gubrium & J. A. Holstein. Thousand Oaks, CA: Sage Publications.

Morris, M., & L. Jacobs. 2000. You got a problem with that? Exploring evaluators' disagreements about ethics. *Evaluation Review* 24: 384–406.

Morrow, R. 1994. *Critical theory and methodology.* Thousand Oaks, CA: Sage.

Morrow-Howell, N., M. Jonson-Reid, S. McCrary, Y. Lee, & E. Spitznagel. 2009. Evaluation of experience corps student reading outcomes. Report from the Center for Social Development, Washington University, St. Louis. Retrieved on June 30, 2009, from http://csd.wustl.edu/Publications/Documents/RP09-01.pdf

Mossakowski, K. 2008. Dissecting the influence of race/ethnicity and socioeconomic status on mental health in young adulthood. Paper presented at the Annual Meeting of the American Sociological Association, Boston, August 1–4.

Mosteller, F. 1955. Use as evidenced by an examination of wear and tear on selected sets of ESS. In *A study of the need for a new encyclopedic treatment of the social sciences*, edited by K. Davis. Unpublished manuscript.

Muir, K. B., & T. Seitz. 2004. Machismo, misogyny, and homophobia in a male athletic subculture: Participant-observation study of deviant rituals in collegiate rugby. *Deviant Behavior* 25: 303–327.

Nathan, R. 2005. *My freshman year: What a professor learned by becoming a student.* Ithaca, NY: Cornell University Press.

National Academy of Sciences. 1993. Methods and values in science. In *The "racial" economy of science*, 341–343, edited by S. Harding. Bloomington: Indiana University Press.

National Alliance for Caregiving. 2005. Young caregivers in the U.S. findings from a national study. Retrieved on January 22, 2006, from http://www.uhfnyc.org/usr_doc/Young_Caregivers_Study_083105.pdf

National Center for Health Statistics. 2009. National Health Interview Survey (NHIS). Retrieved on May 29, 2009, from http://www.cdc.gov/nchs/about/major/nhis/hisdesc.htm

National Science Foundation. 2009. 45 CFR Part 690: Federal policy for the protection of human subjects. Retrieved on August 17, 2009, from http://www.nsf.gov/bfa/dias/policy/docs/45cfr690.pdf

Neto, F., & J. Barros. 2000. Psychosocial concomitants of loneliness among students of Cape Verde and Portugal. *Journal of Psychology* 134: 503–514.

NHS. 2008. History of the Nurses' Health Study. Retrieved on August 30, 2009, from http://www.channing.harvard.edu/nhs/history/index.shtml

NICHD ECCRN. 2008. Social competence with peers in third grade: Associations with earlier peer experiences in childcare. *Social Development* 17: 419–453.

Norusis, M. J. 2009. *The SPSS guide to data analysis.* Chicago: SPSS Inc.

Nyden, P. A., A. Figert, M. Shibley, & D. Burrows. 1997. *Building community.* Thousand Oaks, CA: Pine Forge.

O'Neill, K. 2008. Public sociology and participatory action research in rural sociology. *Footnotes* 36(9). Retrieved on July 2, 2009, from http://www.asanet.org/footnotes/dec08/rural.html

Oakley, A. 1981. Interviewing women: A contradiction in terms. In *Doing feminist research*, 30–61, edited by H. Roberts. London: Routledge & Kegan Paul.

Olson, J. E., I. H. Frieze, S. Wall, B. Zdaniuk, A. Ferligoj, T. Kogovšek, J. Horvat, N. Šarlija, E. Jarošová, D. Pauknerová, L. A. N. Luu, M. Kovacs, J. Miluska, A. Orgocka, L. Erokhina, O. V. Mitina, L. V. Popova, N. Petkevičiūtė, M. Pejic-Bach, S. Kubušová, & M. R. Makovec. 2007. Beliefs in equality for women and men as related to economic factors in Central and Eastern Europe and the United States. *Sex Roles* 56: 297–308.

Olson, S. 1976. *Ideas and data: The process and practice of social research.* Homewood, IL: Dorsey.

Ono, H., & H.-J. Tsai. 2008. Race, parental socioeconomic status, and computer use time outside of school among young American children, 1997 to 2003. *Journal of Family Issues* 29: 1650–1672.

Oransky, M., & C. Fisher. 2009. The development and validation of meanings of adolescent masculinity scale. *Psychology of Men & Masculinity* 10(1): 57–72.

Oriel, K., M. B. Plane, & M. Mundt. 2004. Family medicine residents and the impostor phenomenon. *Family Medicine* 36: 248–252.

Ozbay, O., & Y. Ozcan. 2008. A test of Hirschi's social bonding theory: A comparison of male and female delinquency. *International Journal of Offender Therapy and Comparative Criminology* 52(2): 134–157.

Paluck, E. L. 2009. Reducing intergroup prejudice and conflict using the media: A field experiment in Rwanda. *Journal of Personality and Social Psychology* 96: 574–587.

Palys, T., & J. Lowman. 2001. Social research with eyes wide shut: The limited confidentiality dilemma. *Canadian Journal of Criminology* 43: 255–267.

Palys, T., & J. Lowman. 2006. Protecting research confidentiality: Towards a research-participant shield law. *Canadian Journal of Law and Society* 21: 163–185.

Paradis, E. K. 2000. Feminist and community psychology ethics in research with homeless women. *American Journal of Community Psychology* 28: 839–854.

Park, J., & A. E. Zeanah. 2005. An evaluation of voice recognition software for use in interview-based research: A research note. *Qualitative Research* 5: 245–251.

Park, P. 1993. What is participatory research? A theoretical and methodological perspective. In *Voices of change, participatory research in the United States and Canada*, edited by P. Park, M. Brydon-Miller, B. Hall, & T. Jackson. Westport, CT: Bergin & Garvey 1–20.

Pattman, R., S. Frosh, & A. Phoenix. 2005. Constructing and experiencing boyhoods in research in London. *Gender and Education* 17(5): 555–561.

Paver, D. 2007. The use of field experiments for studies of employment discrimination: Contributions, critiques, and directions for the future. *Annals of the American Academy of Political and Social Science* 609: 104–133.

Perrone, K. M., C. L. Civiletto, L. K. Webb, & J. C. Fitch. 2004. Perceived barriers to and supports of the attainment of career and family goals among academically talented individuals. *International Journal of Stress Management* 11: 114–131.

Perry, C. L., K. A. Komro, S. Veblen-Mortenson, & L. M. Bosma. 2000. The Minnesota DARE PLUS Project: Creating community partnerships to prevent drug use and violence. *Journal of School Health* 70: 84–88.

Pew Research Center for People & the Press. 2009. Economy, jobs trump all other policy priorities in 2009. Retrieved on May 20, 2009, from http://people-press.org/report/485/economy-top-policy-priority

Plutzer, D., & M. Berkman. 2005. The graying of America and support for funding of the nation's schools. *Public Opinion Quarterly* 69(1): 66–86.

Portes, A., J. M. Clarke, & R. Manning. 1985. After Mariel: A survey of the resettlement experiences of 1980 Cuban refugees in Miami. *Cuban Studies* 15: 37–59.

Pothier, M. 2009. Pop goes the market. The strange connection between billboard hits and stock prices. *Boston Globe*, February 22. Retrieved on February 22, 2009, from http://www.boston.com/bostonglobe/ideas/articles/2009/02/22/pop_goes_the_market/

Powell, R. A., H. M. Single, & K. R. Lloyd. 1996. Focus groups in mental health research: Enhancing the validity of user and provider questionnaires. *International Journal of Social Psychiatry* 42: 193–206.

Prior, M. 2006. The incumbent in the living room: The rise of television

and the incumbency advantage in U.S. house elections. *Journal of Politics* 3: 657–673.

Ptacek, J. 1988. Why do men batter their wives? In *Feminist perspectives on wife abuse*, edited by K. Yllo & M. Bograd. Newbury Park, CA: Sage.

Public Agenda. 2002. Right to die: Major proposals. *Public Agenda Online*. Retrieved on April 18, 2002, from http://www.publicagenda.org/issues/major-proposals_details.dfm?

Punch, M. 1994. Politics and ethics of qualitative research. In *Handbook of qualitative research*, 83–97, edited by N. Denzin & Y. Lincoln. Thousand Oaks, CA: Sage.

Putnam, R. D. 2000. *Bowling alone: The collapse and revival of American community*. New York: Simon & Schuster.

Puwar, N. 1997. Reflections on interviewing women MPs. *Sociological Research Online* 2. Retrieved on September 16, 2009, from http://www.socresonline.org.uk/2/1/4.html

Pyrczak, F., & R. R. Bruce. 2007. *Writing empirical research reports: A basic guide for students of the social and behavioral sciences*, 6th edition. Los Angeles, CA: Pyrczak.

Racher, F. E., J. M. Kaufert, & B. Havens. 2000. Conjoint research interviews with frail elderly couples: Methodological implications. *Journal of Nursing Research* 6: 367–379.

Radley, A., D. Hodgetts, & A. Cullen. 2005. Visualizing homelessness: A study in photography and estrangement. *Journal of Community & Applied Social Psychology* 15(4): 273–295.

Rasiniski, K. A., D. Mingay, & N. M. Bradburn. 1994. Do respondents really "mark all that apply" on self-administered questions? *Public Opinion Quarterly* 58: 400–408.

Rasmussen Reports. 2005. 72% say they're willing to vote for woman president. Retrieved on June 23, 2009, from http://legacy.rasmussenreports.com/2005/Woman%20President.htm

Rasmussen Reports. 2009. 44% say global warming due to planetary trends, not people. Retrieved on May 20, 2009, from http://www.rasmussenreports.com/public_content/politics/issues2/articles/44_say_global_warming_due_to_planetary_trends_not_people

Rathje, W. L., & C. Murphy. 1992. *Rubbish! The archaeology of garbage*. New York: HarperCollins.

RealClearPolitics. 2008. General election: McCain vs. Obama. Retrieved on February 2, 2009, from http://www.realclearpolitics.com/epolls/2008/president/us/general_election_mccain_vs_obama

Reger, J. 2004. Organizational "emotion work" through consciousness-raising: An analysis of a feminist organization. *Qualitative Sociology* 27(2): 205–222.

Reinharz, S. 1992. *Feminist methods in social research*. New York: Oxford University Press.

Rew, L., S. D. Horner, L. Riesch, & R. Cauvin. 2004. Computer-assisted survey interviewing of school-age children. *Advances in Nursing Science* 27(2): 129–137.

Reynolds, A. J., L. C. Mathieson, & J. W. Topitzes. 2009. Do early childhood interventions prevent child maltreatment? A review of research. *Child Maltreatment* 14: 182–206.

Reynolds, A. J., & D. Robertson. 2003. School-based early intervention and later child maltreatment in the Chicago Longitudinal Study. *Child Development* 74(1): 3–26.

Reynolds, A. J., J. A. Temple, S. Ou, D. L. Robertson, J. P. Mersky, J. W. Topitzes, & M. D. Niles. 2007. Effects of a school-based, early childhood intervention on adult health and well-being; A 19-year follow-up of low-income families. *Archives of Pediatrics & Adolescent Medicine* 161: 370–739.

Reynolds, A. J., J. A. Temple, D. L. Robertson, & E. A. Mann. 2002. Age 21 cost-benefit analysis of the title I Chicago child-parent centers. *Educational Evaluation and Policy Analysis* 24: 267–303.

Richards, T. J., & L. Richards. 1994. Using computers in qualitative research. In *Handbook of qualitative research*, 445–462, edited by N. K. Denzin & Y. S. Lincoln. Thousand Oaks, CA: Sage.

Richman, W., S. Kiesler, S. Weisband, & F. Drasgow. 1999. A meta-analytic study of social desirability distortion in computer-administered questionnaires, traditional questionnaires, and interviews. *Journal of Applied Psychology* 84: 754–775.

Riemer, J. W. 1977. Varieties of opportunistic research. *Urban Life* 5: 467–477.

Ringheim, K. 1995. Ethical issues in social science research within special reference to sexual behavior research. *Social Science and Medicine* 40: 1691–1697.

Roberts, S. 2007. 51% of women are now living without spouse. *New York Times*, January 16: 1ff.

Robinson, J., & S. Martin. 2008. What do happy people do? *Social Indicators Research* 89(3): 565–571.

Rochlan, A., G. Good, & T. Carver. 2009. Predictors of gender-related barriers, work, and life satisfaction among men in nursing. *Psychology of Men & Masculinity* 10(1): 44–56.

Rogers, D. 2008. 2010 census: Higher cost, lower tech. Retrieved on March 11, 2009, from http://www. politico.com/news/stories/0408/ 9376.html

Roig, M., & L. DeTommaso. 1995. Are college cheating and plagiarism related to academic procrastination? *Psychological Reports* 77(2): 691–698.

Rollins, J. 1985. *Between women*. Philadelphia: Temple University Press.

Rose, D. S., S. D. Sidle, & K. H. Griffith. 2007. A penny for your thoughts, monetary incentives improve response rates for company-sponsored employee surveys. *Organizational Research Methods* 10: 225–240.

Rosenberg, M. 1968. *The logic of survey analysis*. New York: Basic Books.

Rosengren, K. E. 1981. Advances in Scandinavian content analysis. In *Advances in content analysis*, 9–19, edited by K. E. Rosengren. Beverly Hills, CA: Sage.

Rosenhan, D. L. 1971. On being sane in insane places. *Science* 179: 250–258.

Rossi, P. H., & H. E. Freeman. 1993. *Evaluation: A systematic approach*. Newbury Park, CA: Sage.

Rossi, P. H., & J. D. Wright. 1984. Evaluation research: An assessment. In *Annual review of sociology*, vol. 10, 332–352, edited by R. H. Turner & J. F. Short. Palo Alto, CA: Annual Reviews.

Rubin, L. B. 1983. *Intimate strangers: Men and women together*. New York: Harper & Row.

Rubin, L. B. 1994. *Families on the fault line: America's working class speaks about the family, the economy, race, and ethnicity*. New York: HarperCollins.

Runcie, J. F. 1980. *Experiencing social research*. Homewood, IL: Dorsey.

Rutherford, M. B. 2004. Authority, autonomy, and ambivalence: Moral choice in twentieth-century commencement speeches. *Sociological Forum* 19: 583–609.

Salmela-Aro, K., N. Kiuru, E. Leskinen, & J. Nurmi. 2009. School Burnout Inventory (SBI): Reliability and validity. *European Journal of Psychological Assessment* 25(1): 48–57.

Sampson, R. J., & S. W. Raudenbush. 1999. Systematic social observation of public spaces: A new look at disorder in urban neighborhoods. *American Journal of Sociology* 105: 603–651.

Sanchez, J., C. Lomeli-Loibl, & A. A. Nelson. 2009. Sacramento's LGBTQ youth: Youth-led participatory action research for mental health justice with Youth in Focus. *Focal Point* 23: 6–8. Retrieved on July 2, 2009, from http://www.rtc. pdx.edu/PDF/fpS0902.pdf

Saucier, P. K. 2008. *We eat Cachupa, not clam Chowder: Mapping second generation Cape Verdean youth identity in the greater Boston area*. Doctoral dissertation at Northeastern University.

Scarce, R. 1995. Scholarly ethics and courtroom antics: Where researchers stand in the eyes of the law. *American Sociologist* 26(1): 87–112.

Scarce, R. 2005. *Contempt of court: A scholar's battle for free speech from behind bars*. Lanham, MD: AltaMira Press.

Schaeffer, N. C., & S. Presser. 2003. The science of asking questions. *Annual Review of Sociology* 29: 65–88.

Schieman, S. 2001. Age, education, and the sense of control: A test of the cumulative advantage hypothesis. *Research on Aging* 23: 153–178.

Schlosser, J. A. 2008. Issues in interviewing inmates navigating the methodological landmines of prison research. *Qualitative Inquiry* 14: 1500–1525.

Schumaker, J. F., & J. D. Shea. 1993. Loneliness and life satisfaction in Japan and Australia. *Journal of Psychology* 127: 65–71.

Schuman, H., & S. Presser. 1981. *Questions and answers in attitude surveys*. New York: Academic Press.

Schwarz, N., & H. Hippler. 1987. What responses may tell your respondents: Informative functions of response alternatives. In *Social information processing and survey method*, 163–178, edited by H. Hippler, N. Schwarz, & S. Sudman. New York: Springer.

SDA Archive. 2009. GSS cumulative datafile 1972–2008. Retrieved on April 2009 from http://sda.berkeley. edu/archive.htm

Seale, C., J. Charteris-Black, C. Dumelow, L. Locock, & S. Ziebland. 2008. The effect of joint interviewing on the performance of gender. *Field Methods* 20: 107–128.

Seidman, I. 1991. *Interviewing as qualitative research*. New York: Teachers College, Columbia University.

Seidman, I. 1998. *Interviewing as qualitative research*, 2nd edition. New York: Teachers College, Columbia University.

Seiter, E. 1993. *Sold separately: Children and parents in consumer culture*. New Brunswick, NJ: Rutgers University Press.

Serrano-Garcia, I. 1990. Implementing research: Putting our values to work. In *Researching community psychology*, 171–182, edited by P. Tolan, C. Keys, F. Chertock, & L. Jason. Washington, DC: American Psychological Association.

Shadish, W. R., Jr., T. D. Cook, & L. C. Leviton. 1991. *Foundations of program evaluation*. Newbury Park, CA: Sage.

Shaw, C. R., & H. D. McKay. 1931. Social factors in juvenile delinquency. *Reports on the causes of crime*, vol. II. National Commission on Law Observance and Enforcement, Report No. 13. Washington, DC: U.S. Government Printing Office.

Sheets, R., & J. Mohr. 2009. Perceived social support from friends and family and psychosocial functioning in bisexual young adult college students. *Journal of Counseling Psychology* 56(1): 152–163.

Shepard III, E. M. 2001. The economic costs of D.A.R.E. Unpublished paper. Retrieved on July 3, 2009, from http://www.drugpolicy.org/ docUploads/DAREfinalRP.pdf

Sherwood, J. H. 2004. Talk about country clubs: Ideology and the reproduction of privilege. Ph.D. dissertation, North Carolina State University. http://www.lib.ncsu.edu/ theses/available/etd-04062004-083555

Shih, T., & X. Fan. 2008. Comparing response rates from web and mail surveys: A meta-analysis. *Field Methods* 20: 249–271.

Silver, H. J. 2006. Science and politics: The uneasy relationship. *Footnotes* 34(2): 1 and 5.

Silverman, D. 2004. *Doing qualitative research*, 2nd edition. London: Sage Publications.

Sims, R. C. 2007. Web surveys: Applications in denominational research. *Review of Religious Research* 49: 69–80.

Singer, E., & F. J. Levine. 2003. Protection of human subjects and research: Recent developments and future prospects for the social sciences. *Public Opinion Quarterly* 67(1): 148–164.

Singer, P. 2005. Anti-fraud research rules take effect. *National Journal* 37(28): 2210.

Singleton, R., Jr., B. C. Straits, M. M. Straits, & R. J. McAllister. 1988. *Approaches to social research*. New York: Oxford University Press.

Singleton, R. A., Jr., & B. C. Straits. 2001. Survey interviewing. In *Handbook of interview research*, 59–82, edited by J. F. Gubrium & J. A. Holstein. Thousand Oaks, CA: Sage Publications.

Sloane, P. D., L. W. Cohen, T. R. Konrad, C. S. Williams, J. G. Schumacher, & S. Zimmerman. 2008. Physician interest in volunteer service during retirement. *Annals of Internal Medicine* 149: 317–322.

Small, S. A., S. M. Cooney, & C. O'Connor. 2009. Evidence-informed program improvement: Using principles of effectiveness to enhance the quality and impact of family-based prevention programs. *Family Relations* 58: 1–13.

Smith, A. E., J. Sim, T. Scharf, & C. Phillipson. 2004. Determinants of quality of life amongst older people in deprived neighborhoods. *Ageing and Society* 24: 793–814.

Smith, P. M., & B. B. Torrey. 1996. The future of the behavioral and social sciences. *Science* 271: 611–612.

Smith, T. W. 1989. That which we call welfare by any other name would smell sweeter: An analysis of the impact of question wording on response patterns. In *Survey research methods*, 99–107, edited by E. Singer & S. Presser. Chicago: University of Chicago Press.

Smith, T. W. 2005. Generation gaps in attitudes and values from the 1970s to the 1990s. In *On the frontier of adulthood: Theory, research and public policy*, edited by R. A. Settersen, F. F. Furstenberg Jr., & R. G. Rumbaut. Chicago: University of Chicago Press.

Smith, T. W., J. Kim, A. Koch, & A. Park. 2005. Social-science research and the general social surveys. *ZUMA-Nachrichten* 56: 68–77.

Smyth, J. D., D. A. Dillman, L. M. Christian, & M. J. Stern. 2006. Comparing check-all and forced-choice question formats in web surveys. *Public Opinion Quarterly* 70: 66–77.

Snell, D., & D. Hodgetts. 2007. Heavy metal, identity and the social negotiation of a community of practice. *Journal of Community & Applied Social Psychology* 17: 430–445.

Socolar, R. S., D. K. Runyan, & L. Amaya-Jackson. 1995. Methodological and ethical issues related to studying child maltreatment. *Journal of Family Issues* 16: 565–586.

Sokolow, J. A. 1999. The darker side of persuasion: Stanley Milgram's experiments on obedience to authority. *Proposal Management* 31–36.

Soule, S., & B. King. 2008. Competition and resource partitioning in three social movement industries. *American Journal of Sociology* 113(6): 1568–1610.

Southgate, D., & V. Roscigno. 2009. The impact of music on childhood and adolescent achievement. *Social Science Quarterly* 90(1): 4–21.

Spaaij, R. 2008. Violence, masculinity, and collective identity in football hooliganism. *Journal of Sports & Social Issues* 32(4): 369–392.

Srivastava, S., M. Tamir, K. M. McGonigal, O. P. John, & J. J. Gross. 2009. The social costs of emotional suppression: A prospective study of the transition to college. *Journal of Personality and Social Psychology* 96(4): 883–897.

Stake, R. E. 2003. Case studies. In *Strategies of qualitative inquiry*, 2nd edition, 134–164, edited by N. K. Denzin & Y. S. Lincoln. Thousand Oaks, CA: Sage.

Stalker, K., J. Carpenter, C. Connors, & R. Phillips. 2004. Ethical issues in social research: Difficulties encountered gaining access to children in hospital for research. *Child: Care,*

Health & Development 30: 377–383.

Stark, L. 2007. Victims in our own minds? IRBs in myth and practice. *Law & Society Review* 41: 777–786.

Starr, P. 1987. The sociology of official statistics. In *The politics of numbers*, 7–60, edited by W. Alonso & P. Starr. New York: Russell Sage Foundation.

Stevanovic, P., & P. A. Rupert. 2009. Work-family spillover and life satisfaction among professional psychologists. *Professional Psychology: Research an Practice* 40(1): 62–68.

Stevic, C., & R. M. Ward. 2008. Initiating personal growth: The role of recognition and life satisfaction on the development of college students. *Social Indicators Research* 89(3): 523–534.

Stille, A. 2001. New attention for the idea that abortion averts crime. *New York Times*, April 14, Arts and Ideas Section.

Stouffer, S. A. 1950. Some observations on study design. *American Journal of Sociology* 55: 355–361.

Strauss, A. 1987. *Qualitative analysis for social scientists*. Cambridge: Cambridge University Press.

Streb, M. J., B. Burrell, B. Frederick, & M. A. Genovese. 2008. Social desirability effects and support for a female American president. *Public Opinion Quarterly* 72: 76–89.

Subedi, B., & J. Rhee. 2008. Negotiating collaboration across differences. *Qualitative Inquiry* 14: 1070–1092.

Suchman, L., & B. Jordan. 1992. Validity and the collaborative construction of meaning in face-to-face surveys. In *Questions about questions*, 241–270, edited by J. M. Tanur. New York: Russell Sage Foundation.

Sudman, S. 1967. *Reducing the cost of surveys*. Chicago: Aldine.

Sudman, S., & N. Bradburn. 1982. *Asking questions: A practical guide to questionnaire design*. San Francisco: Jossey-Bass.

Sulik, G. A. 2007. The balancing act: Care work for the self and coping with breast cancer. *Gender & Society* 21: 857–877.

Sullivan, M., R. Bhuyan, K. Senturia, S. Shiu-Thornton, & S. Ciske. 2005. Participatory action research in practice: A case study in addressing domestic violence in nine cultural communities. *Journal of Interpersonal Violence* 20(8): 977–995.

Szinovacz, M., & A. Davey. 2006. Effects of retirement and grandchild care on depressive symptoms. *International Journal of Aging and Human Development* 62(1): 1–20.

Tan, J., & Y. Ko. 2004. Using feature films to teach observation in undergraduate research methods. *Teaching Sociology* 32: 109–118.

Tavris, C. 1996. Mismeasure of women. In *The meaning of difference: American constructions of race, sex and gender*, edited by K. E. Rosenblum & T. Travis. New York: McGraw Hill.

Taylor, L. 2005. All for him: Articles about sex in American lad magazines. *Sex Roles* 52: 152–163.

Teisl, M. F., B. Roe, & M. Vayda. 2006. Incentive effects on response rates, data quality, and survey administration costs. *International Journal of Public Opinion Research* 218: 364–373.

The Disaster Center. 2008. UCR crime statistics. Retrieved on February 28, 2009, from http://www.disaster center.com/crime/uscrime.htm

Thurlow, M. B. 1935. An objective analysis of family life. *Family* 16: 13–19.

Tomaskovic-Devey, D., C. P. Wright, R. Czaja, & K. Miller. 2006. Self-reports of police speeding stops by race: Results from the North Carolina reverse record check survey. *Journal of Quantitative Criminology* 22: 279–297.

Topitzes, J., O. Godes, J. P. Mersky, S. Ceglarek, & A. J. Reynolds. 2009. Educational success and adult health: Findings from the Chicago longitudinal study. *Prevention Science* 10: 175–195.

Touliatos, J., B. F. Perlmutter, & M. A. Straus. 2001. *Handbook of family measurement techniques.* Thousand Oaks, CA: Sage.

Tourangeau, R., E. Singer, & S. Presser. 2003. Context effects in attitude surveys: Effects on remote items and impact on predictive validity. *Sociological Methods & Research* 31(4): 486–513.

Traver, A. E. 2007. Home(land) Décor: China adoptive parents' consumption of Chinese cultural objects for display in their homes. *Qualitative Sociology* 30: 201–220.

Trochian, W. M. K. 2006. Research methods: Knowledge base. Retrieved on January 28, 2009, from http://www.socialresearchmethods.net/kb/index.php

Trochim, W. M., S. E. Marcus, L. C. Mâsse, R. P. Moser, & P. C. Weld. 2008. The evaluation of large research initiatives. *American Journal of Evaluation* 29: 8–28.

Tseng, V. 2004. Family interdependence and academic adjustment in college: Youth from immigrant and U.S. born families. *Child Development* 75(3): 966–983.

Turabian, K. L., revised by J. Grossman and A. Bennett. 1996. *A manual for writers of term papers, theses, and dissertations*, 6th edition. Chicago: University of Chicago Press.

Turner, J. A. 1991. *The structure of sociological theory*, 5th edition. Belmont, CA: Wadsworth

Turney, L., & C. Pocknee. 2005. Virtual focus groups: New frontiers in research. *International Journal of Qualitative Methods* 4(2), Article 3. Retrieved on January 14, 2006, from http://www.ualberta.ca/~ijqm/backissues/4_2/pdf/turney.pdf

Tutt, D. 2008. Where the interaction is: Collisions of the situated and the mediated in living room interaction. *Qualitative Inquiry* 14(7): 1157–1179.

Ulrich, L. T. 1990. *A midwife's tale: The life of Martha Ballard, based on her diary, 1785–1812.* New York: Vintage Books.

Umaña-Taylor, A. J., & M. Y. Bámaca. 2004. Conducting focus groups with Latino populations: Lessons from the field. *Family Relations* 53(3): 261–272.

United Nations Statistics Division. 2000. The world's women 2000: Trends and statistics. Table 5.D: Indicators of economic activity. Retrieved on May 17, 2004, from http://unstats.un.org/unsd/demographic/ww2000/

Urbaniak, G. C., & S. Plous. 2009. *Research randomizer.* Retrieved on March 9, 2009, from http://randomizer.org

U.S. Census Bureau. 2000. Census bureau director says 92 percent of U.S. households accounted for. *United States Department of Commerce News*, May 31. Retrieved on October 2005 from www.census.gov/Press-Release/www/200041.html

U.S. Census Bureau. 2009a. *Statistical abstract: The national data book.* Retrieved on February 13, 2009, from http://www.census.gov/compendia/statab/tables/09s1294.pdf

U.S. Census Bureau. 2009b. What is the census? US census 2010. Retrieved on June 23, 2009, from http://2010.census.gov/2010census/

U.S. Census Bureau. 2009c. Internet use triples in decades, Census Bureau reports. Retrieved on June 23, 2009, from http://www.census.gov/Press-Release/www/releases/archives/communication_industries/013849.html

U.S. Commission on Immigration Reform. 1994. *Restoring credibility.* Washington, DC: U.S. Commission on Immigration Reform.

U.S. Government Printing Office. 1949. Trials of war criminals before the Nuremberg military tribunals under Control Council Law No. 10, Vol. 2, pp. 181–182. Washington, DC: U.S. Government Printing Office. Retrieved on March 11, 2009, from http://ohsr.od.nih.gov/guidelines/nuremberg.html

Van Leeuwen, T., & C. Jewitt. 2001. *Handbook of visual analysis.* Thousand Oaks, CA: Sage.

Vera, E., C. Thakral, R. Gonzales, M. Morgan, W. Conner, E. Caskey, A. Bauer, L. Mattera, S. Clark, K. Bena, & L. Dick. 2008. Subjective well-being in urban adolescents of color. *Cultural Diversity and Ethnic Minority Psychology* 14(3): 224–233.

Verhovek, S. H. 1997. In poll, Americans reject means but not ends of racial diversity. *New York Times*, December 14, 1–32.

Vidich, A. J., & J. Bensman. 1964. The Springdale case: Academic bureaucrats and sensitive townspeople. In *Reflections on community studies*, edited by A. Vidich, J. Bensman, & M. R. Stein. New York: Wiley.

Villalobos, A. 2008. Compensatory connection: A new explanation for the modern intensification of mothering. Paper presented at the Annual Meeting of the American Sociological Association, Boston, August 1–4.

Voloshin, I. 2008. Adolescent work patterns: Impact of family resources on employment sector and intensity of male and female high school seniors. Paper presented at the Annual Meeting of the American Sociological Association, Boston, August 1–4.

Von Hippel, W., J. W. Schooler, K. J. Preacher, & G. A. Radvansky. 2005. Coping with stereotype threat: Denial as an impression management strategy. *Journal of Personality and Social Psychology* 89: 22–35.

Wallace, W. 1971. *The logic of science in sociology*. Chicago: Aldine.

Walton, J. 1992. Making the theoretical case. In *What is a case? Exploring the foundations of social inquiry*, 121–138, edited by C. C. Ragin & H. S. Becker. New York: Cambridge University Press.

Ward, J., & Z. Henderson. 2003. Some practical and ethical issues encountered while conducing tracking research with young people leaving the "care" system. *International Journal of Social Research Methodology* 6(3): 255–259.

Warr, M., & C. G. Ellison. 2000. Rethinking social reactions to crime: Personal and altruistic fear in family households. *American Journal of Sociology* 106: 551–578.

Warren, C. A. B., T. Barnes-Brus, H. Burgess, L. Wiebold-Lippisch, J. Hackney, G. Harkness, V. Kennedy, R. Dingwall, P. C. Rosenblatt, A. Ryen, & R. Shuy. 2003. After the interview. *Qualitative Sociology* 26(1): 93–110.

Wax, R. H. 1971. *Doing fieldwork: Warnings and advice*. Chicago: University of Chicago.

Weaver, D. 1994. The work and retirement decisions of older women: A literature review. *Social Security Bulletin* 57: 3–25.

Webb, E. J., D. T. Campbell, R. D. Schwartz, & L. Sechrest. 2000. *Unobtrusive measures*, revised edition. Thousand Oaks, CA: Sage.

Weber, M. 1947. *The theory of social and economic organization*. New York: Free Press.

Weer, C. 2006. The role of maternal employment, role-altering strategies, and gender in college students'

expectation of work-family conflict. *Sex Roles* 55(7/8): 535–544.

Weisberg, H. F. 2005. *The total survey error approach*. Chicago: The University of Chicago Press.

Weisberg, H. F., J. A. Krosnick, & B. D. Bowen. 1989. *An introduction to survey research and data analysis*. Glenview, IL: Scott, Foresman.

Weiss, C. H., E. Murphy-Graham, & S. Birkeland. 2005. An alternate route to policy influence how evaluations affect D.A.R.E. *American Journal of Evaluation* 26: 12–30.

Weiss, R. S. 1994. *Learning from strangers: The art and method of qualitative interview studies*. New York: Free Press.

Weitz, R. 1991. *Life with AIDS*. New Brunswick, NJ: Rutgers University Press.

Weitzman, E., & M. B. Miles. 1995. *Computer programs for qualitative analysis*. Thousand Oaks, CA: Sage.

Weitzman, L. J. 1985. *The divorce revolution*. New York: Free Press.

Westervelt, S. D., & K. J. Cook. 2007. Feminist research methods in theory and practice: Learning from death row exonerees. In *Criminal justice research and practice: Diverse voices from the field*, 21–37, edited by S. Miller. Boston: University Press of New England.

Westkott, M. 1979. Feminist criticism of the social sciences. *Harvard Educational Review* 49: 422–430.

Whyte, W. F. 1955. *Street corner society: The social structure of an Italian slum*. Chicago: University of Chicago Press.

Whyte, W. F. 1997. *Creative problem solving in the field: Reflections on a career*. Walnut Creek, CA: AltaMira Press.

Williamson, T., & A. Long. 2005. Qualitative data analysis using data displays. *Nurse Researcher* 12: 7–19.

Wilson, D. C., D. W. Moore, P. F. McKay, & D. R. Avery. 2008. Affirmative action programs for women and minorities; expressed support affected by question order. *Public Opinion Quarterly* 72: 514–522.

Wolcott, H. F. 1982. Differing styles for on-site research, or, "If it isn't ethnography, what is it?" *Review Journal of Philosophy and Social Science* 7: 154–169.

Wolkomir, M. 2001. Emotion work, commitment, and the authentication of the self. *Journal of Contemporary Ethnography* 30: 305–334.

Wong, J. D., & M. A. Hardy. 2009. Women's retirement expectations: How stable are they? *Journals of Gerontology Series B: Psychological Sciences and Social Sciences* 64: 77–86.

Wysong, E., R. Aniskiewicz, & D. Wright. 1993. Truth and DARE: Tracking drug education to graduation and as symbolic politics. *Social Problems* 31: 448–472.

Yassour-Borochowitz, D. 2004. Reflections on the researcher-participant relationship and the ethics of dialogue. *Ethics and Behavior* 14: 175–186.

Yin, R. Y. 2009. *Case study research: Design and methods*. Thousand Oaks, CA: Sage Publications.

Zhao, S., S. Grasmuck, & J. Martin. 2008. Identity construction on Facebook: Digital empowerment in anchored relationships. *Computers in Human Behavior* 24(5): 1816–1836.

Zipp, J. F., & J. Toth. 2002. She said, he said, they said: The impact of spousal presence in survey research. *Public Opinion Quarterly* 66(2): 177–208.

INDEX

Credits

This page constitutes an extension of the copyright page. We have made every effort to trace the ownership of all copyrighted material and to secure permission from copyright holders. In the event of any question arising as to the use of any material, we will be pleased to make the necessary corrections in future printings. Thanks are due to the following authors, publishers, and agents for permission to use the material indicated.

Chapter 2. 34: Adapted from Walter Wallace (1971), The Logic of Science in Sociology. ©1971 by Walter L. Wallace. Used by permission of Walter or Aldine de Gruyter.

Chapter 3. 45: Desiree Ciambrone, "Ethical Concerns and Researcher Responsibilities in Studying Women with HIV/AIDS". Reprinted by permission of the author.

Chapter 4. 75: This article was written for this text by Michele Hoffnung, Ph.D., Quinnipiac University, and is published with permission. **83:** Jessica Holden Sherwood is a graduate student in the Department of Sociology, North Carolina State University. This article is published with permission. **93:** Reprinted by permission of the author.

Chapter 5. 105: From PEW Research Center for teh People and the Press by Scott Keeter, Michael Dimock, and Leah Christian. Reprinted by permission of Scott Keeter.

Chapter 6. 136: The data was collected by Michele Hoffnung, Professor of Psychology, Quinnipac University as part of her study of women's lives and is published with permission. **137:** Joseph R. Ferrari, "Impostor Tendencies and Academic Dishonesty," published in SBP in the 2005, Vol 33 (No.1) edition, pp.11-18. Reprinted by permission of Social Behavior and Personality.

Chapter 7. 180: Susan E. Chase, "A Case Study of City University". Reprinted by permission of the author.

Chapter 8. 198: Chris Caldeira, "Student Learning: An Experiment in the Classroom," reprinted by permission of the author.

Chapter 9. 214: K. Brandon Lang and Christopher W. Podeschi, "Environmentalism Among College Students: A Sociological Investigation". Reprinted by permission of the authors.

Chapter 10. 255: Sandra Enos, "Interviewing and Confidentiality" Reprinted by permission of the author. **255:** This article has been revised by Sandra Enos from her paper of the same name that appeared in the first edition of this book and is published with permission. **284:** Nicole Banton, "On My Team?" Why Intragroup